ADVANCES IN ASYMMETRIC AUTOCATALYSIS AND RELATED TOPICS

National Academy of Sciences, Literature and Arts
(Modena)

Chain of Studies 43

Accademia Nazionale di Scienze, Lettere e Arti
(Modena)

Collana di studi 43

ADVANCES IN ASYMMETRIC AUTOCATALYSIS AND RELATED TOPICS

Editors

GYULA PÁLYI
University of Modena and Reggio Emilia, Modena, Italy

RÓBERT KURDI
University of Pannonia, Veszprém, Hungary

CLAUDIA ZUCCHI
University of Modena and Reggio Emilia, Modena, Italy

ELSEVIER

ACADEMIC PRESS
An imprint of Elsevier

Academic Press is an imprint of Elsevier
125 London Wall, London EC2Y 5AS, United Kingdom
525 B Street, Suite 1800, San Diego, CA 92101-4495, United States
50 Hampshire Street, 5th Floor, Cambridge, MA 02139, United States
The Boulevard, Langford Lane, Kidlington, Oxford OX5 1GB, United Kingdom

Notices

Knowledge and best practice in this field are constantly changing. As new research and experience broaden our understanding, changes in research methods, professional practices, or medical treatment may become necessary.

Practitioners and researchers must always rely on their own experience and knowledge in evaluating and using any information, methods, compounds, or experiments described herein. In using such information or methods they should be mindful of their own safety and the safety of others, including parties for whom they have a professional responsibility.

To the fullest extent of the law, neither the Publisher nor the authors, contributors, or editors, assume any liability for any injury and/or damage to persons or property as a matter of products liability, negligence or otherwise, or from any use or operation of any methods, products, instructions, or ideas contained in the material herein.

British Library Cataloguing-in-Publication Data
A catalogue record for this book is available from the British Library

Library of Congress Cataloging-in-Publication Data
A catalog record for this book is available from the Library of Congress

ISBN: 978-0-12-812824-4

For Information on all Academic Press publications
visit our website at https://www.elsevier.com/books-and-journals

Publisher: Cathleen Sether
Acquisition Editor: Emily McCloskey
Editorial Project Manager: Emily Thomson
Production Project Manager: Paul Prasad Chandramohan
Designer: Mark Rogers

Typeset by MPS Limited, Chennai, India

CONTENTS

List of Contributors *xi*
Preface *xv*

1. Asymmetric Autocatalysis and the Origins of Homochirality of Organic Compounds. An Overview **1**

Kenso Soai, Arimasa Matsumoto and Tsuneomi Kawasaki

1.1 Introduction 1
1.2 Examination of the Origins of Homochirality by Using Asymmetric Autocatalysis with Amplification of Chirality 4
1.3 Asymmetric Autocatalysis Initiated by Hydrogen, Carbon, and Oxygen Isotope Chirality and by Cryptochirality: Highly Sensitive Chiral Discrimination by Asymmetric Autocatalysis 15
1.4 Structure of Asymmetric Autocatalyst Determined by Single Crystal X-Ray Crystallography 20
1.5 Summary 22
References 23

2. Soai Autocatalyst as Conveyor of Chirality Coupling Autocatalysis With Asymmetric Induction **31**

Matias Funes-Maldonado, Carlo Romagnoli, Bora Sieng and Mohamed Amedjkouh

2.1 Introduction 31
2.2 Results and Discussion 34
2.3 Conclusion 48
References 49

3. Graph Theoretical and Statistical Analysis of the Impact of Soai Reaction on Natural Sciences **53**

Béla Barabás and Ottilia Fülöp

3.1 Introduction 53
3.2 Soai Graph 54
3.3 Main Communities of the Soai Graph 57
3.4 Disciplines of Journals and Books Citing Soai's Paper 65
References 66

4. Kinetic and Structural Aspects of Mirror-Image Symmetry Breaking in the Soai Reaction 71

Thomas Buhse, María E. Noble-Terán, José-Manuel Cruz, Jean-Claude Micheau and Christophe Coudret

4.1 Introduction 71
4.2 A Brief History of the Discovery of Soai Reaction 73
4.3 Structural Aspects of the Soai Reaction Intermediates 81
4.4 Kinetic Analysis of the Soai Reaction 86
4.5 Theoretical Models of the Soai Reaction 88
4.6 Kinetic Modeling of the Additive Effect, *ee* Amplification During the Stepwise Soai Reaction, Aldehyde Concentration and Apparent Kinetic Order 96
4.7 Conclusion 100
Acknowledgments 102
References 102

5. Macromolecular Versus Micromolecular World 111

Luciano Caglioti and Valeria Righi

5.1 Introduction 111
5.2 Results and Discussion 114
5.3 Conclusion 116
References 117

6. The Importance of the Idea of "Parachirality" in Life Science 119

Noriko Fujii, Norihiko Fujii, Takumi Takata and Hiroaki Sakaue

6.1 Introduction 120
6.2 Localization of D-Asp Residues in Proteins of Living Tissues 121
6.3 Mechanism of D-Asp Formation in Proteins Under Physiological Conditions 121
6.4 The Influence of Asp Isomerization on Protein Structure and Function 123
6.5 New Method of Analysis for Bound Form of D-Amino Acids 124
6.6 Towards Parachirality 128
References 129

7. Graph Theoretical Tools in Two- and Three-Dimensional Chirality Problems 133

Ottilia Fülöp and Béla Barabás

7.1 Introduction 133
7.2 Chemistry and Graph Theory: A Beautiful Connection 134
7.3 Basic Notions 135

7.4 2D Results 138
7.5 Chirality in R^3 143
7.6 Conclusion 147
References 147
Further Reading 148

8. Synthesis of Pyrimidine-Terminated Chiral Large Molecular Architectures With Functions of Self-Replication and Self-Improvement by Asymmetric Autocatalysis 149

Tsuneomi Kawasaki, Arimasa Matsumoto, Itaru Sato and Kenso Soai

8.1 Introduction 149
8.2 Self-Replication of Hexakis(5-pyrimidyl alkanol)hexaalkylsilane With Self-Improvement of Its Enantiomeric and Diastereomeric Purity 151
8.3 Self-Replication and Self-Improvement of Enantiomeric and Diastereomeric Ratio of Tetrakis(pyrimidyl alkanol)tetraalkylsilane 158
8.4 Summary 162
Acknowledgment 162
References 162

9. Model Equivalency in Interpreting the Stochastic Enantiomer Distribution in the Soai Reaction 167

Gábor Lente

9.1 Introduction 167
9.2 Results and Discussion 170
9.3 Conclusions 177
References 178

10. Structural Study of Asymmetric Autocatalysis by X-Ray Crystallography 183

Arimasa Matsumoto, Tsuneomi Kawasaki and Kenso Soai

10.1 Introduction 183
10.2 Results and Discussion 186
10.3 Conclusion 195
Acknowledgments 196
References 196

11. On the Relative Measures of the Degrees of Chirality and General Asymmetry 203

Paul G. Mezey

11.1 Introduction 203
11.2 A General 3D Symmetry Deficiency and Chirality Measure Approach 205

11.3 Some Natural Choices for the Molecular Representations X, Suitable for the Application of the R-Set—Based 3D Symmetry Deficiency and Chirality Measures 210
11.4 A Chemically Motivated, Configuration Space Approach Applied to Some Symmetry Deficiency and Chirality Measure Problems 215
11.5 Choices for Chirality Measures and Descriptors in Terms of a "Symmetry Deficiency Spectrum" 218
Acknowledgment 219
References 219
Further Reading 222

12. Asymmetric Reduction of Ketones to Chiral Platform Molecules 223

Bálint Fridrich, József M. Tukacs and László T. Mika

12.1 Introduction 223
12.2 Conclusion 237
Acknowledgment 237
References 237

13. Searching for Spontaneous Mirror-Symmetry Breaking in Organoautocatalytic Reactions 241

Guillem Valero and Albert Moyano

13.1 Introduction 241
13.2 Results and Discussion 244
13.3 Conclusions 256
Acknowledgments 257
References and Notes 257

14. Isotope Chirality in the Reagents and Intermediates of the Soai Reaction 259

Béla Barabás, Róbert Kurdi and Gyula Pályi

14.1 Introduction 259
14.2 Results and Discussion 260
14.3 Conclusions 271
References and Notes 272

15. Chiral Polythiophenes 277

Francesca Parenti, Luisa Schenetti and Francesco Tassinari

15.1 Introduction 277
15.2 Results and Discussion 278

15.3 Alkylsulfanyl and Related Polythiophenes 285
15.4 Application of Chiral Polythiophenes 291
References 295

16. Abiotic Emergence of Biological Homochirality 299

Josep M. Ribó, Joaquim Crusats, Zoubir El-Hachemi, David Hochberg and Albert Moyano

16.1 Introduction 299
16.2 The Spontaneous Mirror-Symmetry Breaking Scenario 302
16.3 Reaction Network 303
16.4 System Thermodynamic Architectures 311
16.5 From Stochastic to Seterministic Spontaneous Mirror-Symmetry Breaking 312
References and Notes 313

17. Catalytic Applications of Supported Ionic Liquid Phases 317

Csaba Fehér, Máté Papp, Béla Urbán and Rita Skoda-Földes

17.1 Introduction 317
17.2 Results and Discussion 319
17.3 Conclusions 334
Acknowledgments 334
References 334

18. Asymmetric Autocatalysis Initiated by Crystal Chirality of Achiral Compounds 337

Arimasa Matsumoto, Yoshiyasu Kaimori, Tsuneomi Kawasaki and Kenso Soai

18.1 Introduction 337
18.2 Results and Discussion 339
18.3 Conclusion 350
Acknowledgments 350
References 350

19. Amino Acids as Food Quality Factors in Parmigiano Reggiano Hard Cheese 357

Franco Bellesia, Adriano Pinetti, Livia Simon-Sarkadi, Claudia Zucchi, János Csapó, Bart Weimer, Luciano Caglioti and Gyula Pályi

19.1 Introduction 357
19.2 Amino-Acid Analysis 358

19.3 Results and Discussion 359
19.4 Conclusions 365
Acknowledgments 366
References 366

Index *369*

LIST OF CONTRIBUTORS

Mohamed Amedjkouh
Department of Chemistry, University of Oslo, Oslo, Norway

Béla Barabás
Institute of Mathematics, Department of Stochastics, Budapest University of Technology and Economics, Budapest, Hungary

Franco Bellesia
Department of Chemistry, University of Modena and Reggio Emilia, Modena, Italy

Thomas Buhse
Centro de Investigaciones Químicas, Universidad Autónoma del Estado de Morelos, Cuernavaca, Morelos, Mexico

Luciano Caglioti
Department of Chemistry of Bioactive Compounds, University "La Sapienza"-Roma, Roma, Italy

Christophe Coudret
Laboratoire des IMRCP, Université de Toulouse, CNRS UMR 5623, Université Paul Sabatier, 118 route de Narbonne 31062 Toulouse Cedex 9, France

Joaquim Crusats
Department Organic Chemistry, University of Barcelona, Catalonia, Spain; Institute of Cosmos Science (IEEC-UB), University of Barcelona, Catalonia, Spain

José-Manuel Cruz
Centro de Investigaciones Químicas, Universidad Autónoma del Estado de Morelos, Cuernavaca, Morelos, Mexico

János Csapó
Faculty of Animal Science, University of Kaposvár, Kaposvár, Hungary

Zoubir El-Hachemi
Department Organic Chemistry, University of Barcelona, Catalonia, Spain; Institute of Cosmos Science (IEEC-UB), University of Barcelona, Catalonia, Spain

Csaba Fehér
University of Pannonia, Institute of Chemistry, Department of Organic Chemistry, Veszprém, Hungary

Bálint Fridrich
Department of Chemical and Environmental Process Engineering, Budapest University of Technology and Economics, Budapest, Hungary

Norihiko Fujii
Radioisotope Research Center, Teikyo University, Kaga, Itabashi, Tokyo, Japan

Noriko Fujii
Research Reactor Institute, Kyoto University, Osaka, Japan

Ottilia Fülöp
Institute of Mathematics, Department of Differential Equations, Budapest University of Technology and Economics, Budapest, Hungary

Matías Funes-Maldonado
Department of Chemistry, University of Oslo, Oslo, Norway

David Hochberg
Department of Molecular Evolution, Centro de Astrobiología (CSIC-INTA), Madrid, Spain

Yoshiyasu Kaimori
Department of Applied Chemistry, Tokyo University of Science, Tokyo, Japan

Tsuneomi Kawasaki
Department of Materials Science and Engineering, University of Fukui, Bunkyo, Fukui, Japan; Research Center for Chirality, Research Institute for Science and Technology (RIST), Tokyo University of Science, Kagurazaka, Shinjuku-ku, Tokyo, Japan

Róbert Kurdi
Institute of Environmental Engineering, University of Pannonia, Veszprém, Hungary

Gábor Lente
Department of Inorganic and Analytical Chemistry, University of Debrecen, Debrecen, Hungary

Arimasa Matsumoto
Research Center for Chirality, Research Institute for Science and Technology (RIST), Tokyo University of Science, Kagurazaka, Shinjuku-ku, Tokyo, Japan; Department of Applied Chemistry, Tokyo University of Science, Kagurazaka, Shinjuku-ku, Tokyo, Japan

Paul G. Mezey
Canada Research Chair in Scientific Modelling and Simulation, Department of Chemistry and Department of Physics and Physical Oceanography Memorial University of Newfoundland, St. John's, NL, Canada

Jean-Claude Micheau
Laboratoire des IMRCP, Université de Toulouse, CNRS UMR 5623, Université Paul Sabatier, 118 route de Narbonne 31062 Toulouse Cedex 9, France

László T. Mika
Department of Chemical and Environmental Process Engineering, Budapest University of Technology and Economics, Budapest, Hungary

Albert Moyano
Department of Inorganic and Organic Chemistry, University of Barcelona, Barcelona, Catalonia, Spain

María E. Noble-Terán
Centro de Investigaciones Químicas, Universidad Autónoma del Estado de Morelos, Cuernavaca, Morelos, Mexico

Gyula Pályi
Department of Chemistry, University of Modena and Reggio Emilia, Modena, Italy; Department of Life Sciences, University of Modena and Reggio Emilia, Modena, Italy

Máté Papp
University of Pannonia, Institute of Chemistry, Department of Organic Chemistry, Veszprém, Hungary

Francesca Parenti
Dipartimento di Scienze Chimiche e Geologiche, Università di Modena e Reggio Emilia, Modena, Italy

Adriano Pinetti
Department of Chemistry, University of Modena and Reggio Emilia, Modena, Italy

Josep M. Ribó
Department Organic Chemistry, University of Barcelona, Catalonia, Spain; Institute of Cosmos Science (IEEC-UB), University of Barcelona, Catalonia, Spain

Valeria Righi
Department of Sciences for the Quality of Life, University of Bologna, Rimini, Italy

Carlo Romagnoli
Department of Chemistry, University of Oslo, Oslo, Norway

Hiroaki Sakaue
International University of Health and Welfare, Ohtawara, Tochigi, Japan

Itaru Sato
Graduate School of Science and Engineering, Ibaraki University, Bunkyo, Ibaraki, Japan

Luisa Schenetti
Dipartimento di Scienze della Vita, Università di Modena e Reggio Emilia, Modena, Italy

Bora Sieng
Department of Chemistry, University of Oslo, Oslo, Norway

Livia Simon-Sarkadi
Institute of Biochemistry and Food Chemistry, Technical University of Budapest, Budapest, Hungary

Rita Skoda-Földes
University of Pannonia, Institute of Chemistry, Department of Organic Chemistry, Veszprém, Hungary

Kenso Soai
Research Center for Chirality, Research Institute for Science and Technology (RIST), Tokyo University of Science, Kagurazaka, Shinjuku-ku, Tokyo, Japan; Department of Applied Chemistry, Tokyo University of Science, Kagurazaka, Shinjuku-ku, Tokyo, Japan

Takumi Takata
Department of Biochemistry, School of Pharmacy, Tokyo University of Pharmacy and Life Sciences, Hachioji, Tokyo, Japan

Francesco Tassinari
Dipartimento di Scienze della Vita, Università di Modena e Reggio Emilia, Modena, Italy

József M. Tukacs
Department of Chemical and Environmental Process Engineering, Budapest University of Technology and Economics, Budapest, Hungary

Béla Urbán
University of Pannonia, Institute of Chemistry, Department of Organic Chemistry, Veszprém, Hungary

Guillem Valero
Department of Inorganic and Organic Chemistry, University of Barcelona, Barcelona, Catalonia, Spain

Bart Weimer
Nutrition and Food Sciences, Utah State University, Logan, UT, United States

Claudia Zucchi
Department of Chemistry, University of Modena and Reggio Emilia, Modena, Italy

PREFACE

Really fundamental discoveries are rare in Natural Sciences, and thus also in Chemistry. One of these was published in 1995, in the highly prestigious periodical, *Nature*, by a group of scientists from the Tokyo University of Science (Japan), led by *Kenso SOAI*. This publication opened new perspectives in preparative organic chemistry and in several related fields of Natural Sciences. Soai and co-workers described the first well-documented example of *asymmetric autocatalysis*, a new type of chemical reactions, in which the reaction product not only catalyzes its own formation, but also promotes the prevalence of formation of molecules that have the same spatial structure (chirality). This highly interesting result has very practical consequences in several fields, such as, for example, the pharmaceutical industry or the structural chemistry of certain polymers. It also enables the study of theoretically highly important topics, such as absolute enantioselecive synthesis as well as the origin of biological chirality.

Our Academy, the (Italian) National Academy of Sciences, Literature and Arts (Modena), recognized fairly early the importance of this research awarding in 2003 the Medal of Scientific Merit to Prof. Kenso Soai and 2 years later electing him to *honorary member* of the Academy.

An international group of scientists decided to make an attempt at summarizing the new results of the *asymmetric autocatalysis* and some related fields, where progress could be expected. The fruit of their work is the present book, which was prepared under the patronage of the National Academy of Sciences, Literature and Arts (Modena), as well as of the University of Modena and Reggio Emilia (I), the University of Technology and Economics (Budapest, H) and the University of Pannonia (Veszprém, H). Our Academy adopted the book as Volume 43 of the *Collana di studi*.

It is an honor for me to present this outstanding collection of papers as *Advances in Asymmetric Autocatalysis and Related Topics* to the international community of natural scientists.

Ernesto Milano

National Academy of Sciences, Literature and Arts (Modena)

Modena, June 6, 2016

CHAPTER 1

Asymmetric Autocatalysis and the Origins of Homochirality of Organic Compounds. An Overview

Kenso Soai[1,2], Arimasa Matsumoto[1,2] and Tsuneomi Kawasaki[2,3]
[1]Department of Applied Chemistry, Tokyo University of Science, Kagurazaka, Shinjuku-ku, Tokyo, Japan
[2]Research Center for Chirality, Research Institute for Science and Technology (RIST), Tokyo University of Science, Kagurazaka, Shinjuku-ku, Tokyo, Japan
[3]Department of Materials Science and Engineering, University of Fukui, Bunkyo, Fukui, Japan

Abstract

Asymmetric autocatalysis is a reaction in which a chiral product acts as a chiral catalyst for its own production. 5-Pyrimidyl alkanol acts as an asymmetric autocatalyst with amplification of chirality in the enantioselective addition of diisopropylzinc to pyrimidine-5-carbaldehyde. 2-Alkynyl-5-pyrimidyl alkanol with an extremely low enantiomeric excess (*ca.* 0.00005% ee) acts as an asymmetric autocatalyst to afford the same alkanol with significantly amplified ee ($>$99.5%). Theories of the origins of homochirality have been examined by using asymmetric autocatalysis with amplification of ee. Circularly polarized light, chiral inorganic crystals, such as quartz, chiral organic crystals composed of achiral compounds, and spontaneous absolute asymmetric synthesis, have been considered as candidates for the origins of chirality. It was found that asymmetric autocatalysis is triggered by these origins of chirality, affording the pyrimidyl alkanol with high ee and with the corresponding absolute configurations to the chirality of the chiral triggers. Chiral isotopomers of carbon (C^{13}/C^{12}), oxygen (O^{18}/O^{16}), and hydrogen (D/H) were also found to trigger asymmetric autocatalysis. Asymmetric autocatalysis enables us to discriminate the chirality of cryptochiral compounds.

Keywords: Asymmetric autotatalysis; asymmetric amplification; chiral isotopomer; spontaneous absolute asymmetric synthesis; cryptochirality; origin of chirality; circularly polarized light

1.1 INTRODUCTION

Two of the characteristic features of living organisms are (1) self-replication, i.e., autocatalytic multiplication of cells and individuals, and (2) strong single-handedness of their molecular components, such as L-amino acids and D-sugars. This biological homochirality is essential for life. The

Advances in Asymmetric Autocatalysis and Related Topics
DOI: http://dx.doi.org/10.1016/B978-0-12-812824-4.00001-0

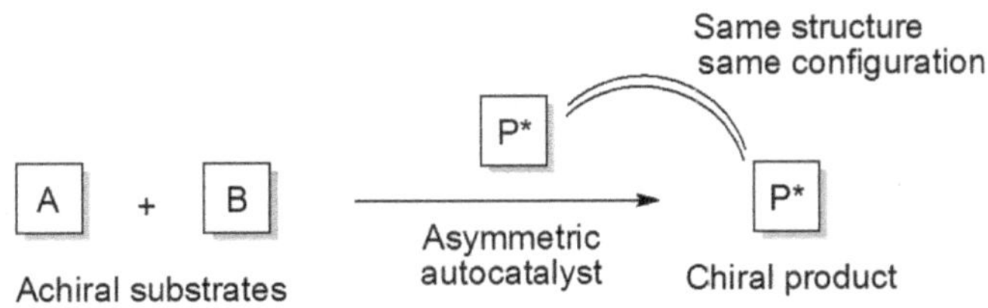

Figure 1.1 Asymmetric autocatalysis.

origin of biological homochirality has attracted much attention for nearly 170 years, ever since the discovery of molecular dissymmetry of sodium ammonium tartrate by Pasteur [1].

Several origins of molecular chirality have been proposed [2]. However, the proposed theories have usually been capable of inducing only very low enantiomeric excesses (ees). Therefore in order to achieve the single-handedness of biological organic compounds, an amplification process of low enantioenrichment is necessary [3]. In this context, asymmetric autocatalysis with amplification of chirality, if it exists, has been proposed as the most straightforward process.

Asymmetric autocatalysis is a reaction in which a chiral product acts as a chiral catalyst for its own production (Fig. 1.1). The process is an automultiplication of the chiral compound. Asymmetric autocatalysis has inherent advantages over the usual nonautocatalytic asymmetric catalysis: (1) high efficiency, because of the process of self-replication; (2) an increase in the amount of catalyst during the reaction, because the product itself becomes the catalyst; and (3) no process is required to separate product from catalyst, because the structures of the catalyst and the product are the same. In an ideal case, the catalytic activity does not decrease due to an increase in the amount of catalyst.

In 1953, Frank proposed a mathematical mechanism of asymmetric autocatalysis without mentioning any chemical structure [4]. However, no actual asymmetric autocatalytic reaction had been presented until our first report, in 1990, on the asymmetric autocatalysis of 3-pyridyl alkanol in the enantioselective addition of dialkylzincs to pyridine-3-carbaldehyde (Fig. 1.2) [5,6].

Ever since our first report on asymmetric autocatalysis, we have continued to work on this topic. In 1995, we finally found the highly enantioselective asymmetric autocatalysis of pyrimidyl alkanol with amplification of ee [7–10] in the enantioselective addition of diisopropylzinc (i-Pr_2Zn) to pyrimidine-5-carbaldehyde.

Pyrimidyl alkanol **1** acts as an asymmetric autocatalyst to afford more of itself with the same absolute configuration with >99.5% ee and in >99%

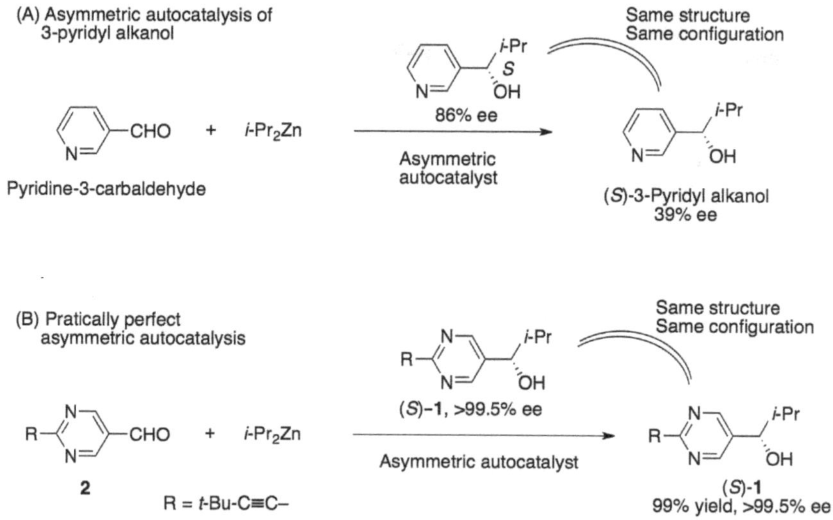

Figure 1.2 Asymmetric autocatalysis in the addition reaction of *i*-Pr_2Zn to nitrogen-containing aromatic aldehydes. (A) The first asymmetric autocatalysis of 3-pyridyl alkanol; (B) practically perfect asymmetric autocatalysis of 5-pyrimidyl alkanol **1**.

yield (Fig. 1.3) [11a,b]. The asymmetric autocatalysis of pyrimidyl alkanol **1** was found to exhibit significant amplification of ee. Thus an extremely low ee of *ca.* 0.00005% of the asymmetric autocatalyst **1** after three consecutive asymmetric autocatalyses amplified the ee of alkanol **1** to >99.5% and the amount by a factor of about 630,000 times [11c]. The amplification of ee by asymmetric autocatalysis is unique because the amplification of ee is made possible without the need for any other chiral auxiliary other than the initial low enantioenrichment of asymmetric autocatalyst **1** itself.

The above asymmetric autocatalysis with amplification of ee proves the existence of an actual chemical reaction in which very slight enantioenrichment can be amplified significantly to almost enantiopurity (>99.5% ee).

We then envisaged that the asymmetric autocatalysis with amplification of chirality could be used to elucidate the origins of homochirality [12].

1.2 EXAMINATION OF THE ORIGINS OF HOMOCHIRALITY BY USING ASYMMETRIC AUTOCATALYSIS WITH AMPLIFICATION OF CHIRALITY

The enantioenrichments induced by the proposed mechanisms of the origin of chirality have usually been very low (as mentioned in the Introduction). The amplification of very low enantioenrichment of organic compounds to

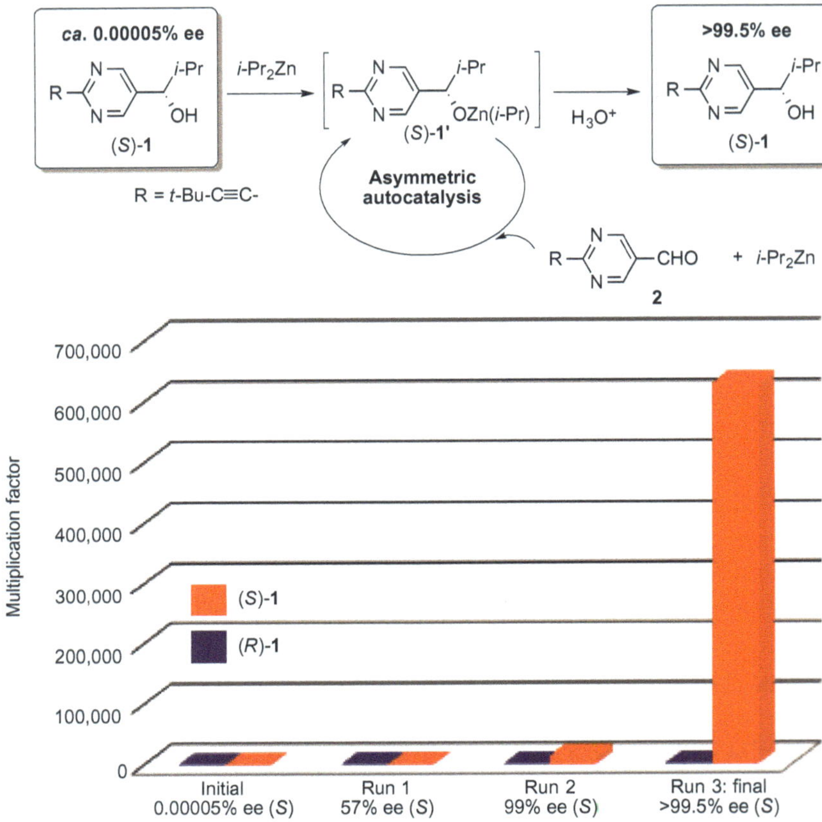

Figure 1.3 Asymmetric autocatalysis of 5-pyrimidyl alkanol with amplification of ee.

enantiopure organic compounds has remained puzzling. Moreover, some of the proposed mechanisms of the origin of chirality have not induced any ee above the detection level. We anticipated that if a proposed origin of chirality could induce a small ee in pyrimidyl alkanol, either directly or indirectly, the subsequent asymmetric autocatalysis with amplification of ee would afford pyrimidyl alkanol of high ee with the absolute configuration corresponding to that of the origin of chirality. We used asymmetric autocatalysis to examine and clarify the relationship between the several proposed mechanisms for the origin of chirality and an organic compound with high ee.

1.2.1 Circularly Polarized Light

Left (*l*) and right (*r*)-handed circularly polarized light (CPL) is a chiral physical force. It has been proposed as the origin of chirality of organic

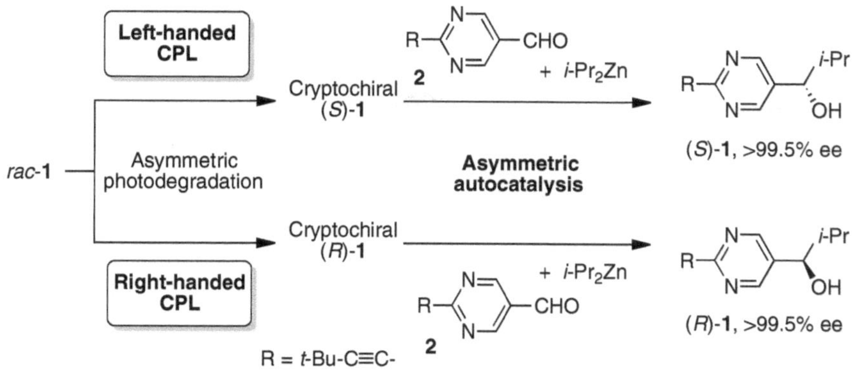

Figure 1.4 Asymmetric autocatalysis triggered by direct irradiation of circularly polarized light.

compounds. The occurrence of relatively strong CPL has been observed in some of the star-forming regions [13]. CPL irradiation of the racemic organic compounds such as leucine and hexahelicene is capable of inducing low enantioenrichment with only *ca.* 2% ee [14]. The induced low ees have, however, not been correlated to the homochirality of organic compounds [15].

The irradiation of *l*-CPL to racemic (*rac*) asymmetric autocatalyst **1** [16] and the subsequent asymmetric autocatalysis with amplification of ee affords (*S*)-alkanol **1** with >99.5% ee (Fig. 1.4) [16b]. On the contrary, irradiation with *r*-CPL affords (*R*)-**1** with >99.5% ee.

The correlation between the handedness of CPL and (*R*)-**1** and (*S*)-**1** can be explained as follows: (*R*)-**1** and (*S*)-**1** exhibit positive and negative Cotton effects, respectively, in their circular dichroism (CD) spectra at 313 nm. This means that the irradiation of *l*-CPL on *rac*-**1** would induce the asymmetric photoreaction of (*R*)-**1** due to its preferential absorbance of *l*-CPL. Therefore the less reactive (*S*)-**1** would become the remaining enantiomer, and the subsequent asymmetric autocatalysis amplifies the ee of (*S*)-**1** to >99.5% ee. This result is the first direct correlation of the handedness of CPL with that of an organic compound of near enantiopurity.

1.2.2 Chiral Inorganic Crystals of Quartz, Sodium Chlorate, Cinnabar, and Retgersite

It is widely accepted that many minerals existed on Earth even before the emergence of life. Chiral natural minerals such as quartz have been

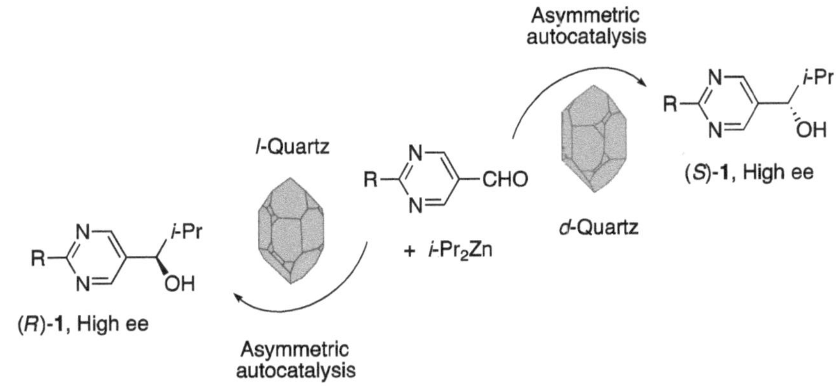

Figure 1.5 *d*- and *l*-Quartz induced asymmetric autocatalysis with amplification of chirality.

considered as one of the possible origins of biological homochirality [17]. Quartz is a chiral single crystal of silicon dioxide and it exhibits enantiomorphism. Although many reports have appeared on the use of quartz to induce chirality of organic compounds, no significant asymmetric induction using quartz has yet been reported.

We envisaged that, if chiral *d*- and *l*-quartz induce any enantiomeric imbalance in an asymmetric reaction, the subsequent asymmetric autocatalysis amplifies significantly the ee of the product [18]. As anticipated, in the asymmetric autocatalysis using pyrimidine-5-carbaldehyde **2** and *i*-Pr_2Zn, triggered by *d*-quartz powder, (*S*)-**1** with 97% ee was obtained in a yield of 95% (Fig. 1.5 and Table 1.1, entry 1). In contrast, *l*-quartz afforded (*R*)-**1** with 97% ee. These results clearly show that *d*- and *l*-quartz trigger asymmetric autocatalysis and acts as the origin of chirality. This result also stands as the first example of the correlation of the chirality of an inorganic crystal to a near enantiopure organic compound.

Sodium chlorate ($NaClO_3$) and sodium bromate ($NaBrO_3$) are chiral inorganic ionic crystals [3d,19]. *d*-$NaClO_3$ triggers asymmetric autocatalysis to afford (*S*)-**1** with high ee, while *l*-$NaClO_3$ affords (*R*)-**1** [20a]. In contrast, *l*-$NaBrO_3$ triggers the formation of (*S*)-**1** while *d*-$NaBrO_3$ triggers the formation of (*R*)-**1** [20b]. It should be noted that $NaClO_3$ and $NaBrO_3$ of the same type of enantiomorph have opposite signs of optical activity (Entries 2 and 3) [20]. Cinnabar is a red chiral crystal of mercury(II) sulfide (HgS). Enantiomorphic *P*- and *M*-crystals of cinnabar are constructed with −S−Hg−S−Hg spiral chains. *P*-HgS acts as a chiral

Table 1.1 Enantioselective synthesis initiated with chiral inorganic crystals in conjunction with asymmetric autocatalysis

Entry	Chiral inorganic crystal			Alkanol 1		Refs.
		Space group	Chirality	ee (%)[a]	Config.	
1	Quartz (SiO_2)	$P3_221$	*d*	97	*S*	[18]
		$P3_121$	*l*	97	*R*	
2	$NaClO_3$	$P2_13$	*d*	98	*S*	[20a,b]
			l	98	*R*	
3	$NaBrO_3$	$P2_13$	*l*	98	*S*	[20b]
			d	98	*R*	
4	Cinnabar (HgS)	$P3_221$	*M*	>99.5	*S*	[21]
		$P3_121$	*P*	>99.5	*R*	
5	Retgersite ($NiSO_4 \cdot 6H_2O$)	$P4_12_12$	CD(+) 390_{Nujol}	>99.5	*S*	[22]
		$P4_32_12$	CD(−) 390_{Nujol}	>99.5	*R*	

[a]The highest ee value reported in the original publications.

initiator of asymmetric autocatalysis to afford (*R*)-**1**, while *M*-HgS affords (*S*)-**1** (Entry 4) [21]. Retgersite ($NiSO_4.6H_2O$) is a chiral mineral. Retgersite of [CD(+)390_{Nujol}] triggers asymmetric autocatalysis to afford (*S*)-**1**, while retgersite of [CD(−)390_{Nujol}] affords (*R*)-**1** [22].

As described, chiral inorganic crystals such as quartz, sodium chlorate, cinnabar, and retgersite trigger asymmetric autocatalysis and act as the origins of chirality to afford, in conjunction with asymmetric autocatalysis, organic compounds of high ee with the correlated absolute configurations.

1.2.3 Chiral Crystals Formed from Achiral Organic Compounds

There are certain classes of achiral organic compounds that form chiral crystals, with each crystal exhibiting one of the two possible enantiomorphs [23]. Some of these chiral organic crystals have been used as reactants in stereospecific reactions [2h]. However, chiral organic crystals consisting of achiral organic compounds have rarely been used as chiral auxiliaries in enantioselective reactions [24]. We examined asymmetric autocatalysis triggered by chiral crystals composed of achiral organic compounds.

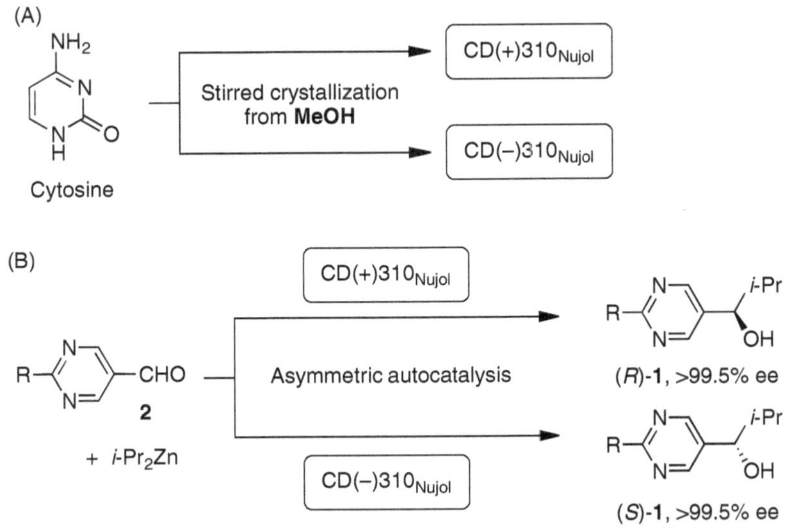

Figure 1.6 (A) Chiral crystallization of cytosine under stirred condition. (B) Asymmetric autocatalysis initiated by chiral crystal of achiral nucleobase cytosine.

Cytosine is an achiral organic compound and acts as an important nucleobase. It is formed under prebiotic conditions [25]. Crystallization of cytosine from methanol yields chiral crystals that exhibit either plus or minus Cotton effects in solid-state CD spectra at *ca.* 310 nm (Fig. 1.6A). Asymmetric autocatalysis was examined by using chiral crystals of cytosine (Fig. 1.6B) [26a]. When a [CD(+)310_{Nujol}]-cytosine crystal was used as a chiral trigger, the reaction between aldehyde **2** and *i*-Pr_2Zn afforded enantioenriched (*R*)-alkanol **1** in conjunction with asymmetric autocatalysis with amplification of ee. In contrast, a [CD(−)310_{Nujol}]-cytosine crystal triggered the production of (*S*)-**1**. Thus, the chiral cytosine crystal is shown to act as the origin of chirality.

When cytosine is crystallized from water, unlike when it is crystallized from methanol, it forms achiral crystals of cytosine monohydride (Fig. 1.7A). It was found that the elimination of crystal water from one of the enantiomeric faces (Fig. 1.7B) of the achiral crystal of cytosine monohydrate, by heating, afforded a chiral crystal of dehydrated cytosine [26b]. The chirality of the dehydrated cytosine formed is nicely correlated to the enantiotopic face of the achiral cytosine monohydrate (Fig. 1.7C). It should be emphasized that the dehydration of crystal water from one of the enantiotopic faces under reduced pressure conditions also afforded the chiral crystal of cytosine, but that the formed crystal chirality is opposite

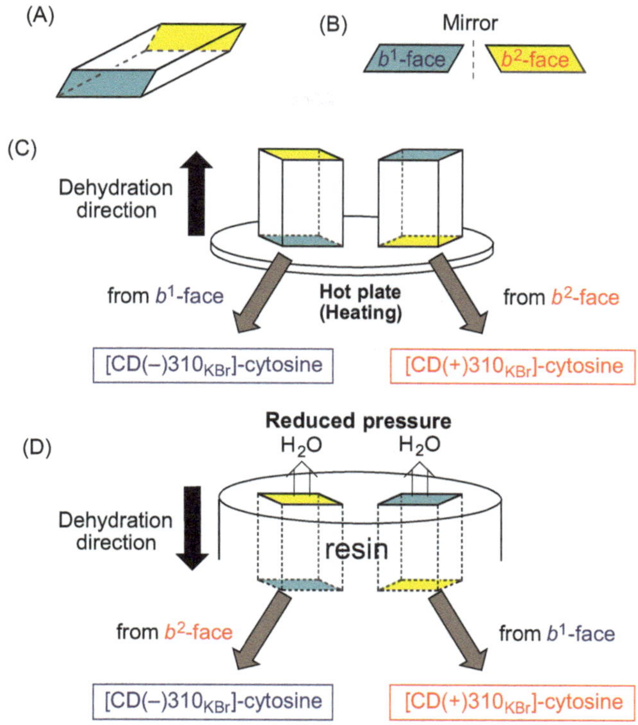

Figure 1.7 Generation of absolute controlled crystal chirality by the removal of crystal water from the achiral crystal of cytosine monohydrate.

to that dehydrated by heating (Fig. 1.7D). It should also be noted that the sign of the Cotton effects of the cytosine chiral crystal observed are opposite to those observed when using a KBr disk and Nujol mull, that is, CD $(+)310_{KBr}$ and $CD(-)310_{Nujol}$ mean the same chirality of cytosine crystal. This is the first example of the formation of chiral crystals with controlled chirality from an achiral crystal [26b,c].

Another example of an achiral nucleobase is adenine. Adenine dinitrate forms chiral crystals. Chiral crystals of adenine dinitrate trigger asymmetric autocatalysis (Table 1.2, Entry 1) [27a]. This indicates that achiral nucleobases such as cytosine and adenine can act as the origin of chirality in their chiral crystalline state.

In addition, enantiomorphous crystals composed of achiral *N*-benzoylglycine, [27b] benzil, [27c] tetraphenylethylene, [27d] 2-thenoylglycine, [27e] aromatic triester, [27f] ethylenediammonium sulfate, [27g] and certain chiral cocrystals consisting of two achiral compounds, [24] can act as chiral triggers of asymmetric autocatalysis (Entries 2–14).

Table 1.2 Stereochemical correlations between the crystals used as chiral initiators and the synthesized 5-pyrimidyl alkanol **1**

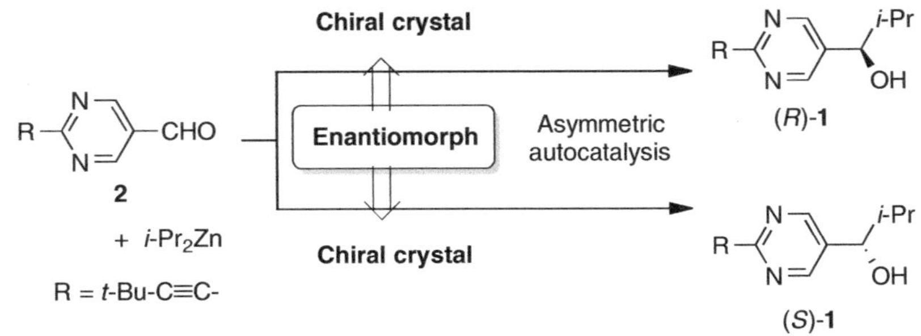

Entry	Crystal	Space group	Helicity (axis)[a]	Solid-state CD	Alkanol **1**		Refs.
					ee[b]	Config.	
1	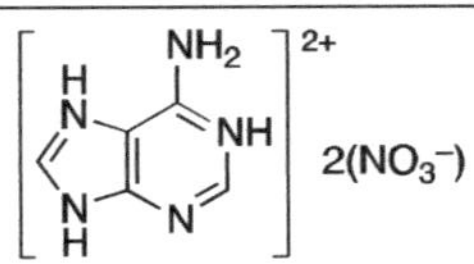	$P2_12_12_1$	*P* (*a*) *M* (*a*)	$CD(-)250_{Nujol}$ $CD(+)250_{Nujol}$	>99.5 >99.5	*S* *R*	[27a]
2	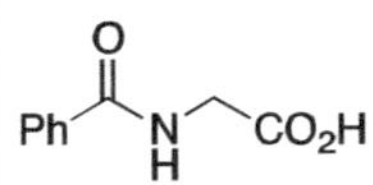	$P2_12_12_1$	– –	$CD(+)260_{Nujol}$ $CD(-)260_{Nujol}$	>99.5 >99.5	*S* *R*	[27b]
3	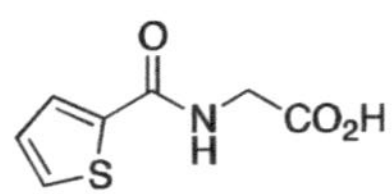	$P2_12_12_1$	*M* (*c*) *P* (*c*)	$CD(+)300_{KBr}$ $CD(-)300_{KBr}$	75 90	*S* *R*	[27e]
4	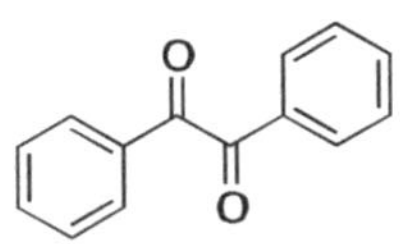	$P3_121$ or $P3_221$	– –	$CD(-)400_{KBr}$ $CD(+)400_{KBr}$	95 96	*S* *R*	[27c]

5	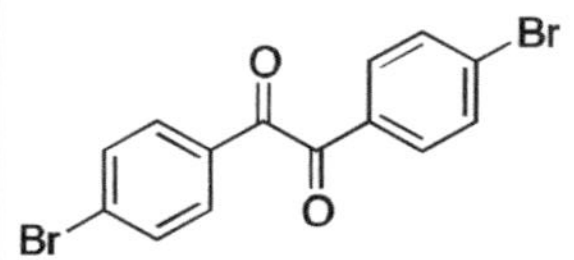	$P2_12_12$	P^c M^c	CD(−)320_{KBr} CD(+)320_{KBr}	97 97	*S* *R*	[27c]
6	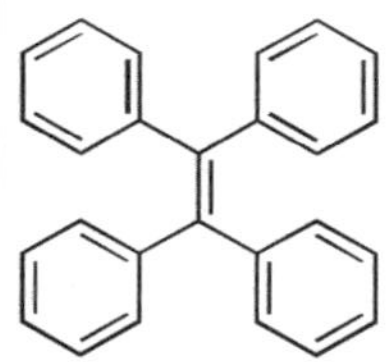	$P2_1$	– –	CD(+)270_{Nujol} CD(−)270_{Nujol}	>99.5 >99.5	*S* *R*	[27d]
7	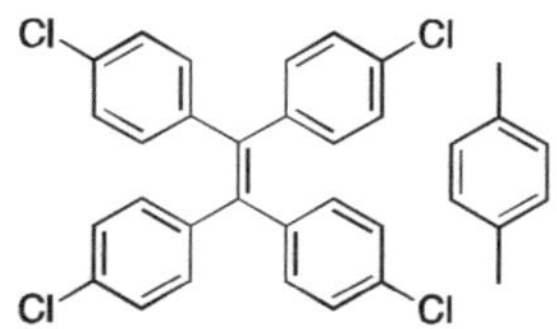	$P2_12_12_1$	M^c P^c	CD(−) 270_{Nujol}[d] CD(+) 270_{Nujol}[d]	>99.5 >99.5	*S* *R*	[27d]
8	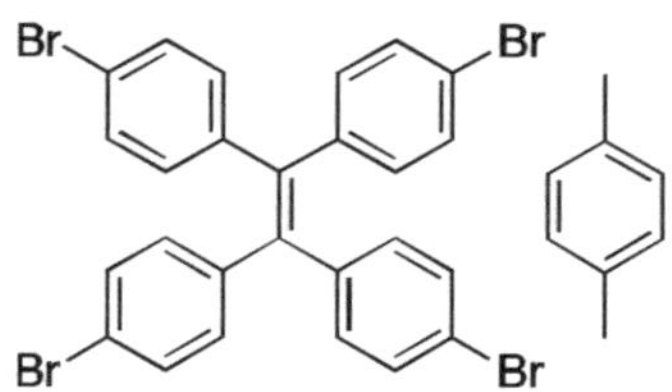	$P2_12_12_1$	M^c P^c	– –	>99.5 >99.5	*S* *R*	[27d]
9	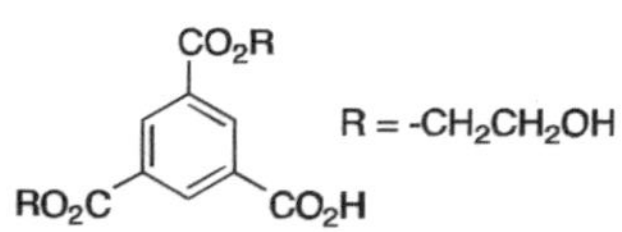	$P6_1$ or $P6_5$	*P* *M*	CD(+)224_{KBr} CD(−)224_{KBr}	>99.5 >99.5	*S* *R*	[27f]

(*Continued*)

Table 1.2 (Continued)

Entry	Crystal	Space group	Helicity (axis)[a]	Solid-state CD	Alkanol 1		Refs.
					ee[b]	Config.	
10	$(CH_2NH_2)_2 \cdot H_2SO_4$	$P4_12_12$	M^c	$(-)^e$	>99.5	*S*	[27g]
		$P4_32_12$	P^c	$(+)^e$	>99.5	*R*	
11	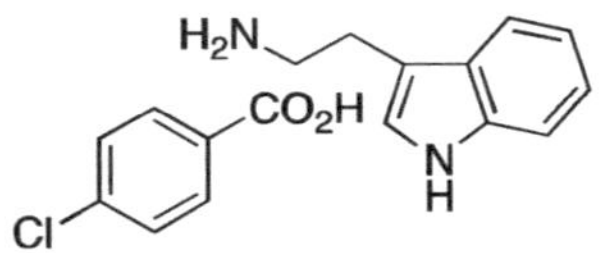	$P2_12_12_1$	*M* (*c*)	$CD(-)225_{Nujol}$	92	*S*	[24]
			P (*c*)	$CD(+)225_{Nujol}$	96	*R*	
12	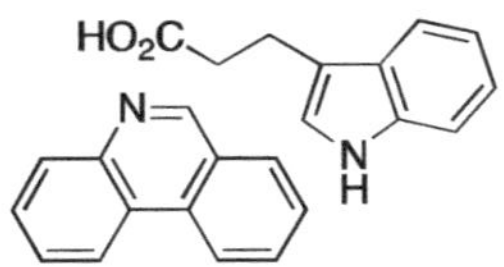	$P2_1$	*M* (*b*)	$CD(-)250_{Nujol}$	97	*S*	[24]
			P (*b*)	$CD(+)250_{Nujol}$	92	*R*	
13	$K[Co(edta)] \cdot 2H_2O$	–	–	$(+)_{545}{}^f$	91	*S*	[27h]
			–	$(-)_{545}{}^f$	94	*R*	
14	$K[Co(trdta)] \cdot 2H_2O$	–	–	$(+)_{545}{}^f$	85	*S*	[27h]
			–	$(-)_{545}{}^f$	88	*R*	

[a]Helicity of the molecular arrangement in crystal is indicated.
[b]The highest ee value reported in the original publications.
[c]The conformational chirality of each molecule is indicated.
[d]Solid-state CD after the removal of *p*-xylene.
[e]Defined by using polarizing microscope.
[f]The sing of the optical rotation at 545 nm is indicated.

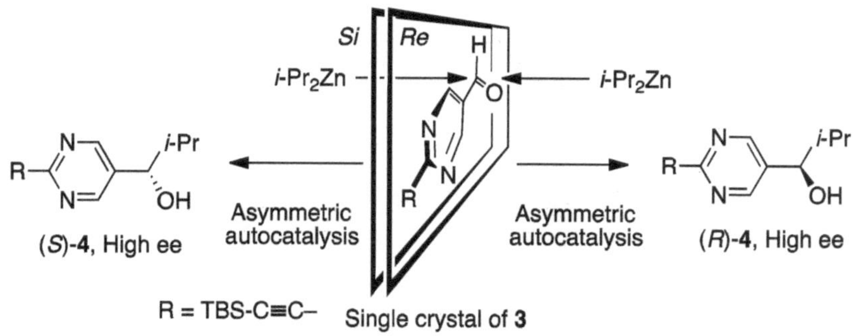

Figure 1.8 Enantioselective addition of diisopropylzinc at the enantiotopic surface of an achiral single crystal of 2-(*tert*-butyldimethylsilylethynyl)pyrimidine-5-carbaldehyde.

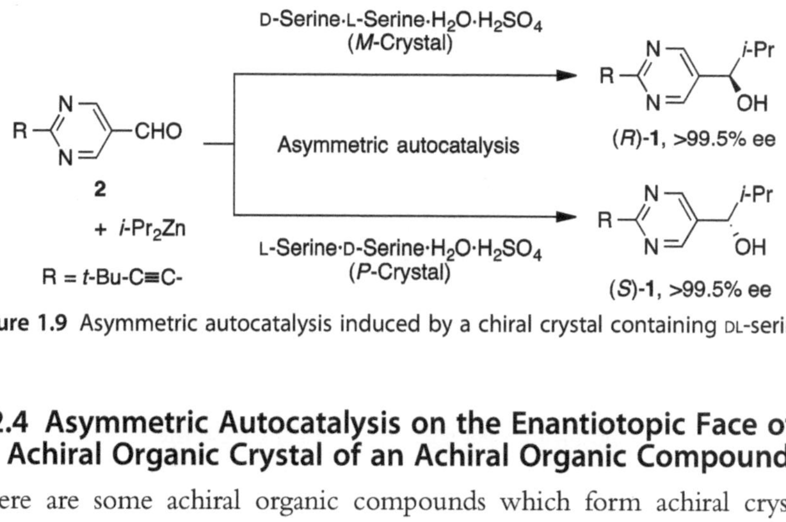

Figure 1.9 Asymmetric autocatalysis induced by a chiral crystal containing DL-serine.

1.2.4 Asymmetric Autocatalysis on the Enantiotopic Face of an Achiral Organic Crystal of an Achiral Organic Compound

There are some achiral organic compounds which form achiral crystals with enantiotopic faces. A crystal of 2-(*tert*-butyldimethylsilylethynyl) pyrimidine-5-carbaldehyde **3** is an example of an achiral crystal (*P*-1) that has enantiotopic faces (Fig. 1.8) [28]. When one of the enantiotopic faces, the *Re*-face, was exposed to *i*-Pr_2Zn vapor, (*R*)-pyrimidyl alkanol **4** was formed. On the other hand, exposure of *i*-Pr_2Zn vapor on the *Si*-face afforded (*S*)-alkanol **4** [28]. The ee of the obtained alkanol **4** was enhanced to >99.5% ee by the subsequent asymmetric autocatalysis.

1.2.5 Chiral Crystal of DL-Serine Sulfate

The crystal of DL-diserinium sulfate hydrate forms the chiral crystal ($P2_12_12_1$). The chiral crystal of DL-diserinium sulfate hydrate was used as a chiral trigger for asymmetric autocatalysis (Fig. 1.9) [29]. The *M*-crystal of DL-diserinium sulfate hydrate triggered the formation of (*R*)-pyrimidyl alkanol **1**, while the *P*-crystal triggered the formation of (*S*)-alkanol **1**. This is the first example showing that chiral crystals formed from a racemic compound act as an origin of chirality.

1.2.6 Spontaneous Absolute Asymmetric Synthesis

The term "absolute asymmetric synthesis" had been used for the asymmetric synthesis without the use of any chiral compound, for example, for the reaction induced by chiral CPL. However, in his excellent commentary on absolute asymmetric synthesis, Mislow proposed a new definition [2g]: absolute asymmetric synthesis is the formation of an enantioenriched compound without the intervention of any chiral factor. The spontaneous absolute asymmetric synthesis has been proposed as one of the origins of homochirality in nature. It was widely believed that without the intervention of any chiral factor the so-called racemic product is formed: the probability of the formation of both *R* and *S* products in a reaction is 50:50. According to the theory of statistics, however, the numbers of *R* and *S* enantiomers are rarely exactly the same, i.e., there are almost always fluctuations in the numbers of enantiomers [2g]. For example, if a coin is flipped 100 times, there is only an 8% probability of exactly 50 heads and 50 tails. In the other 92% of cases, either heads or tails will be in excess: 51:49, 48:52, and so on. Pályi and Caglioti mentioned the statistical fluctuations of ee of various amounts of so-called racemic molecules [30]. When the number of molecules is very low, statistical fluctuation exhibits rather high ee. In an extreme case, when the number of molecules is an odd number, the numbers of each of the enantiomeric molecules can never be the same.

Asymmetric autocatalysis of pyrimidyl alkanol with amplification of ee enables us to enhance extremely low ee to >99.5% ee. We reasoned that the reaction of achiral reactants without the intervention of any chiral factors affords products with low ee as a result of fluctuation in the formation of racemic mixtures, and that the low ee can be amplified by subsequent asymmetric autocatalysis (Fig. 1.10).

Based on the considerations described above, we examined spontaneous absolute asymmetric synthesis using asymmetric autocatalysis. The reactions of **2** [31a] and 2-methylpyrimidine-5-carbaldehyde [31b,c] with *i*-Pr_2Zn without the addition of a chiral substance afforded enantioenriched (*S*)-**1** or (*R*)-**1** and (*S*)- or (*R*)-2-methylpyrimidyl alkanol, respectively [31]. When aldehyde **2** reacted with *i*-Pr_2Zn in a mixed solvent of ether and toluene, zinc alkoxide of pyrimidyl alkanol formed in situ, and

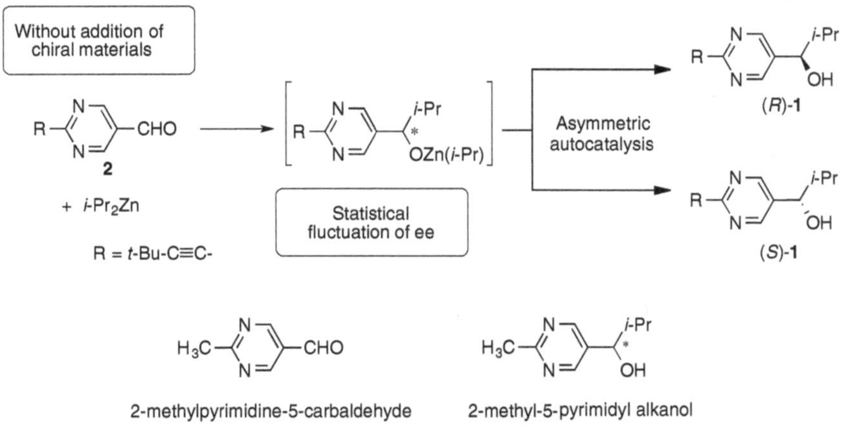

Figure 1.10 Spontaneous absolute asymmetric synthesis by asymmetric autocatalysis with amplification of ee.

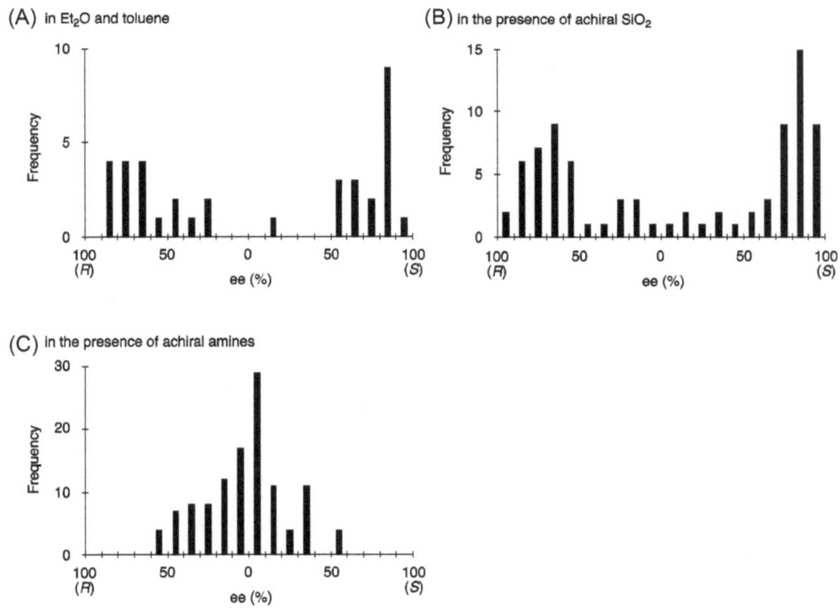

Figure 1.11 Histograms of the absolute configuration and ee of pyrimidyl alkanol **1** produced by spontaneous absolute asymmetric synthesis.

the subsequent asymmetric autocatalysis with amplification of ee afforded enantioenriched (*S*)- or (*R*)-**1** with an ee well above the detection level. In a total of 37 reactions, (*S*)-alkanol **1** formed 19 times and (*R*)-**1** formed 18 times (Fig. 1.11A) [31a]. Thus, the absolute configurations of **1** produced showed an approximately stochastic distribution of *S* and *R* enantiomers. In addition, the absolute asymmetric synthesis between pyrimidine-5-carbaldehyde **2** and *i*-Pr_2Zn in the presence of achiral amorphous silica gel (Fig. 1.11B) [32a] and achiral amines (Fig. 1.11C) [32b] spontaneously afforded (*S*)-**1** or (*R*)-**1** in an approximately stochastic distribution. These results, i.e., the approximately stochastic distribution of the formation of (*S*) or (*R*)-alkanol **1**, fulfill one of the conditions necessary for spontaneous absolute asymmetric synthesis [33].

1.3 ASYMMETRIC AUTOCATALYSIS INITIATED BY HYDROGEN, CARBON, AND OXYGEN ISOTOPE CHIRALITY AND BY CRYPTOCHIRALITY: HIGHLY SENSITIVE CHIRAL DISCRIMINATION BY ASYMMETRIC AUTOCATALYSIS

Asymmetric autocatalysis with amplification of ee can be applied to the discrimination of chirality. When the pyrimidine-5-carbaldehyde **2** is treated with *i*-Pr_2Zn in the presence of cryptochiral compounds, the sense of chirality of the product **1** should be controlled by the configuration of the chiral initiators [12].

1.3.1 Chiral Amino Acids Arising from Hydrogen Isotope Substitution

Glycine and α-methylalanine are achiral amino acids, but they become chiral when one of the hydrogen atoms of the methylene group of glycine or one of the methyl groups of α-methylalanine are deuterated. Chiral isotopomers of glycine-α-*d* and α-methyl-d_3-alanine are used as chiral triggers in asymmetric autocatalysis (Table 1.3) [34b]. The reaction of *i*-Pr_2Zn with pyrimidine-5-carbaldehyde **2** afforded alkanol **1** with high ee in conjunction with asymmetric autocatalysis with amplification of chirality. The absolute configuration of the formed alkanol **1** was correlated with the chirality resulting from hydrogen isotope substitution. In the presence of (*S*)-glycine-α-*d*, the reaction of *i*-Pr_2Zn with aldehyde **2** afforded (*S*)-alkanol **1** with high ee. In contrast, (*R*)-**1** was formed in the presence of (*R*)-glycine-α-*d* (Entry 1). In addition, α-methyl-d_3-alanine acts as chiral initiator in asymmetric autocatalysis with amplification of chirality (Entry 2). Even chiral amino acid derivatives with partially deuterated substituents such as mono- and dideuterated methyl groups ($-CDH_2$ and $-CD_2H$), i.e., *N*-benzoyl-α-methyl-d_1-alanine and *N*-benzoyl-α-methyl-d_2-alanine, act as chiral

Table 1.3 Discrimination of chirality arising from hydrogen isotope substitution of amino acid derivatives

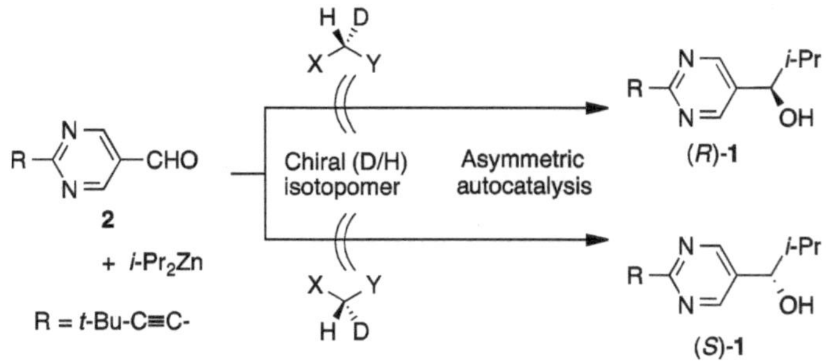

Entry	(D/H)-isotopomer		Alkanol 1			Refs.
	Structure	Config.	Config.	ee[a]	Yield	
1	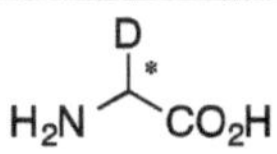	*S* *R*	*S* *R*	96 93	94 94	[34b]
2	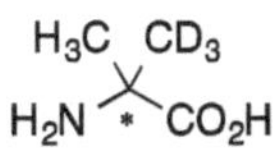	*R* *S*	*S* *R*	99 99	95 93	[34b]
3	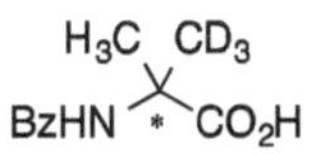	*S* *R*	*S* *R*	>99.5 >99.5	98 97	[34c]
4	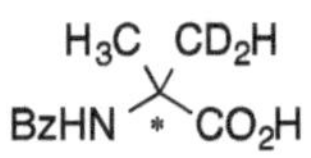	*S* *R*	*S* *R*	>99.5 >99.5	99 94	[34c]
5	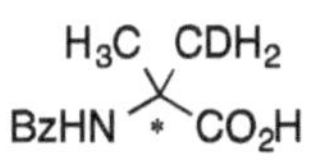	*S* *R*	*S* *R*	>99.5 >99.5	97 95	[34c]
6	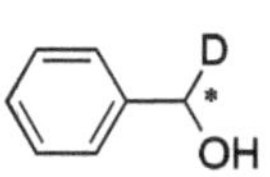	*R* *S*	*S* *R*	95 96	98 95	[34a]
7	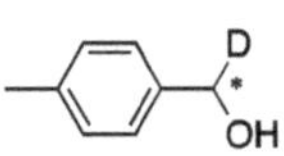	*R* *S*	*S* *R*	95 96	94 92	[34a]
8	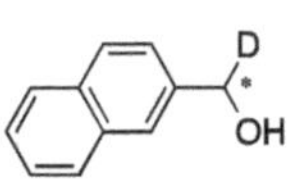	*R* *S*	*S* *R*	92 95	96 96	[34a]
9	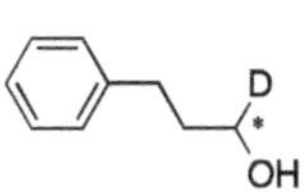	*R* *S*	*S* *R*	92 94	95 98	[34a]

[a]The highest ee value reported in the original publications.

initiators of asymmetric synthesis (Entries 3–5) [34c]. These results provide the first example of a highly enantioselective synthesis induced by chiral amino acids due to deuterium substitution. Furthermore, chirally deuterated primary alcohols act as chiral initiators of asymmetric autocatalysis (Entries 6–9) [34a].

It is known that the ratio of D to H of amino acids in meteorites is much higher than that on Earth [35a]. Glycine has only five hydrogen atoms. Therefore it is an interesting question whether the highly deuterated glycine molecule found in meteorites is a chiral isotopomer. Asymmetric autocatalysis with amplification of ee can serve as an effective tool to detect the chirality of deuterated amino acids as well as any chiral compounds in meteorites [35b].

1.3.2 Chiral Compounds Arising from Carbon and Oxygen Isotopomers as Chiral Initiators of Asymmetric Autocatalysis

Because chirality originates from the much smaller difference in atomic weight between carbon-12/carbon-13 and oxygen-16/oxygen-18 than that between H/D, carbon and oxygen isotopomers should be experimentally more difficult to discriminate than those of H/D isotopomers (Fig. 1.12). Thus, it had not been considered possible that chiral carbon isotopomers could induce chirality in some enantioselective reactions. We found that chiral compounds arising solely from carbon isotope ($^{12}C/^{13}C$) substitution can trigger asymmetric autocatalysis [36].

A chiral carbon isotopomer, i.e., dimethylphenylmethanol acts as a chiral initiator of asymmetric autocatalysis. When aldehyde **2** reacted with *i*-Pr_2Zn in the presence of (*R*)-dimethylphenylmethanol, arising from ^{13}C substitution of the phenyl group, (*R*)-pyrimidyl alkanol **1** with high ee was obtained (Table 1.4, Entry 1). In contrast, (*S*)-dimethylphenylmethanol initiated the formation of (*S*)-pyrimidyl alkanol **1**. The isotopomers of diphenylmethanol (Entry 2) and 1,3-dimethoxy-2-propanol (Entry 3) also served as chiral triggers of asymmetric autocatalysis to afford pyrimidyl alkanol **1** with high ee of the absolute configurations corresponding to those of chiral triggers.

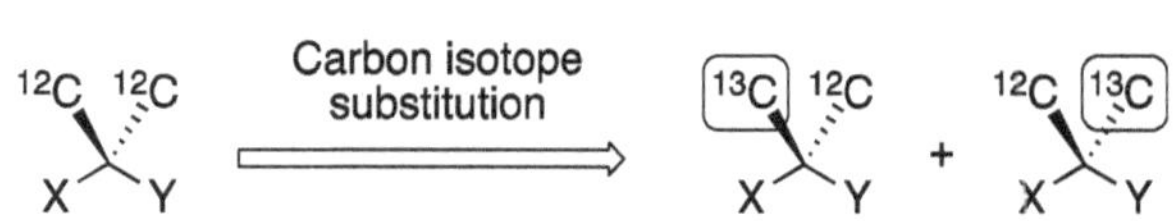

Figure 1.12 Generation of carbon isotope chirality.

Table 1.4 Asymmetric autocatalysis triggered by chiral compounds arising from carbon and oxygen isotope substitution

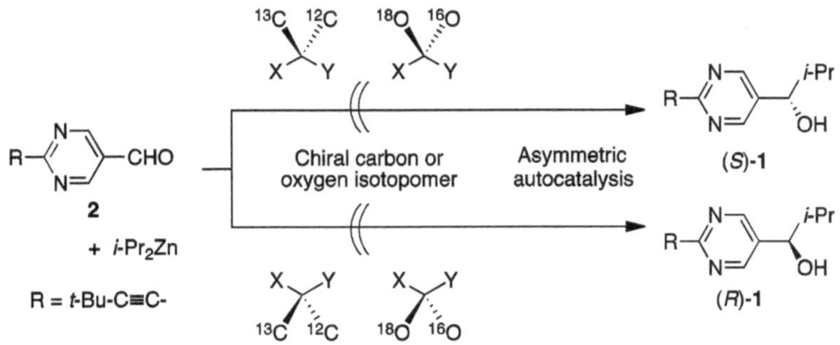

Entry	Isotopomer		Alkanol 2			Refs.
	Structure	Config.	Config.	ee[a]	Yield	
1	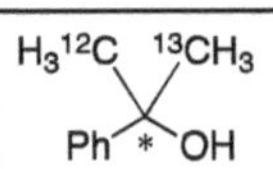	*R* *S*	*S* *R*	94 96	94 95	[36]
2	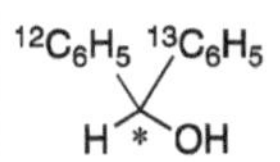	*S* *R*	*S* *R*	97 95	92 90	[36]
3	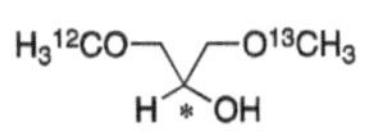	*S* *R*	*S* *R*	95 92	96 96	[36]
4	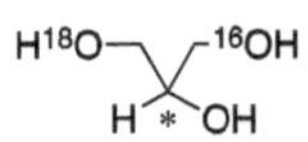	*R* *S*	*S* *R*	>99.5 >99.5	85 89	[37a]
5	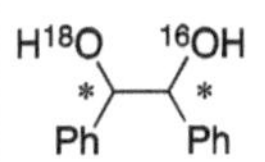	[^{18}O] *R*, [^{16}O] *S* [^{18}O] *S*, [^{16}O] *R*	*S* *R*	98 97	99 97	[37b]

[a]The highest ee value reported in the original publication.

It was also found that asymmetric autocatalysis is initiated by the oxygen ($^{16}O/^{18}O$) isotopomers, i.e., glycerin (Entry 4) and 2,3-butanediol (Entry 5) [37].

These results are the first examples of asymmetric induction in asymmetric synthesis by carbon and oxygen isotopically chiral compounds. Many achiral organic compounds become chiral when carbon and oxygen isotopes are considered. The neglected carbon and oxygen isotope chirality of many organic compounds on Earth can thus be discriminated by asymmetric autocatalysis. Pályi and Barabás describe the theoretical considerations of the natural chiral isotopomers [38].

1.3.3 Discrimination of Cryptochirality by Asymmetric Autocatalysis

Asymmetric autocatalysis with amplification of chirality can be used for the highly sensitive chiral discrimination of cryptochiral compounds. It has been defined that the carbon atom is an asymmetric carbon atom when the four substituents or atoms of a tetrahedral carbon atom are all different. An extreme case is a saturated quaternary hydrocarbon 5-ethyl-5-propylundecane, i.e., (*n*-butyl)ethyl(*n*-hexyl)(*n*-propyl)methane [39]. This compound is chiral because the structures of the four substituents are different. However, the compound does not exhibit any detectable optical rotation because the structures of the substituents are quite similar to each other [39]. Thus the discrimination of chirality of the compound has not been achieved experimentally, by spectroscopy. Mislow termed such hidden chirality that cannot be detected "cryptochirality" [2g,40]. Asymmetric autocatalysis is initiated by (*R*)- and (*S*)-5-ethyl-5-propylundecane to afford (*S*)-pyrimidyl alkanol **1** and (*R*)-**1** with high ee, respectively (Fig. 1.13A) [41a]. We also found that cryptochiral isotactic polystyrene triggers asymmetric autocatalysis (Fig. 1.13B) [41b].

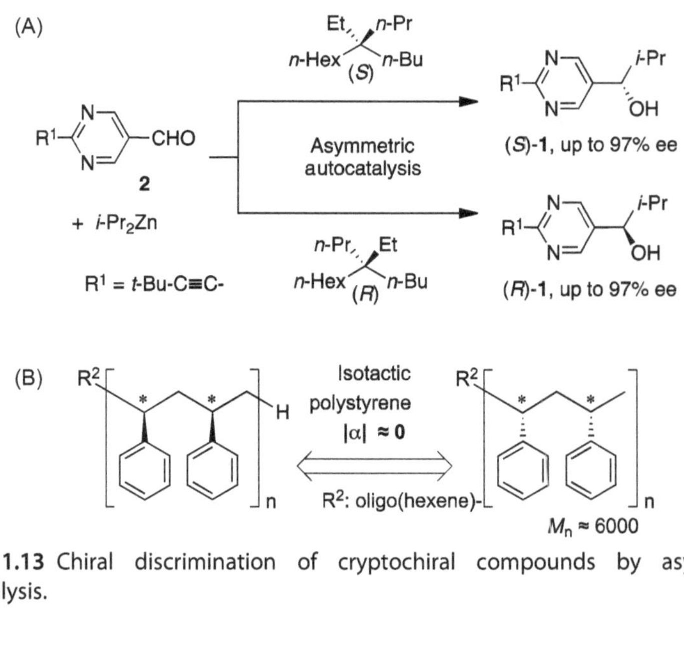

Figure 1.13 Chiral discrimination of cryptochiral compounds by asymmetric autocatalysis.

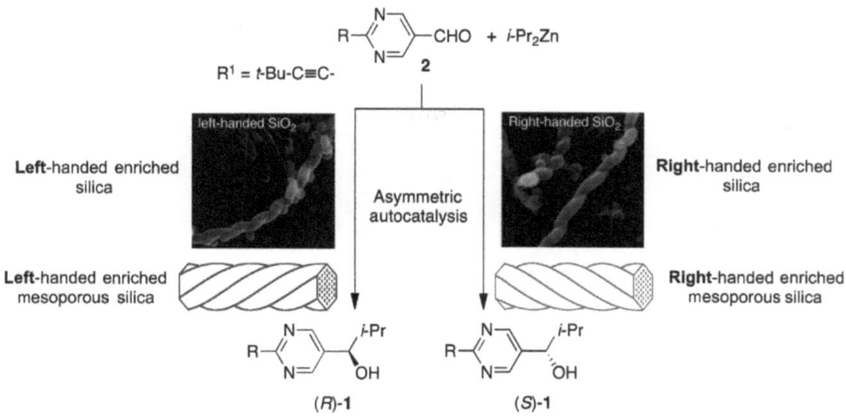

Figure 1.14 Asymmetric autocatalysis initiated with helical silica and helical mesoporous silica.

1.3.4 Chiral Discrimination of Helical Silica and Helical Mesoporous Silica by Asymmetric Autocatalysis

Artificially designed chiral inorganic materials such as helical silica [42] and helical mesoporous silica [43] have been used as chiral triggers of asymmetric autocatalysis. The chiralities of helical silica [44a] and mesoporous silica [44b] have been well discriminated by asymmetric autocatalysis (Fig. 1.14). These results are the first examples of asymmetric induction of high ee by artificially designed chiral inorganic materials.

1.4 STRUCTURE OF ASYMMETRIC AUTOCATALYST DETERMINED BY SINGLE CRYSTAL X-RAY CRYSTALLOGRAPHY

In asymmetric autocatalysis, pyrimidyl alkanol **2** reacts with *i*-Pr_2Zn to form isopropylzinc alkoxide. The crystal structures of isopropylzinc alkoxides derived from either enantiopure alkanol **2** or racemic **2** have been determined by single x-ray crystallography (Fig. 1.15) [45]. When an excess amount of *i*-Pr_2Zn was used, tetrameric crystal structures were obtained from both enantiopure and racemic alkanol **2**. There are, however, large differences in the crystal structures. Details of the crystal structures are described in Chapter 10 of this book.

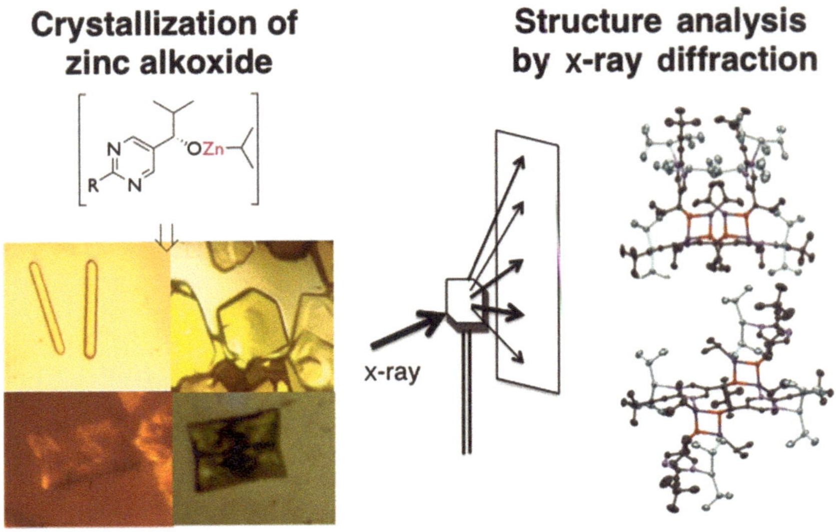

Figure 1.15 Single crystal structure of asymmetric autocatalyst.

1.5 SUMMARY

We have demonstrated asymmetric autocatalysis with amplification of ee in the enantioselective addition of i-Pr_2Zn to pyrimidine-5-carbaldehyde. The ee of (*S*)-2-alkynylpyrimidyl alkanol **1** with ee as low as *ca.* 0.00005% ee was amplified to >99.5% ee during three consecutive rounds of asymmetric autocatalysis. The product was used as an asymmetric autocatalyst for the consecutive rounds. During three consecutive rounds of asymmetric autocatalysis, the amount of the initially major (*S*)-alkanol **1** automultiplied by a factor of about 630,000 times. This amplification of ee by asymmetric autocatalysis is a unique phenomenon in that no chiral factor other than the asymmetric autocatalyst itself, i.e., pyrimidyl alkanol **1**, is necessary.

Theories of the origins of homochirality have been examined by using asymmetric autocatalysis with amplification of ee. The initial low enantioenrichment induced by the origin of chirality is amplified by asymmetric autocatalysis to very high enantioenrichments. The irradiation of right or left-handed CPL to the racemic pyrimidyl alkanol, and the subsequent asymmetric autocatalysis, enabled us to correlate the chirality of CPL with the chirality of the pyrimidyl alkanol **1** with high ee. Quartz, sodium chlorate, cinnabar, and retgersite, i.e., chiral inorganic crystals were found to be chiral triggers of asymmetric autocatalysis.

Cytosine, *N*-benzoylglycine, benzyl, and tetraphenylethylene form chiral organic crystals composed of achiral compounds. These chiral crystals act as chiral initiators of asymmetric autocatalysis. The enantiotopic faces of achiral organic crystals and chiral crystals of DL-serinium sulfate hydrate also act as chiral triggers of asymmetric autocatalysis. Spontaneous absolute asymmetric synthesis, i.e., the formation of apparently enantioenriched compounds without the intervention of any chiral factors, was achieved by using the significant power of amplification of ee by asymmetric autocatalysis of pyrimidyl alkanol. The reaction between pyrimidine-5-carbaldehyde and *i*-Pr_2Zn without the addition of any other chiral factor enables the formation of (*S*)- or (*R*)-alkanol showing a bimodal distribution. Asymmetric autocatalysis is triggered by chiral compounds resulting from carbon ($^{13}C/^{12}C$) and oxygen ($^{16}O/^{18}O$) isotopes. Asymmetric autocatalysis is also triggered by amino acids that are chiral because of H/D isotope substitution, such as glycine-α-*d*. Chiralities of cryptochiral compounds, helical and mesoporous helical silica are discriminated by asymmetric autocatalysis. X-ray crystallographic analysis reveals the structure of asymmetric autocatalyst.

REFERENCES

[1] Pasteur, L. Recherches Sur les Relations Qui Peuvent Exister Entre la Forme Crystalline, la Composition Chimique et le Sens de la Polarisation Rotatoire. *Ann. Chim. Phys.* **1848,** *24*, 442–459.

[2] (a) Inoue, Y. Asymmetric Photochemical Reactions in Solution. *Chem. Rev.* **1992,** *92*, 741–770.

(b) Bolli, M.; Micura, R.; Eschenmoser, A. Pyranosyl-RNA: Chiroselective Self-assembly of Base Sequences by Ligative Oligomerization of Tetranucleotide-2',3'-cyclophosphates (with a Commentary Concerning the Origin of Biomolecular Homochirality). *Chem. Biol.* **1997,** *4*, 309–320.

(c) Siegel, J. S. Homochiral Imperative of Molecular Evolution. *Chirality* **1998,** *10*, 24–27.

(d) Feringa, B. L.; van Delden, R. A. Absolute Asymmetric Synthesis: The Origin, Control, and Amplification of Chirality. *Angew. Chem. Int. Ed.* **1999,** *38*, 3418–3438.

(e) Kondepudi, D. K.; Asakura, K. *Spontaneous Symmetry Breaking, and Stochastic Behavior* **2001,** *34*, 946–954.

(f) Ribó, J. M.; Crusats, J.; Sagués, F.; Claret, J.; Rubires, R. Chiral Sign Induction by Vortices During the Formation of Mesophases in Stirred Solutions. *Science* **2001,** *292*, 2063–2066.

(g) Mislow, K. Absolute Asymmetric Synthesis: A Commentary. *Collect. Czech. Chem. Commun.* **2003,** *68*, 849–864.

(h) Weissbuch, I.; Lahav, M. Crystalline Architectures as Templates of Relevance to the Origins of Homochirality. *Chem. Rev.* **2011,** *111*, 3236–3267.

(i) Saito, Y.; Hyuga, H. Colloquium: Homochirality: Symmetry Breaking in Systems Driven Far from Equilibrium. *Rev. Mod. Phys.* **2013,** *85*, 603–621.

[3] (a) Katagiri, T.; Yoda, C.; Furuhashi, K.; Ueki, K.; Kubota, T. Separation of an Enantiomorph and Its Racemate by Distillation: Strong Chiral Recognizing Ability of Trifluorolactates. *Chem. Lett.* **1996,** *25*, 115–116.

(b) Green, M. M.; Park, J. W.; Sato, T.; Teramoto, A.; Lifson, S.; Selinger, R. L. B., et al. The Macromolecular Route to Chiral Amplification. *Angew. Chem. Int. Ed.* **1999,** *38*, 3138–3154.

(c) Pizzarello, S.; Weber, A. L. Prebiotic Amino Acids as Asymmetric Catalysts. *Science* **2004,** *303*, 1151.

(d) Viedma, C. Chiral Symmetry Breaking During Crystallization: Complete Chiral Purity Induced by Nonlinear Autocatalysis and Recycling. *Phys. Rev. Lett.* **2005,** *94*, 065504.

(e) Klussmann, M.; Iwamura, H.; Mathew, S. P.; Wells, D. H.; Pandya, U.; Armstrong, A. Thermodynamic Control of Asymmetric Amplification in Amino Acid Catalysis. *Nature* **2006,** *441*, 621–623.

(f) Hayashi, Y.; Matsuzawa, M.; Yamaguchi, J.; Yonehara, S.; Matsumoto, Y.; Shoji, M. Large Nonlinear Effect Observed in the Enantiomeric Excess of Proline in Solution and that in the Solid State. *Angew. Chem. Int. Ed.* **2006,** *45*, 4593–4597.

(g) Soloshonok, V. A.; Ueki, H.; Yasumoto, M.; Mekala, S.; Hirschi, J. S.; Singleton, D. A. Phenomenon of Optical Self-purification of Chiral Non-racemic Compounds. *J. Am. Chem. Soc.* **2007,** *129*, 12112–12113.

(h) Satyanarayana, T.; Abraham, S.; Kagan, H. B. Nonlinear Effects in Asymmetric Catalysis. *Angew. Chem. Int. Ed.* **2009,** *48*, 456–494.

(i) Breslow, R.; Cheng, Z.-L. L-amino Acids Catalyze the Formation of an Excess of D-glyceraldehyde, and Thus of Other D Sugars, under Credible Prebiotic Conditions. *Proc. Natl. Acad. Sci. U.S.A.* **2010,** *107*, 5723–5725.

[4] Frank, F. C. On Spontaneous Asymmetric Synthesis. *Biochim. Biophys. Acta* **1953,** *11*, 459–463.

[5] Soai, K.; Niwa, S.; Hori, H. Asymmetric Self-catalytic Seaction. Self-production of Chiral 1-(3-pyridyl)alkanols as Chiral Self-catalysts in the Enantioselective Addition of Dialkylzinc Reagents to Pyridine-3-carbaldehyde. *J. Chem. Soc. Chem. Commun.* **1990,** 982–983.

[6] (a) Soai, K.; Hayase, T.; Shimada, C.; Isobe, K. Catalytic Asymmetric Synthesis of Chiral Diol, Bis[2-(l-hydroxyalkyl)- phenyllether, an Asymmetric Autocatalytic Reaction1. *Tetrahedron: Asymmetry* **1994,** *5*, 789–792.

(b) Soai, K.; Hayase, T.; Takai, K. Catalytic Chirally Self-replicating Molecule. Asymmetric Autocatalytic Reaction of a Zinc Alkoxide of Chiral 1-ferrocenyl-2-methylpropan-1-ol. *Tetrahedron: Asymmetry* **1995,** *6*, 637–638.

[7] Soai, K.; Shibata, T.; Morioka, H.; Choji, K. Asymmetric Autocatalysis and Amplification of Enantiomeric Excess of a Chiral Molecule. *Nature* **1995,** *378*, 767–768.

[8] (a) Soai, K.; Shibata, T.; Sato, I. Enantioselective Automultiplication of Chiral Molecules by Asymmetric Autocatalysis. *Acc. Chem. Res.* **2000,** *33*, 382–390.

(b) Sato, I.; Shimizu, M.; Kawasaki, T.; Soai, K. Asymmetric Autocatalysis Initiated by Chiral Ephedrine Immobilized on Silica Gel in a Diisopropylzinc Addition to Pyrimidine-5-carbaldehyde. *Bull. Chem. Soc. Jpn.* **2004,** *77*, 1587–1588.

(c) Soai, K.; Kawasaki, T. Discovery of Asymmetric Autocatalysis with Amplification of Chirality and Its Implication in Chiral Homogeneity of Biomolecules. *Chirality* **2006,** *18*, 469–478.

(d) Soai, K.; Kawasaki, T. Amplification of Chirality. In *Topics in Current Chemistry;* Soai, K., Ed.; 284; Topics in Current Chemistry; Springer: Berlin, Heidelberg, 2008; pp 1−31.
(e) Kawasaki, T.; Soai, K. Amplification of Chirality as a Pathway to Biological Homochirality. *J. Fluor. Chem* **2010,** *131*, 525−534.
(f) Kawasaki, T.; Soai, K. Asymmetric Induction Arising from Enantiomerically Enriched Carbon-13 Isotopomers and Highly Sensitive Chiral Discrimination by Asymmetric Autocatalysis. *Bull. Chem. Soc. Jpn.* **2011,** *84*, 879−892.
(g) Kawasaki, T.; Soai, K. Asymmetric Autocatalysis Triggered by Chiral Crystals Formed from Achiral Compounds and Chiral Isotopomers. *Isr. J. Chem.* **2012,** *52*, 582−590.
(h) Kawasaki, T.; Matsumoto, A.; Soai, K. Asymmetric Autocatalysis Pathway to the Biological Homochirality. *Chim. Oggi Chem. Today* **2012,** *30* (Suppl), 10−13.
(i) Soai, K.; Kawasaki, T. Asymmetric Autocatalysis−Discovery and State of The Art. In *The Soai Reaction and Related Topic;* Palyi, G., Zucchi, C., Caglioti, C., Eds.; Academia Nationale di Scienze Lettere e Arti Modena: Modena, Edizioni Artestampa, 2012; pp 9−34.
(j) Soai, K.; Kawasaki, T. Asymmetric Autocatalysis of Pyrimidyl Alkanol. *Top Organomet. Chem.* **2013,** *44*, 261−279.
(k) Soai, K.; Kawasaki, T.; Matsumoto, A. The Origins of Homochirality Examined by Using Asymmetric Autocatalysis. *Chem. Rec.* **2014,** *14*, 70−83.
(l) Soai, K.; Kawasaki, T.; Matsumoto, A. Asymmetric Autocatalysis of Pyrimidyl Alkanol and Its Application to the Study on the Origin of Homochirality. *Acc. Chem. Res.* **2014,** *47*, 3643−3654.

[9] (a) Bolm, C.; Bienewald, F.; Seger, A. Asymmetric Autocatalysis with Amplification of Chirality. *Angew. Chem. Int. Ed.* **1996,** *35*, 1657−1659.
(b) Podlech, J.; Gehring, T. New Aspects of Soai's Asymmetric Autocatalysis. *Angew. Chem. Int. Ed.* **2005,** *44*, 5776−5777.
(c) Caglitoti, L.; Zucchi, C.; Pályi, G. Single-molecule Chirality. *Chim. Oggi* **2005,** *38−39*, 42−43.
(d) Gehring, T.; Busch, M.; Schlageter, M.; Weingand, D. A Concise Summary of Experimental Facts about the Soai Reaction. *Chirality* **2010,** *22* (Suppl 1), E173−E182.

[10] (a) Sato, I.; Omiya, D.; Igarashi, H.; Kato, K.; Ogi, Y.; Tsukiyama, K., et al. Relationship between the Time, Yield, and Enantiomeric Excess of Asymmetric Autocatalysis of Chiral 2-alkynyl-5-pyrimidyl Alkanol with Amplification of Enantiomeric Excess. *Tetrahedron: Asymmetry* **2003,** *14*, 975−979.
(b) Blackmond, D. G. Asymmetric Autocatalysis and Its Implications for the Origin of Homochirality. *Proc. Natl. Acad. Sci. U.S.A.* **2004,** *101*, 5732−5736.
(c) Gridnev, I. D.; Serafimov, J. M.; Brown, J. M. Solution Structure and Reagent Binding of the Zinc Alkoxide Catalyst in the Soai Asymmetric Autocatalytic Reaction. *Angew. Chem. Int. Ed.* **2004,** *43*, 4884−4887.
(d) Klankermayer, J. J.; Gridnev, I. D.; Brown, J. M. Role of the Isopropyl Group in Asymmetric Autocatalytic Zinc Alkylations. *Chem. Commun.* **2007,** 3151−3153.
(e) Schiaffino, L.; Ercolani, G. Unraveling the Mechanism of the Soai Asymmetric Autocatalytic Reaction by First-principles Calculations: Induction and Amplification of Chirality by Self-assembly of Hexamolecular Complexes. *Angew. Chem. Int. Ed.* **2008,** *47*, 6832−6835.
(f) Ercolani, G.; Schiaffino, L. Putting the Mechanism of the Soai Reaction to the Test: DFT Study of the Role of Aldehyde and Dialkylzinc Structure. *J. Org. Chem.* **2011,** *76*, 2619−2626.

(g) Gehring, T.; Quaranta, M.; Odell, B.; Blackmond, D. G.; Brown, J. M. Observation of a Transient Intermediate in Soai's Asymmetric Autocatalysis: Insights from ^{1}H NMR Turnover in Real Time. *Angew. Chem. Int. Ed.* **2012,** *51*, 9539–9542.
(h) Gridnev, I. D.; Vorobiev, A. K. Quantification of Sophisticated Equilibria in the Reaction Pool and Amplifying Catalytic Cycle of the Soai Reaction. *ACS Catal* **2012,** *2*, 2137–2149.
(i) Micheau, J.-C.; Coudret, C.; Cruz, J.-M.; Buhse, T. Amplification of Enantiomeric Excess, Mirror-image Symmetry Breaking and Kinetic Proofreading in Soai Reaction Models with Different Oligomeric Orders. *Phys. Chem. Chem. Phys.* **2012,** *14*, 13239–13248.

[11] (a) Shibata, T.; Hayase, T.; Yamamoto, J.; Soai, K. One-pot Asymmetric Autocatalytic Reaction with Remarkable Amplification of Enantiomeric Excess. *Tetrahedron: Asymmetry* **1997,** *8*, 1717–1719.
(b) Shibata, T.; Yonekubo, S.; Soai, K. Practically Perfect Asymmetric Autocatalysis with (2-Alkynyl-5-pyrimidyl)alkanols. *Angew. Chem. Int. Ed.* **1999,** *38*, 659–661.
(c) Sato, I.; Urabe, H.; Ishiguro, S.; Shibata, T.; Soai, K. Amplification of Chirality from Extremely Low to Greater than 99.5 % ee by Asymmetric Autocatalysis. *Angew. Chem. Int. Ed.* **2003,** *42*, 315–317.

[12] Shibata, T.; Yamamoto, J.; Matsumoto, N.; Yonekubo, S.; Osanai, S.; Soai, K. Amplification of a Slight Enantiomeric Imbalance in Molecules Based on Asymmetric Autocatalysis: The First Correlation between High Enantiomeric Enrichment in a Chiral Molecule and Circularly Polarized Light. *J. Am. Chem. Soc.* **1998,** *120*, 12157–12158.

[13] (a) Bailey, J.; Chrysostomou, A.; Hough, J. H.; Gledhill, T. M.; McCall, A.; Clark, S., et al. Circular Polarization in Star-formation Regions: Implications for Biomolecular Homochirality. *Science* **1998,** *281*, 672–674.

[14] (a) Kagan, H.; Moradpour, A.; Nicoud, J. F.; Balavoine, G.; Tsoucaris, G. Photochemistry with Circularly Polarized Light. Synthesis of Optically Active Hexahelicene. *J. Am. Chem. Soc.* **1971,** *93*, 2353–2354.
(b) Nishino, H.; Kosaka, A.; Hembury, G. A.; Aoki, F.; Miyauchi, K.; Shitomi, H., et al. Absolute Asymmetric Photoreactions of Aliphatic Amino Acids by Circularly Polarized Synchrotron Radiation: Critically pH-Dependent Photobehavior. *J. Am. Chem. Soc.* **2002,** *124*, 11618–11627.

[15] Noorduin, W. L.; Bode, A. a C.; van der Meijden, M.; Meekes, H.; van Etteger, A. F.; van Enckevort, W. J. P., et al. Complete Chiral Symmetry Breaking of an Amino Acid Derivative Directed by Circularly Polarized Light. *Nat. Chem.* **2009,** *1*, 729–732.

[16] (a) Sato, I.; Sugie, R.; Matsueda, Y.; Furumura, Y.; Soai, K. Asymmetric Synthesis Utilizing Circularly Polarized Light Mediated by the Photoequilibrium of Chiral Olefins in Conjunction with Asymmetric Autocatalysis. *Angew. Chem. Int. Ed.* **2004,** *43*, 4490–4492.
(b) Kawasaki, T.; Sato, M.; Ishiguro, S.; Saito, T.; Morishita, Y.; Sato, I., et al. Enantioselective Synthesis of near Enantiopure Compound by Asymmetric Autocatalysis Triggered by Asymmetric Photolysis with Circularly Polarized Light. *J. Am. Chem. Soc.* **2005,** *127*, 3274–3275.

[17] (a) Bonner, W. A.; Kavasmaneck, P. R.; Martin, F. S.; Flores, J. J. Asymmetric Adsorption of Alanine by Quartz. *Science* **1974,** *186*, 143–144.
(b) Hazen, R. M.; Filley, T. R.; Goodfriend, G. A. Selective Adsorption of L- and D-amino Acids on Calcite: Implications for Biochemical Homochirality. *Proc. Natl. Acad. Sci. U.S.A.* **2001,** *98*, 5487–5490.

(c) Hazen, R. M.; Sholl, D. S. Chiral Selection on Inorganic Crystalline Surfaces. *Nat. Mater.* **2003,** *2*, 367–374.

[18] Soai, K.; Osanai, S.; Kadowaki, K.; Yonekubo, S.; Shibata, T.; Sato, I. *d*- and *l*-Quartz-Promoted Highly Enantioselective Synthesis of a Chiral Organic Compound. *J. Am. Chem. Soc.* **1999,** *121*, 11235–11236.

[19] (a) Kondepudi, D. K.; Kaufman, R. J.; Singh, N. Chiral Symmetry Breaking in Sodium Chlorate Crystallizaton. *Science* **1990,** *250*, 975–976.

(b) McBride, J. M.; Carter, R. L. Spontaneous Resolution by Stirred Crystallization. *Angew. Chem. Int. Ed.* **1991,** *30*, 293–295.

[20] (a) Sato, I.; Kadowaki, K.; Soai, K. Asymmetric Synthesis of an Organic Compound with High Enantiomeric Excess Induced by Inorganic Ionic Sodium Chlorate. *Angew. Chem. Int. Ed.* **2000,** *39*, 1510–1512.

(b) Sato, I.; Kadowaki, K.; Ohgo, Y.; Soai, K. Highly Enantioselective Asymmetric Autocatalysis Induced by Chiral Ionic Crystals of Sodium Chlorate and Sodium Bromate. *J. Mol. Catal. A Chem.* **2004,** *216*, 209–214.

[21] Shindo, H.; Shirota, Y.; Niki, K.; Kawasaki, T.; Suzuki, K.; Araki, Y., et al. Asymmetric Autocatalysis Induced by Cinnabar: Observation of the Enantioselective Adsorption of a 5-Pyrimidyl Alkanol on the Crystal Surface. *Angew. Chem. Int. Ed.* **2013,** *52*, 9135–9138.

[22] Matsumoto, A.; Ozawa, H.; Inumaru, A.; Soai, K. Asymmetric Induction by Retgersite, Nickel Sulfate Hexahydrate, in Conjunction with Asymmetric Autocatalysis. *New J. Chem.* **2015,** *39*, 6742–6745.

[23] (a) Sakamoto, M. Absolute Asymmetric Synthesis from Achiral Molecules in the Chiral Crystalline Environment. *Chem. Eur. J* **1997,** *3*, 684–689.

(b) Kahr, B.; Gurney, R. W. Dyeing Crystals. *Chem. Rev.* **2001,** *101*, 893–951.

(c) Matsuura, T.; Koshima, H. Introduction to Chiral Crystallization of Achiral Organic Compounds Spontaneous Generation of Chirality. *J. Photochem. Photobiol. C: Photochem. Rev.* **2005,** *6*, 7–24.

[24] Kawasaki, T.; Jo, K.; Igarashi, H.; Sato, I.; Nagano, M.; Koshima, H., et al. Asymmetric Amplification Using Chiral Cocrystals Formed from Achiral Organic Molecules by Asymmetric Autocatalysis. *Angew. Chem. Int. Ed.* **2005,** *44*, 2774–2777.

[25] Robertson, M. P.; Miller, S. L. An Efficient Prebiotic Synthesis of Cytosine and Uracil. *Nature* **1995,** *375*, 772–774.

[26] (a) Kawasaki, T.; Suzuki, K.; Hakoda, Y.; Soai, K. Achiral Nucleobase Cytosine Acts as an Origin of Homochirality of Biomolecules in Conjunction with Asymmetric Autocatalysis. *Angew. Chem. Int. Ed.* **2008,** *47*, 496–499.

(b) Kawasaki, T.; Hakoda, Y.; Mineki, H.; Suzuki, K.; Soai, K. Generation of Absolute Controlled Crystal Chirality by the Removal of Crystal Water from Achiral Crystal of Nucleobase Cytosine. *J. Am. Chem. Soc.* **2010,** *132*, 2874–2875.

(c) Mineki, H.; Kaimori, Y.; Kawasaki, T.; Matsumoto, A.; Soai, K. Enantiodivergent Formation of a Chiral Cytosine Crystal by Removal of Crystal Water from an Achiral Monohydrate Crystal under Reduced Pressure. *Tetrahedron: Asymmetry* **2013,** *24*, 1365–1367.

[27] (a) Mineki, H.; Hanasaki, T.; Matsumoto, A.; Kawasaki, T.; Soai, K. Asymmetric Autocatalysis Initiated by Achiral Nucleic Acid Base Adenine: Implications on the Origin of Homochirality of Biomolecules. *Chem. Commun.* **2012,** *48*, 10538–10540.

(b) Kawasaki, T.; Suzuki, K.; Hatase, K.; Otsuka, M.; Koshima, H.; Soai, K. Enantioselective Synthesis Mediated by Chiral Crystal of Achiral Hippuric Acid in Conjunction with Asymmetric Autocatalysis. *Chem. Commun.* **2006,** *2*, 1869–1871.

(c) Kawasaki, T.; Harada, Y.; Suzuki, K.; Tobita, T.; Florini, N.; Pályi, G., et al. Enantioselective Synthesis Utilizing Enantiomorphous Organic Crystal of Achiral Benzils as a Source of Chirality in Asymmetric Autocatalysis. *Org. Lett.* **2008,** *10*, 4085–4088.
(d) Kawasaki, T.; Nakaoda, M.; Kaito, N.; Sasagawa, T.; Soai, K. Asymmetric Autocatalysis Induced by Chiral Crystals of Achiral Tetraphenylethylenes. *Orig. Life Evol. Biosph.* **2010,** *40*, 65–78.
(e) Carter, D. J.; Rohl, A. L.; Shtukenberg, A.; Bian, S.; Hu, C.; Baylon, L. Prediction of Soai Reaction Enantioselectivity Induced by Crystals of N -(2-Thienylcarbonyl)glycine. *Cryst. Growth Des.* **2012,** *12*, 2138–2145.
(f) Kawasaki, T.; Uchida, M.; Kaimori, Y.; Sasagawa, T.; Matsumoto, A.; Soai, K. Enantioselective Synthesis Induced by the Helical Molecular Arrangement in the Chiral Crystal of Achiral Tris(2-Hydroxyethyl) 1,3,5-Benzenetricarboxylate in Conjunction with Asymmetric Autocatalysis. *Chem. Lett.* **2013,** *42*, 711–713.
(g) Matsumoto, A.; Ide, T.; Kaimori, Y.; Fujiwara, S.; Soai, K. Asymmetric Autocatalysis Triggered by Chiral Crystal of Achiral Ethylenediamine Sulfate. *Chem. Lett.* **2015,** *44*, 688–690.
(h) Sato, I.; Kadowaki, K.; Ohgo, Y.; Soai, K.; Ogino, H. Highly Enantioselective Asymmetric Autocatalysis Induced by Chiral Cobalt Complexes Due to the Topology of the Coordination of Achiral Ligands. *Chem. Commun.* **2001,** 1022–1023.

[28] Kawasaki, T.; Kamimura, S.; Amihara, A.; Suzuki, K.; Soai, K. Enantioselective C-C Bond Formation as a Result of the Oriented Prochirality of an Achiral Aldehyde at the Single-crystal Face Upon Treatment with a Dialkyl Zinc Vapor. *Angew. Chem. Int. Ed.* **2011,** *50*, 6796–6798.

[29] Kawasaki, T.; Sasagawa, T.; Shiozawa, K.; Uchida, M.; Suzuki, K.; Soai, K. Enantioselective Synthesis Induced by Chiral Crystal Composed of DL-serine in Conjunction with Asymmetric Autocatalysis. *Org. Lett.* **2011,** *13*, 2361–2363.

[30] Barabas, B.; Caglioti, L.; Zucchi, C.; Maioli, M.; Gál, E.; Micskei, K., et al. Violation of Distribution Symmetry in Statistical Evaluation of Absolute Enantioselective Synthesis. *J. Phys. Chem. B.* **2007,** *111*, 11506–11510.

[31] (a) Soai, K.; Sato, I.; Shibata, T.; Komiya, S.; Hayashi, M.; Matsueda, Y., et al. Asymmetric Synthesis of Pyrimidyl Alkanol without Adding Chiral Substances by the Addition of Diisopropylzinc to Pyrimidine-5-carbaldehyde in Conjunction with Asymmetric Autocatalysis. *Tetrahedron: Asymmetry* **2003,** *14*, 185–188.
(b) Soai, K.; Shibata, T.; Kowata, Y.: Japan Kokai Tokkyo Koho JP 1997 9-268179. Application date: February 1 and April 18, 1996. An abstract is readily available as JPH09268179 from the European Patent Office (http://worldwide.espacenet.com) The Title of the Patent is "Production of Optically Active Pyrimidylalkyl Alcohol by Spontaneous Absolute Asymmetric Synthesis".
(c) Singleton, D. A.; Vo, L. K. A Few Molecules Can Control the Enantiomeric Outcome. Evidence Supporting Absolute Asymmetric Synthesis Using the Soai Asymmetric Autocatalysis. *Org. Lett.* **2003,** *5*, 4337–4339.

[32] (a) Kawasaki, T.; Suzuki, K.; Shimizu, M.; Ishikawa, K.; Soai, K. Spontaneous Absolute Asymmetric Synthesis in the Presence of Achiral Silica Gel in Conjunction with Asymmetric Autocatalysis. *Chirality* **2006,** *18*, 479–482.
(b) Suzuki, K.; Hatase, K.; Nishiyama, D.; Kawasaki, T.; Soai, K. Spontaneous Absolute Asymmetric Synthesis Promoted by Achiral Amines in Conjunction with Asymmetric Autocatalysis. *J. Syst. Chem.* **2010,** *1*, 5.

[33] (a) Lente, G. Stochastic Kinetic Models of Chiral Autocatalysis: a General Tool for the Quantitative Interpretation of Total Asymmetric Synthesis. *J. Phys. Chem. A* **2005,** *109*, 11058–11063.

(b) Islas, J. R.; Lavabre, D.; Grevy, J.-M.; Lamoneda, R. H.; Cabrera, H. R.; Micheau, J.-C., et al. Mirror-symmetry Breaking in the Soai Reaction: A Kinetic Understanding. *Proc. Natl. Acad. Sci. U.S.A.* **2005,** *102*, 13743–13748.
(c) Micskei, K.; Rábai, G.; Gál, E.; Caglioti, L.; Pályi, G. Oscillatory Symmetry Breaking in the Soai Reaction. *J. Phys. Chem. B* **2008,** *112*, 9196–9200.
(d) Lavabre, D.; Micheau, J.-C.; Rivera Islas, J.; Buhse, T.: Kinetic Insight into Specific Features of the Autocatalytic Soai Reaction. In Amplification Of Chirality; Soai, K. Ed.; 284; Top. Curr. Chem.; 2008; pp 67–96.
(e) Saito, Y.; Hyuga, H., Rate Equation Approaches to Amplification of Enantiomeric Excess and Chiral Symmetry Breaking. In Amplification Of Chirality; Soai, K. Ed.; 284; Top. Curr. Chem.; 2008; pp 97–118.
(f) Crusats, J. Frank Model and Spontaneous Emergence of Chirality in Closed Systems. *ChemPhysChem* **2009,** *10*, 2123–2131.
(g) Barabas, B.; Caglioti, L.; Micskei, K.; Palyi, G. Data-based Stochastic Approach to Absolute Asymmetric Synthesis by Autocatalysis. *Bull. Chem. Soc. Jpn.* **2009,** *82*, 1372–1376.

[34] (a) Sato, I.; Omiya, D.; Saito, T.; Soai, K. Highly Enantioselective Synthesis Induced by Chiral Primary Alcohols Due to Deuterium Substitution. *J. Am. Chem. Soc.* **2000,** *122*, 11739–11740.
(b) Kawasaki, T.; Shimizu, M.; Nishiyama, D.; Ito, M.; Ozawa, H.; Soai, K. Asymmetric Autocatalysis Induced by Meteoritic Amino Acids with Hydrogen Isotope Chirality. *Chem. Commun.* **2009,** 4396–4398.
(c) Kawasaki, T.; Ozawa, H.; Ito, M.; Soai, K. Enantioselective Synthesis Induced by Compounds with Chirality Arising from Partially Deuterated Methyl Groups in Conjunction with Asymmetric Autocatalysis. *Chem. Lett.* **2011,** *40*, 320–321.

[35] (a) Pizzarello, S.; Huang, Y. The Deuterium Enrichment of Individual Amino Acids in Carbonaceous Meteorites: A Case for the Presolar Distribution of Biomolecule Precursors. *Geochim. Cosmochim. Acta* **2005,** *69*, 599–605.
(b) Kawasaki, T.; Hatase, K.; Fujii, Y.; Jo, K.; Soai, K.; Pizzarello, S. The Distribution of Chiral Asymmetry in Meteorites: An Investigation Using Asymmetric Autocatalytic Chiral Sensors. *Geochim. Cosmochim. Acta* **2006,** *70*, 5395–5402.

[36] Kawasaki, T.; Matsumura, Y.; Tsutsumi, T.; Suzuki, K.; Ito, M.; Soai, K. Asymmetric Autocatalysis Triggered by Carbon Isotope ($^{13}C/^{12}C$) Chirality. *Science* **2009,** *324*, 492–495.

[37] (a) Kawasaki, T.; Okano, Y.; Suzuki, E.; Takano, S.; Oji, S.; Soai, K. Asymmetric Autocatalysis: Triggered by Chiral Isotopomer Arising from Oxygen Isotope Substitution. *Angew. Chem. Int. Ed.* **2011,** *50*, 8131–8133.
(b) Matsumoto, A.; Oji, S.; Takano, S.; Tada, K.; Kawasaki, T.; Soai, K. Asymmetric Autocatalysis Triggered by Oxygen Isotopically Chiral Glycerin. *Org. Biomol. Chem.* **2013,** *11*, 2928–2931.

[38] (a) Barabás, B.; Caglioti, L.; Micskei, K.; Zucchi, C.; Pályi, G. Isotope Chirality and Asymmetric Autocatalysis: A Possible Entry to Biological Chirality. *Orig. Life Evol. Biosph.* **2008,** *38*, 317–327.
(b) Barabás, B.; Kurdi, R.; Pályi, G. Natural Abundance Isotopic Chirality in the Reagents of the Soai Reaction. *Symmetry (Basel)* **2016,** *8*, 2.

[39] Wynberg, H.; Houbiers, J.P.M.; Bosch, H.W.: The Optical Activity of Butylethylhexylpropylmethane. *J. Am. Chem Soc.* 1965, 87, 2635–2639.

[40] Mislow, K.; Bickart, P. An Epistemological Note on Chirality. *Isr. J. Chem.* **1976,** *15*, 1–6.

[41] (a) Kawasaki, T.; Tanaka, H.; Tsutsumi, T.; Kasahara, T.; Sato, I.; Soai, K. Chiral Discrimination of Cryptochiral Saturated Quaternary and Tertiary Hydrocarbons by Asymmetric Autocatalysis. *J. Am. Chem. Soc.* **2006,** *128*, 6032–6033.
(b) Kawasaki, T.; Hohberger, C.; Araki, Y.; Hatase, K.; Beckerle, K.; Okuda, J., et al. Discrimination of Cryptochirality in Chiral Isotactic Polystyrene by Asymmetric Autocatalysis. *Chem. Commun.* **2009,** 5621–5623.
[42] Jung, J. H.; Ono, Y.; Hanabusa, K.; Shinkai, S. Creation of Both Right-handed and Left-handed Silica Structures by Sol- Gel Transcription of Organogel Fibers Comprised of Chiral Diaminocyclohexane Derivatives. *J. Am. Chem. Soc.* **2000,** *122*, 5008–5009.
[43] Che, S.; Liu, Z.; Ohsuna, T.; Sakamoto, K.; Terasaki, O.; Tatsumi, T. Synthesis and Characterization of Chiral Mesoporous Silica. *Nature* **2004,** *429*, 281–284.
[44] (a) Sato, I.; Kadowaki, K.; Urabe, H.; Jung, J. H.; Ono, Y.; Shinkai, S., et al. Highly Enantioselective Synthesis of Organic Compound Using Right- and Left-handed Helical Silica. *Tetrahedron Lett.* **2003,** *44*, 721–724.
(b) Kawasaki, T.; Araki, Y.; Hatase, K.; Suzuki, K.; Matsumoto, A.; Yokoi, T., et al. Helical Mesoporous Silica as an Inorganic Heterogeneous Chiral Trigger for Asymmetric Autocatalysis with Amplification of Enantiomeric Excess. *Chem. Commun.* **2015,** *51*, 8742–8744.
[45] Matsumoto, A.; Abe, T.; Hara, A.; Tobita, T.; Sasagawa, T.; Kawasaki, T., et al. Crystal Structure of the Isopropylzinc Alkoxide of Pyrimidyl Alkanol: Mechanistic Insights for Asymmetric Autocatalysis with Amplification of Enantiomeric Excess. *Angew. Chem. Int. Ed.* **2015,** *54*, 15218–15221.

CHAPTER 2

Soai Autocatalyst as Conveyor of Chirality Coupling Autocatalysis With Asymmetric Induction

Matías Funes-Maldonado, Carlo Romagnoli, Bora Sieng and Mohamed Amedjkouh
Department of Chemistry, University of Oslo, Oslo, Norway

Abstract

Herein, we report on the development of a new concept to generate and propagate chirality, which relies on a chemical system combining asymmetric amplification of Soai autocatalyst with parallel/consecutive asymmetric induction in a subsequent alkylation and alkynylation of azaaryl aldehydes. Thus, Soai autocatalyst is amplified in situ from 11% to >98% *ee* and concomitantly catalyzes the addition of $Zn(i\text{-}Pr)_2$ to provide pyridyl alkanols with up to 99% *ee*. Such process displayed unique amplifications of *ee* in one cycle for both Soai autocatalyst and other alkanol compounds produced throughout the reaction. Likewise, in a sequence of autocatalysis-catalysis in asymmetric alkynylation of azaaryl aldehydes, the Soai autocatalyst is generated from minimal enantiopurity by asymmetric autocatalysis in >98% *ee*, to promote addition of alkynylzinc to a variety of pyridyl aldehydes and provide enantioenriched propargylic alcohols in up 86% *ee*. A different and complementary approach allows the remote amplification of chirality. Thus, several chiral alkynols with very low enantiomeric excess are used as triggers for the Soai autocatalyst amplification, which is used as their own chiral catalyst. This way "enantiopoor" alkynols are remotely amplified with retension or inversion of absolute configuration.

Keywords: Systems chemistry; self-replication; chirality; asymmetric amplification; reaction networks

2.1 INTRODUCTION

The chiral homogeneity of biomolecules is considered to be directly related to the origin and evolution of life, and continues to fascinate scientists [1]. Homochirality might have preceded the origin of life, but it remains intriguing how and when biomolecules achieved high enantioenrichment. Once generated, even with a small imbalance, complex chemical systems are required for amplification and propagation of

Advances in Asymmetric Autocatalysis and Related Topics
DOI: http://dx.doi.org/10.1016/B978-0-12-812824-4.00002-2

homochirality. In this context, asymmetric autocatalysis is particularly interesting; as such a chemical system may combine self-replication with amplification of chirality [2]. The Soai reaction offered the first, and to date the only, example of an asymmetric autocatalytic reaction to provide spectacular chiral amplification [3]. However, addressing this matter by direct autocatalytic processes is far from obvious for most chemical reactions.

Nevertheless, the rise of homochirality requires the following three sequential stages: mirror-symmetry breaking (a small enantiomeric imbalance arise) followed by chiral amplification (enantiomeric enrichment occurs) and finally this chirality may be transferred to another set of molecules in a subsequent process (Scheme 2.1).

Moreover, in our developing societies, there is a strong demand to access chiral molecules from the pharmaceutical, flavor, fragrance, and agrochemical industries. Therefore the synthesis of chiral molecules has become a fundamental task for organic chemists in the last 40 years, and has stimulated intensive research in the field of asymmetric catalysis. However, a major challenge in asymmetric catalysis is the design and optimization of new ligands and catalytic systems, which usually suffer from substrate specificity and require very long prospection time (Scheme 2.2). Despite the significant progress in enantioselective transformations, the chiral pool remains the only source of chirality towards optically pure catalysts, since a linear relationship of the *ee* values between the chiral catalyst and the product is anticipated, although there has been a recent surge of interest in chiral amplification as provided by Kagan in term of nonlinear effects (NLE) that are sometimes observed in catalytic reactions [4].

The Soai reaction shows the characteristics of mirror-symmetry breaking by yielding enantiomerically enriched pyrimidyl alkanol in absence of any external chiral influence [5]. As stated by Frank in his theoretical work, autocatalysis alone is not sufficient, but a mutual inhibition process is also necessary to reach high levels of chiral amplification. These two

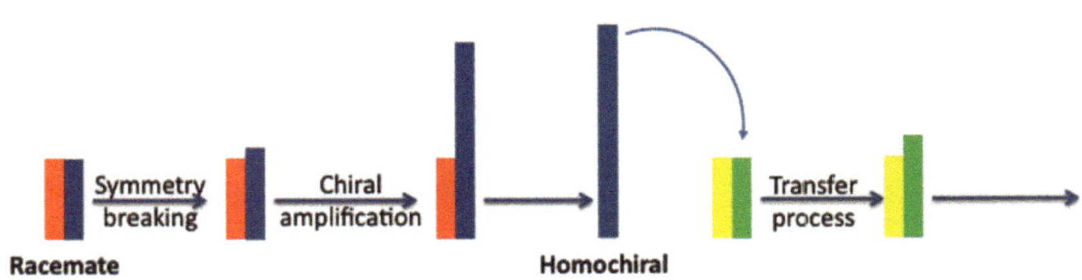

Scheme 2.1 Process for generation and propagation of homochirality.

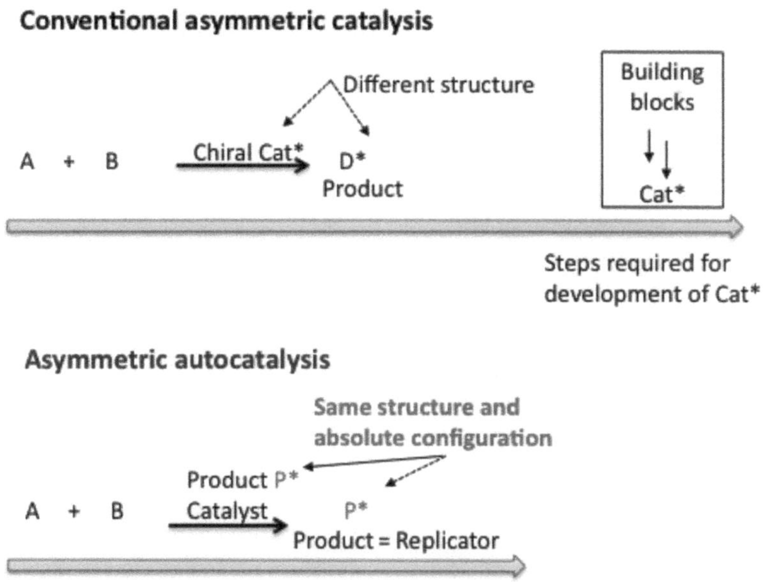

Scheme 2.2 Asymmetric autocatalysis versus conventional catalysis.

features can give rise to the mirror-symmetry breaking phenomenon. Therefore the Soai reaction can be described as a Frank-type reaction. The kinetic model proposed by Blackmond and Brown accurately predicted the reaction rate and the amplification of *ee* for the Soai reaction [6]. Ercolani and Schiaffino proposed for the first time a catalytic cycle detailed at the molecular level for the Soai reaction by density functional theory (DFT) calculations [7].

Buhse and Micheau proposed kinetic models based on numerical integration of coupled rate equations. In their studies, they provided information on the origin of the mirror-symmetry breaking in the Soai reaction [8]. Ribó et al. also used numerical simulations to describe the Soai reaction as a network of coupled reactions taking place in a closed system which gives rise to absolute asymmetric synthesis under kinetic control [9]. These two groups examined different models that involve autocatalysis and mutual inhibition, which were able to reproduce the asymmetric amplification and the mirror-symmetry breaking in a similar way to that observed in the Soai reaction.

Over the last 25 years, the Soai reaction remains the only example of asymmetric autocatalysis with amplification of *ee*. Since this reaction can be used for the construction of a model for understanding the origin of

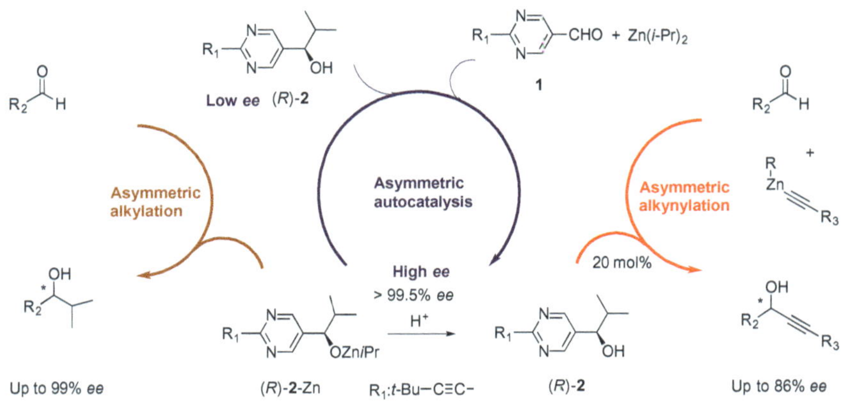

Scheme 2.3 The Soai autocatalyst as a conveyor of amplified chirality.

homochirality on Earth, most of the efforts have been directed towards the elucidation of its mechanism, and much less in finding applications of this reaction. These applications focused on its use as a tool for the detection of asymmetry or determination of low *ee* values. Interestingly, Soai's group also reported an efficient one-pot sequential asymmetric alkylation [10]. Thus, the chiral catalyst first self improves its *ee* by asymmetric autocatalysis and then acts as a chiral catalyst in a subsequent alkylation with significant amplification of *ee* in newly formed products.

Although this preliminary report could provide a new vision for asymmetric catalysis and for the rise and propagation of homochirality, this new concept of asymmetric catalysis went overlooked and no further studies could be found in the literature.

In this chapter, we describe our recent work using the Soai autocatalyst as a conveyor of chirality by coupling asymmetric autocatalysis with concomitant asymmetric alkylation and subsequent alkynylation, as depicted in Scheme 2.3 [11]. Taking a step even further, we also show that Soai autocatalyst provides very high remote asymmetric amplification of target alkynols when they are used as chiral seeds with very low initial *ee*.

2.2 RESULTS AND DISCUSSION

2.2.1 Asymmetric Alkylation: Preliminary Approach

We envisioned a reaction network combining simultaneous enantioselective autocatalysis with asymmetric induction in the alkylation of azaaryl substrates, which are incapable of such asymmetric amplification on their own. The proposed catalytic process consists of an asymmetric autocatalytic loop where the low *ee* pyrimidyl alcohol **2** undergo a self-replication process with amplification of *ee* from aldehyde **1**, in conjunction with a second loop where **2** should act as the catalyst for asymmetric induction in the dialkylzinc addition to a different set of pyridine-3-carboxaldehydes **3–11**. It is important to note that control reactions showed that the selected compounds **12–20** can undergo autocatalysis but with a pronounced loss of *ee*, i.e., no amplification was observed.

Following the procedure described by Soai, chiral pyrimidyl alcohol (*R*)-**2** was achieved in high yield at ca.98% *ee* and then used as a catalyst to promote the asymmetric addition of $Zn(i\text{-}Pr)_2$ to different aldehydes **3–11** [12] as depicted in Table 2.1. Among the compounds investigated, two groups emerged: one group of alkanols **12–16** showed asymmetric induction ranging from very low to good with *R* configuration. Similarly, compounds *(R)*-**17–20** provided good asymmetric induction with very good to excellent chirality transfer between 89% and 93% *ee*, suggesting a conceivable interaction with regard to the structural similarity of **17–20** and **2**. This result clearly shows the beneficial contribution of autocatalyst **2** to achieve high *ee* in systems that are themselves incapable of amplification of *ee*.

Remarkably a strong amplification was observed when (*R*)-**17–20** were generated under a reverse mode of addition of $Zn(i\text{-}Pr)_2$ to precursor aldehydes, as shown in parentheses in Table 2.1. The structural similarity with the original Soai (*R*)-**2** may account for the striking asymmetric induction observed with alkanols **17–20**, which suggests a positive interaction between autocatalysts and the newly formed pyridine alkanols. Then the question rises as to whether it is possible to set up a chemical system coupling asymmetric autocatalysis with asymmetric induction without prior amplification of (*R*)-**2,** as shown in Scheme 2.4. Such a one-pot process would prefigure a powerful synthetic tool if autocatalyst (*R*)-**2** and pyridine alkanol are generated in parallel in the reaction.

2.2.2 Parallel Asymmetric Alkylation by Soai Autocatlayst

This approach proved to be very powerful for the amplification of both autocatalyst **2** and alkanols **17–20** in one catalytic cycle. Moreover, under these conditions (*R*)-**2** showed high autocatalytic activity and was amplified to a final 98% *ee*. Therefore it is plausible that compound **2** is not operating just as an autocatalyst, but might participate in an improved catalytic system combining asymmetric amplification of autocatalyst (*R*)-**2**

Table 2.1 Enantioselective addition of $Zn(i\text{-}Pr)_2$ to aldehydes using (*R*)-pyrimidyl alcohol **2** as chiral catalyst

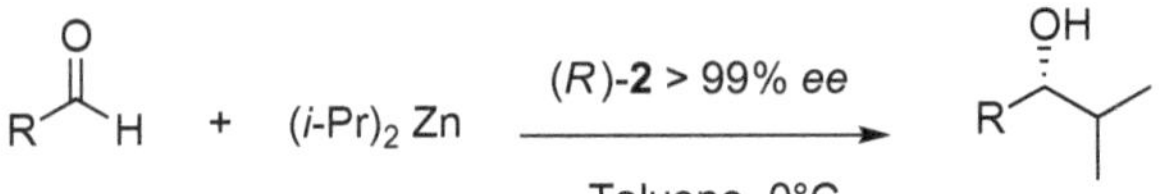

3–11 → **12–20**

Entry[a]	Aldehyde	R	Prod.	Yield (%)[b]	ee (%)[c]
1	**3**	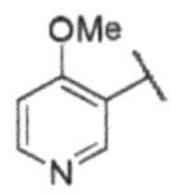	**12**	58	20
2	**4**	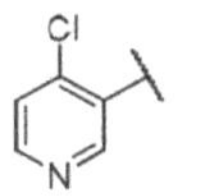	**13**	71	68
3	**5**	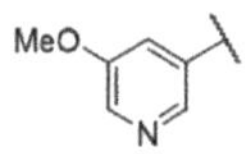	**14**	65	68
4	**6**	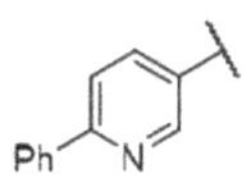	**15**	61	16
5	**7**	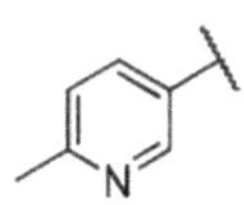	**16**	57	26
6	**8**	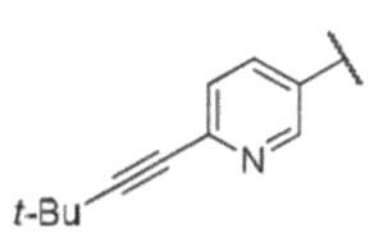	**17**	92	89 (98)
7	**9**	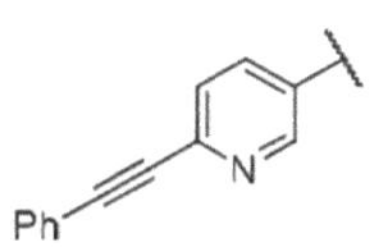	**18**	82	82 (95)
8	**10**	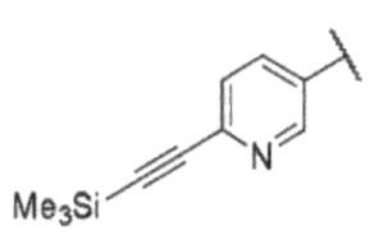	**19**	85	91 (91)
9	**11**	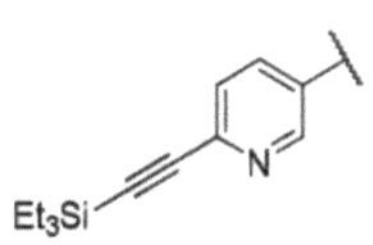	**20**	83	93 (99)

[a]Reactions were performed under nitrogen at 0°C using 2.2 equivalents of $(i\text{-}Pr)_2Zn$ and 10 mol% of (*R*)-**2**.
[b]Isolated product after column chromatography.
[c]ee were determined by HPLC and/or GC analysis on chiral stationary phases, *ee* values in parentheses with reverse addition of dialkylzinc.

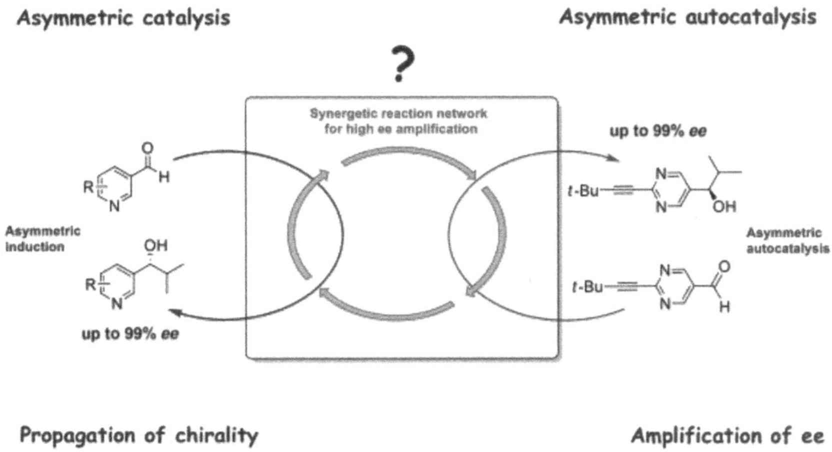

Scheme 2.4 Parallel amplification of Soai autocatalyst and pyridyl alkanols.

Table 2.2 One-pot asymmetric autocatalysis and enantioselective addition of $Zn(i\text{-}Pr)_2$ to aldehydes using low *ee* (*R*)-pyrimidyl alcohol **2** as chiral catalyst

Entry[a]	Aldehyde	Catalyst (*R*)-**2** (%) ee initial	(*R*)-**2** (%) *ee* final	Product (*R*)-**12-20**		
					Yield (%)[b]	ee (%)[c]
1	**3**	15	–[d]	**12**	47	3
2	**4**	15	40	**13**	65	15
3	**5**	11	42	**14**	59	30
4	**6**	11	41	**15**	60	12
5	**7**	11	39	**16**	62	26
6	**8**	11	97	**17**	81	98
7	**9**	11	80	**18**	96	95
8	**10**	11	84	**19**	93	82
9	**11**	11	96	**20**	92	98

[a]Reactions were performed under nitrogen at 0°C using 2.2 equivalents of $(i\text{-}Pr)_2Zn$, 10 mol% of aldehyde **1** and 1 mol% (*R*)-**2**.
[b]Yield determined by ^1H NMR.
[c]ee were determined by HPLC and/or GC analysis on chiral stationary phases.
[d]The soai alcohol **2** could not be isolated.

alongside a catalytic addition of $Zn(i\text{-}Pr)_2$ to aldehydes **8–11**, as shown in Table 2.2.

In this one-pot chemical combination, (*R*)-**2** (1 mol%) performed remarkably well, with high asymmetric amplification even when starting with only 11% *ee*. For example, the reaction provided high yields of (*R*)-**2** in a final >97% *ee* together with (*R*)-**17** in >98% *ee*.

However, aldehydes **3–7** failed to exhibit the expected amplification of *ee*, maybe because of the missing necessary electronic or steric features for the cooperative interaction with the autocatalysis of (*R*)-**2**. Indeed, we clearly observe a negative effect on the amplification of (*R*)-**2**, which was generated with a maximum of 42% *ee*. Subsequently, this has also resulted in poor final *ee* for products **12–16** (Table 2.2, Entries 1–5). Therefore one could consider an inhibition interaction during the reaction of these substrates.

With a striking difference, substrates **8–11** reacted efficiently to afford high yields of **17–20**, and exhibited remarkable amplification of *ee*. Repeatedly, overexpression of autocatalyst (*R*)-**2** reached high values in the range of 80–95% *ee* after just one cycle (Table 2.2, Entries 6–9). Compounds (*R*)-**18–20** were isolated in up to 98% *ee*. Notably, alkanol (*R*)-**18** is isolated with 95% *ee*, although (*R*)-**2** had only reached a final 80% *ee*. So, enantiopure autocatalyst (*R*)-**2** is not critical to achieve an almost perfect chirality transfer in the subsequent alkylation reaction of pyridine aldehydes.

The remarkable efficiency of this new chemical system combining autocatalysis with chiral induction prompted us to examine whether the amplifications of (*R*)-**2** and (*R*)-**17–20** were connected. For example, continuous monitoring of *ee* progress over time in the formation of **2** and **17** revealed a clear time-dependent correlation and pronounced asymmetric amplification between the *ee* of both (*R*)-**2** and (*R*)-**17**. Furthermore, the enantiopurity of newly forming alkanol (*R*)-**17** increased rapidly up to 90% *ee* after 30 min while (*R*)-**2** had reached only 63% *ee* from the initial condition (11% *ee*). The maximum enantiomeric excess of 98% *ee* for both alkanols was reached after 2 h (Fig. 2.1). As can be seen, although it would represent only 10 mol% at a maximum, autocatalyst (*R*)-**2** reacts

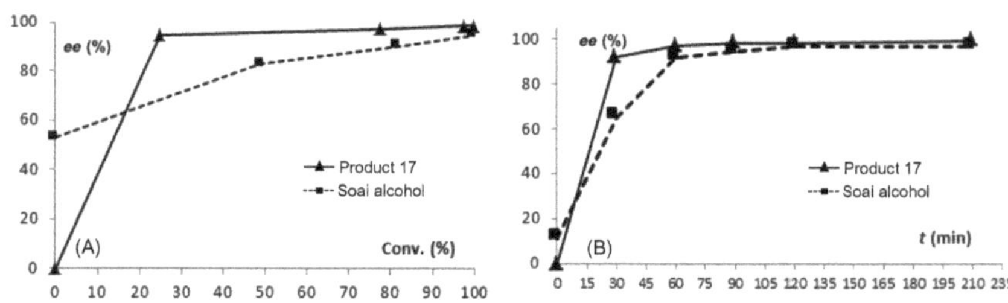

Figure 2.1 Evolution of the *ee* of product 17 (▲) and Soai alcohol 2 (■) during the advancement of the reaction as a function of conversion (A) or reaction time (B).

much faster than (*R*)-**17** increased. While (*R*)-**2** reaches 49% conversion with 83% *ee*, (*R*)-**17** amplifies rapidly to 94% *ee* after only 25% conversion. These plots also reveal two strong NLE for both autocatalysts (*R*)-**2** and (*R*)-**17** probably as a result of an unusual amplification profile [4], and two striking observations could be made. First, the Soai autocatalyst (*R*)-**2** amplifies in a single reaction step by contrast to the sequential procedure in which 3–4 reactions are usually required. Second, this asymmetric autocatalytic system is very powerful for amplifying a tiny *ee* of (*R*)-**2** by promoting the formation of (*R*)-**17** with even higher *ee*, i.e, nonlinear relationship.

Thus interaction between (*R*)-**2** and (*R*)-**17** could promote a reaction network coupling enantioselective autocatalysis to asymmetric induction. Possibly, a fast equilibrium of heterodimers may allow cross catalysis or mutual enantiomeric inhibition.

2.2.3 Coupling Asymmetric Autocatalysis With Asymmetric Alkynylation

The enantioselective alkynylzinc addition to carbonyl compounds, is interesting because a new C–C bond is formed with concomitant creation of a stereogenic center, and the resulting propargylic alcohols are versatile building blocks, serving as precursors for a variety of important chiral targets after further functionalization (Scheme 2.5) [13].

In 1979, Mukaiyama et al. reported the first asymmetric addition of derivatives of lithium acetylide to benzaldehyde in the presence of a chiral amino alcohol [14]. Later in 1990, Niwa and Soai described the use of alkynylzinc reagents to access propargylic alcohols in the range of 7%–35% *ee* using *N,N*-dibutylnorephedrine [15]. Other chiral catalysts including amino alcohols, sulfonamides, oxazolidines, and β-hydroxy amides, have been prepared and tested since then [13]. Suprisingly there are no reports describing alkynylation of pyrimidine- and pyridinecarboxaldehydes with organozinc reagent. This absence of examples might be attributed to the low or no enantioselectivity due to the strong coordination of the nitrogen atom in the substrate to the organozinc reagent.

Moving in this direction, we suggested a new concept coupling asymmetric autocatalysis with asymmetric induction in a sequence of two consecutive reactions (Scheme 2.6). Thus, a chiral catalyst is generated from minimal enantiopurity and provides entries to highly enantioenriched chiral molecules.

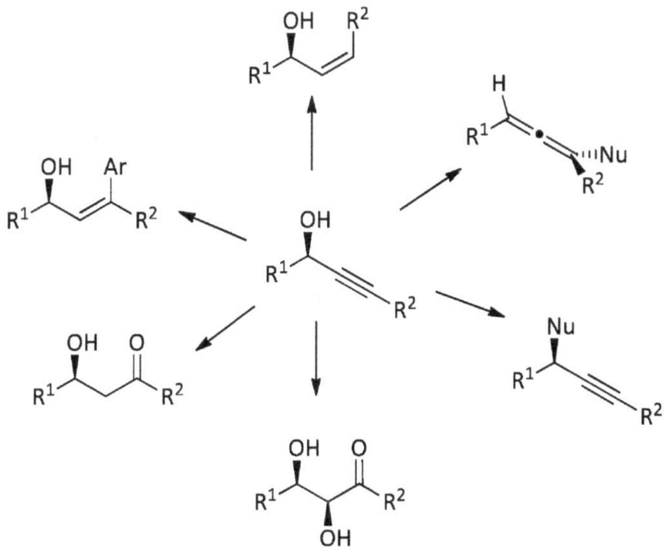

Scheme 2.5 Examples of derivatization of chiral propargylic alcohols.

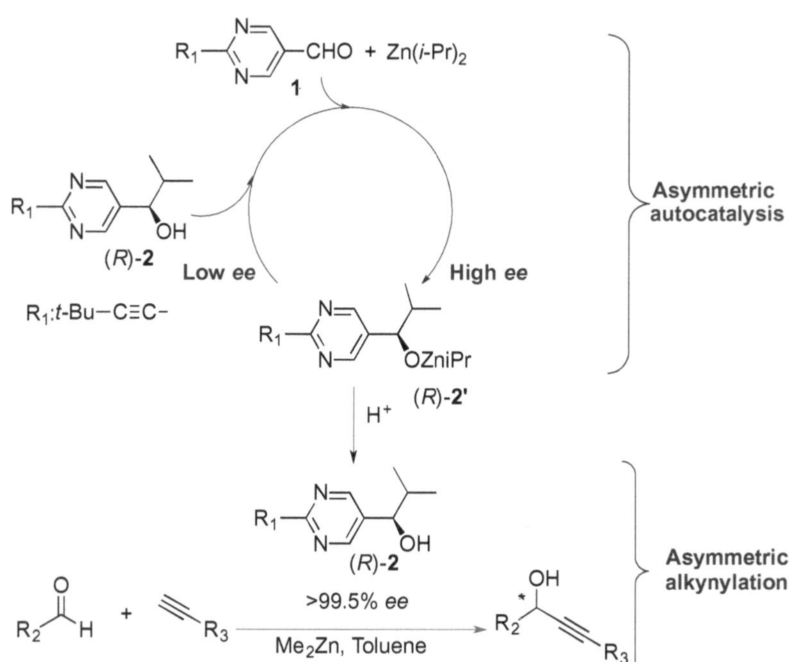

Scheme 2.6 Consecutive asymmetric catalyst amplification and subsequent alkynylation of aldehydes.

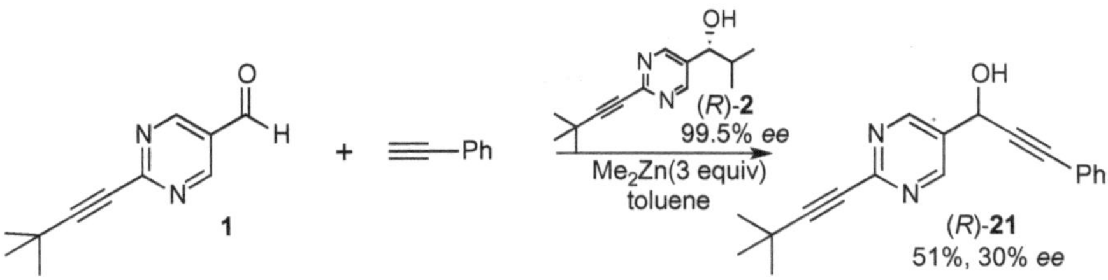

Scheme 2.7 Addition of phenylacetylene to Soai aldehyde **1**.

2.2.3.1 Reaction Optimization

Our first approach was to explore the catalytic properties of the Soai alcohol (*R*)-**2** in the asymmetric alkynylation of aldehyde **1** in the presence of dimethylzinc, as shown in Scheme 2.7. To our surprise and disappointment the reaction produced alkynol (*R*)-**21** product with 30% *ee*.

In stark contrast, when the same procedure was applied to the asymmetric alkynylation of 3-pyridinecarboxaldehyde **22**, the corresponding propargylic alcohol (*S*)-**23** was obtained with 76% *ee* after optimizing the reaction conditions (Table 2.3, Entry 3).

Further studies were carried out to evaluate the influence of other dialkylzinc reagents as precursors of zinc alkynylides. With Et_2Zn not only resulted in lower yield (due to the formation of the alkyl adduct) but also decreased the enantioselectivity (Table 2.3, Entry 8). In contrast, the more reactive *i*Pr_2Zn provided the opposite enantiomer with 42% *ee* and good yield (Table 2.3, Entry 9). Varying the concentration of the catalyst, under the reaction conditions described in Table 2.3, Entry 3, clearly demonstrates that increasing the catalyst loading from 20% to 35% and 55% did not result in improved of selectivity (Fig. 2.2).

To establish the viability of this new concept coupling asymmetric autocatalysis with asymmetric alkynylation, its efficiency must be compared to conventional catalytic procedures available from the chiral pool as reference. The commercially available ProPhenol ligand **24**, synthesized from L-proline, is a well known catalyst for the asymmetric alkynylation of aryl, aliphatic, and α,β-unsaturated aldehydes [16]. However, this catalyst has not been examined with azaaryl aldehydes.

Surprisingly, (*S,S*)-Prophenol **24** provided propargylic alcohol **23** in comparable *ee* to that of autocatalyst **2**, whereas norephedrine derivative (*S,R*)-**25** gave a low asymmetric induction of 38% *ee* (Scheme 2.8) [15]. Since Soai's autocatalyst can be generated from achiral precursors (absolute asymmetric synthesis), this concept provides a very powerful

Table 2.3 Optimization of reaction conditions for the enantioselective addition of phenylacetylene to aldehyde **22**

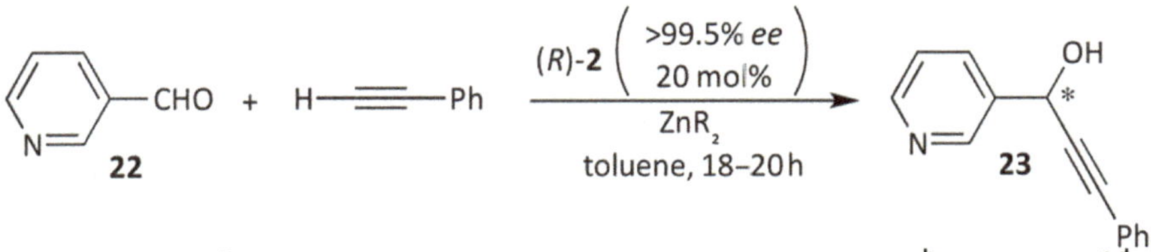

Entry	ZnR_2[a]	ZnR_2/Alkyne	Temp. (°C)	Yield (%)[b]	ee (%)[c,d]
1	Me_2Zn	3:4	0	64	68 (*S*)
2	Me_2Zn	1.4:1.4	0	11	53 (*S*)
3	Me_2Zn	2.5:7	0	91	76 (*S*)
4[e]	Me_2Zn	2.5:7	0	70	74 (*S*)
5	Me_2Zn	2.5:7	r.t.	83	64 (*S*)
6	Me_2Zn	2.5:7	−40	42	62 (*S*)
7[f]	Me_2Zn	2.5:7	0	84	56 (*S*)
8	Et_2Zn	2.5:7	0	54	43 (*S*)
9	*i*Pr_2Zn	2.5:7	0	69	42 (*R*)

[a]All the reactions were carried out in 2.6 mL of toluene under nitrogen, pre-formation of the alkynylzinc specie was performed at 75°C, Me_2Zn (0.77 mL, 1.2 M in toluene) and alkyne (2.6 mmol), 2 h, then r.t. and (*R*)-**2** (20 mol%), 1 h, then aldehyde (0.37 mmol), 18–22 h, 0°C.
[b]Isolated product after column chromatography.
[c]ee values were determined by HPLC with a Daicel Chiralcel OD column.
[d]Absolute configuration was determined by analysis of the corresponding Mosher ester ^{1}H NMR spectrum.
[e]20 mol% of D*i*MPEG 2000 as additive.
[f]10 mol% of (*R*)-2.

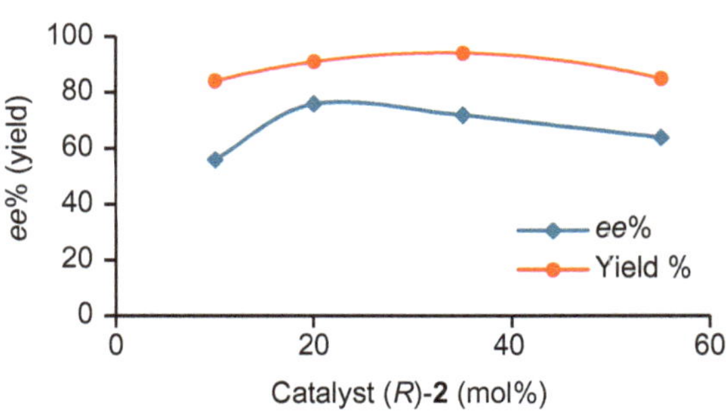

Figure 2.2 Yield (in red) and *ee* (in blue) of product **23** as a function of catalyst loading **2.**

synthetic tool for generating and amplifying chirality in other chemical reactions. The system obtained clearly illustrates the three sequential stages: *mirror-symmetry breaking* followed by *chiral amplification* and finally *chiral transmission process.*

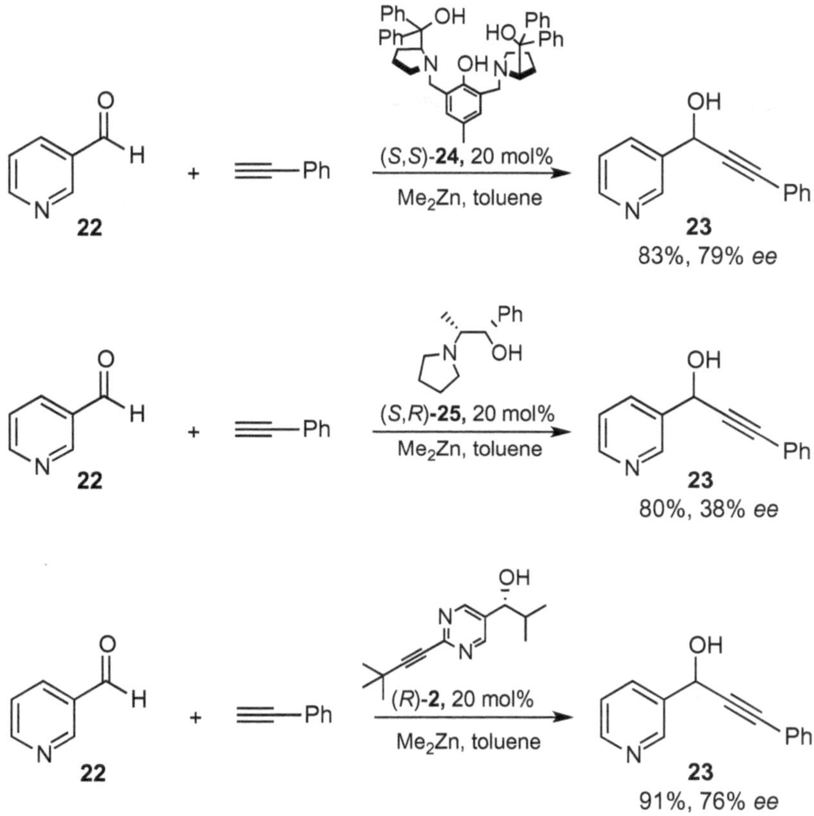

Scheme 2.8 Comparison of selectivity of catalyst **24** and **25**, prepared from the chiral pool, with autocatalyst **2**.

2.2.3.2 Asymmetric Alkynylation of Azaaryl Aldehydes

The results depicted in Table 2.4 demonstrate that alkynylation is compatible with a range of substituents on the substrate, and proceeds with high enantioselectivity, up to 86% *ee* and good yields. Notably, another known aldehyde for the Soai system, pyrimidine-5-carbaldehyde provided a better yield and *ee* of 88% and 74%, respectively, than the more common Soai aldehyde **1** (Table 2.4, Entries 8–9). Thus a variety of highly enantiopure alkynols can be accessed by Soai autocatalyst, which could amplify from very low *ee*.

Interestingly, the same absolute configuration as the autocatalyst (*R*)-**2** was obtained with alkynol **29**. In a similar way, alkynol **30** a reminiscence of the Soai aldehyde, also favored the *R* enantiomer in the presence of

Table 2.4 Asymmetric alkynylation of azaaryl aldehydes

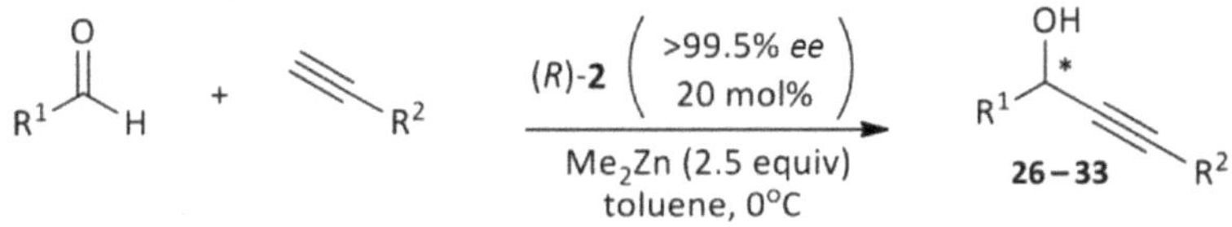

Entry[a]	R^1	R^2	Prod.	Yield (%)[b]	ee (%)[c,d]
1	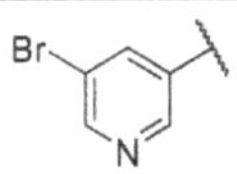	Ph	**26**	77	86 (*S*)
2	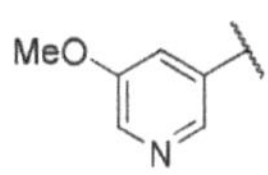	Ph	**27**	72	62 (*S*)
3	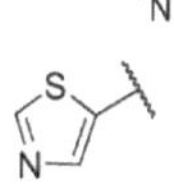	Ph	**28**	66	40 (*S*)
4	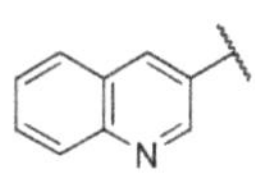	Ph	**29**	71	67 (*R*)
5	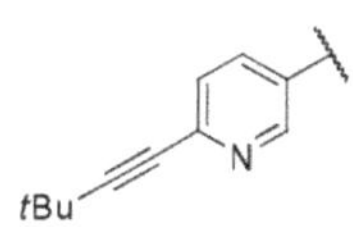	Ph	**30**	76	65 (*R*)
6	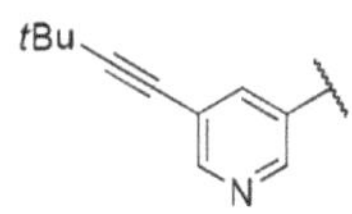	Ph	**31**	40	63 (*S*)
7	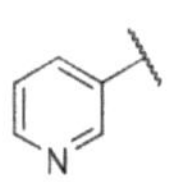	TMS	**32**	65	84 (*R*)
8	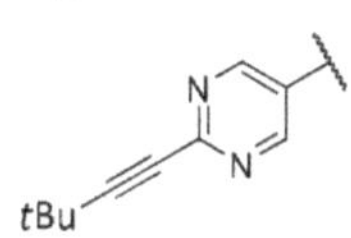	Ph	**21**	71	30 (*R*)
9	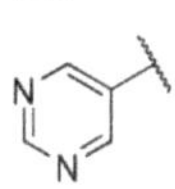	Ph	**33**	88	74 (*S*)

[a]All the reactions were carried out in 2.6 mL of toluene under nitrogen, preformation of the alkynylzinc specie was performed at 75°C, Me_2Zn (0.77 mL, 1.2 M in toluene) and alkyne (2.6 mmol), 2 h, then r.t. and (*R*)-**2** (20 mol%), 1 h, then aldehyde (0.37 mmol), 18–22 h, 0°C.
[b]Isolated product after column chromatography.
[c]ee values were determined by HPLC analysis.
[d]Absolute configurations were determined by analysis of the corresponding Mosher ester ^1H NMR spectrum.

(*R*)-**2**. Despite a very high enantiopurity reached with alkynol **33**, the absolute configuration was opposite to that of Soai autocatalyst.

2.2.3.3 One-Pot Asymmetric Autocatalysis-Catalysis Sequence

Following the successful activity of Soai autocatalyst (*R*)-**2** in transferring the amplified chirality in the alkynylation reaction, perhaps the most challenging next step would consist of an autocatalysis-catalysis sequence without isolation of (*R*)-**2,** as shown in Scheme 2.9. Indeed, starting with 2 mol% of (*R*)-**2**, originally with only 16% *ee*, three consecutive cycles were required to achieve amplification to a practically enantiomerically pure product (>98% *ee*). Consecutively, 20 mol% of the newly formed isopropylzinc alkoxide of (*R*)-**2′** was used as a catalyst in the enantioselective alkynylation of 3-pyrdidinecarboxaldehyde **22**, providing the corresponding propargylic alcohol (*S*)-**23** in 91% yield and 65% *ee*.

The autocatalysis-catalysis sequence provided alcohol **23** in slightly lower *ee* than with the conventional procedure; this might be attributed to a small amount of iPr_2Zn remaining from the amplification step, which might induce changes in the structure of the active zinc species. Despite this decrease in *ee*, we have here a demonstration of an in-situ generated

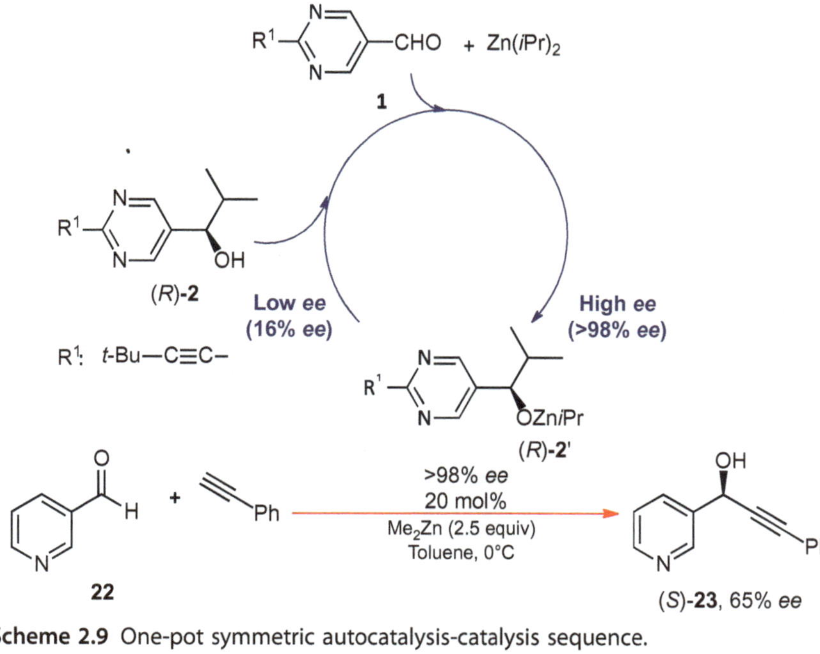

Scheme 2.9 One-pot symmetric autocatalysis-catalysis sequence.

autocatalyst by asymmetric amplification that acts as a conveyor of chirality in a chemically different reaction.

2.2.4 Towards Remote Amplification of Chirality

We mentioned that chiral compounds including alcohols, amino acids, epoxides, and [$Cr(acac)_3$] could serve as external stimuli to induce asymmetric autocatalysis of **2** as already reported by Soai [17].

As just demonstrated, the Soai autocatalyst is not only able to sense and amplify a slight asymmetry from a chiral inducer compound, but is also capable of transferring such amplified chirality.

Therefore based on our recent findings described above, we reasoned that chiral target molecules with very low *ee* can also act as trigger compounds for asymmetric amplification of autocatalyst **2** (asymmetric read-in) and operate as a chiral relay (Scheme 2.10). In other words: target molecule would be able to trigger the formation of their own chiral catalyst for remote amplification of chirality [18].

The formation of alkanol **2** was subjected to the external chiral stimulus of various chiral alkynols. Since the chiral initiators were used with low *ee*, several autocatalytic cycles were necessary to provide almost enantiopure alkanol **2** (>95% *ee*). The amplification process was carried out as illustrated in Scheme 2.11, with chiral initiator/target compounds **23**, **29**, **30**, **32**, and **34**. In the first step, 0.2 mol% of the target molecule with low *ee*, triggers the asymmetric amplification of the Soai autocatalyst **2**,

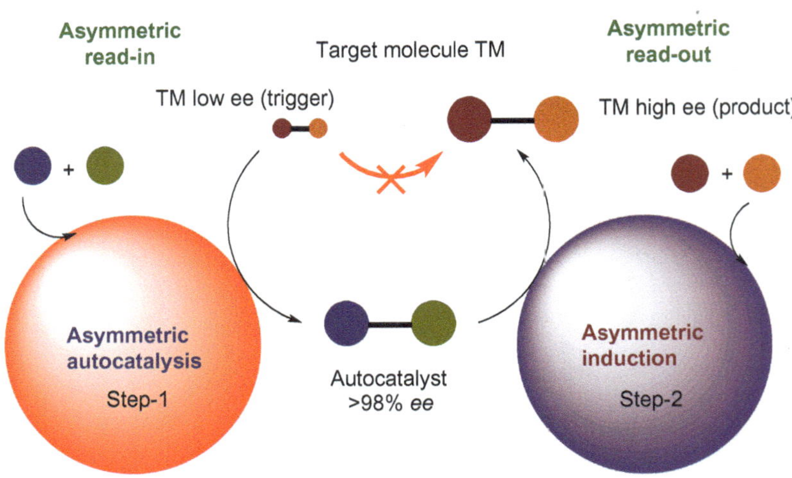

Scheme 2.10 The concept of remote amplification of chirality.

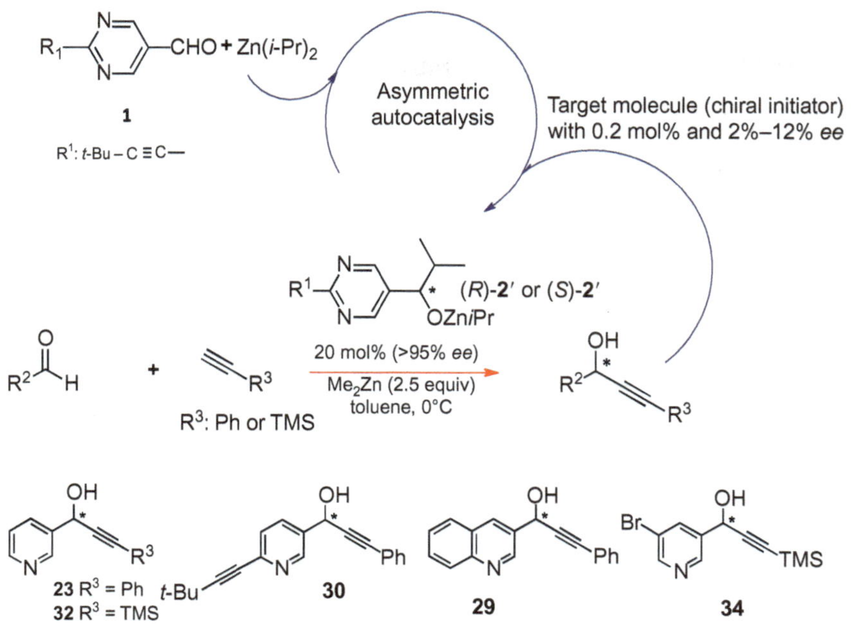

Scheme 2.11 Remote asymmetric amplification of trigger/target alkynols with Soai autocatalyst.

i.e., asymmetric read-in. In the second step, 20 mol% of Soai autocatalyst induces the formation of the target alkynol from its precursors with a higher *ee*, i.e., asymmetric read-out. Since the target compound is not capable of asymmetric amplification by itself, this chemical system represents a good example of an overall process for remote asymmetric amplification. Although, in some cases opposite absolute configuration was obtained along the course of read-in to read-out (Table 2.5).

At the outset, (*R*)-pyridyl alkynol **23** was used as chiral source with 8% *ee* to induce asymmetric autocatalysis of (*R*)-**2**. Near enantiopure pyrimidyl alkanol (*R*)-**2** with 95% *ee* was obtained after four consecutive autocatalytic cycles. In response, the alkoxide (*R*)-**2′** catalyzed alkynylation reaction providing (*S*)-**23** with 67% *ee* and an 84% yield (Table 2.5, Entry 1). In contrast, (*S*)-**23** with only 2% *ee* induced the production of (*S*)-**2**, which provided (*R*)-**23** with 68% *ee* (Table 2.5, Entry 2). In the same manner, (*S*)-**32** (c.10% *ee*) and (*S*)-**34** (c.12% *ee*) can act as a chiral initiator of asymmetric autocatalysis to induce the formation of (*R*)-**2**. The subsequent step provided (*R*)-**32** and (*R*)-**34** with 82% and 84% *ee* respectively (Table 2.5, Entries 3 and 4). It should be noted that this reaction sequence proceeded with inversion of configuration from trigger to product molecules.

Table 2.5 Remote asymmetric amplification through Soai asymmetric autocatalysis

Entry[a]	Chiral trigger	Trigger's initial ee	Abs. conf **2**	Product	
				Yield (%)[b]	ee (%)[c] (conf)[d]
1	(*R*)-**23**	8	(*R*)	84	67 (*S*)-**23**
2	(*S*)-**23**	2	(*S*)	88	68 (*R*)-**23**
3	(*S*)-**32**	10	(*R*)	74	82 (*R*)-**32**
4	(*S*)-**34**	12	(*R*)	58	84 (*R*)-**34**
5	(*R*)-**29**	10	(*R*)	86	67 (*R*)-**29**
6	(*R*)-**30**	11	(*R*)	83	68 (*R*)-**30**
7	(*S*)-**30**	2	(*S*)	88	71 (*S*)-**30**

[a]Reactions were carried out under nitrogen. The molar ratio in the amplification step of chiral trigger aldehyde **2** i-Pr_2Zn = 0.01:1:1.4.
[b]Isolated product after column chromatography.
[c]ee values were determined by HPLC analysis.
[d]Absolute configuration was determined by analysis of the corresponding Mosher ester ^{1}H NMR spectrum.

Remarkably, quinolyl alkynol (*R*)-**29** and alkynol (*R*)-**30** were able to catalyze their own remote amplification, showing retention of configurations and affording excellent yields of 86% and 83% with 67% and 68% *ee*, respectively (Table 2.5, Entries 5 and 6). The reliability of this process was confirmed by translating a very low initial *ee* (e.g., 2%) of (*S*)-**30** into an enantioenriched product with 71% *ee* (Table 2.5, Entry 7). The system shows how a molecule with low *ee*, is able to improve its *ee* by remote amplification of chirality.

The autocatalytic amplification of **2** was monitored by HPLC during each catalytic cycle and showed the amplification in each cycle, triggered by the chiral propagylic alcohols. (*R*)-**23** (8% *ee*) and (*R*)-**29** (10% *ee*) prompted remarkable amplification of (*R*)-**2** to 62% and 74% *ee* within only the first cycle, respectively. At the second cycle this amplification increased up to 82% and 94% *ee*, suggesting that remote replication with high level of *ee* through amplification of (*R*)-**2** may be implemented after only two autocatlytic cycles.

2.3 CONCLUSION

In the present chapter we have confirmed the singularity of the Soai reaction by the discovery of a new reactivity. The Soai's autocatalyst **2**, which is readily available in >99% *ee* by asymmetric amplification from a virtually achiral precursor, is a powerful chemical tool as a catalyst for ensuing enantioselective transformations.

A demonstration is presented for a new chemical system combining asymmetric amplification of autocatalyst **2** alongside a catalytic addition of $Zn(i\text{-}Pr)_2$ to pyridyl aldehydes in a one-pot process. In such a catalytic system **2** and pyridyl alkanols are generated with remarkable amplification up to >98% *ee* in one reaction cycle for both compounds, even when starting from a very low enantiopurity of **2**. Moreover, a strong nonlinear relationship is observed in a cooperative reaction of autocatalysts **2** and pyridyl alkanol.

Notably, in a different set of reactions, highly enantioenriched pyridyl alkynols (up to 86% *ee*) are accessed using the Soai alkanol **2** as a catalyst. The asymmetric induction is also correlated to the degree of amplification of autocatlyst **2**. These results are the realization of novel process in which the Soai autocatalyst is amplified from a minimal *ee* and used to induce asymmetry in a different chemical reaction, i.e., *alkynylation*.

As a further advance, we demonstrate that target alkynols with tiny initial *ee*, and in trace amount, act as chiral triggers to generate their own catalyst, namely the Soai autocatalyst, by asymmetric amplification, which in turn enables the formation of the same alkynols with greater enantiomeric purity. The overall process leads to the copy catalysis of nonautocatalytic target molecules through a "read in" and "read out" for a remote amplification of chirality.

The unique reactivity of Soai autocatalyst **2** described here provides premises for the construction of a mechanistic model for the rise and propagation of chirality, and offers new settings for a novel approach to access enantiomerically pure molecules in the future.

REFERENCES

[1] (a) Frank, F. C. On Spontaneous Ssymmetric Synthesis. *Biochim. Biophys. Acta* **1953,** *11* (0), 459–463.
(b) Calvin, M. Chemical Evolution. *Chem. Br.* **1969,** *5* (1), 22–28.
(c) Feringa, B. L.; van Delden, R. A. Absolute Asymmetric Synthesis: The Origin, Control, and Amplification of Chirality. *Angew. Chem., Int. Ed.* **1999,** *38* (23), 3418–3438.
(d) Ribo, J. M.; Crusats, J.; Sagues, F.; Claret, J.; Rubires, R. Chiral Sign Induction by Vortices During the Formation of Mesophases in Stirred Solutions. *Science* **2001,** *292* (5524), 2063–2066.
(e) Mislow, K. Absolute Asymmetric Synthesis: A Commentary. *Collect. Czech. Chem. Commun.* **2003,** *68* (5), 849–864.
(f) Weissbuch, I.; Lahav, M. Crystalline Architectures as Templates of Relevance to the Origins of Homochirality. *Chem. Rev.* **2011,** *111* (5), 3236–3267.
(g) Micskei, K.; Rabai, G.; Gal, E.; Caglioti, L.; Palyi, G. Oscillatory Symmetry Breaking in the Soai Reaction. *J. Phys. Chem. B.* **2008,** *112* (30), 9196–9200.

(h) Zepik, H.; Shavit, E.; Tang, M.; Jensen, T. R.; Kjaer, K.; Bolbach, G., et al. Chiral Amplification of Oligopeptides in Two-Dimensional Crystalline Self-Assemblies on Water. *Science* **2002,** *295* (5558), 1266–1269.

[2] Bissette, A. J.; Fletcher, S. P. Mechanisms of Autocatalysis. *Angew. Chem., Int. Ed. Engl.* **2013,** *52* (49), 12800–12826.

[3] Soai, K.; Shibata, T.; Morioka, H.; Choji, K. Asymmetric Autocatalysis and Amplification of Enantiomeric Excess of a Chiral Molecule. *Nature* **1995,** *378* (6559), 767–768.

[4] Puchot, C.; Samuel, O.; Dunach, E.; Zhao, S.; Agami, C.; Kagan, H. B. Nonlinear Effects in Asymmetric Synthesis. Examples in Asymmetric Oxidations and Aldolization Reactions. *J. Am. Chem. Soc.* **1986,** *108* (9), 2353–2357.

[5] (a) Gridnev, I. D.; Serafimov, J. M.; Quiney, H.; Brown, J. M. Reflections on Spontaneous Asymmetric Synthesis by Amplifying Autocatalysis. *Org. Biomol. Chem.* **2003,** *1* (21), 3811–3819.

(b) Singleton, D. A.; Vo, L. K. A Few Molecules Can Control the Enantiomeric Outcome. Evidence Supporting Absolute Asymmetric Synthesis Using the Soai Asymmetric Autocatalysis. *Org. Lett.* **2003,** *5* (23), 4337–4339.

(c) Soai, K.; Sato, I.; Shibata, T.; Komiya, S.; Hayashi, M.; Matsueda, Y., et al. Asymmetric Synthesis of Pyrimidyl Alkanol without Adding Chiral Substances by the Addition of Diisopropylzinc to Pyrimidine-5-Carbaldehyde in Conjunction with Asymmetric Autocatalysis. *Tetrahedron: Asymmetry* **2003,** *14* (2), 185–188.

(d) Singleton, D. A.; Vo, L. K. Enantioselective Synthesis without Discrete Optically Active Additives. *J. Am. Chem. Soc.* **2002,** *124* (34), 10010–10011.

[6] (a) Blackmond, D. G.; McMillan, C. R.; Ramdeehul, S.; Schorm, A.; Brown, J. M. Origins of Asymmetric Amplification in Autocatalytic Alkylzinc Additions. *J. Am. Chem. Soc.* **2001,** *123* (41), 10103–10104.

(b) Buono, F. G.; Blackmond, D. G. Kinetic Evidence for a Tetrameric Transition State in the Asymmetric Autocatalytic Alkylation of Pyrimidyl Aldehydes. *J. Am. Chem. Soc.* **2003,** *125* (30), 8978–8979.

[7] (a) Ercolani, G.; Schiaffino, L. Putting the Mechanism of the Soai Reaction to the Test: DFT Study of the Role of Aldehyde and Dialkylzinc Structure. *J. Org. Chem.* **2011,** *76* (8), 2619–2626.

(b) Schiaffino, L.; Ercolani, G. Unraveling the Mechanism of the Soai Asymmetric Autocatalytic Reaction by First-Principles Calculations: Induction and Amplification of Chirality by Self-Assembly of Hexamolecular Complexes. *Angew. Chem., Int. Ed.* **2008,** *47* (36), 6832–6835.

(c) Schiaffino, L.; Ercolani, G. Mechanism of the Asymmetric Autocatalytic Soai Reaction Studied by Density Functional Theory. *Chemistry* **2010,** *16* (10), 3147–3156.

[8] (a) Islas, J. R.; Lavabre, D.; Grevy, J. M.; Lamoneda, R. H.; Cabrera, H. R.; Micheau, J. C., et al. Mirror-Symmetry Breaking in the Soai Reaction: A Kinetic Understanding. *Proc. Natl. Acad. Sci. U.S.A.* **2005,** *102* (39), 13743–13748.

(b) Micheau, J.-C.; Cruz, J.-M.; Coudret, C.; Buhse, T. An Autocatalytic Cycle Model of Asymmetric Amplification and Mirror-Symmetry Breaking in the Soai Reaction. *ChemPhysChem* **2010,** *11* (16), 3417–3419.

(c) Micheau, J.-C.; Coudret, C.; Cruz, J.-M.; Buhse, T. Amplification of Enantiomeric Excess, Mirror-Image Symmetry Breaking and Kinetic Proofreading in Soai Reaction Models with Different Oligomeric Orders. *Phys. Chem. Chem. Phys.* **2012,** *14* (38), 13239–13248.

[9] (a) Crusats, J.; Hochberg, D.; Moyano, A.; Ribó, J. M. Frank Model and Spontaneous Emergence of Chirality in Closed Systems. *ChemPhysChem* **2009,** *10* (12), 2123–2131.

(b) Ribo, J. M.; Blanco, C.; Crusats, J.; El-Hachemi, Z.; Hochberg, D.; Moyano, A. Absolute Asymmetric Synthesis in Enantioselective Autocatalytic Reaction Networks: Theoretical Games, Speculations on Chemical Evolution and Perhaps a Synthetic Option. *Chemistry* **2014,** *20* (52), 17250–17271.

[10] Sato, I.; Urabe, H.; Ishii, S.; Tanji, S.; Soai, K. Asymmetric Synthesis with a Chiral Catalyst Generated from Asymmetric Autocatalysis. *Org. Lett.* **2001,** *3* (24), 3851–3854.

[11] (a) Funes-Maldonado, M.; Sieng, B.; Amedjkouh, M. Enabling Asymmetric Alkynylation of Azaaryl Aldehydes with Soai Autocatalyst. *Eur. J. Org. Chem.* **2015,** *2015* (19), 4081–4086.

(b) Romagnoli, C.; Sieng, B.; Amedjkouh, M. Asymmetric Amplification Coupling Enantioselective Autocatalysis and Asymmetric Induction for Alkylation of Azaaryl Aldehydes. *Eur. J. Org. Chem.* **2015,** *2015* (19), 4087–4092.

(c) Sato, I.; Urabe, H.; Ishii, S.; Tanji, S.; Soai, K. Asymmetric Synthesis with a Chiral Catalyst Generated from Asymmetric Autocatalysis. *Org. Lett.* **2001,** *3* (24), 3851–3854.

[12] Sato, I.; Urabe, H.; Ishiguro, S.; Shibata, T.; Soai, K. Amplification of Chirality from Extremely Low to Greater than 99.5% *ee* by Asymmetric Autocatalysis. *Angew. Chem., Int. Ed.* **2003,** *42* (3), 315–317.

[13] Trost, B. M.; Weiss, A. H. The Enantioselective Addition of Alkyne Nucleophiles to Carbonyl Groups. *Adv. Synth. Catal.* **2009,** *351* (7–8), 963–983.

[14] Mukaiyama, T.; Suzuki, K.; Soai, K.; Sato, T. Enantioselective Addition of Acetylene to Aldehyde. Preparation of Optically Active Alkynyl Alcohols. *Chem. Lett.* **1979,** *8* (5), 447–448.

[15] Niwa, S.; Soai, K. Catalytic Asymmetric-Synthesis of Optically-Active Alkynyl Alcohols by Enantioselective Alkynylation of Aldehydes and by Enantioselective Alkylation of Alkynyl Aldehydes. *J. Chem. Soc. Perkin Trans. 1* **1990,** (4), 937–943.

[16] Trost, B. M.; Bartlett, M. J.; Weiss, A. H.; von Wangelin, A. J.; Chan, V. S. Development of Zn-ProPhenol-Catalyzed Asymmetric Alkyne Addition: Synthesis of Chiral Propargylic Alcohols. *Chemistry* **2012,** *18* (51), 16498–16509.

[17] (a) Sato, I.; Ohgo, Y.; Igarashi, H.; Nishiyama, D.; Kawasaki, T.; Soai, K. Determination of Absolute Configurations of Amino Acids by Asymmetric Autocatalysis of 2-Alkynylpyrimidyl Alkanol as a Chiral Sensor. *J. Organomet. Chem.* **2007,** *692* (9), 1783–1787.

(b) Kawasaki, T.; Omine, T.; Sato, M.; Morishita, Y.; Soai, K. Asymmetric Autocatalysis Induced by Octahedral Tris(acetylacetonato)chromium(III) Complex as a Homogeneous Chiral Initiator. *Chem. Lett.* **2007,** *36* (1), 30–31.

(c) Kawasaki, T.; Shimizu, M.; Suzuki, K.; Sato, I.; Soai, K. Enantioselective Synthesis Induced by Chiral Epoxides in Conjunction with Asymmetric Autocatalysis. *Tetrahedron: Asymmetry* **2004,** *15* (23), 3699–3701.

(d) Shibata, T.; Yamamoto, J.; Matsumoto, N.; Yonekubo, S.; Osanai, S.; Soai, K. Amplification of a Slight Enantiomeric Imbalance in Molecules Based on Asymmetric Autocatalysis: The First Correlation between High Enantiomeric Enrichment in a Chiral Molecule and Circularly Polarized Light. *J. Am. Chem. Soc.* **1998,** *120* (46), 12157–12158.

[18] Funes-Maldonado, M.; Sieng, B.; Amedjkouh, M. Asymmetric Autocatalysis as a Relay for Remote Amplification of Chirality of Target Molecules used as Trigger. *Org. Lett.* **2016,** *18* (11), 2536–2539.

CHAPTER 3

Graph Theoretical and Statistical Analysis of the Impact of Soai Reaction on Natural Sciences*

Béla Barabás[1] and Ottilia Fülöp[2]
[1]Institute of Mathematics, Department of Stochastics, Budapest University of Technology and Economics, Budapest, Hungary
[2]Institute of Mathematics, Department of Differential Equations, Budapest University of Technology and Economics, Budapest, Hungary

Abstract

Asymmetric autocatalysis (Soai reaction) was reported 20 years ago. Since then, this discovery has had an exceptional influence on several branches of natural sciences. In the present paper citations to this milestone article—556 publications of 908 scientists from Google Scholar's database—are analyzed by graph theoretical and statistical tools. The results enable localization of the most active research groups, their priorities and the identification of the distribution of scientific disciplines influenced by the Soai reaction.

Keywords: Soai reaction; asymmetric autocatalysis; chirality; social network; community; graph theory; statistics

3.1 INTRODUCTION

Chirality is one of the central problems of modern chemistry [4–7]. The main goal of practical research focused on problems of chirality is the preparation of single enantiomers in a selective way [8–13]. Theoretical efforts are mainly focused on the origin of the so-called biological chirality (the empirical fact that living organisms utilize chiral biomolecules only in the form of high excess of one of the enantiomers) [14–16]. Several of the unresolved problems in these fields of asymmetric (enantioselective) autocatalysis provide important steps toward the solution.

Asymmetric autocatalysis is found in achiral-to-chiral reactions, where the product catalyzes its own formation (conventional catalysis) and promotes the preferential formation of that enantiomer of the product which

* Some results of the present paper were reported at Symposia [1,2] and described in a preliminary publication [3].

Advances in Asymmetric Autocatalysis and Related Topics
DOI: http://dx.doi.org/10.1016/B978-0-12-812824-4.00003-4

corresponds to its own configuration (similar to the effect of "chiral auxiliaries"). The combination of these two effects produces asymmetric autocatalysis, which had no precedents before its discovery in 1995 by Kenso Soai and coworkers (Tokyo University of Science) [17]. In the "basic" variant of this reaction, however, some initial excess of the chiral product was necessary, even if in very low quantities [18–22]. The Soai reaction (as asymmetric autocatalysis [17] is called now [23]) is operative in a class of $C(sp^3)-C(sp^3)$ bond-making reactions. In the last two decades it has been discovered that it has several important consequences (and through these consequences—applications). The most important of these is the realization of the first well-documented example of absolute asymmetric (enantioselective) synthesis, that is: the Soai reaction without any chiral additive or asymmetric physical field [24–27]. This synthesis is one of the century-old dreams of organic chemists. It was found that the high enantioselectivity of the Soai reaction could be achieved by chiral additives other than the Soai-product (a chiral secondary alcohol). Such additives even include cryptochiral compounds [28,29], or chiral compounds, where the chirality is due exclusively to isotopic substitution ($^{1}H/^{2}H$, $^{12}C/^{13}C$, $^{14}N/^{15}N$ or $^{16}O/^{17}O$, ^{18}O) [30–33]. Chiral surfaces (crystal faces) of achiral compounds can also be used as "chiral auxiliaries" of this particular reaction [34–37]. Beyond the fundamental importance of these "triggering" experiments, such variants of the Soai reaction can be used for structure elucidation of cryptochiral compounds, exploration of details of biological isotope effects or as observation tools in atomic (ionic)-level surface stereochemistry. Several authentic reviews cover various sectors of this relatively new field of chemistry [23,38–42]. The revolutionary new features of the Soai autocatalysis induced new directions of research activity in several nonchemical fields of natural sciences, as it will be shown in the present paper.

3.2 SOAI GRAPH

The representation and analysis of social networks as graphs has a great potential for researchers in many disciplines. First we summarize some of the key graph theoretical and social network analysis (SNA) concepts we use in this chapter.

In general, an undirected graph $G = (V, E)$ is represented by a set of vertices (or nodes) V and a set of edges (or links) E connecting some pairs of vertices. In this chapter we deal only with finite undirected graphs.

We say that two vertices x and y are adjacent if they are linked by at least one edge xy.

A graph $G = (V, E)$ is said to be *complete* if any two vertices from V are adjacent. Multiple edges are two or more edges that link the same vertices. A loop is an edge that links a vertex to itself. A path between x and y is a sequence of successive edges xv_1, v_1v_2, …, v_iy, where vertices x, v_1, v_2, …, v_i, y are distinct from one another.

A clique of the graph is a subset C of the vertices such that every two distinct vertices from C are adjacent, thus any clique is a vertex set of a complete subgraph of G. The vertex set $C \subset V$ is a maximum clique of the graph if there is no clique with more vertices. Vertices x and y are said to be connected (by path) if there exists at least one path between x and y. Otherwise x and y are disconnected. The length of a path is the number of its edges.

A graph $G = (V, E)$ is connected if every two vertices x and y from V are connected. Otherwise G is a disconnected graph. A disconnected graph is formed by components that are maximal (i.e., largest possible) subgraphs in which all vertices are connected to each other by path.

The degree of a vertex x (denoted by $deg(x)$) is the number of edges incident to the vertex x, where loops are counted twice. The maximum degree of a graph G is denoted by ΔG. The geodesic distance between two vertices of a connected undirected graph is the length of the shortest path (geodesic) connecting the vertices.

The eccentricity of a vertex (denoted by $\in (x)$) is the longest geodesic distance between x and any other vertex. The diameter $d(G)$ of a graph.

$G = (V, E)$ the maximum of all eccentricities, i.e., $d = \max_{x \in V} \in (x)$. The overall centrality (or simply centrality) of a node x was introduced by Linton C. Freeman [43] and it can be computed as follows:

$$C_B(x) = \sum_{s \neq x \neq t \in V} \frac{g_{st}(x)}{g_{st}}$$

where g_{st} is the total number geodesics between the nodes s and t and $g_{st}(x)$ is the number of geodesics connecting s and t that contain x. The terms of this sum are the probabilities that vertex x falls on a randomly selected geodesic connecting s and t for all different s, $t \in V \setminus \{x\}$.

For example in Fig. 3.1A all vertices have the overall centrality 0.5 and in an undirected star graph with four vertices and three edges (Fig. 3.1B) the overall centrality of the center vertex is 3 and for all other

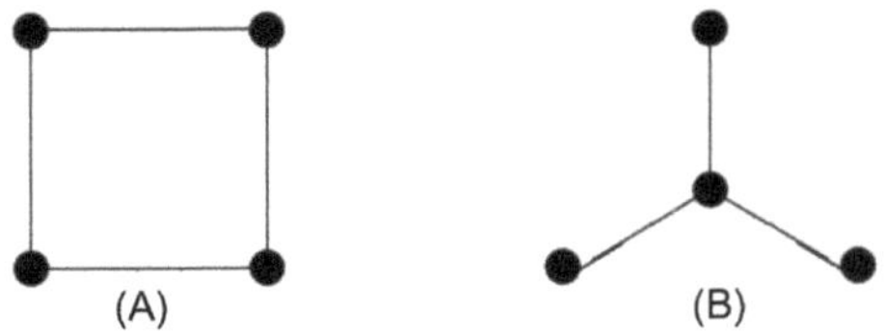

Figure 3.1 Case (A): n = 4, centrality = 0.5 of all vertex. Case (B): n = 4, centrality = 3 of center vertex and centrality = 0 for all other vertex.

vertices it is equal to 0. Freeman [43] proved that the maximum value taken by overall centrality in a graph with n vertices is achieved only by a center vertex in a star graph and it is equal to $(2^{n-1}) = \frac{n^2 - 3n + 2}{2}$.

Definition 1

The average degree of a graph with n vertices and e edges provides information about the number of edges compared to the number of vertices of the graph. Each edge is incident to two vertices and counts in the degree of both nodes, thus the average degree of an undirected graph is $\frac{2e}{n}$.

Definition 2

Social network analysis (SNA) is the mapping of relationships between people, groups, computers and other connected entities. The nodes in the network are the examined people or groups while the links (edges) are the relationships between these vertices.

Definition 3

Let us call any article or book written in the Latin alphabet from Google Scholar's database [44] an *S-paper* if it makes reference to the paper of K. Soai, T. Shibata, H. Morioka, and K. Choji: Asymmetric autocatalysis and amplification of enantiomeric excess of a chiral molecule published on 28 December 1995 in *Nature* [17].

Definition 4

The *Soai graph* is a social network represented as an undirected graph with a node set formed by all scientists who have at least one S-paper. Two nodes are adjacent (so they form an edge) if the corresponding two authors have at least one common S-paper. Edge xy does not mean that the corresponding common S-paper or S-papers have only two authors (x and y), in case of an edge xy the possibility of other co-authors is not excluded.

Figure 3.2 Soai graph.

The Soai graph has 908 nodes, 60 of them being isolated because they correspond to individual authors of S-papers who have no S-papers in common with others. We used the loops incident for all isolated nodes (and only isolated nodes), represented here as at least one S-paper written by the corresponding (individual) author. To simplify our social network, loops were left out for individual authors who have other S-papers with co-authors. There are no multiple edges in the graph, so in cases of two adjacent nodes the number of common S-papers cannot be seen. The graph has 2534 edges, 60 of them being loops. The average degree of the Soai graph is 5.58.

Thus the Soai graph shows the social interactions between authors of S-papers, i.e., scientists influenced by the work of K. Soai et al. [17]. The Soai graph, as defined above, is shown in Fig. 3.2. It was made using the "Igraph" package, which utilizes the library of the R programming language for network analysis [45].

3.3 MAIN COMMUNITIES OF THE SOAI GRAPH

Analyzing the Soai graph we used mathematical algorithms to derive communities in the social network. A network is said to have a community structure if the nodes of the network can be grouped into sets of nodes, such that each set of nodes is internally densely connected. Overlapping communities are allowed. The more general definition is based on the principle that pairs of nodes are more likely to be connected if they are both members of the same community, and less likely to be connected if they do not share communities [46].

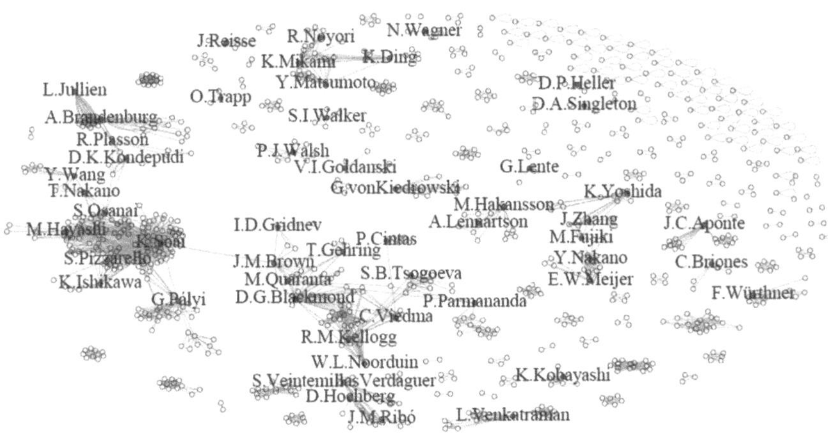

Figure 3.3 Articulation points.

Two different mathematical algorithms were chosen. One of them is called the "Walktrap Algorithm" [47]. This is a stochastic algorithm based on random walks. The idea is that short random walks tend to stay in the same community.

The Walktrap Algorithm found 121 communities in the Soai graph, plus 60 individual authors (of S-papers) who have no S-papers in common with the co-authors. The average size of these communities was ≈ 7.1.

We also used another algorithm that is called "Fastgreedy". This is a deterministic algorithm that follows the heuristic problem-solving process of making the locally optimal choice at each stage with the hope of finding a global optimum [48]. Fastgreedy Algorithm found 114 communities plus 60 individual authors. The average size of these communities was ≈ 7.4.

It is important to notice that the two algorithms based on completely different ideas have produced very similar community structures on the Soai graph.

Modularity is a measure of the structure of a network. Networks with high modularity have dense connections between the vertices within communities (modules), but sparse connections between vertices from different communities.

The Soai network has a high modularity: 0.8835 if we consider the individual authors, and 0.9057 without them.

Vertices that make connections between communities are called articulation points. If someone deletes them from the graph, the number of connected components increases.

In Fig. 3.3 *red* (dark gray in print versions) vertices represent the articulation points.

Table 3.1 Size of the four largest communities

	Soai's community	Blackmond's community	Pályi's community	Noyori's community
Walktrap	135	58	20	26
Fastgreedy	130	67	33	26

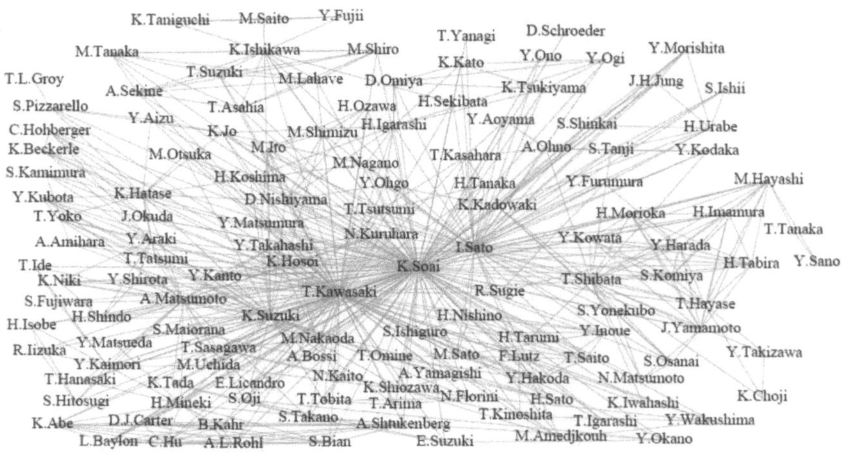

Figure 3.4 K. Soai's community using the Walktrap Algorithm.

We named the four largest communities. Table 3.1 highlights the number of their vertices by both algorithms.

The community structure of the network is as follows.

The largest community is the so-called *Soai's community* illustrated in Fig. 3.4. It contains 135 scientists (vertices) and 631 edges (pairs of co-authors) by the Walktrap Algorithm and 130 scientists by the Fastgreedy Algorithm.

A great portion of vertices of a network has a low (overall) centrality score. Those with high centrality are the most important since they assert more control over the network. Using the Walktrap Algorithm, the three highest scores are K. Soai (5126.98), T. Kawasaki (1881.21), and I. Sato (781.74). It means that the center of this community is Kenso Soai.

This fact is indicates that:

- even today Kenso Soai is the leading personality in the research of asymmetric autocatalysis
- Soai's school of scientists is numerous enough to perform very laborious experiments and thereby influence the work of an unusually high number of scientists.

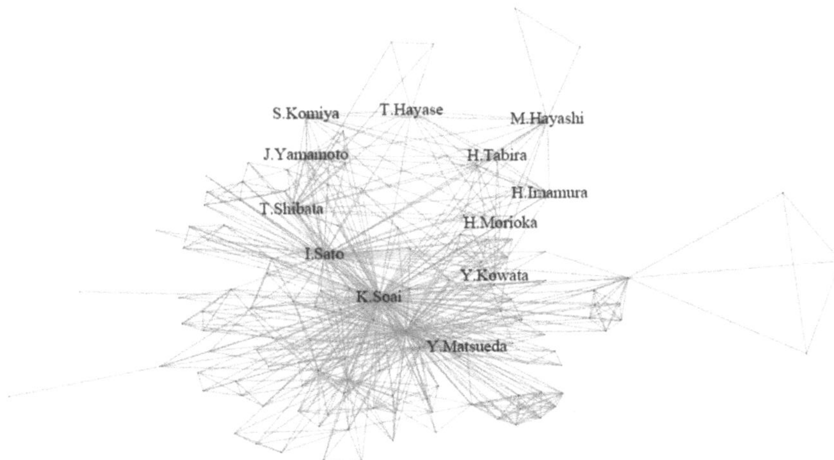

Figure 3.5 Maximum clique in Soai's community.

Further, the high number of edges might indicate a good collegial atmosphere between researchers dealing with the Soai reaction.

The chemical content of the papers [17–20,24–26,28–42] from the Soai Research Group from Tokyo University of Science shows that the main goal of scientists cooperating with Kenso Soai is to obtain preparative information about asymmetric autocatalysis. Doubtless, this is the most laborious branch of chemistry and requires a high number of co-workers, reflected by the high number of nodes (135).

To reveal the closest relations between co-workers one can search cliques in the graph. Both mathematical algorithms show that the maximum clique in the Soai's community has 12 members. See Fig. 3.5.

The second largest community is called D. G. Blackmond's community since Donna Blackmond has the highest centrality score in this community. (The three highest scores: D. G. Blackmond (961.53), W. L. Noorduin (297.00), and R. M. Kellogg (285.23).) This community with 58 scientists is shown in Fig. 3.6.

It is interesting to see that the only link between the two largest communities is J. M. Brown, who is a co-author with both K. Soai and D. G. Blackmond. The scientific content of the publications from this group is the exploration of the mechanism of the Soai autocatalysis, mostly with kinetic and sometimes with spectroscopic methods. Efforts at extending the range of chemistry involved in asymmetric autocatalysis were also reported [49–52].

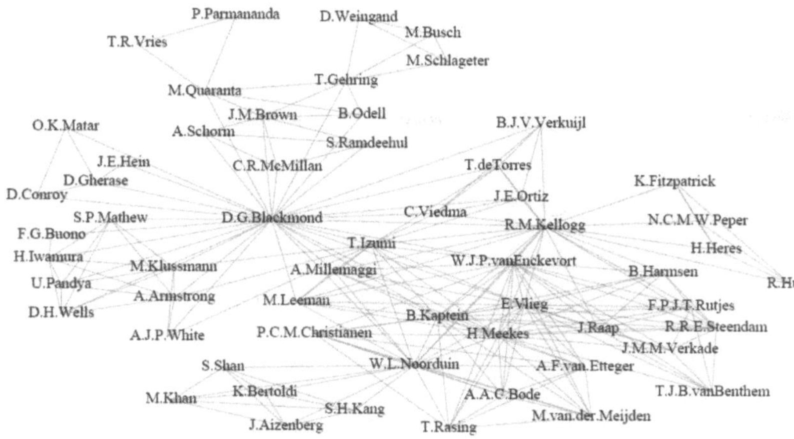

Figure 3.6 D. G. Blackmond's community using the Walktrap Algorithm.

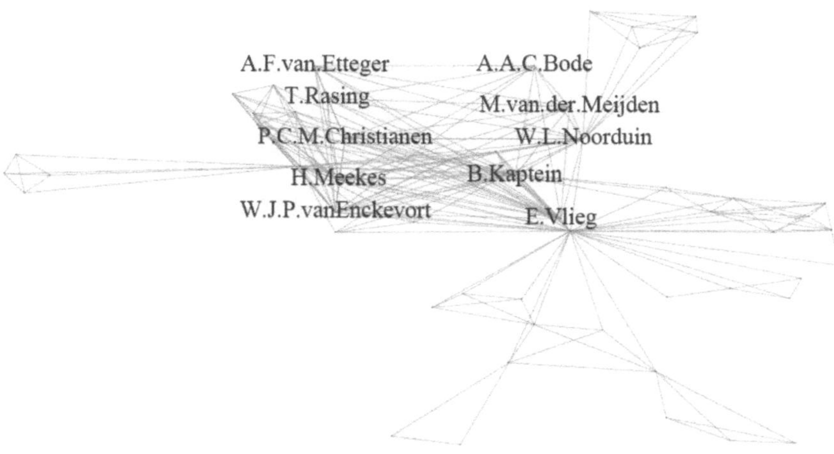

Figure 3.7 Maximum clique in D. G. Blackmond's community using the Walktrap Algorithm.

The Igraph software found two maximum cliques in the Blackmond's community. Both have 10 members.

They are illustrated in Figs. 3.7 and 3.8.

(The three highest scores: D. G. Blackmond (1190.03), R. M. Kellogg (754.03), and W. L. Noorduin (351.00).)

The third largest community shown by the Walktrap Algorithm is G. Pályi's community with 20 members (Fig. 3.9) or 33 members by the Fastgreedy Algorithm (Fig. 3.10).

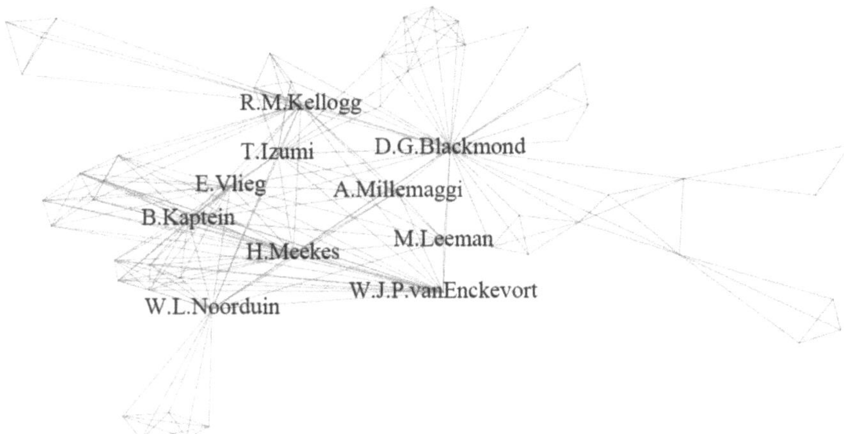

Figure 3.8 Maximum clique in D. G. Blackmond's community using the Fastgreedy Algorithm.

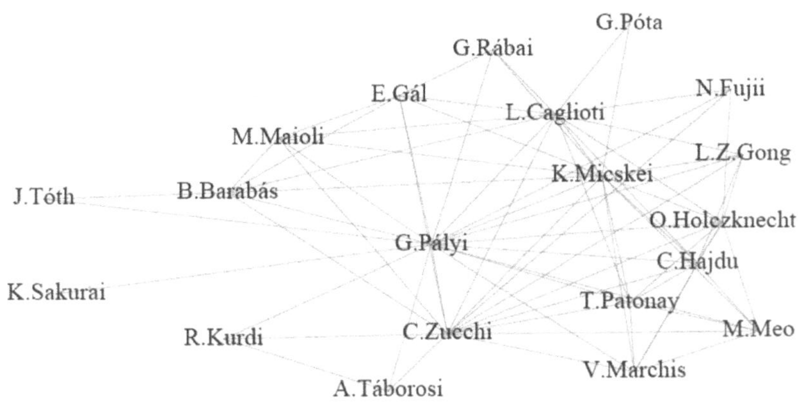

Figure 3.9 G. Pályi's community using the Walktrap Algorithm.

The centrality scores using Fastgreedy Algorithm are as follows:
G. Pályi (327.71), P. Lazzaretti (93.83), and F. Faglioni (93.66).
G. Pályi's group made efforts at exploring the limits of the Soai autocatalysis using empirical and stochastic methods. This group also made preparative attempts at developing a "designed" asymmetric autocatalytic system. Typical examples are given in Refs. [21,22,53–56].
The link between the K. Soai's community and the G. Pályi's community is G. Pályi and N. Florini. They published papers with K. Soai and

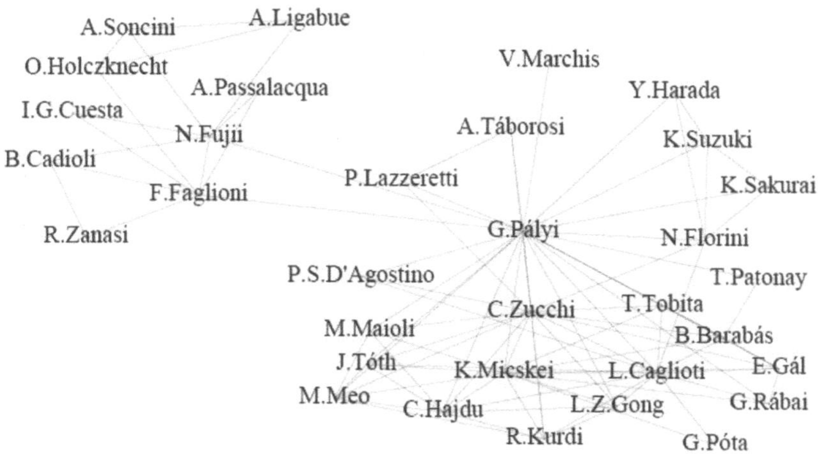

Figure 3.10 G. Pályi's community using the Fastgreedy Algorithm.

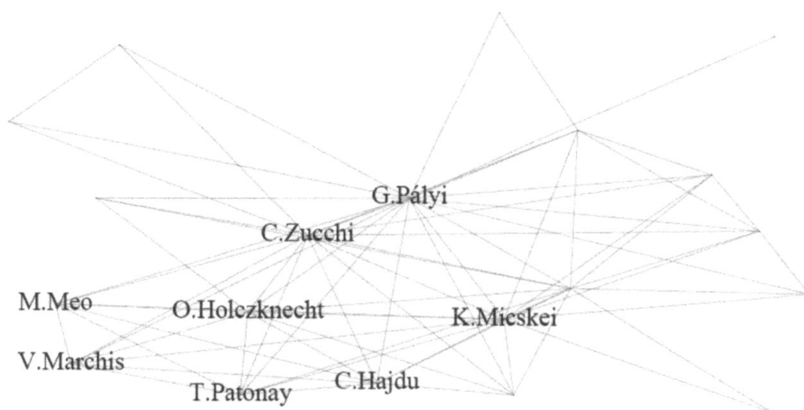

Figure 3.11 The maximum clique in Pályi's community using the Fastgreedy Algorithm.

other five scientists from the Soai community. There is, however, no direct link between the Soai's community and the Noyori's community.

The maximum clique in Pályi's community shown by the Walktrap Algorithm has eight members and it is shown in Fig. 3.11.

The maximum clique in Pályi's community shown by the Fastgreedy Algorithm has seven members and it is shown in Fig. 3.12.

The next community is that of R. Noyori with 26 nodes and 86 pairs of scientists with an average degree of 6.692 (see Fig. 3.13). Both the

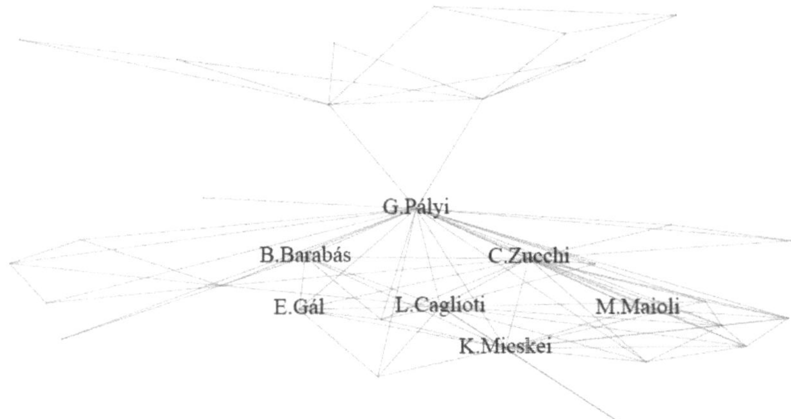

Figure 3.12 Maximum clique in Pályi's community using the Fastgreedy Algorithm.

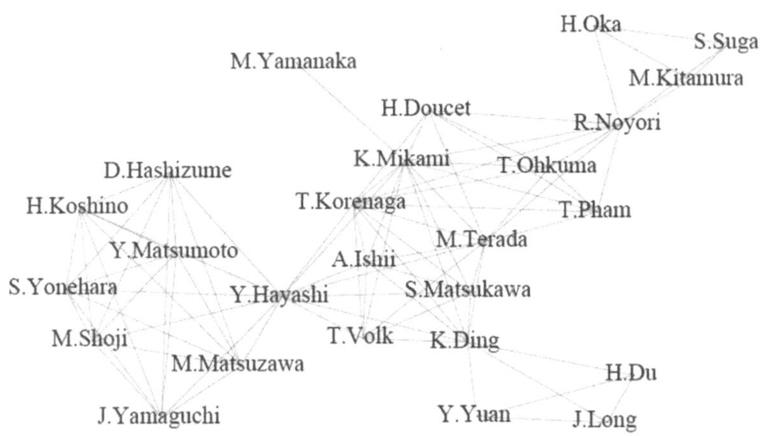

Figure 3.13 R. Noyori's community.

Walktrap and Fastgreedy Algorithm reveals the same community. Noyori's group is the leading authority in classical asymmetric catalysis [57,58]. It is noteworthy, that the highest centrality score has Y. Matsumoto (126), R. Noyori (66), and K. Ding (66). The reason we named this community after R. Noyori is that he is a particularly distinguished person within this group since he won Nobel Prize in chemistry for asymmetric catalysis, in 2001.

3.4 DISCIPLINES OF JOURNALS AND BOOKS CITING SOAI'S PAPER

The distribution of journals and books citing the milestone paper [17] of Soai et al. is shown in Fig. 3.14. The broad spectrum of these publications gives additional support to our initial statement that the discovery of the Soai reaction not only influenced research in several branches of chemistry, but that it also has a significant ongoing impact on all disciplines of natural sciences.

Actually 556 publications were analyzed (few of them belong to two or more categories) (Table 3.2).

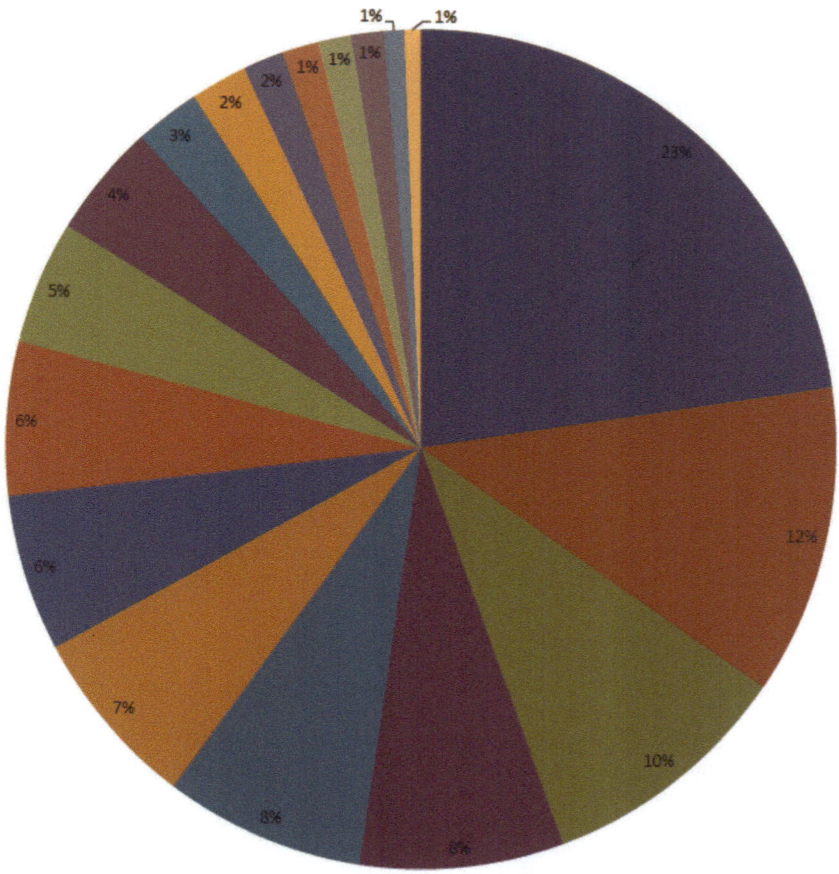

Figure 3.14 Distribution of the disciplines of journals and books citing paper [17].

Table 3.2 Distribution of journals and books citing paper

Chemistry, Journals of general coverage (JACS, Angew. Chem., etc.)	143	23%
Various (theses, web sites, etc.)	75	12%
Chemistry, stereochemistry, chirality	62	10%
Chemistry, physical chemistry	50	8%
Physics	49	8%
Biochemistry	44	7%
Cosmology, cosmochemistry, astrobiology	38	6%
Chemistry, organic chemistry	37	6%
Chemistry, reviews	30	5%
General journals (Nature, Science, PNAS, etc.)	28	4%
Chemistry, catalysis	16	3%
Biology, biophysics	14	2%
Systems theory, mathematical chemistry	10	2%
Book, chemistry	9	1%
Chemistry, inorganic chemistry	8	1%
Chemistry, organometallic chemistry	8	1%
Book, biology, biochemistry	5	1%
Chemistry, colloids, interface Chemistry, chromatography, engineering	4	1%

Source: Soai, K.; Shibata, T.; Morioka, H.; Choji, K. Asymmetric Autocatalysis and Amplification of Enantionmeric Excess of a Chiral Molecule. *Nature* **1995**, *378* (6559), 767–768.

REFERENCES

[1] Barabás, B.; Fülöp, O. Impact of the Soai-Autocatalysis on Natural Sciences. In *Third International Symposium on the Soai Reaction and Related Topic*; Felsőmocsolád, 2015.

[2] Barabás, B. Analysis of Social Network Researchers Working on Soai-Reaction and Related Topics. In: *14th Symposium on Chemical Approaches to Chirality*; Tokyo, Japan, 2015.

[3] Barabás, B.; Fülöp, O. Impact of the Soai-Autocatalysis on Natural Sciences. *J. Math. Chem.* **2016,** *54*, 10–17.

[4] Eliel, E. L.; Wilen, S. H.; Mander, L. N. *Stereochemistry of Organic Compounds;* Wiley: Chichester, 1994.

[5] Markó, L. *Organic Chemistry;* Veszprémi Egyetemi Kiadó: Veszprém, 2000.

[6] Christmann, M.; Bräse, S. *Asymmetric Syntheses.* In *The Essentials;* Wiley-VCH: Weinheim, 2007.

[7] Pályi, G.; Zucchi, C.; Caglioti, L. In *Organometallic Chirality;* Mucchi, Ed.; Accad. Nazl. Sci. Lett. Arti: Modena, 2008.

[8] Noyori, R. *Asymmetric Catalysis in Organic Synthesis;* Wiley: New York, 1994.

[9] Jacobsen, E. N.; Pfaltz, A.; Yamamoto, H. *Comprehensive Asymmetric Catalysis*, Vol. I–III. Springer: Heidelberg, 1999.

[10] Ojima, I. *Catalytic Asymmetric Synthesis*, 2nd. ed.; Wiley: Chichester, 2000.

[11] Horváth, I. T. *Encyclopedia of Catalysis;* Wiley-Interscience: Hoboken, NJ, 2003.
[12] Berkessel, A.; Gröger, H. *Asymmetric Organocatalysis;* Wiley-VCH: Weinheim, 2005.
[13] Mikami, K.; Lautens, M. *New Frontiers in Asymmetric Catalysis;* Wiley: Chichester, 2007.
[14] Pályi, G.; Zucchi, C.; Caglioti, L. *Progress in Biological Chirality;* Elsevier: Oxford, 2004.
[15] Fujii, N. D-Amino Acids in Living Higher Organisms. *Orig. Life Evol. Biosph.* **2002,** *32*, 103–127.
[16] Pavlov, V. A.; Klabunovski, E. I. Homochirality. Origin in Nature: Possible Versions. *Curr. Org. Chem.* **2014,** *18*, 93–114.
[17] Soai, K.; Shibata, T.; Morioka, H.; Choji, K. Asymmetric Autocatalysis and Amplification of Enantionmeric Excess of a Chiral Molecule. *Nature* **1995,** *378* (6559), 767–768.
[18] Sato, I.; Urabe, H.; Ishiguro, S.; Shibata, T.; Soai, K. Amplification of Chirality from Extremely Low Level to Greater than 99.5% ee. *Angew. Chem., Int. Ed.* **2003,** *42*, 315–317.
[19] Shibata, T.; Morioka, H.; Hayase, T.; Choji, K.; Soai, K. Highly Enantioselective Catalytic Asymmetric Automultiplication of Chiral Pyrimidyl Alcohol. *J. Am. Chem. Soc.* **1996,** *118*, 471–472.
[20] Shibata, T.; Yonekubo, S.; Soai, K. Practically Perfect Asymmetric Autocatalysis with (2-Alkynyl-5-pyrimidyl)alkanols. *Angew. Chem., Int. Ed.* **1999,** *38*, 659–661.
[21] Micskei, K.; Maioli, M.; Zucchi, C.; Caglioti, L.; Pályi, G. Generalization Possibilities of Autocatalytic Absolute Enantioselective Synthesis. *Tetrahedron: Asymmetry* **2006,** *17*, 2960–2962.
[22] Caglioti, L.; Micskei, K.; Pályi, G. Chirality of the Very First Molecule in Absolute Enantioselective Synthesis. *Viva Orig.* **2007,** *35*, 82–84.
[23] Pályi, G.; Zucchi, C.; Caglioti, L. *The Soai Reaction and Related Topic;* Accad. Nazl. Sci. Lett. Arti - Artestampa: Modena, 2012.
[24] Soai, K.; Shibata, T.; Kowata, Y. Production of Optically Active Pyrimidylalkyl Alcohol by Spontaneous Asymmetric Synthesis. Japan Kokai, Tokkyo Koho, JP, 9-268179, **1997**.
[25] Soai, K.; Sato, I.; Shibata, T.; Komiya, S.; Hayashi, M.; Matsueda, Y., et al. Asymmetric Synthesis of Pyrimidyl Alkanol Without Adding Chiral Substances by the Addition of Diisopropylzinc to Pyrimidine-5-Carbaldehyde in Conjunction with Asymmetric Autocatalysis. *Tetrahedron: Asymmetry* **2003,** *14*, 185–188.
[26] Kawasaki, T.; Suzuki, K.; Shimizu, M.; Ishikawa, K.; Soai, K. Spontaneous Absolute Asymmetric Synthesis in the Presence of Achiral Silica Gel in Conjunction with Asymmetric Autocatalysis. *Chirality* **2006,** *18*, 479–482.
[27] Mislow, K. Absolute Asymmetric Synthesis: A Commentary. *Collect. Czech. Chem. Commun.* **2003,** *68*, 849–864.
[28] Kawasaki, T.; Tanaka, H.; Tsutsumi, T.; Kasahara, T.; Sato, I.; Soai, K. Chiral Discrimination of Cryptochiral Saturated Quaternary and Tertiary Hydrocarbons by Asymmetric Autocatalysis. *J. Am. Chem. Soc.* **2006,** *128*, 6032–6033.
[29] Kawasaki, T.; Hohberger, C.; Araki, Y.; Hatase, K.; Beckerle, K.; Okuda, J., et al. Discrimination of Cryptochirality in Chiral Isotactic Polystyrene by Asymmetric Autocatalysis. *Chem. Commun.* **2009,** 5621–5623.
[30] Sato, I.; Omiya, D.; Saito, T.; Soai, K. Highly Enantoselective Synthesis Induced by Chiral Primary Alcohols Due to Deuterium Substitution. *J. Am. Soc.* **2000,** *122*, 11739–11740.
[31] Kawasaki, T.; Matsumura, Y.; Tsutsumi, T.; Suzuki, K.; Ito, M.; Soai, K. Asymmetric Autocatalysis Triggered by Carbon Isotope (13C/12C) Chirality. *Science* **2009,** *324*, 492–495.

[32] Kawasaki, T.; Okano, Y.; Suzuki, E.; Takano, S.; Oji, S.; Soai, K. Asymmetric Autocatalysis: Triggered by Chiral Isotopomer Arising from Oxygen Isotope Substitution. *Angew. Chem., Int. Ed.* **2011,** *50*, 8131–8133.
[33] Ozaki, H.; Tada, K.; Harada, S.; Ayugase, T.; Ozawa, H.; Kawasaki, T., et al. Asymmetric Autocatalysis Triggered by Nitrogen Isotope Chirality. In Third International Symposium on the Soai Reaction and Related Topic, Felsömocsolád (H), 2015, pp 24–25.
[34] Soai, K.; Osanai, S.; Kadowaki, K.; Yonekubo, S.; Shibata, T.; Sato, I. d- and l-Quartz-Promoted Highly Enantioselective Synthesis of a Chiral Organic Compound. *J. Am. Chem. Soc.* **1999,** *121*, 11235–11236.
[35] Sato, I.; Kadowaki, K.; Soai, K. Asymmetric Synthesis of an Organic Compound with High Enantiomeric Excess Induced by Inorganic Ionic Sodium Chlorate. *Angew. Chem., Int. Ed.* **2000,** *39*, 1510–1512.
[36] Kawasaki, T.; Hakoda, Y.; Mineki, H.; Suzuki, K.; Soai, K. Generation of Absolute Controlled Crystal Chirality by the Removal of Crystal Water from Achiral Crystal of Nucleobase Cytosine. *J. Am. Chem. Soc.* **2010,** *132*, 2874–2875.
[37] Mineki, H.; Hanasaki, T.; Matsumoto, A.; Kawasaki, T.; Soai, K. Asymmetric Autocatalysis Initiated by Achiral Nucleic Acid Base Adenine: Implications on the Origin of Homochirality of Biomolecules. *Chem. Commun.* **2012,** 10538–10540.
[38] Soai, K.; Shibata, T.; Sato, I. Enantioselective Automultiplication of Chiral Molecules by Asymmetric Autocatalysis. *Acc. Chem. Res.* **2000,** *33*, 382–390.
[39] Soai, K. Asymmetric Autocatalysis, Absolute Asymmetric Synthesis and Origin of Homochirality of Biomolecules. In *Progress in Biological Chirality;* Palyi, G., Zucchi, C., Caglioti, L., Eds.; Elsevier: Oxford, 2004; pp 355–364.
[40] Soai, K.; Kawasaki, T. Discovery of Asymmetric Autocatalysis with Amplification of Chirality and its Implication in Chiral Homogeneity of Biomolecules. *Chirality* **2006,** *18*, 469–478.
[41] Soai, K.; Kawasaki, T. Asymmetric Autocatalysis of Pyrimidyl Alkanol. Top. *Organomet. Chem* **2013,** *44*, 261–280.
[42] Soai, K.; Kawasaki, T.; Matsumoto, A. Asymmetric Autocatalysis of Pyrimidyl Alkanol and its Application to the Study on the Origin of Homochirality. *Acc. Chem. Res.* **2014,** *47*, 3643–3654.
[43] Freeman, L. C. Centrality in Social Networks I: Conceptual Classification. *Soc. Networks* **1978/79,** *1*, 215–293.
[44] www.scholar.google.com (accessed May 15, 2015).
[45] http://igraph.org (accessed May 20, 2015).
[46] https://en.wikipedia.org/wiki/Community_structure (accessed May 15, 2015).
[47] Pons, P.; Latapy, M. Computing Communities in Large Networks Using Random Walks. *Lect. Notes Comput. Sci.* **2005,** *3733*, 284–293.
[48] https://www.cs.princeton.edu/wayne/kleinberg-tardos/pdf/04GreedyAlgorithmsI.pdf (accessed May 15, 2015).
[49] Blackmond, D. G.; McMillan, C. R.; Ramdeehul, S.; Schorm, A.; Brown, J. M. Origins of Asymmetric Amplification in Autocatalytic Alkylzinc Additions. *J. Am. Chem. Soc.* **2001,** *123*, 10103–10104.
[50] Buono, F. G.; Blackmond, D. G. Kinetic Evidence for a Terameric Transition State in the Asymmetric Autocatalytic Alkylation of Pyrimidyl Aldehydes. *J. Am. Chem. Soc.* **2003,** *125*, 8978–8979.
[51] Quaranta, M.; Gehring, T.; Odell, B.; Brown, J. M.; Blackmond, D. G. Unusual Temperature Dependence on Reaction Rate in the Asymmetric Autocatalityc Alkylation of Pyrimidyl Aldehydes. *J. Am. Chem. Soc.* **2010,** *132*, 15104–15107.

[52] Gehring, T.; Quaranta, M.; Odell, B.; Blackmond, D. G. Observation of Transient Intermediate in Soai's Asymmetric Autocatalysis: Insights from 1H NMR Turnover in Real Time. *Angew. Chem., Int. Ed. Eng.* **2012,** *51* (38), 9539–9542.
[53] Micskei, K.; Rábai, G.; Gál, E.; Caglioti, L.; Pályi, G. Oscillatory Symmetry Breaking in the Soai Reaction. *J. Phys. Chem. B* **2008,** *112*, 9196–9200.
[54] Micskei, K.; Patonay, T.; Caglioti, L.; Pályi, G. Amino Acid Chirality for Enantioselective Syntheses. *Chem. Biodivers.* **2010,** 7, 1660–1669.
[55] Pályi, G.; Micskei, K.; Zékány, L.; Zucchi, C.; Caglioti, L. Racemates and the Soai Reaction. *Magy. Kém. Lapja* **2005,** *58*, 218–223.
[56] Barabás, B.; Caglioti, L.; Micskei, K.; Pályi, G. Data-Based Stochastic Approach to Absolute Asymmetric Synthesis by Autocatalysis. *Bull. Chem. Soc. Jpn.* **2009,** *82*, 1372–1375.
[57] Noyori, R.; Suga, S.; Kitamura, M. Self and Nonself Recognition of Chiral Catalysts: The Origin of Nonlinear Effects in the Amino-Alcohol Catalyzed Asymmetric Addition of Diorganozincs to Aldehydes. *Chem. Rec.* **2001,** *1*, 85–100.
[58] Ohkuma, T.; Doucet, H.; Pham, T.; Mikami, K.; Korenaga, T.; Terada, M., et al. Asymmetric Activation of Racemic Ruthenium (II) Complexes for Enantioselective Hydrogenation. *J. Am. Chem. Soc.* **1998,** *120*, 1086–1087.

CHAPTER 4

Kinetic and Structural Aspects of Mirror-Image Symmetry Breaking in the Soai Reaction

Thomas Buhse[1], María E. Noble-Terán[1], José-Manuel Cruz[1], Jean-Claude Micheau[2] and Christophe Coudret[2]
[1]Centro de Investigaciones Químicas, Universidad Autónoma del Estado de Morelos, Cuernavaca, Morelos, Mexico
[2]Laboratoire des IMRCP, Université de Toulouse, CNRS UMR 5623, Université Paul Sabatier, 118 route de Narbonne 31062 Toulouse Cedex 9, France

Abstract

The mechanism of the Soai reaction remains a challenge for experimentalists and theorists. It appears to be much more complex than simple models describing asymmetric autocatalyis in generalized terms. Numerous oligomer species obeying the principles of chiral combinatorics are likely involved. A number of these intermediates have been either isolated or recognized by nuclear magnetic resonance (NMR) studies. A combination of kinetic analyses and density functional theory (DFT) calculations could shed some light on the deeper understanding of this asymmetric autocatalysis. A recent breakthrough was provided by the identification of X-ray structures of various oligomers. To rationalize its mechanism, kinetic modeling of the Soai reaction has been developed in two complementary directions: a structural approach validated by DFT calculations and a functional approach by numerical studies and the reproduction of experimental features. Recently, we introduced a new mechanistic feature related to the concept of kinetic proofreading. We showed that whatever the values of the kinetics parameters, an increase in the catalytic oligomer size from one to three lead to a higher tolerance to poorer chiral recognition between the diastereoisomers, which is a key element for asymmetric amplification and spontaneous symmetry breaking. In the present contribution, we also show that numerical simulation is useful to understand the *ee* amplification during the stepwise addition of aldehyde and organozinc in a typical Soai reaction experimental procedure. Also, we present a first modeling approach in respect to the observed high sensitivity of the Soai reaction towards chiral additives. Our model predicts that the aldehyde concentration plays the role of a bifurcation parameter inducing a transition from a racemic state to an optically active one. Our results also confirm that in such complex systems, kinetic orders must be handled with care.

Keywords: Asymmetric autocatalysis; Soai reaction, mechanism of; Soai reaction, NMR studies of; Soai reaction, DFT studies of; Soai reaction, kinetic studies of; Soai reaction, modeling approach to

Advances in Asymmetric Autocatalysis and Related Topics
DOI: http://dx.doi.org/10.1016/B978-0-12-812824-4.00004-6

4.1 INTRODUCTION

When a three-dimensional object is not superimposable onto its mirror image, it is said to be chiral. In organic chemistry, when a tetragonal carbon atom is surrounded by four different molecular residues it takes one of the two possible mirror-image enantiomeric configurations. One is called the R form (*rectus*), and the other S (*sinister*). When chiral organic substances are synthesized from achiral reactants, usually a mixture of R and S enantiomers in a 50:50 ratio is obtained. This mixture is racemic. Otherwise, the enantiomeric ratio [R]/[S] defines the enantiomeric excess $ee = ([R]/[S] - 1)/([R]/[S] + 1)$. Chiral products are characterized by an enantiomeric ratio $0 < [R]/[S] < \infty$, i.e., $-1 < ee < +1$. In most studies, it is the absolute value of the *ee*, which is considered to be $(0 < |ee| < 1)$.

One of the characteristic features of life is the strong one-handedness of its molecular components, such as L-amino acids and D-sugars. The origin of biological homochirality has been a subject of considerable attention.

The first theoretical model for the spontaneous emergence of homochirality was proposed by Frank in 1953 [1]. His model consisted of an autocatalytic production of enantiomers with mutual destruction. The achievement of such a situation, called *absolute asymmetric synthesis*, is the dream of all synthetic organic chemists [2]. However, until now, only one sufficiently documented experimental example of absolute asymmetric synthesis is known: this is the Soai reaction.

In the 1995 seminal paper of the Soai group [3], it is described how a chiral product acts as a chiral catalyst for its own production. Namely, in the alkylation of certain pyrimidinyl aldehydes with diisopropyl zinc, the chiral pyrimidinic alkanol accelerates its own formation and promotes the prevalence of its own configuration. The process gives rise to an auto-multiplication of a chiral compound (Scheme 4.1).

The widespread interest in the mechanism of the Soai reaction is mostly fueled by its possible connection to the emergence of homochirality in the biosphere [4]. Moreover, the recently established crystal structure of the isopropylzinc alkoxide of pyrimidyl alkanol aggregates has boosted the attractiveness of this system, since it is believed that these aggregate structures may be the keys for delivering the high enantiomeric amplification and spontaneous symmetry breaking in the Soai reaction.

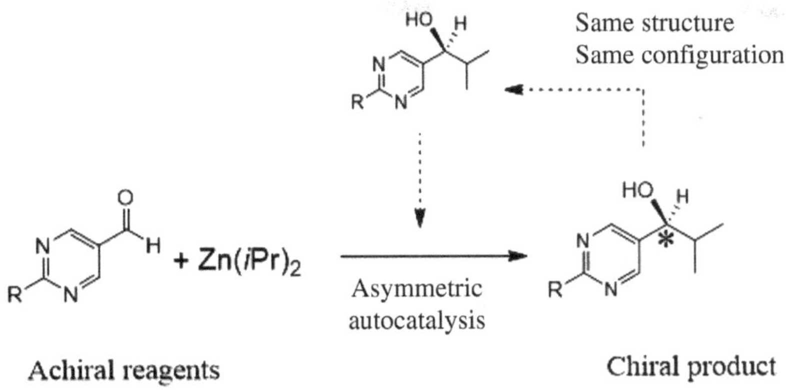

Scheme 4.1 Principle of chiral autocatalysis in the Soai reaction.

4.2 A BRIEF HISTORY OF THE DISCOVERY OF SOAI REACTION

The first chiral autocatalysis was reported by Soai et al. in 1990 [5]. They reported the enantioselective addition of diisopropylzinc ($i\text{Pr}_2\text{Zn}$) to pyridine-3-carbaldehyde using a catalytic amount of (*S*)-pyridylalkanol. During the course of this reaction the newly formed (*S*)-pyridylalkanol product played the role of a chiral catalyst, but an attenuation of the *ee* from 86% to 35% was witnessed. An important breakthrough was achieved in 1995 by the use of pyrimidinic compounds [3]. These chiral autocatalytic systems are able to perform valuable *ee* amplification as predicted by the simple theoretical Frank model [1]. Table 4.1 compares the *ee* amplification performances of the various systems. Some of them have been submitted to consecutive autocatalysis, i.e., a sequence of runs where the product of the previous run serves as the catalyst for the next one, but here we have isolated what we believe to be the steps where the *ee* amplification was at its maximum.

Table 4.1 illustrates the *ee* amplification power of the various systems ranging from a simple autocatalysis with *ee* attenuation to a highly selective autocatalysis. These properties are likely to be related to the presence of N-atoms in the aromatic part of the aldehyde. With only one nitrogen atom, such as in the case of the pyridine carbaldehyde, there is no chiral amplification. On the contrary, moderate chiral amplification appears in the case of quinoline and pyrimidine carbaldehydes likely to be related to the presence of two nucleophilic centers. Methyl, alkenyl, and vinyl substituent are favorable since higher chiral amplification is witnessed.

Table 4.1 *ee* **Amplifications after one step alkylation of various aldehydes by diisopropylzinc**

Aldehyde structure	ee_i (%)	ee_j (%)	Amplification	Refs.
Pyridine	86	35	0.16	[5]
Quinoline	63	82	2.3	[6]
Pyrimidine	10	57	3	[3]
5-Carbamoyl-3-pyridine	14	74	5	[7,8]
Ferrocenylethynyl-pyrimidine	8	73	5.5	[9]
2-Alkenyl and 2-vinyl-5-pyrimidine	36	93.4	13	[10]
Me-pyrimidine	0.28	87	14.3	[11]
Adamantyl-ethynyl-pyrimidine	7.2	89	15	[12]
t-Bu-ethynyl-pyrimidine	57	99	54	[13,14]

ee_i is the enantiomeric excess of the autocatalyst; ee_j is the enantiomeric excess of the newly formed carbinol after the next step; steps giving rise to the higher amplification have been selected; amplification $= (R/S)_i/(R/S)_j = [(1 + ee_i)(1 - ee_j)]/[(1 - ee_i)(1 + ee_j)]$.

Trimethylsilyl and phenyl substituents have also been recommended but without any *ee* amplification measurements [15]. The highest amplification occurs with the *t*-butyl ethynyl substitution, while other bulky radicals (ferrocenyl, adamantyl) are slightly less favorable than the *t*-butyl. That is the reason why most of the results about the Soai reaction have been obtained with the *t*-butyl ethynyl pyrimidinyl aldehyde, which is now known as the "Soai's aldehyde." Moreover, the importance of the structure of the alkyl substituent on the organo-zinc also has been pointed-out [16].

4.2.1 Extreme Sensitivity, Absolute Asymmetric Synthesis and Mirror-Symmetry Breaking

Among the most remarkable chiral amplification performed by the Soai reaction is the amplification of an extremely low *ee*, namely from around 0.00005% *ee* to 57% in a single run [14]. The Soai reaction is so sensitive to any chiral environment that it has been proposed as a general enantioenrichment detector [17–19].

Moreover, the first experimental evidence of absolute asymmetric synthesis was claimed by Soai et al. [20]. In 37 experiments that were carried out without adding any chiral substance they found 19 times the formation of the *S* and 18 times the formation of the *R* pyrimidinic alkanols. Singleton and Vo [21,22] confirmed independently these results

Table 4.2 Compilation of the various experimental mirror-image symmetry breakings and spontaneous asymmetric synthesis in the Soai asymmetric autocatalytic reaction performed without adding any chiral substance

Aldehyde (a)	Conditions	Runs	(*R*)-	(*S*)-	Ref.
Methyl	Ether (b)	45	22	23	[22]
t-Bu-ethynyl	Toluene/ether (b)	37	18	19	[20]
Me_3Si-ethynyl	Toluene (c)	20	8	4 (d)	[23]
t-Bu-ethynyl	Toluene + achiral silica (e)	84	39	45	[24]
Adamantyl-ethynyl	Toluene + 2% rac-carbinol (c)	4	1	3	[12]
t-Bu-ethynyl	Toluene + achiral amines (f)	115	56	59	[25]

Number of independent runs and chirality ((*R*)- or (*S*)- of the produced carbinol. (a), all aldehydes are 2-substituted pyrimidinic carbaldehydes; (b), stepwise addition of iPr_2Zn on aldehyde (two steps); (c), addition of iPr_2Zn on aldehyde in toluene (one step); (d), 8/20 were racemic; (e), stepwise addition of iPr_2Zn and aldehyde (three steps); (f), addition of iPr_2Zn on aldehyde in toluene, then aldehyde on iPr_2Zn in reacting mixture (two steps)).

by showing a clear random distribution of the *R* and *S* enantiomers in some of their 54 experiments. Each experimental run provided a prevalence of either the *R* or the *S* enantiomer, with 27 runs with the *R* enantiomer and 27 with the *S*. Several other papers indicate a stochastic behavior in respect of the outcoming chirality sign when achiral starting conditions were used. Gridnev et al. [23] carried out 20 experiments of which eight resulted in (*R*)-product, four resulted in (*S*)-product and eight were racemic. The random formation of (*S*)- and (*R*)-5-pyrimidyl alkanol in the presence of achiral silica gel was also reported by Soai et al. [24]. Furthermore a similar stochastic tendency was witnessed by Gehring et al. [12]. The Soai group also reported the stochastic formation of chiral (*S*)- and (*R*)-pyrimidyl alkanols in the presence of achiral amines, such as *N,N'*-dimethylpiperazine, *N,N'*-diethylpiperazine or *N*-methyl-morpholine [25]. Table 4.2 gathers all of the already-published experiments where the Soai reaction has been carried out without any addition of chiral initiator.

These results advocate a mirror-symmetry breaking event during the Soai reaction. With the exception of few crystallization processes [26,27], such behavior is quasi-unprecedented in organic synthesis [28,29] and can be considered as an example of "absolute asymmetric synthesis" [30]. In classical organic chemistry it is expected that without the intervention of any chiral factor, the product is racemic. However, when a system involves asymmetric autocatalysis with an amplification of *ee*, an initial bias in *ee* resulting from a fluctuation in the racemate would be enhanced to produce an enantio-enriched product because the racemic state becomes unstable. Fluctuations are sufficient to drive the system away from an unstable state. The origin of such fluctuations has been discussed. According to Mills [31], statistical deviations from the racemic state become significant when the population of the chiral species is small since the expected *ee* of a tiny sample is proportional to the reciprocal of the square root of the number of molecules ($ee \div N^{-1/2}$). The possible effect of natural abundance isotopic chirality in the reagents of the Soai reaction has been recently discussed [32]. It is concluded that the chiral isotopic species formed by natural isotopic substitution in achiral reagents could provide possible chiral induction in the Soai reaction. Moreover, the 84 experiments [24] have been submitted to various sophisticated statistical analysis and a tiny but significant bias was demonstrated towards the (*S*)-carbinol. While the exact origin of such a bias remains unknown, it was assumed that it could be either fortuitous [33] or the manifestation of a chiral external force such as parity-violating energy difference [34]. In summary, all these results show that in the Soai reaction very small chiral fluctuations induce a slight initial enantiomeric imbalance that is enlarged by consecutive asymmetric autocatalysis. As the Soai reaction is able to perform mirror-image symmetry breaking and spontaneous asymmetric synthesis it is expected that it will be very responsive to any kind of cryptochiralities. This statement is the basis of a remarkable series of fascinating demonstrations carried out by the Soai's group.

4.2.2 Effect of Circularly Polarized Light

Circularly polarized light (CPL) has been used to induce chirality in organic compounds [35]. However, due to the very small anisotropy factor exhibited by organic compounds, only very low enantioenrichments have been achieved. However, when these low *ee* are used as chiral inducers in the Soai reaction, highly enantioenriched carbinols have been generated. Kawasaki et al. [36] have shown that the asymmetric photodecomposition of racemic pyrimidyl alkanol with CPL and the subsequent asymmetric autocatalysis afforded a practically enantiopure ($ee > 99.5\%$) pyrimidyl alkanol. Moreover, the absolute configuration of the carbinol was in accordance with the handedness of the CPL. Because,

(*R*)- and (*S*)-carbinols respectively exhibit positive and negative circular dichroism spectra in the UV region, the correlation was rationalized by considering that the irradiation of the racemic carbinol by left-handed-CPL induces the photodegradation of the (*R*)- faster than the (*S*)- leaving the (*S*)- with a tiny *ee*. Symmetrically, right-handed-CPL afforded the (*R*)- in the same conditions. This work appears to be the first demonstration of an asymmetric autocatalysis triggered directly by a chiral physical force with an unequivocal correlation between the handedness of the CPL and the absolute configuration of a quasi enantiopure organic compound.

4.2.3 Effect of Chiral Additives

The asymmetric autocatalysis of pyrimidylalkanol has been examined under the influence of various chiral additives. The following tables show that almost every imaginable chiral inductor including all kinds of functional groups are able to direct the Soai reaction specifically. However, they must be divided into two main groups: those which are soluble in the reaction mixture giving rise to chiral recognition in solution and those which are not soluble, yielding to chiral surface adsorption.

4.2.3.1 Easily Soluble Additives

Hydrocarbons

Among the easily soluble additives the hydrocarbon cases deserve to be regarded separately because their molecular interactions are limited in the absence of any heteroatom. Table 4.3 gathers the main amplification experiments using soluble hydrocarbon (C_xH_y) chiral additives.

While there is no clear-cut relationship between the configurations of the various additives and the carbinols, results show that in all cases chiral hydrocarbons play the role of chiral initiators to give enantioenriched pyrimidyl alkanols. Therefore, the Soai reaction provides a highly sensitive discrimination between enantiomers whose hidden chirality cannot be differentiated by spectroscopy. However, being pure hydrocarbons they provide only a few molecular interactions that could give rise to some chiral recognition in solution. In our opinion, these effects constitute an example of an amplification of some London dispersion forces.

Table 4.3 List of Soai reaction experiments carried out in presence of chiral hydrocarbons

Additive	$ee_{additive}$ (%)	$ee_{carbinol}$ (%)	Ref.
(*P*)-(+)- and (*M*)-(−)-[5]- and [6]-helicenes	0.13 to >99.5	(*S*)-: 56−95	[37]
(*P*) → *S*; (*M*) → *R*		(*R*)-: 62−94	
(S_p)- & (R_p)-hydrocarbon[2.2]-paracyclophane	88 to >99.5	(*S*)-: 93−99	[38]
(S_p) → *S*; (R_p) → *R*		(*R*)-: 93−99	
(−)-(*R*) & (+)-(*S*)-1,3-disubstituted allenes	43 to >99.5	(*S*)-: 94−98	[39]
(*R*) → *S*; (*S*) → *R*		(*R*)-: 90−97	
(*R*)- & (*S*)-chiral olefins	56 to 99	(*S*)-: 90−96	[40]
(*R*) → *S*; (*S*) → *R*		(*R*)-: 91−96	
(*R*)- & (*S*)-cryptochiral saturated quaternary hydrocarbons	86 to 91	(*S*)-: 91−97	[41]
(*R*) → *S*; (*S*) → *R*		(*R*)-: 94−97	
(*S*)- & (*R*)-cryptochiral saturated tertiary hydrocarbons	82 to 99.5	(*S*)-: 93−99	[41]
(*S*) → *S*; (*R*) → *R*		(*R*)-: 89−98	
(*M*)- & (*P*)-(12,8)-[4]$CC_{2,8}$ finite single-wall carbon nanotube	−	(*S*)-: 89−91	[42]
(*M*) → *S*; (*P*) → *R*		(*R*)-: 87−91	
Cryptochiral isotactic polystyrene; Δ,*S*,*S* → *S*; Λ,*R*,*R* → *R*	−	(*S*)-: 67−94 (*R*)-: 63−92	[43]

Heteroatomic Additives

On the other side, stronger molecular interations and some catalysis are likely to occur in presence of heteroatoms. Table 4.4 collects the Soai reaction amplification tests employing chiral heteroatomic additives which were assumed to be soluble in the Soai reaction mixture (i.e., in toluene).

4.2.3.2 Nonsoluble Additives

We have gathered in Table 4.5 the main amplification experiments using additives which were assumed to be insoluble. Surface reactions have also been collected.

The detailed mechanism of the asymmetric induction with chiral crystal additives is still under investigation. However, examination of these

Table 4.4 List of Soai reaction experiments carried out in presence of soluble chiral heteroatomic additives

Additive	$ee_{additive}$ (%)	$ee_{carbinol}$ (%)	Refs.
(*P*)-(+)- and (*M*)-(−)-tetrathia-[7]-helicene; (*P*) → *S*; (*M*) → *R*	2	83	[44]
(*R*)- & (*S*)-4-carboxy[2.2]-paracyclophane	2.5–99	(*S*)-: 89–97	[45]
(*R*) → *S*; (*S*) → *R*		(*R*)-: 94–95	
Chiral Cr(acac)$_3$ complex; Δ → *S*; Λ → *R*	9	58	[46]
Chiral diols (a)	(b)	99	[47]
Chiral secondary alcohol (c)	8	96	[48]
Propylene oxide, styrene oxide	3	79	[49]
(*R*) → *S*; (*S*) → *R*	5	80	
Dialkyl norephedrines (d)	>99.5	98.8	[50–53]
^{13}C/^{12}C chiral alcohol (e)	86	95	[54]
D-labeled chiral primary alcohols; (*R*) → *S*; (*S*) → *R*	>95	96	[55]
Deuterated *N*-Bz-alpha-methylalanine; (*S*) → *S*; (*R*) → *R*	(b)	90	[56]
^{18}O-labeled hydrobenzoin; (*R*) → *S*; (*S*) → *R*	>95	98	[57]
^{18}O-isotopically chiral glycerin; (*R*) → *S*; (*S*) → *R*	>99 (f)	97	[58]

Situations showing the stronger *ee* amplification have been selected. (a), depend on the diol and the presence of achiral additives; (b), assumed to be enantiomerically pure; (c), based on configuration correlations, the substituents of the secondary alcohols were compared with the phenylgroup in term of bulkiness; (d), depend on the norephedrine and the presence of achiral additives; (e), depend on the alcohol; (f), estimated from the chiral precursor.

results shows that the configuration of the carbinol products is accurately determined by the chirality of the additives. Three main effects can be suggested to explain the enantioselectivity observed in these reactions: (1) an asymmetric activation of *i*Pr$_2$Zn by the chiral surfaces could induce the formation of a chiral zinc species enabling an enantioselective addition to the aldehyde; (2) an enantiotopic face selective adsorption of the aldehyde to the chiral surface of the additives may promote the addition of *i*Pr$_2$Zn onto the preferred face; or (3) an enantioselective adsorption of the initially formed carbinol zinc alkoxides could also induce some

Table 4.5 List of Soai reaction experiments carried out in presence of insoluble chiral additives

Chiral crystals	Chiral relationship	$ee_{carbinol}$ (%) (a)	Refs.
CPL enriched leucine and valine	L → *R*; D → *S* *ee* = 1% → 51%	(b)	[59]
α-methyl-d_3-alanine	(*R*) → *S*; (*S*) → *R* *ee* = 69% → 99%	(c)	[60]
Helical mesoporous silica	*P* → *S*; *M* → *R* *ee* = 19% → 97%	97 (d)	[61]
d- and *l*-quartz	*l* → *R*; *d* → *S*	97 (d)	[62]
Sodium chlorate and bromate	*l* → *R*; *d* → *S* (ClO_3^-) *l* → *S*; *d* → *R* (BrO_3^-)	98 (d)	[63,64]
Cinnabar	*P* → *R*; *M* → *S*	92 (d)	[65]
Retgersite, nickel sulphate hexahydrate	(−) → *R*; (+) → *S*	95 (c)	[66]
DL-Serine	*P* → *R*; *M* → *S*	95 (d)	[67]
Cytosine	(+) → *R*; (−) → *S* (−) → *R*; (+) → *S* (e)	96 (c)	[68–70]
Adenine dinitrate	(−) → *R*; (+) → *S* *P* → *R*; *M* → *S*	93	[71]
Thienylcarbonylglycine	*P* → *R*; *M* → *S*	90 (c)	[72]
Hippuric acid	(+) → *R*; (−) → *S*	89 (a)	[73]
Achiral benzils	*P* → *S*; *M* → *R*	97 (c)	[74]
Achiral tetraphenylethylenes	*P* → *R*; *M* → *S*	96 (c)	[75]
Salts between achiral acids and achiral amines	*P* → *R*; *M* → *S*	97 (c)	[76]
Ethylenediamine sulfate	(+) → *R*; (−) → *S*	93 (d)	[77]
Achiral tris(2-hydroxyethyl) 1,3,5-benzenetricarboxylate	(−) → *R*; (+) → *S*	94 (c)	[78]
Prochirality of an achiral aldehyde at the single-crystal face	001 → *R*; $00\underline{1}$ → *S*	69 (f, g)	[79]

(a), Experiments run in three or four separate portions; (b), valine; (c), crystals were pre-mixed with iPr_2Zn; (d), crystals were pre-mixed with aldehyde; (e), chiral cytosine crystals obtained after dehydration of achiral cytosine hydrate; (f), reaction of iPr_2Zn on aldehyde enantiotopic crystal faces; (g), there is no clear-cut correlation between the alkylation yield and the final *ee*.

enantioimbalanced asymmetric autocatalyst. Subsequent asymmetric autocatalysis would enhance the small chirality to afford the highly enantioenriched carbinol with an absolute configuration corresponding to the chirality of the additive used.

4.3 STRUCTURAL ASPECTS OF THE SOAI REACTION INTERMEDIATES

4.3.1 Macrocycle

The origin of asymmetric amplification in the Soai reaction has been investigated by Blackmond et al. [80]. The reaction rate was monitored by micro-calorimetry. The reaction exhibits a maximum rate and is faster when the added catalyst is enatiomerically pure. Another prominent result is that the enantiomeric excess of the product increased significantly during the course of the reaction. In its conclusion, this paper provides an unprecedented insight into the structural features of the Soai reaction: a simple dimeric catalyst model is assumed. This is a 12-membered macrocyclic structure where the Zn atom of one alkoxide is coordinated with one of the pyrimidinic nitrogen. It is assumed that such a scaffold offers N, O coordination for the pyrimidinic aldehyde thus supplying a template for attack of $i\text{Pr}_2\text{Zn}$ (Scheme 4.2).

Two years later, an NMR systematic study of the autocatalysis in the Soai reaction was carried out by Gridnev et al. [23]. Their paper provides several direct NMR observations of the chiral symmetry breaking. Two nice NMR kinetics of the aldehyde consumption are displayed. However, there is no more structural investigations and it is assumed that the active catalyst is a product dimer but it is argued in the footnote that the previously macrocyclic dimer does not possess the required lowest energy. Then, Gridnev and Brown [81] carried out DFT calculations at the B3LYP function level with a 6−31G* basis set for all atoms. Their purpose was to define some feasible oligomeric structures that could play a role in the autocatalytic mechanism.

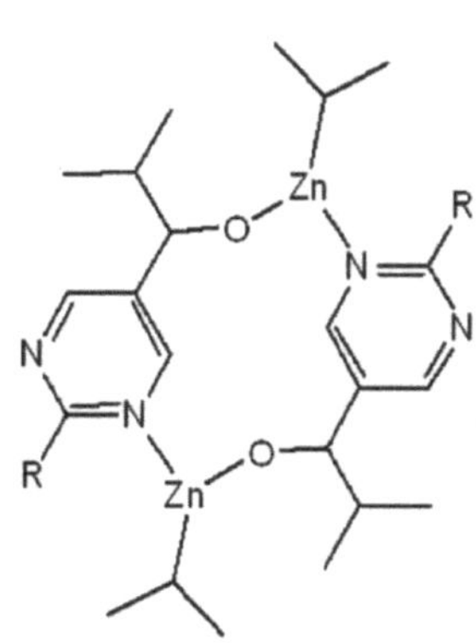

Scheme 4.2 Structure of the 12-membered macrocycle.

4.3.2 Monomer and Dimers

The successive bindings of alkylzinc with aldehyde, alkylation and further binding on the alkoxide have been investigated. In complexation to the aldehyde, it has been shown that N-coordination is more stable than O-coordination. However, on the alkoxide, a bridging agostic alkyl is preferred over the expected N-coordinated structure.

Two low-energy dimer structures have been computed. The square dimer appears to be more stable than the macrocyclic dimer by more than 8 kcal/mol. Further N-coordination of another Zn alkoxide has been witnessed.

In order to test if a dimer could be the catalyst, several trimeric structures including a dimer, an aldehyde and an alkyl Zn have been investigated. However, the precise reaction pathway is not easy to define because further coordinations are needed to build an enantioselective arrangement (Scheme 4.3).

4.3.3 Tetramers and Oligomers

With regard to the tetramers, several clever arrangements were investigated. Among them, the so-called ladder and barrels are particularly attractive since a transition state based on a macrobicyclic trimeric catalyst leading to a barrel has been speculated. However, only the basic ring connectivities have been considered in order to limit the structural complexity of the tetramers.

The main conclusion of this study is on one side the possibility of a tetrameric transition state and on the other side the refutation of any strong aldehyde–ZnR2 binding [82]. Klankermayer et al. carried out more extensive computational work in order to analyze the role of the isopropyl group in asymmetric autocatalytic zinc alkylations [16]. Four energetically favorable tetramers have been confirmed: the cube, the barrel, the ladder and the so-called SMS (sqare-meacrocycle-square). It is

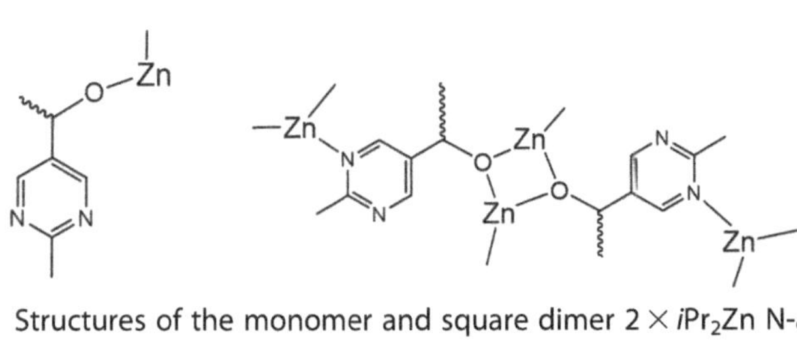

Scheme 4.3 Structures of the monomer and square dimer 2 × $i\mathrm{Pr}_2\mathrm{Zn}$ N-adduct.

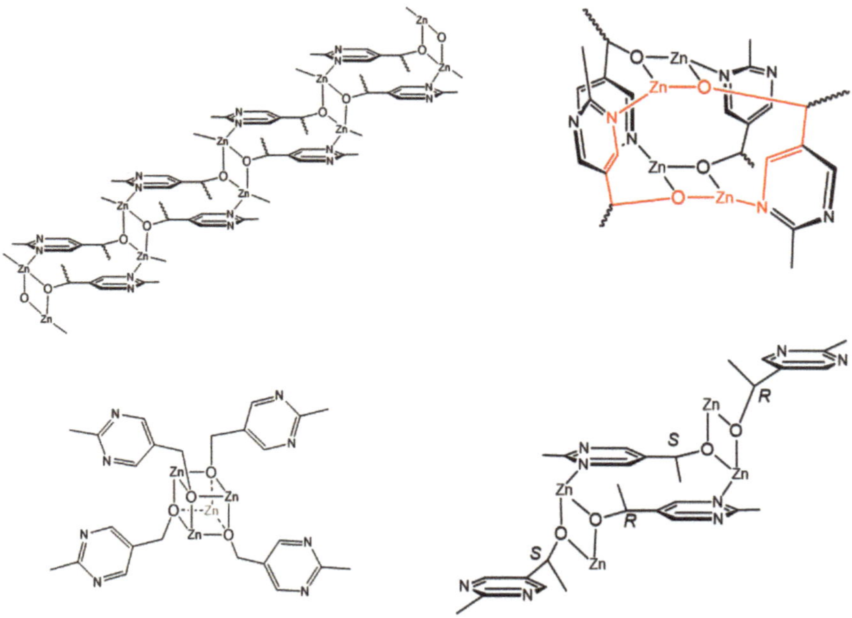

Scheme 4.4 Structures of the ladder, barrel, cube and racemic square-macrocycle-square.

argued that the SMS-tetramer holding two trigonal Zn centers could play the role of catalytic sites and could also give rise to ladder-like oligomeric fragments (Scheme 4.4).

4.3.4 Adducts and Transition State

Independently, an Italian team [83–85] started on the same problem, namely to unknot the mechanism of the Soai asymmetric autocatalytic reaction by first-principles calculations. In addition to the macrocyle and the square dimers, they identified respectively their organozinc O- and N-adducts (Scheme 4.5).

Moreover, they also recognized three more tetrameric structures: a two-aldehyde adduct on the O-Zn macrocyclic dimer, a one-aldehyde adduct on a macrocyclic square trimer and a barrel-like tetramer. It is interesting to note that this one-aldehyde adduct on a macrocyclic square trimer is structurally reminiscent of the noncomputed aldehyde addition transition state found in Gridnev and Brown [81] (Scheme 4.6).

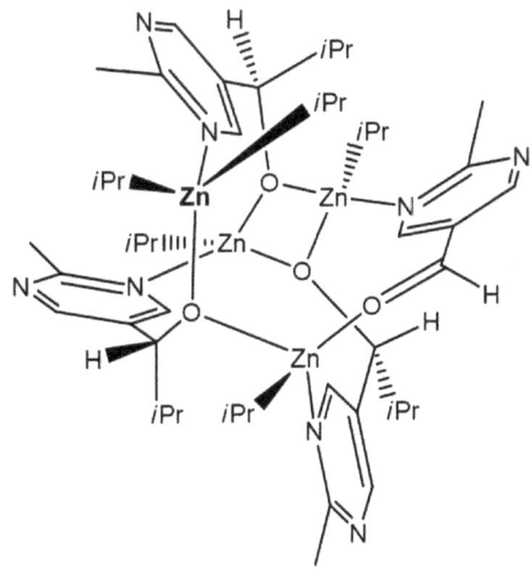

Scheme 4.5 Structure of a one-aldehyde adduct on a macrocyclic square trimer allowing an *i*Pr transfer from iPr_2Zn.

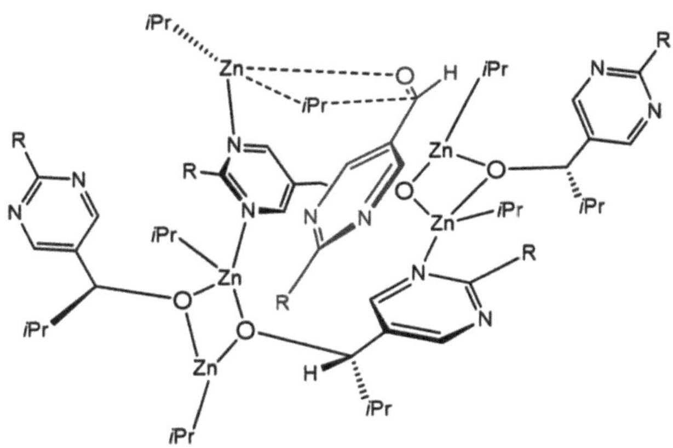

Scheme 4.6 Gridnev computed the SMS-like transition state for the stereospecific alkylation of the aldehyde.

In their next paper, the same authors [86] used DFT calculations to support an autocatalytic mechanism in which dimer catalysts serve as templates for the reaction of two molecules of dialkylzinc with two molecules of aldehyde, so that they are duplicated. It was shown that the observation of asymmetric amplification in the Soai reaction results from the stronger influence of homochiral over heterochiral catalytic cycles. This predominance is due to the lower energy and lower reactivity of the heterochiral species versus their homochiral counterparts. It is finally concluded that this decisive property is insured by a fine-tuning of the structure of the aldehyde and of the dialkylzinc reagent and that higher aggregates could possibly be involved in various competitive catalytic cycles.

Gridnev and Vorobiev [87] performed a combined DFT computational and NMR experimental study 1 year later in order to quantify the sophisticated equilibria occurring in the Soai reaction. It was shown that the tetramers formed by the dimerization of the homochiral square dimers get a SMS-like conformation with orthogonal pyrimidinyl rings. Their most important result is that the homochiral tetramer forms an ideal chiral cavity for the enantioselective alkylation of the aldehyde giving rise to a Zn alkoxycarbinol of the same handedness. On the contrary, in the heterochiral tetramer the pocket is significantly less spacious and disappears upon alkyl Zn coordination, thus excluding the heterochiral tetramers from the catalysis flow [88]. It is finally pointed-out that the realization of the chiral amplification requires an adjusted reversibility of different steps and a tuned relative abundance of homo- and heterochiral species. A tiny formation ($<2\%$ at $0^{\circ}C$) of structurally ill-defined acetal species was also indicated [89,90]. However, it was concluded that the observed species cannot be the precursor of the reaction products.

4.3.5 X-Ray of Isolated Crystals

The year 2015 witnessed an important breakthrough in the Soai reaction knowledge, namely the single-crystal X-ray analysis of the asymmetric autocatalyst [91]. Single crystals of the zinc alkoxide were obtained my mixing pyrimidylalkanol and $i Pr_2Zn$. Five different kinds of single crystals were isolated depending on the chirality of the pyrimidinylalkanol (enantiopure or racemic), the number of $i Pr_2Zn$ equivalents, and the absence of the presence of a solvent (toluene or THF). The single crystals appeared after keeping the solution or after the slow evaporation of the solvent (Scheme 4.7).

Two SMS-like tetrameric structures were confirmed in which two alkoxides form Zn_2O_2 square dimers and two dimers coordinate with each other to form a 12-membered macrocycle through nitrogen coordination with the zinc atom in the squares. In the chiral tetramer, six $i Pr_2Zn$ moieties coordinate with the peripherical nitrogen atoms. In the racemic tetramer, (*R*)- and (*S*)-alkoxides are encountered in both squares and macrocycle and only four $i Pr_2Zn$ are coordinated. Preliminary DFT calculations on both tetramer structures based on this crystal structure

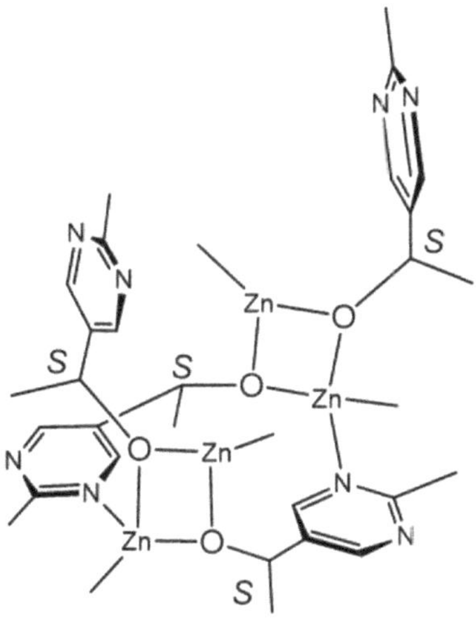

Scheme 4.7 Structure of the homochiral SMS-like tetramer (isolated as a $6 \times i\mathrm{Pr}_2\mathrm{Zn}$ N-adduct).

indicate that the racemic tetramer is more stable than its enantiopure counterpart by 4.8 kcal/mol. In addition, higher oligomers were also observed upon changing the crystallization conditions. Their ladder-like structures depend on the enantiopurity of the pyrimidinylalkanol, and amount of $i\mathrm{Pr}_2\mathrm{Zn}$ and solvent. While it is well known that the reactive species in the solution could be different from the isolated crystallized structures, it is assumed that these coordinative tetramers may be the basis for the huge enantiomeric excess amplification and mirror-image symmetry breaking in the Soai reaction.

4.4 KINETIC ANALYSIS OF THE SOAI REACTION

4.4.1 The Dimer Catalysis Period

Analysis of the Soai reaction started when Blackmond et al. [80] claimed that a simple dimeric catalyst model was able to describe some features of asymmetric autocatalysis. Less than 1 month later, Sato et al. [92] carried out a kinetic analysis of the relationship between the reaction time and the yields of the product and approved a second kinetic order of the catalyst in the enantioselective addition of diisopropylzinc to pyrimidine-5-carbaldehyde using chiral pyrimidyl alkanol as a catalyst. Blackmond [93] criticized this approach in a paper describing the conditions for asymmetric amplification in autocatalytic reactions. She concluded that a mechanism for inhibition of the minor enantiomer must be considered. She also pointed-out that both rate and enantioselectivity must be taken into account in developing any kinetic model for asymmetric autocatalytic reactions with *ee* amplification. Finally, on the basis of kinetic orders, Buono et al. [82] argued that they have found kinetic evidence for a tetrameric transition state in the Soai reaction. Using reaction heat flow techniques, they analyzed a Soai reaction sequence carried out by consecutive addition of aliquots of a methyl substituted aldehyde into an excess of $i\mathrm{Pr_2Zn}$ in toluene in the presence of a catalyst of 10% enantiomerically enriched pyrimidinylalkanol [94]. The rate of the reaction was observed to stabilize while the *ee* increased and a precipitation occurred during the course of the reaction. At last, it was concluded that precipitation may play a significant role only for reactions carried out at lower temperatures or higher concentrations. In a later paper, Blackmond [95] claimed again that the Soai reaction is catalyzed by only the homochiral dimers. The presence of a tetrameric transition state was also proposed, with the concession that further experimental work is needed to describe its structure. While the importance of the analysis of the kinetic shapes is pointed-out, refinements of the previously published models [96–98] are strongly suggested. A few years later, a re-examination of the Soai reaction on the basis of a modified Kagan ML2 model was published [99]. On the basis of a modified Kagan model, it was shown that the pyrimidinylalkoxycarbinols forms dimer species stochastically, and only the homochiral dimers play the role of autocatalysts. Reaction rates and *ee* evolution have been accurately reproduced. However, it was admitted that particular features of the nonlinear effects in the Soai reaction deserved further examination. This last point is of importance, since it is clear that the most striking characteristics cannot be reproduced by such a model; for instance, the huge chiral amplification and the mirror-image symmetry breaking or spontaneous asymmetric synthesis.

4.4.2 The Refinement Period

Quaranta et al. [100] reported an induction period and an anti-Arrhenius effect in the Soai reaction. Moreover, NMR investigations demonstrated an effective tendency for pyrimidinyl-alkoxide to aggregate with preferential formation of unsymmetrical tetramers. It was also admitted that the active catalyst is derived from a tetrameric structure. Such a paper appears as an important refinement in the vision of the Soai reaction since tetrameric structures were observed for the first time.

4.4.3 Bibliographic Survey of the Soai Reaction

The evolution of the knowledge, the performances and understanding of the Soai reaction can be followed by considering the various review papers published from 2000 to 2014 [101–119].

4.5 THEORETICAL MODELS OF THE SOAI REACTION

4.5.1 Deterministic Models

As soon as new experimental results were published, theoretical models of the Soai reaction were suggested. Just looking at the papers' reception dates shows an impressive acceleration in the race at the beginning of 2003, when two pioneering kinetic papers were launched in less than 3 months. First, Buhse [96] proposed a tentative kinetic model for chiral amplification in autocatalytic alkylzinc additions. Six weeks later, the Soai group [97] described independently an experimental and kinetic analysis of the relationship between the time, yield and enantiomeric excess of asymmetric autocatalysis with amplification of enantiomeric excess.

4.5.1.1 Kinetic Curve Fittings and Parameter Identification

In 2005, we proposed the use of kinetic schemes derived from the Frank model, including both stereospecific autocatalysis and mutual inhibition [98]. We were able to reproduce for the first time the experimentally observed giant chiral amplification from an initial enantiomeric excess of $<10^{-6}\%$ to $>60\%$, high sensitivity and positive response to the presence of minute amounts of chiral initiator at concentrations $<10^{-14}$ M, and spontaneous absolute asymmetric synthesis from achiral starting conditions. Assuming either monomer or dimer catalysis, we have fitted the most complete kinetic data of the Soai reaction available [23]. The extracted rate parameters confirm that the mutual inhibition between enantiomers originates from the presence of dimerization equilibria. Higher stability of the heterochiral dimers gave rise to mirror-symmetry breaking. From the kinetic viewpoint, simple autocatalysis involving monomers as catalysts was consistent with all reported Soai reaction experiments.

For instance, an enantiomerically pure catalyst was twice as active as its racemic counterpart, whereas the overall rate profile versus extent of reaction remained independent of the catalyst ee. While, the exact nature of the autocatalytic stereoselective process is still to be revealed, we think

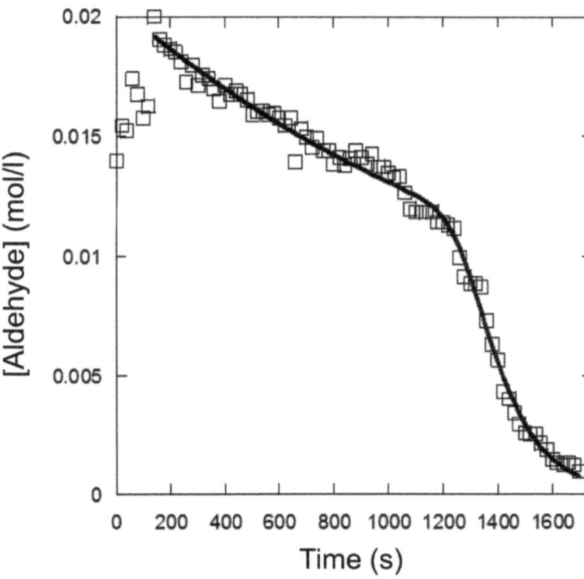

Figure 4.1 Evolution of the aldehyde concentration during a typical Soai reaction [23]. Data points: nuclear magnetic resonance monitoring; solid-line fitting by a monomer catalysis model [98] using the following parameters: uncatalyzed reaction: 5.2×10^{-3}; homo-dimerization: 4.8×10^{3}; homo-dimer splitting: 21; hetero-dimerization: 4.5×10^{5}; hetero-dimer splitting: 5.2×10^{-2}; monomer catalysis: 69 ($M^{-2}\,s^{-1}$); hetero- versus homo-dimers energy difference: 6.3 kcal/mol (hetero more stable); stereoselectivity of the catalytic step: 100%. No initial catalyst added.

that the kinetic order-based arguments [80,82,95,99,119] are not sufficient to conclude that the catalyst is dimeric or even multimeric and that the transition state is tetrameric. On the other side, looking at the fitted parameters, it appears that the monomeric model needs a homodimer versus heterodimer diastereoisomeric energy difference as high as 6.3 kcal/mol while for the dimeric model the requisite is only 2.7 kcal/mol (heterodimer more stable). This second value being closer to our PM-3 calculations (2 kcal/mol), we have consequently refined our previous models accordingly (Fig. 4.1).

4.5.1.2 Model Improvements With Autocatalytic Cycles

These considerations lead us to suggest an autocatalytic cycle model of asymmetric amplification and mirror-symmetry breaking in the Soai reaction [120]. In this regard, recent DFT calculations by Schiaffino and Ercolani [83–85] underlined the presence of alkoxide dimers, trimers and

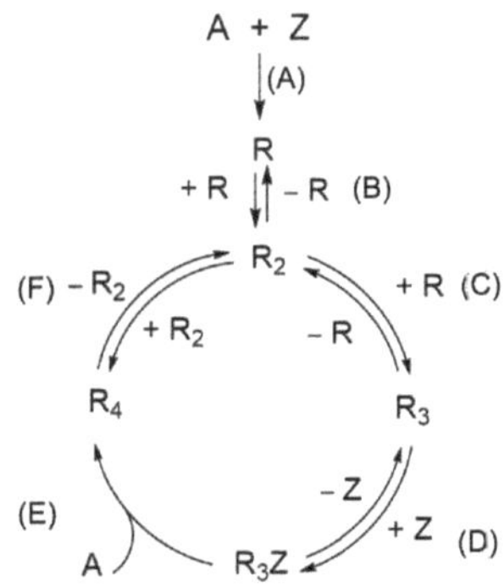

Scheme 4.8 List of main steps (chirality is not displayed) (A) uncatalyzed formation of racemic alkoxide monomers from iPr_2Zn and aldehyde; (B) dimerization; (C) trimer formation; (D) trimer activation; (E) aldehyde addition and alkylation with tetramer formation; (F) tetramer splitting. As trimers are built from three alkoxide monomers, they provide auto-catalysis.

tetramers in an autocatalytic cycle. Scheme 4.8 shows the basic structure of our suggested autocatalytic network.

All possible diastereomeric combinations were later considered, giving rise to a total of 44 processes and 20 species. This model reproduces the remarkable enlargement of the *ee* from $ee_0 = 0.00001\%$ to $ee_{\text{final}} = 39\%$ similar to that observed by Soai et al. [14] (from 0.00005% to 59%). In addition, our computed kinetic results are in apparent agreement with the experimental observations made by Blackmond [82,95], namely second-order kinetics in the aldehyde, first-order kinetics in the homochiral dimers, and zeroth-order kinetics in the organozinc compound. However, while the aldehyde appears as second order, binding of two aldehydes is never considered in our model. This nonintuitive effect is the result of the complexity of the whole kinetic network. Furthermore, the mirror-image symmetry breaking, occurring without any initial addition of chiral additives has been reproduced by considering a slight energetic advantage for the heterochiral dimer (0.2 kcal/mol) and a more noticeable energy difference in favor of heterochiral tetramer (2.7 kcal/mol) and by assuming that the homochiral trimer is 100 times more active than its heterochiral counterpart. Such spontaneous asymmetric synthesis is impossible to presume within the framework of single-rate laws classical kinetics. Considering the results of the X-ray analysis [91], this model is particularly appealing since it predicts that the tetramers accumulate, the heterochiral being slightly more concentrated than the homochiral one.

4.5.1.3 *Kinetic Proofreading in the Soai Reaction*

We then analyzed the amplification of enantiomeric excess, mirror-image symmetry breaking and kinetic proofreading in Soai reaction models with different oligomeric orders [121]. Three prototypical autocatalytic cycle models able to simulate the Soai reaction were considered. The three models vary by their monomeric, dimeric or trimeric order of the assumed catalytic species. Using a numerical integration of differential equations and random grid parameter scanning, a mirror-image symmetry breaking test was imposed on the three models. It was shown that whatever the values of the other kinetics parameters, an increase in the catalytic oligomer size from one to three lead to a higher tolerance to poorer chiral recognition between the diastereoisomers. For instance, typical diastereoisomeric (hetero-oligomer vs homo-oligomer) energy differences where 6.5, 4.5, and 2.9 kcal/mol for respectively monomeric, dimeric and trimeric autocatalysts. The robustness of the higher oligomers was understood as a particular case of so-called kinetic proofreading in asymmetric autocatalysis.

4.5.1.4 *Frank Models in Closed Systems*

At the same time, Crusats et al. [122] showed that a Frank-like model in a closed system yielded absolute asymmetric synthesis even in the absence of any chiral polarization. The model was based on a monomeric irreversible enantioselective autocatalysis coupled to a mutual inhibition from dimerization. Spontaneous mirror-image symmetry breaking was due to the very high chiral amplifications of the initial small statistical deviations from an ideal racemic composition. The need of a mutual inhibition step leading to a heterochiral dimer RS (heterodimerization) more exergonic than the homodimerization leading to R_2 and S_2 together with a long racemization time scale was pointed-out. It was therefore concluded that a similar type of network is at the heart of the Soai reaction because this simple hypothetical model already explains the main experimental trends.

4.5.2 Quantum Chemistry Models

DFT was used to compute the energy level of both ground and transition states of free and alkyl Zn complexed dimer (macrocyclic and square) [84]. Trailing the previous Blackmond's claims [82] a hexamolecular complex including two aldehyldes, two alkoxy pyrimidinyl carbinols and two alkyl Zn was computed.

This assumption was rather fruitful since it was found that at 25°C the homochiral catalyst is 135-times more efficient than the heterochiral one. While a full catalytic cycle was designed, there was no attempt to reproduce any of the specific features of the Soai reaction.

In a subsequent article the same authors [85], proposed a refinement of the dimer catalysis model by taking into account the formation of tetramers. The remarkable *ee* amplification of the Soai reaction has been reproduced assuming that the dimers are isoergonic, that the heterodimer is not catalytic and that heterochiral tetramers are more stable than the homochiral ones. This last point was confirmed by a DFT study on barrel-like tetramers showing higher stability of the heterochiral ones (Schemes 4.9 and 4.10).

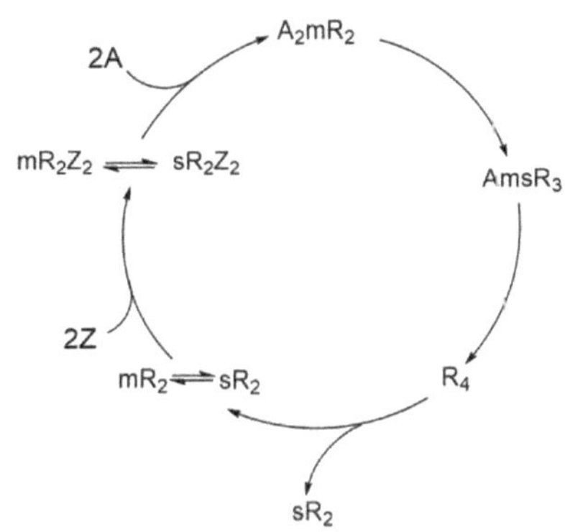

Scheme 4.9 Schiaffino's et al. catalytic cycle. A, aldehyde; A_2mR_2, di-aldhyde adduct on a macrocyclic dimer; $AmsR_3$, mono-aldehyde adduct of a square-macrocyclic trimer; mR_2, macrocyclic dimer; R_4, tetramer; sR_2, square dimer; Z, organo-zinc.

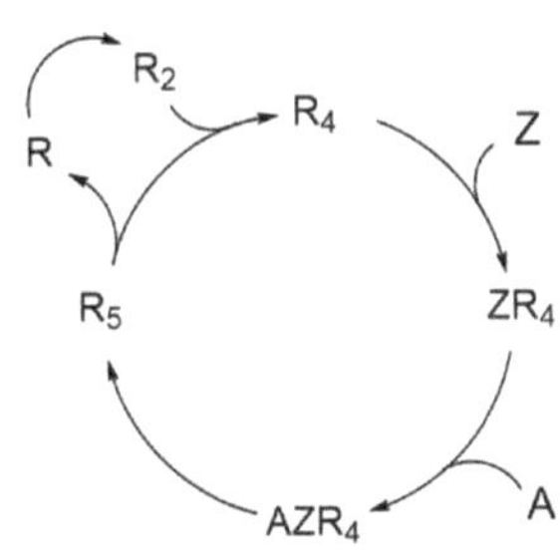

Scheme 4.10 Gridnev's et al. catalytic cycle for the alkylation of aldehyde. A, aldehyde; AZR_4, transition state with aldehyde adduct to activated tetramer; R, monomer; R_2, dimer; R_4, tetramer; R_5, pentamer; Z, organozinc; ZR_4, activated tetramer.

Their third article [83] was still based on the (two-aldehyde + two-alkoxy) transition state assumptions. Calculations were carried out at the M05-2X/6−31G(d) level of theory, by considering 10 energetically distinct paths. This computational method has been recommended for the accurate energetic description of organic compounds containing zinc. As expected, the fully enantioselective catalytic cycle of the homochiral dimers was found to be the dominant mechanism. Two other minor cycles were also recognized. However, there was no attempt to simulate any experimental fact. It was shown, 1 year later, that, depending on the interplay of kinetic constants, a two-aldehyde iso-ergonic dimer catalysis irreversible model can give rise to either chiral amplification or depletion [86]. Calculations of the kinetic constants were carried out from the energetic diagrams in a Gibbs-like approach in order to evaluate the substrate and reactants structural effects. The specific combination of structural and kinetic factors of the Soai reaction was stressed.

Gridnev et al. [89] designed a catalytic cycle taking into account that the three-dimensional cavity observed in the $i\mathrm{Pr_2Zn}$ adduct of the homochiral tetramer forms an ideal chiral pocket for the coordination of the aldehyde. Then, enantioselective alkylation yields monomeric alcoholate of the same handedness as the tetrameric catalyst.

In order to reproduce the anti-Arrhenius effect of the Soai reaction and the dependence of the enantiomeric excess versus conversion at different relative rates of the noncatalytic reaction, several assumptions were necessary, namely: the reaction product exists in solution mainly as the dimer and the concentrations of monomers, tetramers, or higher oligomers are negligible, homo and hetero dimers are formed in a statistical proportion, dimers and tetramers bind only one molecule of diisopropylzinc, the strength of this association varies with the temperature and the aldehyde binding to the homochiral tetramers is weak. In their conclusion, the authors cleverly recognize that numerous requirements for reversibility of different stages, relative abundance of homo- and heterochiral species, and so forth must come together to realize the amplification of chirality.

4.5.3 Stochastic, Statistical and Empirical Models

Such an approach is promoted by Lente [123]. While several different mathematical approaches have been developed under the name of stochastic kinetics, the continuous time discrete state (CDS) stochastic approach is probably closest to the accepted particle-based view of matter. The analysis of the cumulative distribution functions for the random generation of enantiomers in the Soai reaction [20] concluded to a second-order autocalysis like behavior. More recently, a newly developed stochastic Monte Carlo kinetic method combined with deterministic continuation and symmetrization has been applied again to the analysis of the mirror-image symmetry breaking in the Soai reaction [124]. The utility of the method was demonstrated by predicting the enantiomer distribution based on the pioneering Buhse model [97] of the Soai reaction. It is argued that the method opens the way to validate any future models of absolute asymmetric synthesis by direct comparisons of model predictions with measured data.

A data-based stochastic approach to absolute asymmetric synthesis in the Soai reaction has been achieved by Barabás et al. [125,126]. It was concluded that the Soai-autocatalysis might operate by three concerted cooperating catalytic cycles.

Micskei et al.'s [127,128] simple empirical formula enabled the quantitative description of chiral autocatalysis. This formula was used for the prediction of the number of consecutive autocatalytic reaction cycles needed to obtain a high enantiomeric excess in absolute asymmetric synthesis. Its empirical equation relates initial and final enantiomeric excesses:

$$ee_{prod} = \frac{ee_{max}ee_{start}}{B + ee_{start}}$$

where parameter B is indicative of the *ee* amplification power of the system under consideration. If B is very small, only a few cycles could result in a high excess, while if $B = ee_{max}$ no chiral amplification is expected, and if $B > ee_{max}$ racemization occurs. For the Soai's *t*Bu-ethynyl-pyrimidinyl aldehyde parameter B was estimated at around 5×10^{-5}.

4.5.4 Comparison Between the Various Published Models

All these examples show the vigor of the domain. However, as we are mostly interested by the development of chemically realistic models of the Soai reaction, we have gathered in Table 4.6 all of the known attempts to construct a kinetic model involving well-defined species and processes.

Table 4.6 Collection of the various Soai reaction kinetic models

Intermediate species	Number of processes	Comments	Refs.
Homo- and heterodimers	8	(a, b, c)	[80]
Monomer and dimer	3	(b, d, e)	[92]
Monomers, dimers and trimers	22	(b, c, f)	[96]
Monomers, dimers and trimers	10	(b, c, d, e, f)	[97]
Monomer, dimer, tetramer and aldehyde alkyl Zn complex	5	(a, b, c)	[82]
Dimers; activated dimers; tetramers and TS	9	(b, e, g, h)	[84]
Dimers and tetramers	16	(b, c, d, i)	[85]
Dimers; activated dimers; tetramers and TS	26	(b, e, g, h)	[83]
Dimers	7	(b, d, e, g)	[86]
Monomers and dimers	9 (10)	(b, e, f, j, k)	[98]
Monomers and dimers	18	(e, f, i, j, k)	[122,129,130]
Monomers; dimers; free and activated trimers and tetramers	44	(h, i, j)	[120]
Monomers, dimers and trimers	18	(b, j, l)	[126]
Monomers, free and activated dimers and trimers; tetramers	16; 32; 48	(b, f, h, i, j, k)	[121]
Dimers; free and activated tetramers; TS	12	(c, m)	[87]

Ciphers indicate the number of processes; both the direct and reverse steps have been considered. (a), Use of an unique rate law; (b), dimeric catalyst; (c), reproduction of *ee* amplification; (d), algebraic integration; (e), trimolecular steps; (f), formal model without specification of the detailed structures; (g), limited to structural analysis, no attempt to reproduce any experimental feature of the Soai reaction; (h), trimeric catalyst; (i), with stereoselective steps (limited enantioselectivity); (j), reproduce the mirror-image symmetry breaking and the giant *ee* amplification; (k), monomeric catalyst; (l), cf. Buhse model; (m), tetrameric catalyst; TS, transition state.

Table 4.6 shows that kinetic modeling of the Soai reaction has been developed in two complementary directions. A structural direction was promoted independently by Gridnev et al. while a functional direction was promoted by Buhse et al. The structural approach is validated by DFT, while the functional one by numerical integration of coupled non-linear equations and reproduction of experimental features.

4.6 KINETIC MODELING OF THE ADDITIVE EFFECT, *ee* AMPLIFICATION DURING THE STEPWISE SOAI REACTION, ALDEHYDE CONCENTRATION AND APPARENT KINETIC ORDER

In our previous review [131], we have carried out several demonstrative numerical simulations of some of the most striking Soai reaction features using a newly designed autocatalytic cycle model [120].

This trimer catalysis model takes into account all the possible diastereoisomeric interactions in dimers, trimers and tetramers. Table 4.7 provides in a compact way the list of species and processes while the model skeleton is displayed on Scheme 4.8.

Using a realistic parameter set, the model reproduced the huge *ee* amplification (from 0.00005% to 59%) and the mirror-image symmetry breaking. Another specific feature of the Soai reaction is a reversal phenomenon of enantioface selectivity by the cooperative operation of two catalysts [47,51–53]. The opposite enantiomeric product was formed on using a mixture of two chiral catalysts that possess individual enantioselectivity (a mixture of *N,N*-dimethyl-norephedrine and 2-[(1-phenylethyl) amino]-ethanol). It was assumed that this reversal phenomenon originated from specific interactions between catalytic species. Our modeling approach was shown to nicely reproduce this effect [132]. The double-reversal occurs when among two pro-(*R*) additives X and Y, one plays the role of R-catalyst while the other is mostly an S quencher, depending of their relative reactivities according to the processes described in Table 4.8.

Table 4.7 List of species and processes

Homo-S	Hetero	Hetero	Homo-R
$A + Z \rightarrow S$			$A + Z \rightarrow R$
$S + S \leftrightarrow S_2$	$S + R \leftrightarrow RS$	$R + S \leftrightarrow SR$	$R + R \leftrightarrow R_2$
$S + S_2 \leftrightarrow S_3$	$R + S_2 \leftrightarrow RS_2$	$S + R_2 \leftrightarrow SR_2$	$R + R_2 \leftrightarrow R_3$
	$RS + S \leftrightarrow RS_2$	$SR + R \leftrightarrow SR_2$	
$Z + S_3 \leftrightarrow ZS_3$	$Z + RS_2 \leftrightarrow ZRS_2$	$Z + SR_2 \leftrightarrow ZSR_2$	$Z + R_3 \leftrightarrow ZR_3$
$A + ZS_3 \rightarrow S_4$	$A + ZRS_2 \rightarrow R_2S_2$	$A + ZSR_2 \rightarrow S_2R_2$	$A + ZR_3 \rightarrow R_4$
$A + ZS_3 \rightarrow RS_3$			$A + ZR_3 \rightarrow SR_3$
$S_4 \leftrightarrow S_2 + S_2$	$R_2S_2 \leftrightarrow R_2 + S_2$	$S_2R_2 \leftrightarrow S_2 + R_2$	$R_4 \leftrightarrow R_2 + R_2$
	$RS_3 \leftrightarrow RS + S_2$	$SR_3 \leftrightarrow SR + R_2$	

A: aldehyde; R, S: alkoxycarbinol; $R_2S_2 \equiv S_2R_2$: mesotetramers; $RS \equiv SR$: heterodimer; RS_2, SR_2: heterotrimers; RS_3, SR_3: heterotetramers; S_2, R_2: homodimers; S_3, R_3: homotrimers; S_4, R_4: homotetramers; Z: organozinc; ZRS_2, ZSR_2: activated heterotrimers; ZS_3, ZR_3: activated homotrimers; alkylation reactions are assumed to be irreversible. Note the uncatalyzed formation of racemic alkoxide monomers, dimerization, trimer formation, trimer activation, aldehyde addition and alkylation with tetramer formation and tetramer splitting.

Table 4.8 List of processes assumed to occur in presence of X and Y additives

X-side	XY-side	Y-side
X + X ←→ XX	X + Y ←→ XY	Y + Y ←→ YY
X + R ←→ XR		Y + R ←→ YR
X + S ←→ XS		Y + S ←→ YS
XX + Z ←→ XXZ	XY + Z ←→ XZ	YY + Z ←→ YYZ
XR + Z ←→ XRZ		YR + Z ←→ YRZ
XS + Z ←→ XSZ		YS + Z ←→ YSZ
A + XXZ → R(or S) + XX	A + XYZ → R(or S) + XY	A + YYZ → R(or S) + YY
A + XRZ → R(or S) + XR		A + YRZ → R(or S) + YR
A + XSZ → R(or S) + XS		A + YSZ → R(or S) + YS

A: aldehyde; R, S: alkoxycarbinol; X, Y: chiral or nonchiral additives; Z: organozinc. For the sake of simplicity catalytic oligomers higher than dimers have been neglected.

Table 4.8 displays all kinds of dimeric interactions that can occur when two additives X and Y are added to a Soai reaction model. First, the self- and nonself-interactions between the additives create activable catalytic dimers. Second, additive-carbinol interactions, either sequestrate the autocatalytic carbinols or generate specific activable dimers. The possible effect of the additives onto the noncatalyzed step can be modeled by unbalancing the corresponding otherwise symmetrical spontaneous alkylation rate constants. By using only one side (either X or Y), Table 4.8 can be used as a basis for modeling all the additive effects of the Soai reaction.

Finally, our model also predicts that the aldehyde concentration plays the role of a bifurcation parameter inducing a transition from a racemic state to an optically active one. If the aldehyde is too concentrated, the spontaneous formation of racemic pyrimidinyl alkanol exceeds the stereoselective generation of enantiomers and symmetry breaking does not occur. On the other hand, if the aldehyde is too diluted, the formation of autocatalytic oligomers is not promoted, and the final result remains racemic. Fig. 4.2 displays the aldehyde concentration range where *ee* amplification is occurring.

Moreover, numerical simulation is useful to understand the *ee* amplification during the stepwise addition of aldehyde and organozinc in a typical Soai reaction procedure [14]. As an illustration, we provide a simulation of the step-by-step addition typical procedure. The result of a previous step acts as catalyst for the next step. Table 4.9 gathers the results of various simulations where the same amount of aldehyde (0.6 M) and the same amount of organozinc (1.2 M) were used but carried out by varying the aldehyde initial concentration and the next step concentration increase.

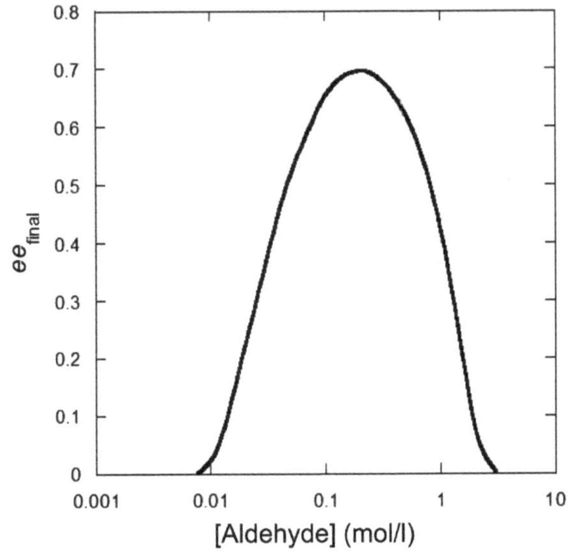

Figure 4.2 Numerical simulation of the *ee* amplification versus initial aldehyde concentration. [Z] = 2*[aldehyde]. Other parameters are the same as in the legend of Table 4.9.

Table 4.9 Numerical simulation of the model using the following parameters: uncatalyzed reaction: 5×10^{-5}; dimerization: 5×10^{3}; dimer splitting: 3; trimerization: 200; trimer splitting: 2×10^{4}; trimer activation: 10^{6}; trimer de-activation: 50; trimer catalysis: 150; tetramer formation: 2×10^{4}; tetramer splitting: 300; hetero- versus homo-oligomers energy difference: 2.3 kcal/mol (hetero more stable); stereoselectivity of the catalytic step: 0.99

Reactants	Step 1	Step 2	Step 3	ee_1	ee_2	ee_3
$[A]_a$; $[Z]_a$	0.2; 0.4	0.2; 0.4	0.2; 0.4	0.875	0.927	0.944
$[A]_b$; $[Z]_b$	0.1; 0.2	0.2; 0.4	0.3; 0.6	0.783	0.912	0.945
$[A]_c$; $[Z]_c$	0.05; 0.1	0.2; 0.4	0.35; 0.7	0.659	0.910	0.950
$[A]_d$; $[Z]_d$	0.05; 0.1	0.1; 0.2	0.45; 0.9	0.659	0.866	0.950
$[A]_e$; $[Z]_e$	0.02; 0.04	0.2; 0.4	0.38; 0.76	0.511	0.926	0.960
$[A]_f$; $[Z]_f$	0.02; 0.04	0.08: 0.16	0.5; 1	0.511	0.868	0.959
$[A]_g$; $[Z]_g$	0.005; 0.01	0.05; 0.1	0.545; 1.09	0.0000154	0.661	0.946

Catalytic charge in pyrimidinylcarbinol units for step 1: 10^{-6}%; ee_0: 4×10^{-8}%. The initial concentrations of steps n and the corresponding *ee* was computed according to the final concentrations of steps $n - 1$.

The ee_3 (after step 3) seems relatively independent of the concentration increments used for the different stepwise simulations. Best amplification occurs in run (*e*). In that case, the effect of the accumulation of the wrong enantiomer is more limited since it is mainly produced during the

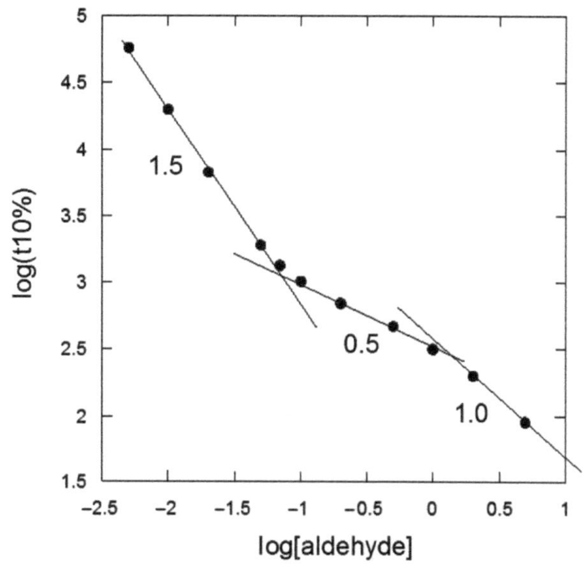

Figure 4.3 Numerical simulation of a log-log plot of the 10% reaction progress time versus [aldehyde] showing an unexpected variation of the apparent aldehyde kinetic order [Z] = 2*[aldehyde]. Other parameters are the same as in the legend of Table 4.9.

low concentration steps, therefore it remains rather diluted. Another persuasive advantage of our modeling approach is to illustrate the complexity of the Soai reaction system. For instance, as shown in Fig. 4.3, it is predicted that the apparent kinetic order of the aldehyde could depend on its concentration range.

Such a result proves that kinetic orders measurements [82] must be handled with care. For instance, when it was accepted that a second-order in aldehyde indicated a two aldehydes molecularity, it was assumed that the corresponding transition state was rate-determining without any other details related to the apparent simultaneous incorporation of two aldehydes [83,84]. The complexity of the Soai reaction mechanism must be borne in mind to avoid inadequate conclusions. Multiple equilibria situations are often nonintuitive.

Although involving only mono- and bi-molecular steps, our present model is certainly perfectible since its does not take into account all the last crystal structures [91] and the most recent DFT calculations [87]. Namely, while tetramers are involved in our model, their organozinc complexes and ladder polymers have not been considered. Moreover, the size of the catalytic species has been limited to trimers. In addition, until now, this model has not been carefully parametrized by experimental kinetic curves fitting, therefore all its kinetic rate constants and relative diastereoisomeric (homo vs hetero) energy differences have been assumed. Actually, it is not easy to compare the extracted thermodynamic parameters with their DFT computed counterparts for two main reasons: (1) parameters could be model dependent, and (2) computed energies are sensitive to the functionals.

4.7 CONCLUSION

The most salient feature of the Soai reaction is its ability to perform mirror-image symmetry breaking or spontaneous absolute asymmetric synthesis i.e., the formation of enantiomerically enriched pyrimidyl alkanol from the reaction of pyrimidine-5-carbaldehyde and $i\mathrm{Pr_2Zn}$ without adding a chiral substance. The approximate stochastic distribution of the absolute configurations of pyrimidyl alkanols was statistically analyzed and a clear S preference was found [34]. The reason for the observed S preference remains complicate. Is it a chiral contamination of biological or chemical origin or the manifestation of a parity violating energy difference effect? None of these two assumptions is satisfactory since independent calculations conclude on one side that the parity-violating energy difference cannot cause the asymmetry of the distribution [133] and on the other side that the computed parity-violation energy is not expected to have any effect on the direction of spontaneous symmetry breaking [134,135]. An alternative hypothesis has been suggested by Barabás [136] about the possibility that the natural abundance of isotopic substitution in isotopically prochiral groups of otherwise achiral molecules could provide stochastically formed enantiomeric excesses exceeding the sensitivity threshold of the Soai reaction. This clever assumption is supported by the numerous cryptochiral experiments published by the Soai group. In all these experiments, a tiny enantiomeric excess of zinc alkoxide must be amplified, and for this purpose consecutive asymmetric autocatalysis must be carried out.

The underlying mechanism of the Soai reaction is still the subject of ongoing research and appears much more complex than the simple Frank model. It is likely to involve numerous oligomeric species obeying the chiral combinatorics principles [137]. Several intermediates have been either isolated or recognized by NMR spectroscopy. A combination of kinetic analysis and DFT calculations had shed some light on our understanding of this asymmetric autocatalysis until the recent real breakthrough that was provided by the publication of X-ray structures of various oligomers. Further mechanistic refinements must exploit the full consequences of these results. Nevertheless, it should be noted that a catalytic intermediate is not necessarily the most abundant species. Moreover, these model improvements must be supported by more kinetically significant results. Although we are aware of the experimental difficulties, we would encourage the Soai reaction community to publish several NMR kinetic monitorings by changing the various experimental constraints: concentration of the reactants and additives.

The following question is worth discussion: Why is the Soai reaction an islet in the organometallic ocean? A preliminary response could be that this rareness arises from the structures of the reactants. The Soai reaction occurs in aromatic solvents with 2-ethynyl pyrimidinyl aldehydes bearing a medium-sized substituent and operates only with diisopropyl zinc, thus exhibiting delicately pitched reactivities. It means that only a few molecules would exhibit the required reactivities. From this point of view, the Soai reaction is certainly interesting, but it is, and will remain, an exception. If no independent chiral autocatalysis are discovered (if we except the reversible aldol and Mannich reactions) this opinion is unquestionable.

This answer is not productive as it closes the door of future investigations. An opposite point of view could be to survey the *ee* amplification literature scrutinizing high and strong asymmetric amplification in both organometallic chemistry [138,139] and organo-catalysis [140]. If organometallic reactions are generally irreversible due to their high reactivity, organocatalysis could benefit from being pushed far from equilibrium conditions by opening the system either to matter (continuous extraction) or to energy (photochemistry). Hunting for unexpected results in this domain could be a promising target; for instance, the recently published reversal of enantioselectivity of mixed chiral amino-alcohol catalysts in $i\mathrm{Pr_2Zn}$ addition to benzaldehydes [53] let us suppose that interactions between catalytic species could exist in situations where monomeric catalysis is accepted. Macromolecules and supramolecular assemblies could be another hopeful objective. Therefore it is possible that researchers have actually encountered many more enantioselective autocatalytic reactions, but that the stochastic nature of the results was mistaken for irreproducibility.

Another point that deserves to be regarded more carefully is the role of water and air traces; it is well known that organozincs are highly hydrolysable and sensitive to oxygen [141–143]. Under these conditions ZnO (wurtzite-like) or cubane-like nanoparticles could be present providing some catalytic effects. Their presence, size and shape depending on the experimental conditions. DLS controls could be useful in checking for their presence.

ACKNOWLEDGMENTS

The research was conducted in the framework of the French-Mexican International Laboratory (LIA) LCMMC. M.E.N.-T. and T.B. acknowledge financial support by the National Council of Science and Technology (CONACYT), Mexico, grant 153699. J.M.C. acknowledges financial support by CONACYT grant 262653.

REFERENCES

[1] Frank, F. C. On Spontaneous Asymmetric Synthesis. *Biochim. Biophys. Acta* **1953,** *11*, 459–463.
[2] Feringa, B. L.; van Delden, R. A. Absolute Asymmetric Synthesis: The Origin, Control and Amplification of Chirality. *Angew. Chem., Int. Ed.* **1999,** *38*, 3418–3438.
[3] Soai, K.; Shibata, T.; Morioka, H.; Choji, K. Asymetric Autocatalysis and Amplification of Enantiomeric Excess of a Chiral Molecule. *Nature* **1995,** *378*, 767–768.
[4] Fülöp, O.; Barabás, B. Impact of the Soai-Autocatalysis on Natural Sciences. *J. Math. Chem.* **2016,** *54*, 10–17.
[5] Soai, K.; Niwa, S.; Hori, H. Asymmetric Self-Catalytic Reaction. Self-Production of Chiral 1-(3-Pyridyl)alkanols as Chiral Self-Catalysts in the Enantioselective Addition of Dialkylzinc Reagents to Pyridine-3-carbaldehyde. *J. Chem. Soc., Chem. Commun.* **1990,** 982–983.
[6] Shibata, T.; Choji, K.; Hayase, T.; Aizu, Y.; Soai, K. Asymmetric Autocatalytic Reaction of 3-Quinolylalkanol with Amplification of Enantiomeric Excess. *Chem. Commun.* **1996,** 1235–1236.
[7] Tanji, S.; Kodaka, Y.; Ohno, A.; Shibata, T.; Sato, I.; Soai, K. Asymmetric Autocatalysis of 5-Carbamoyl-3-pyridyl Alkanols with Amplification of Enantiomeric Excess. *Tetrahedron: Asymmetry* **2000,** *11*, 4249–4253.
[8] Shibata, T.; Morioka, H.; Tanji, S.; Hayase, T.; Kodaka, Y.; Soai, K. Enatioselective Synthesis of Chiral 5-Carbamoyl-3-pyridyl Alcohols by Asymmetric Autocatalytic Reaction. *Tetrahedron Lett.* **1996,** *37*, 8783–8786.
[9] Lutz, F.; Kawasaki, T.; Soai, K. Asymmetric Autocatalysis of a Ferrocene-Containing Chiral Compound with Amplification of Chirality. *Tetrahedron: Asymmetry* **2006,** *17*, 486–490.
[10] Sato, I.; Yanagi, T.; Soai, K. Highly Enantioselective Asymmetric Autocatalysis of 2-Alkenyl and 2-Vinyl-5-pyrimidinyl Alkanols with Significant Amplification of Enantiomeric Excess. *Chirality* **2002,** *14*, 166–168.

[11] Shibata, T.; Hayase, T.; Yamamoto, J.; Soai, K. One-Pot Asymmetric Autocatalytic Reaction with Remarkable Amplification of Enantiomeric Excess. *Tetrahedron: Asymmetry* **1997,** *8*, 1717–1719.

[12] Busch, M.; Schlageter, M.; Weingand, D.; Gehring, T. Systematic Studies Using 2-(1-Adamantylethynyl)pyrimidine-5-carbaldehyde as a Starting Material in Soai's Asymmetric Autocatalysis. *Chem. Eur. J.* **2009,** *15*, 8251–8258.

[13] Shibata, T.; Yonekubo, S.; Soai, K. Practically Perfect Asymmetric Autocatalysis with (2-Alkynyl-5-pyrimidyl) Alkanols. *Angew. Chem., Int. Ed.* **1999,** *38*, 659–661.

[14] Urabe, H.; Ishiguro, S.; Shibata, T.; Soai, K. Amplification of Chirality from Extremely Low to Greater than 99.5% ee by Asymmetric Autocatalysis. *Angew. Chem., Int. Ed.* **2003,** *42*, 315–317.

[15] Gridnev, I. D.; Serafimov, J. M.; Brown, J. M. Solution Structure and Reagent Binding of the Zinc Alkoxide Catalyst in the Soai Asymmetric Autocatalytic Reaction. *Angew. Chem., Int. Ed.* **2004,** *43*, 4884–4887.

[16] Klankermayer, J.; Gridnev, I. D.; Brown, J. M. Role of the Isopropyl Group in Asymmetric Autocatalytic Zinc Alkylations. *Chem. Commun.* **2007,** 3151–3153.

[17] Kawasaki, T.; Hatase, K.; Fujii, Y.; Jo, K.; Soai, K.; Pizzarello, S. The Distribution of Chiral Asymmetry in Meteorites: An Investigation Using Asymmetric Autocatalytic Chiral Sensors. *Geochim. Cosmochim. Acta* **2006,** *70*, 5395–5402.

[18] Welch, C. J.; Biba, M.; Sajonz, P. Fast Methods of Enantiopurity Determination for the Soai Reaction: Towards a General Enantioenrichment Detector? *Chirality* **2007,** *19*, 34–43.

[19] Sato, I.; Ohgo, Y.; Igarashi, H.; Nishiyama, D.; Kawasaki, T.; Soai, K. Determination of Absolute Configurations of Amino Acids by Asymmetric Autocatalysis of 2-Alkynylpyrimidyl Alkanol as a Chiral Sensor. *J. Organomet. Chem.* **2007,** *692*, 1783–1787.

[20] Soai, K.; Sato, I.; Shibata, T.; Komiya, S.; Hayashi, M.; Matsueda, Y., et al. Asymmetric Synthesis of Pyrimidyl Alkanol without Adding Chiral Substances by the Addition of Diisopropylzinc to Pyrimidine-5-carbaldehyde in Conjunction with Asymmetric Autocatalysis. *Tetrahedron: Asymmetry* **2003,** *14*, 185–188.

[21] Singleton, D. A.; Vo, L. K. Enantioselective Synthesis without Discrete Optically Active Additives. *J. Am. Chem. Soc.* **2002,** *124*, 10010–10011.

[22] Singleton, D. A.; Vo, L. K. A Few Molecules Can Control the Enantiomeric Outcome. Evidence Supporting Absolute Asymmetric Synthesis Using the Soai Asymmetric Autocatalysis. *Org. Lett.* **2003,** *5*, 4337–4339.

[23] Gridnev, I. D.; Serafimov, J. M.; Quiney, H.; Brown, J. M. Reflections on Spontaneous Asymmetric Synthesis by Amplifying Autocatalysis. *Org. Biomol. Chem.* **2003,** *1*, 3811–3819.

[24] Kawasaki, T.; Suzuki, K.; Shimizu, M.; Ishikawa, K.; Soai, K. Spontaneous Absolute Asymmetric Synthesis in the Presence of Achiral Silica Gel in Conjunction with Asymmetric Autocatalysis. *Chirality* **2006,** *18*, 479–482.

[25] Suzuki, K.; Hatase, K.; Nishiyama, D.; Kawasaki, T.; Soai, K. Spontaneous Absolute Asymmetric Synthesis Promoted by Achiral Amines in Conjunction with Asymmetric Autocatalysis. *J. Syst. Chem.* **2010,** *1*, 5.

[26] Sögütoglu, L.-C.; Steendam, R. R. E.; Meekes, H.; Vlieg, E.; Rutjes, F. P. J. T. Viedma Ripening: A Reliable Crystallisation Method to Reach Single Chirality. *Chem. Soc. Rev.* **2015,** *44*, 6723–6732.

[27] Steendam, R. R. E.; Verdake, J. M. M.; van Benthem, T. J. B.; Meekes, H.; van Enckevort, W. J. P.; Raap, J., et al. Emergence of Single-Molecular Chirality from Achiral Reactants. *Nat. Comm.* **2014,** *5*, 5543.

[28] Mauksch, M.; Tsogoeva, S. B.; Wei, S.; Martynova, I. M. Demonstration of Spontaneous Chiral Symmetry Breaking in Asymmetric Mannich and Aldol Reactions. *Chirality* **2007,** *19*, 816–825.
[29] Mauksch, M.; Wei, S.; Freund, M.; Zamfir, A.; Tsogoeva, S. B. Spontaneous Mirror Symmetry Breaking in the Aldol Reaction and its Potential Relevance in Prebiotic Chemistry. *Orig. Life Evol. Biosph.* **2010,** *40*, 79–91.
[30] Feringa, B. L.; van Delden, R. A. Absolute Asymmetric Synthesis: The Origin, Control, and Amplification of Chirality. *Angew. Chem., Int. Ed.* **1999,** *38*, 3418–3438.
[31] Mills, W. H. Some Aspects of Stereochemistry. *Chem. Ind. (London)* **1932,** *51*, 750–759.
[32] Barabás, B.; Kurdi, R.; Pályi, G. Natural Abundance Isotopic Chirality in the Reagents of the Soai Reaction. *Symmetry* **2016,** *8*, 1–13.
[33] Lente, G. Stochastic Interpretation of the Asymmetry of Enantiomeric Distribution Observed in the Absolute Asymmetric Soai reaction. *Tetrahedron: Asymmetry* **2011,** *22*, 1595–1599.
[34] Barabás, B.; Caglioti, L.; Zucchi, C.; Maioli, M.; Gal, E.; Micskei, K., et al. Violation of Distribution Symmetry in Statistical Evaluation of Absolute Enantioselective Synthesis. *J. Phys. Chem. B* **2007,** *111*, 11506–11510.
[35] Moradpour, A.; Nicoud, J. F.; Balavoine, G.; Kagan, H.; Tsoucaris, G. Photochemistry with Circularly Polarized Light. The Synthesis of Optically Active Hexahelicene. *J. Am. Chem. Soc.* **1971,** *93*, 2353–2354.
[36] Kawasaki, T.; Sato, M.; Ishiguro, S.; Saito, T.; Morishita, Y.; Sato, I., et al. Enantioselective Synthesis of Near Enantiopure Compound by Asymmetric Autocatalysis Triggered by Asymmetric Photolysis with Circularly Polarized Light. *J. Am. Chem. Soc.* **2005,** *127*, 3274–3275.
[37] Sato, I.; Yamashima, R.; Kadowaki, K.; Yamamoto, J.; Shibata, T.; Soai, K. Asymmetric Induction by Helical Hydrocarbons: [6]- and [5]Helicenes. *Angew. Chem., Int. Ed.* **2001,** *40*, 1096–1098.
[38] Sato, I.; Ohno, A.; Aoyama, Y.; Kasahara, T.; Soai, K. Asymmetric Autocatalysis Induced by Chiral Hydrocarbon [2.2] Paracyclophanes. *Org. Biomol. Chem.* **2003,** *1*, 244–246.
[39] Sato, I.; Matsueda, Y.; Kadowaki, K.; Yonekubo, S.; Shibata, T.; Soai, K. Highly Enantioselective Asymmetric Autocatalysis of Pyrimidin-5-yl Alkanol Induced by Chiral 1,3-Disubstituted Hydrocarbon Allenes. *Helv. Chim. Acta* **2002,** *85*, 3383–3387.
[40] Sato, I.; Sugie, R.; Matsueda, Y.; Furumura, Y.; Soai, K. Asymmetric Synthesis Utilizing Circularly Polarized Light Mediated by the Photoequilibrium of Chiral Olefins in Conjunction with Asymmetric Autocatalysis. *Angew. Chem., Int. Ed.* **2004,** *43*, 4490–4492.
[41] Kawasaki, T.; Tanaka, H.; Tsutsumi, T.; Kasahara, T.; Sato, I.; Soai, K. Chiral Discrimination of Cryptochiral Saturated Quaternary and Tertiary Hydrocarbons by Asymmetric Autocatalysis. *J. Am. Chem. Soc.* **2006,** *128*, 6032–6033.
[42] Hitosugi, S.; Matsumoto, A.; Kaimori, Y.; Iizuka, R.; Soai, K.; Isobe, H. Asymmetric Autocatalysis Initiated by Finite Single-Wall Carbon Nanotube Molecules with Helical Chirality. *Org. Lett.* **2014,** *16*, 645–647.
[43] Kawasaki, T.; Hohberger, C.; Araki, Y.; Hatase, K.; Beckerle, K.; Okuda, J., et al. Discrimination of Cryptochirality in Chiral Isotactic Polystyrene by Asymmetric Autocatalysis. *Chem. Commun.* **2009,** 5621–5623.
[44] Kawasaki, T.; Suzuki, K.; Licandro, E.; Bossi, A.; Maiorana, S.; Soai, K. Enantioselective Synthesis Induced by tetrathia-[7]-helicenes in Conjunction with Asymmetric Autocatalysis. *Tetrahedron: Asymmetry* **2006,** *17*, 2050–2053.

[45] Tanji, S.; Ohno, A.; Sato, I.; Soai, K. Asymmetric Autocatalysis of a Pyrimidyl Alkanol Induced by Chiral Monosubstituted [2.2] Paracyclophanes. *Org. Lett.* **2001,** *3*, 287–289.
[46] Kawasaki, T.; Omine, T.; Sato, M.; Morishita, Y.; Soai, K. Asymmetric Autocatalysis Induced by Octahedral tris(acetylacetonato)chromium(III) Complex as a Homogeneous Chiral Initiator. *Chem. Lett.* **2007,** *36*, 30–31.
[47] Shibata, T.; Tarumi, H.; Kawasaki, T.; Soai, K. Reversal in Enantioselectivity Induced by Achiral Alcohols in Asymmetric Autocatalysis in the Presence of a Chiral Diol as a Chiral Initiator. *Tetrahedron: Asymmetry* **2012,** *23*, 1023–1027.
[48] Shibata, T.; Iwahashi, K.; Kawasakib, T.; Soai, K. Chiral Secondary Alcohol-Induced Asymmetric Autocatalysis: Correlation Between the Absolute Configuration of the Chiral Initiators and the Product. *Tetrahedron: Asymmetry* **2007,** *18*, 1759–1762.
[49] Kawasaki, T.; Shimizu, M.; Suzuki, K.; Sato, I.; Soai, K. Enantioselective Synthesis Induced by Chiral Epoxides in Conjunction with Asymmetric Autocatalysis. *Tetrahedron: Asymmetry* **2004,** *15*, 3699–3701.
[50] Lutz, F.; Sato, I.; Soai, K. The Asymmetric Power of Chiral Ligands Determined by Competitive Asymmetric Autocatalysis. *Org. Lett.* **2004,** *6*, 1613–1616.
[51] Lutz, F.; Igarashi, T.; Kawasaki, T.; Soai, K. Small Amounts of Achiral â-Amino Alcohols Reverse the Enantioselectivity of Chiral Catalysts in Cooperative Asymmetric Autocatalysis. *J. Am. Chem. Soc.* **2005,** *127*, 12206–12207.
[52] Kawasaki, T.; Wakushima, Y.; Asahina, M.; Shiozawa, K.; Kinoshita, T.; Lutz, F., et al. A Reversal Phenomenon of Enantioface Selectivity by the Cooperative Operation of Two Chiral Catalysts. *Chem. Commun.* **2011,** *47*, 5277–5279.
[53] Lutz, F.; Igarashi, T.; Kinoshita, T.; Asahina, M.; Tsukiyama, K.; Kawasaki, T., et al. Mechanistic Insights in the Reversal of Enantioselectivity of Chiral Catalysts by Achiral Catalysts in Asymmetric Autocatalysis. *J. Am. Chem. Soc.* **2008,** *130*, 2956–2958.
[54] Kawasaki, T.; Matsumura, Y.; Tsutsumi, T.; Suzuki, K.; Ito, M.; Soai, K. Asymmetric Autocatalysis Triggered by Carbon Isotope (13C/12C) Chirality. *Science* **2009,** *324*, 492–495.
[55] Sato, I.; Omiya, D.; Saito, T.; Soai, K. Highly Enantioselective Synthesis Induced by Chiral Primary Alcohols Due to Deuterium Substitution. *J. Am. Chem. Soc.* **2000,** *122*, 11739–11740.
[56] Kawasaki, T.; Ozawa, H.; Ito, M.; Soai, K. Enantioselective Synthesis Induced by Compounds with Chirality Arising from Partially Deuterated Methyl Groups in Conjugation with Asymmetric Autocatalysis. *Chem. Lett.* **2011,** *40*, 320–321.
[57] Kawasaki, T.; Okano, Y.; Suzuki, E.; Takano, S.; Oji, S.; Soai, K. Asymmetric Autocatalysis: Triggered by Chiral Isotopomer Arising from Oxygen Isotope Substitution. *Angew. Chem., Int. Ed.* **2011,** *50*, 8131–8133.
[58] Matsumoto, A.; Oji, S.; Takano, S.; Tada, K.; Kawasaki, T.; Soai, K. Asymmetric Autocatalysis Triggered by Oxygen Isotopically Chiral Glycerine. *Org. Biomol. Chem.* **2013,** *11*, 2928–2931.
[59] Shibata, T.; Yamamoto, J.; Matsumoto, N.; Yonekubo, S.; Osanai, S.; Soai, K. Amplification of a Slight Enantiomeric Imbalance in Molecules Based on Asymmetric Autocatalysis: The First Correlation between High Enantiomeric Enrichment in a Chiral Molecule and Circularly Polarized Light. *J. Am. Chem. Soc.* **1998,** *120*, 12157–12158.
[60] Kawasaki, T.; Shimizu, M.; Nishiyama, D.; Ito, M.; Ozawaa, H.; Soai, K. Asymmetric Autocatalysis Induced by Meteoritic Amino Acids with Hydrogen Isotope Chirality. *Chem. Commun.* **2009,** *29*, 4396–4398.
[61] Kawasaki, T.; Araki, Y.; Hatase, K.; Suzuki, K.; Matsumoto, A.; Yokoi, T., et al. Helical Mesoporous Silica as an Inorganic Heterogeneous Chiral Trigger for

Asymmetric Autocatalysis with Amplification of Enantiomeric Excess. *Chem. Commun.* **2015,** *51*, 8742–8744.

[62] Soai, K.; Osanai, S.; Kadowaki, K.; Yonekubo, S.; Shibata, T.; Sato, I. d- and l-Quartz-Promoted Highly Enantioselective Synthesis of a Chiral Organic Compound. *J. Am. Chem. Soc.* **1999,** *121*, 11235–11236.

[63] Sato, I.; Kadowaki, K.; Soai, K. Asymmetric Synthesis of an Organic Compound with High Enantiomeric Excess Induced by an Inorganic Sodium Chlorate. *Angew. Chem., Int. Ed.* **2000,** *39*, 1510–1512.

[64] Sato, I.; Kadowaki, K.; Ohgo, Y.; Soai, K. Highly Enantioselective Asymmetric Autocatalysis Induced by Chiral Ionic Crystals of Sodium Chlorate and Sodium Bromate. *J. Mol. Catal. A: Chem.* **2004,** *216*, 209–214.

[65] Shindo, H.; Shirota, Y.; Niki, K.; Kawasaki, T.; Suzuki, K.; Araki, Y., et al. Asymmetric Autocatalysis Induced by Cinnabar: Observation of the Enantioselective Adsorption of a 5-Pyrimidyl Alkanol on the Crystal Surface. *Angew. Chem., Int. Ed.* **2013,** *52*, 9135–9138.

[66] Matsumoto, A.; Ozawa, H.; Inumaru, A.; Soai, K. Asymmetric Induction by Retgersite, Nickel Sulphate Hexahydrate, in Conjunction with Asymmetric Autocatalysis. *New J. Chem.* **2015,** *39*, 6742–6745.

[67] Kawasaki, T.; Sasagawa, T.; Shiozawa, K.; Uchida, M.; Suzuki, K.; Soai, K. Enantioselective Synthesis Induced by Chiral Crystal Composed of DL-Serine in Conjunction with Asymmetric Autocatalysis. *Org. Lett.* **2011,** *13*, 2361–2363.

[68] Kawasaki, T.; Suzuki, K.; Hakoda, Y.; Soai, K. Achiral Nucleobase Cytosine Acts as an Origin of Homochirality of Biomolecules in Conjunction with Asymmetric Autocatalysis. *Angew. Chem., Int. Ed.* **2008,** *47*, 496–499.

[69] Kawasaki, T.; Hakoda, Y.; Mineki, H.; Suzuki, K.; Soai, K. Generation of Absolute Controlled Crystal Chirality by the Removal of Crystal Water from Achiral Crystal of Nucleobase Cytosine. *J. Am. Chem. Soc.* **2010,** *132*, 2874–2875.

[70] Mineki, H.; Kaimori, Y.; Kawasaki, T.; Matsumoto, A.; Soai, K. Enantiodivergent Formation of a Chiral Cytosine Crystal by Removal of Crystal Water from an Achiral Monohydrate Crystal Under Reduced Pressure. *Tetrahedron: Asymmetry* **2013,** *24*, 1365–1367.

[71] Mineki, H.; Hanasaki, T.; Matsumoto, A.; Kawasaki, T.; Soai, K. Asymmetric Autocatalysis Initiated by Achiral Nucleic Acid Base Adenine: Implications on the Origin of Homochirality of Biomolecules. *Chem. Commun.* **2012,** *48*, 10538–10540.

[72] Carter, D. J.; Rohl, A. L.; Shtukenberg, A.; Bian, S.; Hu, C.; Baylon, L., et al. Prediction of Soai Reaction Enantioselectivity Induced by Crystals of N-(2-Thienylcarbonyl)glycine. *Cryst. Growth Des.* **2012,** *12*, 2138–2145.

[73] Kawasaki, T.; Suzuki, K.; Hatase, K.; Otsuka, M.; Koshima, H.; Soai, K. Enantioselective Synthesis Mediated by Chiral Crystal of Achiral Hippuric Acid in Conjunction with Asymmetric Autocatalysis. *Chem. Commun.* **2006,** *17*, 1869–1871.

[74] Kawasaki, T.; Harada, Y.; Suzuki, K.; Tobita, T.; Florini, N.; Palyi, G., et al. Enantioselective Synthesis Utilizing Enantiomorphous Organic Crystal of Achiral Benzils as a Source of Chirality in Asymmetric Autocatalysis. *Org. Lett.* **2008,** *10*, 4085–4088.

[75] Kawasaki, T.; Nakaoda, M.; Kito, N.; Sasagawa, T.; Soai, K. Asymmetric Autocatalysis Induced by Chiral Crystals, of Achiral Tetraphenylethylenes. *Orig. Life Evol. Biosph.* **2010,** *40*, 65–78.

[76] Kawasaki, T.; Jo, K.; Igarashi, H.; Sato, I.; Nagano, M.; Koshima, H., et al. Asymmetric Amplification Using Chiral Cocrystals Formed from Achiral Organic Molecules by Asymmetric Autocatalysis. *Angew. Chem., Int. Ed.* **2005,** *44*, 2774–2777.

[77] Matsumoto, A.; Ide, T.; Kaimori, Y.; Fujiwara, S.; Soai, K. Asymmetric Autocatalysis Triggered by Chiral Crystal of Achiral Ethylenediamine Sulfate. *Chem. Lett.* **2015,** *44*, 688–690.
[78] Kawasaki, T.; Uchida, M.; Kaimori, Y.; Sasagawa, T.; Matsumoto, A.; Soai, K. Enantioselective Synthesis Induced by the Helical Molecular Arrangement in the Chiral Crystal of Achiral tris(2-hydroxyethyl) 1,3,5-benenetricarboxylate in Conjunction with Asymmetric Autocatalysis. *Chem. Lett.* **2013,** *42*, 711–713.
[79] Kawasaki, T.; Kamimura, S.; Amihara, A.; Suzuki, K.; Soai, K. Enantioselective C = C Bond Formation as a Result of the Oriented Prochirality of an Achiral Aldehyde at the Single-Crystal Face upon Treatment with a Dialkyl Zinc Vapor. *Angew. Chem., Int. Ed.* **2011,** *50*, 6796–6798.
[80] Blackmond, D. G.; McMillan, C. R.; Ramdeehul, S.; Schorm, A.; Brown, J. M. Origins of Asymmetric Amplification in Autocatalytic Alkylzinc Additions. *J. Am. Chem. Soc.* **2001,** *123*, 10103–10104.
[81] Gridnev, I. D.; Brown, J. M. Asymmetric Autocatalysis: Novel Structures, Novel Mechanism? *PNAS* **2004,** *101*, 5727–5731.
[82] Buono, F. G.; Blackmond, D. G. Kinetic Evidence for a Tetrameric Transition State in the Asymmetric Autocatalytic Alkylation of Pyrimidyl Aldehydes. *J. Am. Chem. Soc.* **2003,** *125*, 8978–8979.
[83] Schiaffino, L.; Ercolani, G. Mechanism of the Asymmetric Autocatalytic Soai Reaction Studied by Density Functional Theory. *Chem. Eur. J.* **2010,** *16*, 3147–3156.
[84] Schiaffino, L.; Ercolani, G. Unraveling the Mechanism of the Soai Asymmetric Autocatalytic Reaction by First-Principles Calculations: Induction and Amplification of Chirality by Self-Assembly of Hexamolecular Complexes. *Angew. Chem., Int. Ed.* **2008,** *47*, 6832–6835.
[85] Schiaffino, L.; Ercolani, G. Amplification of Chirality and Enantioselectivity in the Asymmetric Autocatalytic Soai Reaction. *ChemPhysChem* **2009,** *10*, 2508–2515.
[86] Ercolani, G.; Schiaffino, L. Putting the Mechanism of the Soai Reaction to the Test: DFT Study of the Role of Aldehyde and Dialkylzinc Structure. *J. Org. Chem.* **2011,** *76*, 2619–2626.
[87] Gridnev, I. D.; Vorobiev, A. Kh Quantification of Sophisticated Equilibria in the Reaction Pool and Amplifying Catalytic Cycle of the Soai Reaction. *ACS Catal.* **2012,** *2*, 2137–2149.
[88] Gridnev, I. D. Recent Experimental and Computational Studies of the Mechanisms of Enantioselection in Asymmetric Catalytic Reactions. *J. Synth. Org. Chem. Jpn.* **2014,** *72*, 1250–1264.
[89] Gridnev, I. D.; Vorobiev, A. K. On the Origin and Structure of the Recently Observed Acetal in the Soai Reaction. *Bull. Chem. Soc. Jpn.* **2015,** *88*, 333–340.
[90] Gehring, T.; Quaranta, M.; Odell, B.; Blackmond, D. G.; Brown, J. M. Observation of a Transient Intermediate in Soai's Asymmetric Autocatalysis: Insights from 1H NMR Turnover in Real Time. *Angew. Chem., Int. Ed.* **2012,** *51*, 9539–9542.
[91] Matsumoto, A.; Abe, T.; Hara, A.; Tobita, T.; Sasagawa, T.; Kawasaki, T., et al. Crystal Structure of the Isopropylzinc Alkoxide of Pyrimidyl Alkanol: Mechanistic Insights for Asymmetric Autocatalysis with Amplification of Enantiomeric Excess. *Angew. Chem., Int. Ed.* **2015,** *54*, 15218–15221.
[92] Sato, I.; Omiya, D.; Tsukiyama, K.; Ogi, Y.; Soai, K. Evidence of Asymmetric Autocatalysis in the Enantioselective Addition of Diisopropylzinc to Pyrimidine-5-carbaldehyde using Chiral Pyrimidyl Alkanol. *Tetrahedron: Asymmetry* **2001,** *12*, 1965–1969.
[93] Blackmond, D. G. Description of the Condition for Asymmetric Amplification in Autocatalytic Reactions. *Adv. Synth. Catal.* **2002,** *344*, 156–158.

[94] Buono, G. F.; Iwamura, H.; Blackmond, D. G. Physical and Chemical Rationalization for Asymmetric Amplification in Autocatalytic Reactions. *Angew. Chem., Int. Ed.* **2004,** *43*, 2099–2103.
[95] Blackmond, D. G. Mechanistic Study of the Soai Autocatalytic Reaction Informed by Kinetic Analysis. *Tetrahedron: Asymmetry* **2006,** *17*, 584–589.
[96] Buhse, T. A Tentative Kinetic Model for Chiral Amplification in Autocatalytic Alkylzinc Additions. *Tetrahedron: Asymmetry* **2003,** *14*, 1055–1061.
[97] Sato, I.; Omiya, D.; Igarashi, H.; Kato, K.; Ogi, Y.; Tsukiyama, K., et al. Relationship Between the Time, Yield, and Enantiomeric Excess of Asymmetric Autocatalysis of Chiral 2-Alkynyl-5-pyrimidyl Alkanol with Amplification of Enantiomeric Excess. *Tetrahedron: Asymmetry* **2003,** *14*, 975–979.
[98] Rivera Islas, J.; Lavabre, D.; Grevy, J. M.; Hernández Lamoneda, R.; Rojas, H.; Micheau, J. C., et al. Mirror-Symmetry Breaking in the Soai Reaction: A Kinetic Understanding. *PNAS* **2005,** *102*, 13743–13748.
[99] Blackmond, D. G. Kinetic Aspects of Non-linear Effects in Asymmetric Synthesis, Catalysis, and Autocatalysis. *Tetrahedron: Asymmetry* **2010,** *21*, 1630–1634.
[100] Quaranta, M.; Gehring, T.; Odell, B.; Brown, J. M.; Blackmond, D. G. Unusual Inverse Temperature Dependence on Reaction Rate in the Asymmetric Autocatalytic Alkylation of Pyrimidyl Aldehydes. *J. Am. Chem. Soc.* **2010,** *132*, 15104–15107.
[101] Soai, K.; Shibata, T.; Sato, I. Enantioselective Automultiplication of Chiral Molecules by Asymmetric Autocatalysis. *Acc. Chem. Res.* **2000,** *33*, 382–390.
[102] Soai, K.; Sato, I.; Shibata, T. Asymmetric Autocatalysis and the Origin of Chiral Homogeneity in Organic Compounds. *Chem. Rec.* **2001,** *1*, 321–332.
[103] Soai, K.; Sato, I. Asymmetric Autocatalysis and its Implication to Chiral Discrimination. *Chirality* **2002,** *14*, 548–554.
[104] Soai, K.; Shibata, T.; Sato, I. Discovery and Development of Asymmetric Autocatalysis. *Bull. Chem. Soc. Jpn.* **2004,** 77, 1063–1073.
[105] Soai, K.; Kawasaki, T. Discovery of Asymmetric Autocatalysis with Amplification of Chirality and Its Implication in Chiral Homogeneity of Biomolecules. *Chirality* **2006,** *18*, 469–478.
[106] Soai, K.; Kawasaki, T. Asymmetric Autocatalysis with Amplification of Chirality. *Top. Curr. Chem.* **2008,** *284*, 1–33.
[107] Kawasaki, T.; Soai, K. Amplification of Chirality as a Pathway to Biological Homochirality. *J. Fluorine Chem.* **2010,** *131*, 525–534.
[108] Kawasaki, T.; Soai, K. Asymmetric Induction Arising from Enantiomerically Enriched Carbon-13 Isotopomers and Highly Selective Chiral Discrimination by Asymmetric Autocatalysis. *Bull. Chem. Soc. Jpn.* **2011,** *84*, 879–892.
[109] Kawasaki, T.; Soai, K. Asymmetric Autocatalysis Triggered by Chiral Crystals Formed from Achiral Compounds and Chiral Isotopomers. *Isr. J. Chem.* **2012,** *52*, 582–590.
[110] Soai, K.; Kawasaki, T. Asymmetric Autocatalysis of Pyrimidyl Alkanol. *Top. Organomet. Chem.* **2013,** *44*, 261–280.
[111] Soai, K.; Kawasaki, T.; Matsumoto, A. The Origins of Homochirality Examined by Using Asymmetric Autocatalysis. *Chem. Rec.* **2014,** *14*, 70–83.
[112] Soai, K.; Kawasaki, T.; Matsumoto, A. Asymmetric Autocatalysis of Pyrimidyl Alkanol and Its Application to the Study on the Origin of Homochirality. *Acc. Chem. Res.* **2014,** *47*, 3643–3654.
[113] Todd, M. H. Asymmetric Autocatalysis: Product Recruitment for the Increase in the Chiral Environment (PRICE). *Chem. Soc. Rev.* **2002,** *31*, 211–222.
[114] Podlech, J.; Gehring, T. New Aspects of Soai's Asymmetric Autocatalysis. *Angew. Chem., Int. Ed.* **2005,** *44*, 5776–5777.

[115] Brown, J. M.; Gridnev, I.; Klankermayer, J. Asymmetric Autocatalysis with Organozinc Complexes; Elucidation of the Reaction Pathway. *Top. Curr. Chem* **2008,** *284*, 35–65.
[116] Lavabre, D.; Micheau, J. C.; Rivera Islas, J.; Buhse, T. Kinetic Insight into Specific Features of the Autocatalytic Soai Reaction. *Top. Curr. Chem.* **2008,** *284*, 67–96.
[117] Gehring, T.; Busch, M.; Schlageter, M.; Weingand, D. A Concise Summary of Experimental Facts About the Soai Reaction. *Chirality* **2010,** *22*, E173–E182.
[118] Soai, K.; Kawasaki, T. Asymmetric Autocatalysis—Discovery and the State of the Art. In *The Soai Reaction and Related Topic;* Pályi, G., Zucchi, C., Caglioti, L., Eds.; Accademia Nazionale di Scienze Lettere e Arti: Modena, 2012; pp 9–34.
[119] Blackmond, D. G. Asymmetric Autocatalysis and its Implications for the Origin of Homochirality. *PNAS* **2004,** *101*, 5732–5736.
[120] Micheau, J.-C.; Cruz, J.-M.; Coudret, C.; Buhse, T. An Autocatalytic Cycle Model of Asymmetric Amplification and Mirror-Symmetry Breaking in the Soai Reaction. *ChemPhysChem* **2010,** *11*, 3417–3419.
[121] Micheau, J.-C.; Coudret, C.; Cruz, J.-M.; Buhse, T. Amplification of Enantiomeric Excess, Mirror-Image Symmetry Breaking and Kinetic Proofreading in Soai Reaction Models with Different Oligomeric Orders. *Phys. Chem. Chem. Phys.* **2012,** *14*, 13239–13248.
[122] Crusats, J.; Hochberg, D.; Moyano, A.; Ribo, J. M. Frank Model and Spontaneous Emergence of Chirality in Closed Systems. *ChemPhysChem* **2009,** *10*, 2123–2131.
[123] Lente, G. Stochastic Kinetic Models of Chiral Autocatalysis: A General Tool for the Quantitative Interpretation of Total Asymmetric Synthesis. *J. Phys. Chem. A* **2005,** *109*, 11058–11063.
[124] Doka, E.; Lente, G. Mechanism-Based Chemical Understanding of Chiral Symmetry Breaking in the Soai Reaction. A Combined Probabilistic and Deterministic Description of Chemical Reactions. *J. Am. Chem. Soc.* **2011,** *133*, 17878–17881.
[125] Barabás, B.; Caglioti, L.; Micskei, K.; Pályi, G. Data-Based Stochastic Approach to Absolute Asymmetric Synthesis by Autocatalysis. *Bull. Chem. Soc. Jpn.* **2009,** *82*, 1372–1376.
[126] Barabás, B.; Zucchi, C.; Maioli, M.; Micskei, K.; Pályi, G. Stochastic and Empirical Models of the Absolute Asymmetric Synthesis by the Soai-autocatalysis. *J. Mol. Model.* **2015,** *21*, 33.
[127] Micskei, K.; Maioli, M.; Zucchi, C.; Caglioti, L.; Palyi, G. Generalization Possibilities of Autocatalytic Absolute Enantioselective Synthesis. *Tetrahedron: Asymmetry* **2006,** *17*, 2960–2962.
[128] Micskei, K.; Pota, G.; Caglioti, L.; Palyi, G. Empirical Description of Chiral Autocatalysis. *J. Phys. Chem. A* **2006,** *110*, 5982–5984.
[129] Blanco, C.; Crusats, J.; El-Hachemi, Z.; Moyano, A.; Hochberg, D.; Ribo, J. M. Spontaneous Emergence of Chirality in the Limited Enantioselectivity Model: Autocatalytic Cycle Driven by an External Reagent. *ChemPhysChem* **2013,** *14*, 2432–2440.
[130] Ribo, J. M.; Blanco, C.; Crusats, J.; El-Hachemi, Z.; Hochberg, D.; Moyano, A. Absolute Asymmetric Synthesis in Enantioselective Autocatalytic Reaction Networks: Theoretical Games, Speculations on Chemical Evolution and Perhaps a Synthetic Option. *Chem. Eur. J.* **2014,** *20*, 17250–17271.
[131] Micheau, J. C.; Coudret, C.; Buhse, T. Systems Chemistry of the Soai Reaction. In *The Soai Reaction and Related Topic;* Pályi, G., Zucchi, C., Caglioti, L., Eds.; Collana di studi dell'Accademia Nazionale di Scienze Lettere e Arti: Modena, 2012; pp 169–196.

[132] Lavabre, D.; Micheau, J. C.; Rivera Islas, J.; Buhse, T. Enantioselectivity Reversal by Achiral Additives in the Soai Reaction: A Kinetic Understanding. *J. Phys. Chem. A* **2007,** *111*, 281–286.
[133] Lente, G. The Effect of Parity Violation on Kinetic Models of Enantioselective Autocatalysis. *Phys. Chem. Chem. Phys.* **2007,** *9*, 6134–6141.
[134] Faglioni, F.; Lazzeretti, P.; Palyi, G. Parity Violation Energy of 5-Pyrimidyl Alkanol, a Chiral Autocatalytic Molecule. *Chem. Phys. Lett.* **2007,** *435*, 346–349.
[135] Lente, G. Stochastic Analysis of the Parity-Violating Energy Differences between Enantiomers and Its Implications for the Origin of Biological Chirality. *J. Phys. Chem. A* **2006,** *110*, 12711–12713.
[136] Barabás, B.; Caglioti, L.; Micskei, K.; Zucchi, C.; Pályi, G. Isotope Chirality and Asymmetric Autocatalysis: A Possible Entry to Biological Chirality. *Orig. Life Evol. Biosph.* **2008,** *38*, 317–327.
[137] Rivera Islas, J.; Pimienta, V.; Micheau, J. C.; Buhse, T. Kinetic Analysis of Artificial Peptide Self-replication. Part II: The Heterochiral Case. *Biophys. Chem.* **2003,** *103*, 201–211.
[138] Funes-Maldonado, M.; Sieng, B.; Amedjkouh, M. Enabling Asymmetric Alkynylation of Azaaryl Aldehydes with Soai Autocatalyst. *Eur. J. Org. Chem.* **2015,** *2015*, 4081–4086.
[139] Romagnoli, C.; Sieng, B.; Amedjkouh, M. Asymmetric Amplification Coupling Enantioselective Autocatalysis and Asymmetric Induction for Alkylation of Azaaryl Aldehydes. *Eur. J. Org. Chem.* **2015,** *2015*, 4087–4092.
[140] Mauksch, M.; Tsogoeva, S. B.; Martynova, I. M.; Wei, S. Evidence of Asymmetric Autocatalysis in Organocatalytic Reactions. *Angew. Chem., Int. Ed.* **2007,** *46*, 393–396.
[141] Kahn, M. L.; Glaria, A.; Pages, C.; Monge, M.; Saint Macary, L.; Maisonnat, A., et al. Organometallic Chemistry: An Alternative Approach towards Metal Oxide Nanoparticles. *J. Mater. Chem.* **2009,** *19*, 4044–4060.
[142] Bury, W.; Krajewska, E.; Dutkiewicz, M.; Sokolowski, K.; Justyniak, I.; Kaszkur, Z., et al. Tert-Butylzinc Hydroxide as an Efficient Predesigned Precursor of ZnO Nanoparticles. *Chem. Commun.* **2011,** *47*, 5467–5469.
[143] Nchari, L. N.; Hembury, G. A.; Beesley, A. M.; Meehan, D. J.; Tsapatsaris, N.; Hudson, M., et al. Continuous Tubular Flow Reactor for XAFS Studies of Organometallic Reactions: Possibilities and Limitations for Studies of the Soai Reaction. *JPCS* **2009,** *190*, 012156.

CHAPTER 5

Macromolecular Versus Micromolecular World

Luciano Caglioti[1] and Valeria Righi[2]
[1]Department of Chemistry of Bioactive Compounds, University "La Sapienza"-Roma, Roma, Italy
[2]Department of Sciences for the Quality of Life, University of Bologna, Rimini, Italy

Abstract

The origin of life is still a mystery with many unsolved puzzles. How were molecules created? How did they assemble into large structures? Among the many mysteries, the "chirality" phenomenon upon which amino acids and sugars form is particularly exciting. Single chiral molecules in any chemical system play a very particular role, and these are the only systems where one can be sure that they display 100% enantiomer excess. If this kind of system is a sub-sytem of reaction mixtures where highly sensitive autocatalytic reactions may develop, as for example in the case of the Soai reaction, the presence of this initial chiral molecule can decisively influence the subsequent stereochemical developments. In the work presented here we try to understand how the phenomenon of chirality is related to single molecules, and how this starting situation could influence the outcome of reactions where the product(s) is(are) of high molecular weight, such as synthetic or biological polymers.

Keywords: Origin of life; prebiotic chirality; biotic chirality; Soai reaction; single chiral molecule

5.1 INTRODUCTION

The origin of life is still a mystery with many puzzles still unsolved. The question regarding the origin of life remains open and a number of its different aspects have been studied thoroughly in the past years [1—4].

The worlds of prebiotic chemistry and primitive biology lie on opposite sides of the crucial moment for life, when Darwinian evolution was first recognized. Before that time, chemical processes may have led to a substantial level of complexity. Depending on the nature of the prebiotic environment, available building blocks may have included amino acids, hydroxy acids, sugars, purines, pyrimidines and fatty acids. These could have combined to form polymers of largely random sequence and mixed stereochemistry [5].

Prebiotic chemistry proceeds in a multimolecular context, in which different small molecules (both *inorganic* as ammonia, hydrogen chloride,

Advances in Asymmetric Autocatalysis and Related Topics
DOI: http://dx.doi.org/10.1016/B978-0-12-812824-4.00005-8

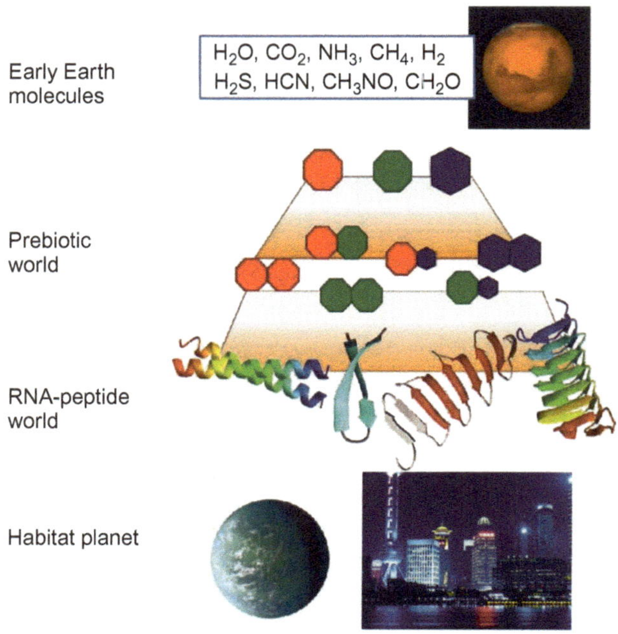

Figure 5.1 Events pertaining to the early history of life on Earth to present.

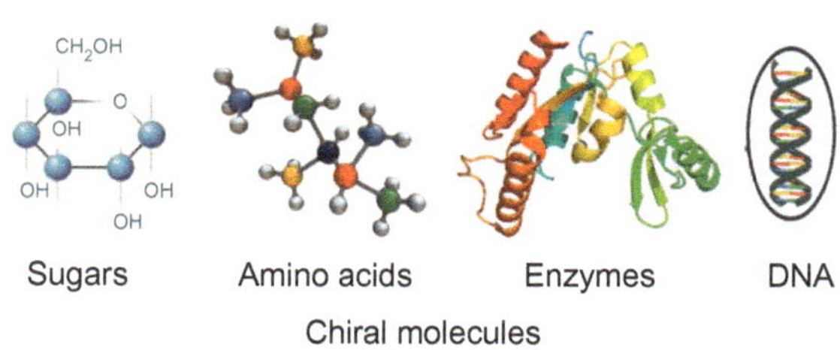

Figure 5.2 Chiral molecules: sugars, amino acids, enzyme and DNA.

carbon, nitrogen and sulfur oxides, phosphates, etc., and *organic* compounds as carbohydrates, amino acids, etc.) are involved under the action of solar and geothermal energy. The complexity of these molecules gradually increases on the basis of this energy supply (Fig. 5.1).

Some of these polymers, that are biological compounds, present chirality such as glucose or ribose (e.g., in ribonucleic acid (RNA) or deoxyribonucleic acid (DNA)), amino acids and enzymes (Fig. 5.2). In living beings sugars are found exclusively in their D-form, whereas other structures such as proteins contain only the L-form of amino acids.

Chirality is a property of individual molecules. If a molecule is chiral then we know that the molecule has two enantiomeric forms (which are almost identical but are actually different molecules). If a molecule is chiral

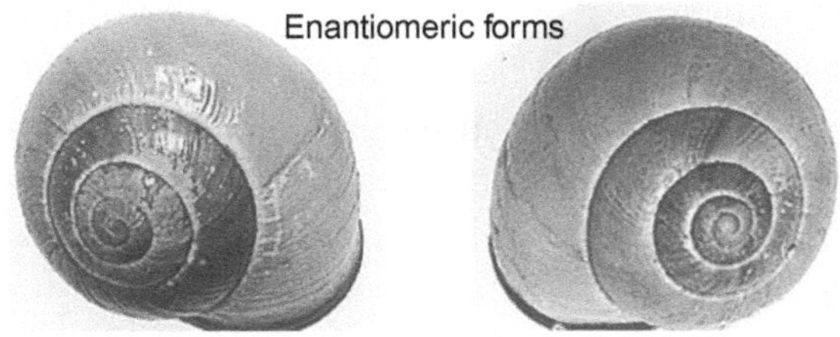

Figure 5.3 Example of twin, or an enantiomer.

then we know that it has an enantiomer (a twin). If a molecule is achiral then it is unique and does not have a twin. However, we can know if a molecule or any other object has a twin just by looking at it (Fig. 5.3).

Enantiomers of chiral molecules have very different properties. For example, about 50 years ago a drug called Thalidomide, which produced as racemates, was developed as a cure for morning sickness. One of its enantiomers had a favorable effect; however, the other enantiomer form caused birth defects. It was found to be strongly teratogenic. European women have used this medication (this drug was not approved in USA) during pregnancy and gave birth to babies with defects (as phocomelic children). For such practical and theoretical reasons chemical reactions have been studied and relevant information about the prebiotic planet in relation to life's origin has been provided.

For example, alanine is a chiral molecule and if it is generated alone in a certain microvolume it will correspond (practically) to requirements of a *single chiral molecule* [6–9].

Obviously this "solitary" alanine can produce additional reactions leading to more complex products. Additional chemical events depend critically on *autocatalysis* and, of course, on energy supply. The products of these second-, third- and following-generation reactions usually *preserve* the chirality of the "parent" alanine and if they are of autocatalytic character the "original chirality" of the solitary alanine will be *amplified* in the sense of generating the same configuration [10].

Caglioti and coworkers report that the origin(s) of life [11] and biological chirality were singular events and, moreover, these did not occur under "sterile" (laboratory-like) conditions [10].

Fuss reported that a chance mechanism can give rise to much higher initial enantiomeric excess (up to 100%) than would be produced by a deterministic mechanism alone, and that it can also explain why not all biological molecules are chirally uniform [12].

Considering all aspects, the conclusion will be that neither the origin of the biological (molecular) chirality, nor the question of the origin of macroscopic asymmetry have been clarified by a satisfactory level of scientific rigor, until today [13]. Considering the importance of these problems, it is understandable that several hypotheses had been formed. Even if many of these theories are of high intellectual quality, we doubt whether the problem could ever be solved with 100% certainty. On the other hand, we believe, that explanations that are consistent with all observations can and should be conceived. We were deeply involved in such efforts in the past decade and we would like to outline here one of our theories [6–9], which, we believe, is sufficiently reduced to elementary facts, to be considered as a starting point for future research and speculations.

5.2 RESULTS AND DISCUSSION

5.2.1 Macromulecular and Micromolecular World

The macromolecular living world is costituted mainly of proteins. Proteins are the most versatile macromolecules in living systems and serve crucial functions in essentially all biological processes. They function as catalysts, they transport and store other molecules such as oxygen, they provide mechanical support and immune protection, they generate movement, they transmit nerve impulses, and they control growth and differentiation. Indeed, much of this text will focus on understanding what proteins do and how they perform these functions.

But, if we think about living individuals, they are molecules or aggregates of distinctly interlocked molecules that can reproduce themselves in the given environment and evolve, as a form, in appropriate achievable environments into forms of continuously increasing complexity and functional intricacy.

In the world micromolecular small molecules have the same functions as macromolecules. The important thing is the sample, or better the number of molecules that costitute the sample. One mole (or mass fraction) as a number of molecules so high that they are stochastically distributed in the sample. The term *stochastic* describes events or systems that are unpredictable due to the influence of a random variable. Therefore the distribution of these molecules follows the probabilistic theory (50% in a way and 50% in another way). (Fig. 5.4). At that point the questions are: Why does one of the two enantiomeric forms prevail? In the sugar D with respect to the L form or in amino acid L with respect to the D form? When is the energy for their formation the same? Instead of going from big to small, if you reverse the reasoning, from small to big, the

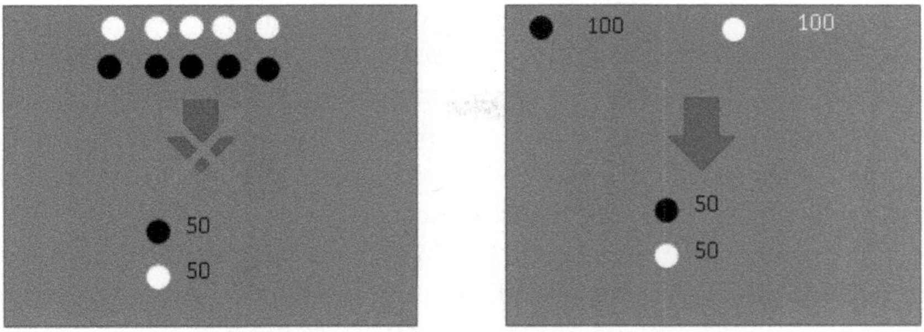

Figure 5.4 Example of statistical distribution of molecule numbers.

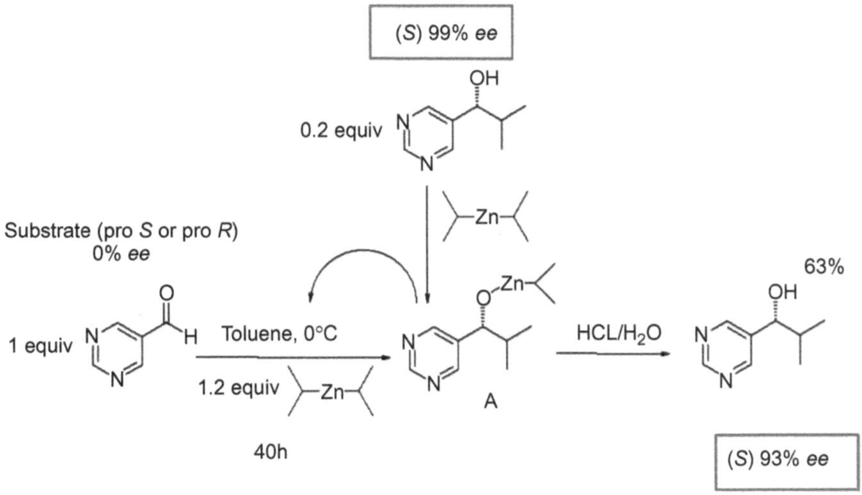

Figure 5.5 Examples of the Soai-reaction [15–18].

probability that you obtain 50% of the two enantiomeric forms is no longer true (Fig. 5.4).

For five black molecules and five white molecules after a few years we may have three white and seven black molecules, because one of the two form takes precedence over the other. Calculations about probability distribution in small numbers of molecules have already been performed [14a,b], for example, with 10 molecules the probability that only one enantiomer is formed is about 0.2%. The statistical calculation is, however, only valid for *even* numbers of molecules, while if an *odd* number of molecules is present in a sample, at least one molecule enantiomeric excess must be present. In exact terms, this means that in 50% of all cases a "racemate" cannot be a true racemate [14]. If we add, in this case, the multiplicity factor due to reactions similar to the Soai reaction (Fig. 5.5), where catalyst and product are the same, we can easily obtain a tremendous (exponential) increase of the initial enantiomer excess (*ee*) [14]. This particular reaction has two important features: the product of the reaction not only speeds up the reaction of its own formation (chemical autocatalysis), but it promotes its own configuration (stereochemical autocatalysis) in the subsequent quantities of the product.

The relative *ee* created initially (i.e., before amplification) by the deterministic mechanisms is very small. In contrast, *ee* is large in the chance mechanism, if the number of initial molecules is small; with a single molecule, it would even imply $ee = 1$ (=100%), so that no amplification is required [19].

There is a fundamental difference in chiral induction by the first (or initial few) molecule(s) in chemical reactions producing "monomolecular" or polymeric products. Initially the first molecule can exert its inducing power only if it gets incorporated in the critical transition state (as is proposed by the Soai reaction) and this happens only in very fortunate cases. In polymerization reactions, however, the "first" molecule is necessarily present at the formation of the "second", while these two units are linked together by chemical bonds, and, naturally, this is true for the formation of the third, fourth, etc., polymeric units [20].

Our original idea [19] can be represented in a simple way: if a chiral molecule could be formed alone and that this process could be amplified by, for example, the Soai reaction, this chiral could multiply exponentially, ultimately prevailing over the other enantiomeric form.

5.3 CONCLUSION

The problems of gaining an understanding of the origins of terrestrial life and the biological chirality are far from being clarified. A chance mechanism, based on small initial numbers, has the advantage that fewer assumptions can be made than in deterministic mechanisms. Model experiments exploring several aspects of these problems are, however, highly important from theoretical and practical points of view [21–24], but their results can be applied only very cautiously to hypotheses regarding the real processes that were displayed some billions of years ago.

Life depends on biopolymer sequences as catalysts and as genetic material. A key step in the origin of life is the emergence of an autocatalytic system leading to homochiral biopolymers.

REFERENCES

[1] Lahav, N. *Biogenesis: Theories on Life's Origin;* Oxford University Press: New York, 1999.
[2] Brack, A. *The Molecular Origins of Life: Assembling the Pieces of the Puzzle;* Cambridge University Press: New York, 1998.
[3] Wills, C.; Bada, J. *The Spark of Life*: Darwin and the Primeval Soup*;* Oxford University Press: Oxford, 2000.
[4] Fry, I. *The Emergence of Life on Earth: A Historical and Scientific Overview;* Rutgers University Press: New Brunswick, NJ, 2000.
[5] Joyce, F. The Antiquity of RNA-Based Evolution Gerald. *Nature* **2002,** *418*, 214–221.
[6] Caglioti, L.; Zucchi, C.; Pályi, G. Single-Molecule Chirality. *Chem. Today* **2005,** *23*, 38–43.
[7] Caglioti, L.; Micskei, K.; Pályi, G. Chirality of the Very First Molecule in Absolute Enantioselective Synthesis. *Viva Origino* **2007,** *35*, 82–84.
[8] Caglioti, L.; Pályi, G. Chiral Chemistry of Single Molecules. *Chem. Today* **2008,** *26*, 41–42.
[9] Caglioti, L.; Micskei, K.; Pályi, G. First Molecules, Biological Chirality, Origin(s) of Life. *Chirality* **2001,** *23*, 65–68.
[10] Caglioti, L.; Pályi, G. Single Chiral Molecule as Possible Starting Element of Complex Chiral Systems. *Rend. Lincei Sci. Fis. Nat.* **2013,** *24*, 191–196.
[11] Pályi, G.; Zucchi, C.; Caglioti, L., Eds. *Fundamentals of Life;* Elsevier: Paris, 2002.
[12] Fuss, W. Does Life Originate from a Single Molecule? *Chirality* **2009a,** *21*, 299–304.
[13] Wu, M.; Walker, S. I.; Higgs, P. G. Autocatalytic Replication and Homochirality in Biopolymers: Is Homochirality a Requirement of Life or a Result of It? *Astrobiology* **2012,** *12*, 818–829.
[14] (a) Pályi, G.; Micskei, K.; Zékány, L.; Zucchi, C.; Caglioti, L. Racemates and the Soai Reaction. *Magy. Kém. Lapja* **2005,** *60*, 17–24.
(b) Barabás, B.; Tóth, J.; Pályi, G. Stochastic Aspects of Asymmetric Autocatalysis and Absolute Asymmetric Synthesis. *J. Math. Chem.* **2010,** *48*, 457–489.
[15] Soai, K.; Shibata, T.; Morioka, H.; Choji, K. Asymmetric Autocatalysis and Amplification of Enantiomeric Excess of a Chiral Molecule. *Nature* **1995,** *378*, 767–768.
[16] Soai, K.; Shibata, T.; Sato, I. Enantioselective Automultiplication of Chiral Molecules by Asymmetric Autocatalysis. *Acc. Chem. Res.* **2000,** *33*, 382–390.
[17] Soai, K.; Shibata, T.; Sato, I. Discovery and Development of Asymmetric Autocatalysis. *Bull. Chem. Soc. Jpn.* **2004,** 77, 1063–1073.
[18] Soai, K.; Kawasaki, T. Asymmetric Autocatalysis with Amplification of Chirality. *Top. Curr. Chem.* **2008,** *284*, 1–33.
[19] Fuss, W. Biological Homochirality as Result from a Single Event. *Colloids Surf. B Biointerfaces* **2009,** *74*, 498–503.
[20] Kimura, J.; Kitadai, N. Polymerization of Building Blocks of Life on Europa and Other Icy Moons. *Astrobiology* **2015,** *15*, 430–441.
[21] Sheldon, R. A. *Chirotechnology;* Dekker: New York and Basel, 1993.
[22] Jacobsen, E. N.; Pfaltz, A.; Yamamoto, H., Eds. *Comprehensive Asymmetric Catalysis II;* Springer: Berlin, 1999.
[23] Jacobsen, E. N.; Pfaltz, A.; Yamamoto, H., Eds. *Comprehensive Asymmetric Catalysis, Supplement 1;* Springer: Berlin, 2003.
[24] Ojima, I., Ed. *Catalytic Asymmetric Synthesis;* 3rd ed. Wiley-VCH: New York, 2010.

CHAPTER 6

The Importance of the Idea of "Parachirality" in Life Science

Noriko Fujii[1], Norihiko Fujii[2], Takumi Takata[3] and Hiroaki Sakaue[4]
[1]Research Reactor Institute, Kyoto University, Osaka, Japan
[2]Radioisotope Research Center, Teikyo University, Kaga, Itabashi, Tokyo, Japan
[3]Department of Biochemistry, School of Pharmacy, Tokyo University of Pharmacy and Life Sciences, Hachioji, Tokyo, Japan
[4]International University of Health and Welfare, Ohtawara, Tochigi, Japan

Abstract

Homochirality is essential for the development and maintenance of life. Therefore biochemists have not paid much attention to the presence and function of biologically uncommon D-amino acids in living organisms. However, recent improvements in analytical techniques have facilitated the accurate analysis of amino acid enantiomers. Consequently, D-amino acids have been detected in a variety of living higher organisms in the form of free, peptide-bound and protein-bound amino acids.

D-aspartyl (Asp) residues present as protein-bound D-amino acids have been widely detected in proteins obtained from various tissues of elderly individuals. The presence of D-Asp in aged tissues of living organisms is a result of the spontaneous racemization of Asp residues during aging. This racemization does not occur uniformly but does so at specific residues on the basis of sequence context or structural considerations that make these sites more susceptible to racemization than others. It is therefore informative to determine the nature of the Asp residues at racemization sites within particular proteins. To date, however, the detection of D-amino acids in proteins has been complex and difficult. In this review, we show the methods for the analysis of protein-bound D-amino acids based on LC-MS/MS that can identify the specific sites of D-Asp in proteins. Furthermore, we show where and how D-amino acids are generated in proteins of living tissues under physiological conditions, and describe the effects of D-amino acid formation on proteins using lens crystallins as an example.

Elucidation of the mechanisms underlying the onset of age-related diseases such as cataract and Alzheimer's disease may be achieved by consideration of the idea of "parachirality", where D-amino acids are incorporated into the L-amino acid arena of life.

Keywords: D-amino acids; D-aspartyl (Asp) residues; isoaspartate (β-aspartate); protein; cataract; aging; age-related diseases; LC-MS/MS

Advances in Asymmetric Autocatalysis and Related Topics
DOI: http://dx.doi.org/10.1016/B978-0-12-812824-4.00006-X

6.1 INTRODUCTION

The homochirality of living organisms on earth might have been the combination of L-amino acids and of the D-sugars before the emergence of life. Nobody knows why, where or how nature selected L-amino acids and D-sugars, or whether their selection was logical or if it was a chance occurrence. However, it is clear that the selection of only one of the enantiomers is necessary to maintain the higher order structures of the protein and RNA/DNA. Although the reasons why the combination of D-amino acids and L-sugars was eliminated before the emergence of life are not known, homochirality is essential for the development and maintenance of life. Once the L-amino acid and D-sugars world was established, their combination was excluded from living systems. Therefore biochemists have not paid much attention to the presence and function of biologically uncommon D-amino acids and L-sugars in living organisms. The presence of L-sugars in living organisms has not been demonstrated but D-amino acids have been found in living organisms in the form of free amino acids, peptides and proteins. Free D-amino acids have important physiological functions in mammals [1–3]. Furthermore, there are many studies of D-amino acid oxidase, D-aspartate oxidase, and racemases that regulate the amount of free D-amino acids in the living body. Advances in studies of free D-amino acids and their related enzymes in vivo have been recently published in an excellent review [4,5]. Short peptides containing a single D-amino acid have been found in various vertebrates and invertebrates. The first known D-amino acid-containing peptide, Dermorphin, was isolated from the skin of a frog (*Phyllomedusinae* from South and Middle America) and is a opioid peptide with the sequence Tyr-D-Ala-Phe-Gly-Tyr-Pro-Ser-NH_2 [6]. Its activity, which is about 1000 times greater than that of morphine, is lost by substituting L-Ala for D-Ala [7]. Many D-amino acid–containing peptides with various physiological functions have since been identified and are described in other excellent reviews [8–10]. In proteins, D-aspartyl (Asp) residues have been widely detected in proteins obtained from various tissues of elderly individuals taking advantage of the recent improvements in analytical techniques. The presence of D-Asp is a result of the spontaneous racemization of Asp residues during aging. In this review, we briefly summarize where and how D-Asp residues are generated in proteins of living tissues under physiological conditions, and describe the effects of D-Asp formation on the structure and functions of proteins using lens crystallins as an example.

6.2 LOCALIZATION OF D-ASP RESIDUES IN PROTEINS OF LIVING TISSUES

In the past few decades, D-Asp has been widely detected in various tissues such as eye lens [11–15], retina [16], conjunctiva [17], cornea [18], teeth [19,20], bone [21–23], aorta [24], ligament [25], brain [26–29], and skin [25,30,31] of elderly individuals (Table 6.1). The presence of D-Asp in aged tissues of living organisms is a result of the spontaneous racemization of Asp residues during aging. Therefore it has been considered that D-Asp seems to be present in the metabolically inert tissues associated with age-related diseases.

However, as shown in Table 6.1, D-Asp seems to be present not only in metabolically inert proteins but also in keratin [31], histone H2B [32] and lysozyme [33], which are not inert. The results suggest that D-Asp does not occur uniformly but does so at specific residues that are more susceptible to racemization than others.

6.3 MECHANISM OF D-ASP FORMATION IN PROTEINS UNDER PHYSIOLOGICAL CONDITIONS

Asp residues in proteins are susceptible to isomerization because Asp has a carboxyl group in its side chain. Inversion and isomerization of Asp residues in proteins is considered to proceed via a succinimide intermediate.

As shown in Fig. 6.1, the simultaneous inversion and isomerization of Asp residues in proteins can be explained as follows: (1) when the lone-pair electron of the nitrogen atom of the amino-acid residue following the Asp residue attacks the carboxyl group of the side chain of the Lα-Asp residue, L-succinimide is generated by intramolecular cyclization; (2) L-succinimide can be converted to D-succinimide via an intermediate through keto–enol tautomerism; and (3) D-succinimide is then hydrolyzed at either side of its two carbonyl groups to form Dα- and Dβ-Asp—similarly, L-succinimide is hydrolyzed to form Lα- and Lβ-Asp. Thus, four isomers, Lα-Asp, Lβ-Asp, Dα-Asp, and Dβ-Asp, are simultaneously formed in the protein. The rate of succinimide formation is considered to depend on the residue neighboring the Asp residue. When the neighboring amino acid has a small side chain, as found in glycine (Gly), alanine (Ala) or serine (Ser), succinimide forms easily because there is no steric hindrance [34,35]. In addition to the effects of the adjacent residues, Asp may also be susceptible to racemization when the residues are located in flexible regions.

Table 6.1 The presence of D-aspartyl residues in proteins from various tissues

Tissue	Protein	Related disease	Specific sites of Asp/Asn	Refs.	Analysis
Lens	αA-crystallin	Cataract	Asp58,76,84,151	[12,15]	CA, LC-MS
Lens	αB-crystallin	Cataract	Asp36, 62,96	[13,15]	CA, LC-MS
Lens	βB2-crystallin	Cataract	Asp4	[14]	CA, LC-MS
Retina	?	AMD	?	[16]	IH
Conjunctiva	?	Pinguecula	?	[17]	IH
Cornea	?	CDK	?	[18]	IH
Tooth	Phosphophoryn	?	?	[20]	CA
Bone	Osteocalcin	?	?	[21]	CA
Bone	Type I collagen C-terminal tellopeptide	Osteoporosis of Paget's disease	Asp1211	[23]	CA
Aorta	Elastin	Arteriosclerosis	?	[24]	IH, CA
Ligament	Elastin	?	?	[25]	IH, CA
Brain	Myelin	?	?	[26]	CA
Brain	β-amyloid	Alzheimer disease	Asp1,7, 23	[29]	CA
brain	Histone H2B	?	Asn 25	[32]	CA,
Skin	Collagen	?	?	UP	LC-MS
Skin	Elastin	?	?	[30]	IH, CA
Skin	Kelatin	?	?	[31]	IH, CA
Body fulid	Lysozyme	?	Asn 127	[33]	CA

AMD, age-related macular diseases; CA, chemical analysis; IH, immunohistochemistry; UP, Unpublished data.

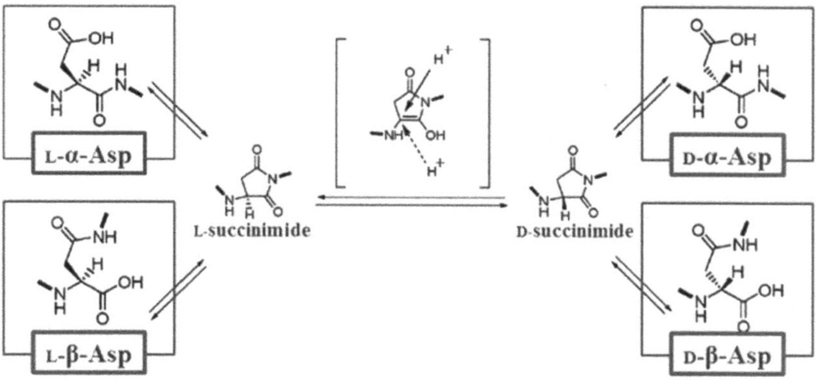

Figure 6.1 Reaction pathway for spontaneous inversion and isomerization of aspartyl residues in proteins.

6.4 THE INFLUENCE OF ASP ISOMERIZATION ON PROTEIN STRUCTURE AND FUNCTION

The appearance of D-Asp isomers in a protein can cause major changes in the 3D structure because the different side-chain orientation may induce an abnormal peptide backbone. In addition to D-Asp formation, the β-linkage of Asp may affect the quaternary structure because the main chain of the protein is elongated.

Human eye lens proteins are mainly composed of α-, β-, and γ-crystallins. Alpha-crystallin exists as a polydisperse hetero-polymer with a molecular mass of 800–1000 kDa, which is composed of αA- and αB-crystallin, each of which has a molecular mass of 20 kDa. The β/γ-crystallin superfamily comprises oligomeric β-crystallin (2–6 subunits) and monomeric γ-crystallin. These crystallins interact stably and function in lens transparency. However, when these lens crystallins undergo isomerization with aging they can aggregate, become insoluble and dissociate, and lose the chaperone function and the transparency of lens.

Therefore the presence of these isomers may be one of the triggers for abnormal aggregation. In fact, αA-crystallin of eye lenses containing large amounts of D-β-Asp obtained from donors of ~80 years of age has been shown to undergo abnormal aggregation to form massive and heterogeneous aggregates [36]. In addition, its chaperone activity which is an

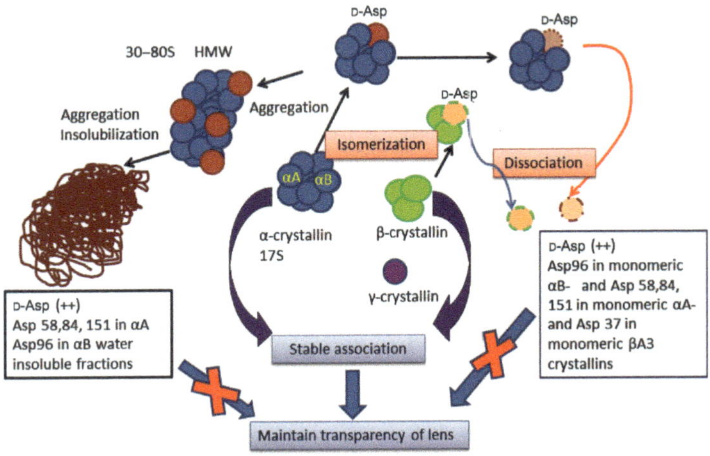

Figure 6.2 The influence of isomerization of Asp on lens protein structure and function.

important function of α-crystallin and prevents the aggregation and insolubilization of other lens proteins, is reduced by 60%, compared to the chaperone activity of young α-crystallin [36]. In contrast, we recently found that isomerization of Asp-96 of αB-crystallin and Asp-37 of βA3-crystallin [37], and Asp 58, 84, 151 of αA-crystallin [38] induced the dissociation from the normal associated states of α or β-crystallin to abnormal monomeric states. Fig. 6.2 shows a summary of the influence of Asp isomerization on lens crystallins, suggesting that the isomerization of Asp might disrupt the higher order polymeric state of α-crystallin and oligomeric β-crystallin, resulting in decreased solubility and function.

6.5 NEW METHOD OF ANALYSIS FOR BOUND FORM OF D-AMINO ACIDS

Under physiological conditions, Asp residues are more easily inverted from the L-form to the D-form compared to other amino-acid residues because the stereoinversion of Asp residues occurs via a succinimide intermediate as shown in Fig. 6.1. The inversion of Asp residues in proteins does not occur uniformly but does so at specific Asp residues on the basis of the sequence context or structural considerations that make the specific residues more susceptible to reaction than others. Therefore it is necessary to determine the nature of the Asp residues at specific sites within particular proteins. The separation of the optical isomers of amino acids previously had been considered to be difficult, because the physical and chemical properties of the optical isomers are the same. In addition to this, enantioseparation of the bound form of amino acids requires the hydrolysis of the protein/peptide before the analysis of the enantiomers. Conventional enantioseparation of amino acids has been performed using gas-chromatography (GC) or RP-HPLC. GC analysis requires nanomole levels of sample while picomole levels are required for RP-HPLC analysis. GC analysis is direct enantioseparation through the use of a chiral capillary column, while RP-HPLC analysis is indirect enantioseparation based on the analysis of the diastereoisomeric derivatives of the amino-acid samples produced by chiral derivatizing reagents. Both methods require the appropriate amino-acid derivatization or preparation in advance for the analysis, the former requiring the samples to be changed into the gaseous state before injection into the GC and the latter requiring production of diastereoisomeric derivatives in the case of the nonchiral column. The process is very complex for free D-amino acid analysis. In addition, in order to analyze the specific sites of D-amino acids in proteins, more complicated steps are required on top of the free D-amino acid analysis. These are: (1) digestion of the protein with an appropriate enzyme; (2) separation of the resulting peptides by RP-HPLC; (3) identification of the peptides by mass analysis and/or protein sequencing; (4) determination of the α or β isomeric state of the identified peptides by the Edman degradation reaction; (5) determination of the D/L ratio of the identified peptides after hydrolysis with 6 N HCl and derivatization; and (6) analysis of the diastereoisomers by RP-HPLC and the D/L ratio of amino acids determined by analysis of the respective peak areas. The resulting analysis of the isomerization of Asp residues in a protein can be accurate but it is a technically demanding process. Consequently, there has been little study of the presence and function of D-amino acids in living organisms.

Recently, we proposed a new accurate and quick LC-MS-based analysis for determining the specific sites having Asp isomers and quantifying the amounts of Asp isomers at the individual sites of all lens crystallins in the water-insoluble (WI) and the water-soluble (WS) fractions without the need for complicated purification from the lens tissues (Fig. 6.3) [15]. This technique involves the human lens sample being homogenized in an appropriate buffer and the supernatant (WS) and precipitate (WI) separated by centrifugation. Both fractions are digested with trypsin overnight, and then the resulting peptides applied to the LC-MS. Generally, each peptide would be expected to elute as one peak with one mass number; however, peptides that contain Asp isomers were separated into

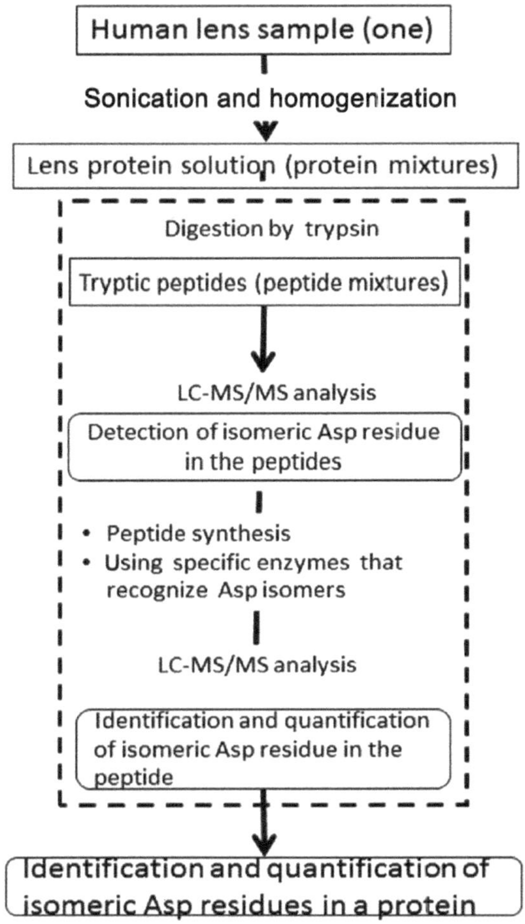

Figure 6.3 New analysis for biologically uncommon Asp isomers in proteins by LC-MS/MS.

multiple peaks and they eluted at different retention times during the LC-MS run even though they had entirely the same sequences. Using this property, we are able to identify peptides that have isomeric Asp residues.

The Asp isomer containing peptides were identified by comparison with the retention time of synthetic peptides [15] or using the commercial enzymes, endoprotease Asp-N (Asp-N), L-isoaspartyl methyltransferase (PIMT) and paenidase, which specifically recognize Lα-Asp, Lβ-Asp, and Dα-Asp residues, respectively [39]. For example, the peptide predicted to correspond to positions 55–65 of αA-crystallin (αA 55–65;

TVLDSGISEVR: $[M + 2H]^{2+} = 588.3$) as identified by the database was mainly separated into four different peaks that eluted at different times, as shown in Fig. 6.4A and B. Fig. 6.4C and D show the LC-MS chromatogram of αB 57–69 (APSWFDTGLSEMR: $[M + 2H]^{2+} = 748.8$) from the WI and WS fractions respectively. This peptide was also separated into several peaks. The number of peptides from αA 55–65 and αB 57–69 was greater from WI protein (Fig. 6.4A and C) than from WS protein (Fig. 6.4B and D). A similar multiple separation of the various peptides containing Asp residues was obtained from all crystallins, that is αA-, αB-, βA3-, βA4-, βB1-, βB2-, and γS-crystallin in both WI and WS protein with the amounts of isomeric peptides in the WI fractions being greater than in the WS fractions. As an example, the amounts for the four Asp isomers of αA- and αB- crystallins from the WI and WS fractions are shown in Fig. 6.5. The amount of normal L-α-Asp is dramatically decreased while the other isomeric ratios increased at all Asp sites in the WI fraction compared to the WS fraction.

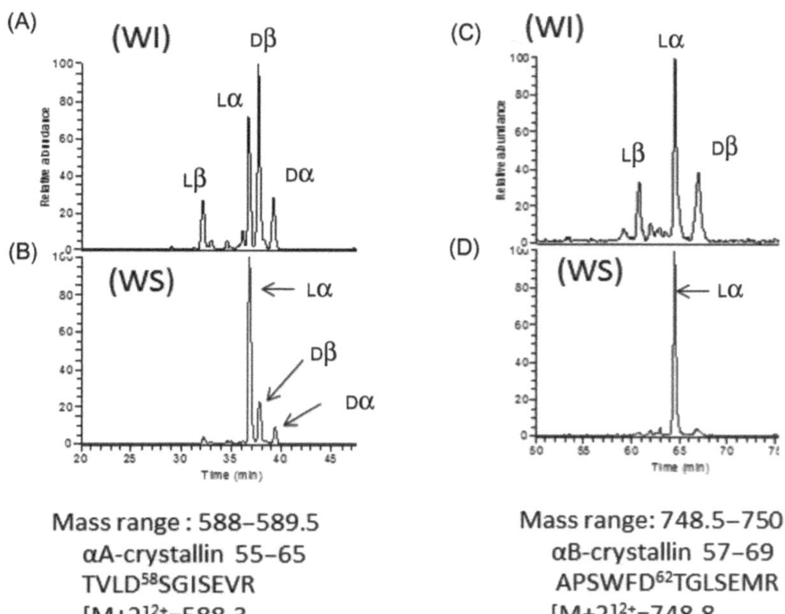

Figure 6.4 LC-MS chromatogram of the tryptic peptides of water-insoluble (WI) and water-soluble (WS) fractions of lens proteins from elderly donors. (A and B): MS range 588–589.5 *m/z* (αA-crystallin 55–65) of WI and WS fractions. (C and D): MS range: 748.5–750 *m/z* (αB-crystallin 57–69) of WI and WS fractions. The peaks were identified by comparison with the retention time of the synthetic peptides. or using the commercial enzymes that recognize the Asp isomer residues.

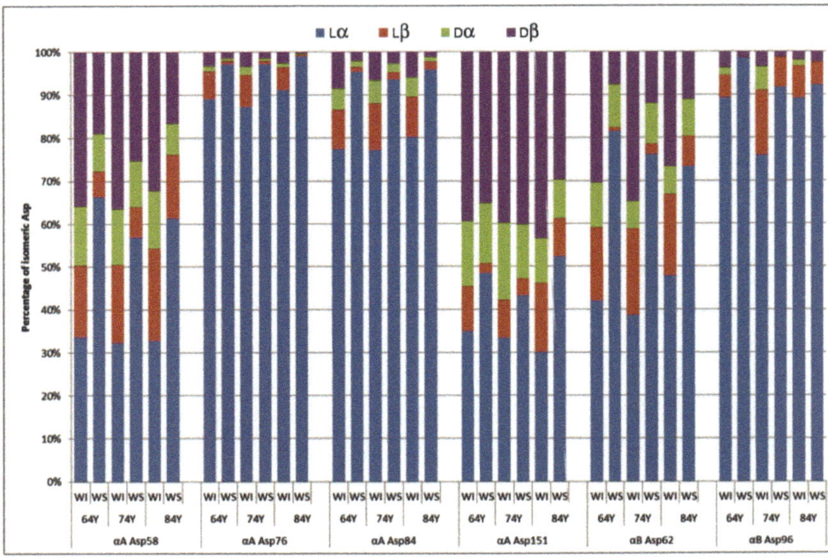

Figure 6.5 The relative amounts of Asp isomers at Asp 58, Asp 76, Asp 84, and Asp 151 of αA-crystallin and Asp 62 and Asp 96 of αB-crystallin from the WI and WS fractions of lenses from 64-, 74-, and 84-year-old donors.

6.6 TOWARDS PARACHIRALITY

Until 40 years ago, D-amino acids were defined as biologically uncommon amino acids with little attention paid to their presence. However, it is now clear that D-amino acids are not so rare in living systems. Developments in analytical methods are advancing our knowledge of the effects of Asp isomerization in proteins. For example, studies have shown that lens crystallins are initially water soluble, but gradually become water insoluble owing to aggregation and insolubilization with aging or with the formation of cataracts. Although the reason for this aggregation is not known, it may be triggered by an increase in D-β-, L-β-, and D-α-Asp isomers because these isomers directly affect protein structure either by inversion of the side chain in the D-form or through formation of the β-linkage. In support of this possibility, αA-crystallin containing large amounts of D-β-Asp obtained from donors aged 80 years forms abnormal and heterogeneous aggregates [36].

Although Asp isomerization has been widely observed in long-lived proteins of various tissues (Table 6.1), there have been few quantitative and site-specific studies of bound protein isomers, mainly because of the lack of a quick and precise method for distinguishing between D- and L-Asp in a peptide. The LC-MS/MS method described in this review provides a solution to this problem. This new approach, which is comprehensive, rapid and low-cost, requires considerably less effort than previous approaches. With the development of this new technique, the identification of D-Asp in proteins in living tissues should become an easy task, and studies of the isomerization and inversion of Asp residues can proceed alongside those of other nonenzymatic PTMs such as deamidation, oxidation and glycation. The importance of D-amino acids in proteins is not limited to lenses, but also to age-related diseases such as Alzheimer's disease, age-related macular degeneration (AMD), and other unknown disorders of protein unfolding. Thus, characterization of these isomers will be a useful marker that will contribute to understanding the etiology of these diseases.

So far, many studies have investigated the relationship between age-related diseases such as cataract and Alzheimer's disease, and genetic abnormality. However, there are few patients with cataracts and Alzheimer's disease identified on the basis of genetic abnormalities. Almost all of these patients have "normal" genes and suffer cataracts and Alzheimer's disease with age. These age-related diseases may not be caused by genetic abnormalities but by protein aggregation. The causes of protein aggregation are not well known. However, elucidation of the mechanisms underlying the onset of age-related diseases may be achieved by consideration of the idea of "parachirality," where D-amino acids are incorporated in the L-amino acid arena of life. The rate of isomerization of Asp residues is faster than we expected under physiological conditions. These findings indicate that abnormal Asp residues are present widely and arise due to racemization of amino acids in various proteins during the lifespan of the individual [40]. Furthermore, it might open up a new scientific field including the origin of biological chirality.

REFERENCES

[1] Furuchi, T.; Homma, H. Free D-aspartate in Mammals. *Biol. Pharm. Bull.* **2005,** *28*, 1566–1570.

[2] Boehning, D.; Snyder, S. H. Novel Neural Modulators. *Annu. Rev. Neurosci.* **2003,** *26*, 105–131.

[3] Hashimoto, A.; Nishikawa, T.; Konno, R.; Niwa, A.; Yasumura, Y.; Oka, T., et al. Free D-serine, D-aspartate and D-alanine in Central Nervous System and Serum in Mutant Mice Lacking D-amino Acid Oxidase. *Neurosci. Lett.* **1993,** *152*, 33–36.

[4] Konno, R.; Hamase, K.; Maruyama, R.; Zaitsu, K. Mutant Mice and Rats Lacking D-amino Acid Oxidase. *Chem. Biodivers.* **2010,** 7, 1450–1458.

[5] Katane, M.; Homma, H. D-aspartate Oxidase: The Sole Catabolic Enzyme Acting on Free D-aspartate in Mammals. *Chem. Biodivers.* **2010,** *7*, 1435–1449.

[6] Montecucchi, P. C.; de Castiglione, R.; Piani, S.; Gozzini, L.; Erspamer, V. Amino Acid Composition and Sequence of Dermorphin, a Novel Opiate-Like Peptide from the Skin of Phyllomedusa Sauvagei. *Int. J. Pept. Protein Res.* **1981,** *17*, 275–283.

[7] Broccardo, M.; Erspamer, V.; Falconieri Erspamer, G.; Improta, G.; Linari, G.; Melchiorri, P., et al. Pharmacological Data on Dermorphins, a New Class of Potent Opioid Peptides from Amphibian Skin. *Br. J. Pharmacol.* **1981,** *73*, 625–631.

[8] Jolles, P. *D-amino Acids in Sequences of Secreted Peptides of Multicellular Oganisms;* Birkhauser Verlag: Basel, Switzerland, 1998; p 192.

[9] Bai, L.; Sheeley, S.; Sweedler, J. V. Analysis of Endogenous D-amino Acid-Containing Peptides in Metazoa. *Bioanal. Rev.* **2009,** *1*, 7–24.

[10] Ollivaux, C.; Soyez, D.; Toullec, J. Y. Biogenesis of D-amino Acid Containing Peptides/Proteins: Where, When and How?. *J. Pept. Sci.* **2014,** *20*, 595–612.

[11] Masters, P. M.; Bada, J. L.; Zigler, J. S., Jr. Aspartic Acid Racemisation in the Human Lens During Ageing and in Cataract Formation. *Nature* **1977,** *268*, 71–73.

[12] Fujii, N.; Satoh, K.; Harada, K.; Ishibashi, Y. Simultaneous Stereoinversion and Isomerization at Specific Aspartic Acid Residues in Alpha A-crystallin from Aged Human Lens. *J. Biochem.* **1994,** *116*, 663–669.

[13] Fujii, N.; Ishibashi, Y.; Satoh, K.; Fujino, M.; Harada, K. Simultaneous Racemization and Isomerization at Specific Aspartic Acid Residues in Alpha B-crystallin from the Aged Human Lens. *Biochim. Biophys. Acta* **1994,** *1204*, 157–163.

[14] Fujii, N.; Kawaguchi, T.; Sasaki, H. Simultaneous Stereoinversion and Isomerization at the Asp-4 Residue in βB2-crystallin from the Aged Human Eye Lenses. *Biochemistry* **2011,** *50*, 8628–8635.

[15] Fujii, N.; Sakaue, H.; Sasaki, H.; Fujii, N. A Rapid, Comprehensive Liquid Chromatography-Mass Spectrometry (LC-MS)-Based Survey of the Asp Isomers in Crystallins from Human Cataract Lenses. *J. Biol. Chem.* **2012,** *287*, 39992–40002.

[16] Kaji, Y.; Oshika, T.; Takazawa, Y.; Fukayama, M.; Takata, T.; Fujii, N. Localization of D-beta-aspartic Acid-containing Proteins in Human Eyes. *Invest. Ophthalmol. Vis. Sci.* **2007,** *48*, 3923–3927.

[17] Kaji, Y.; Oshika, T.; Okamoto, F.; Fujii, N. Immunohistochemical Localisation of D-beta-aspartic Acid in Pingueculae. *Br. J. Ophthalmol.* **2009,** *93*, 974–976.

[18] Kaji, Y.; Oshika, T.; Takazawa, Y.; Fukayama, M.; Fujii, N. Immunohistochemical Localisation of D-beta-aspartic Acid-containing Proteins in Climatic Droplet Keratopathy. *Br. J. Ophthalmol.* **2009,** *93*, 977–979.

[19] Helfman, P. M.; Bada, J. L. Aspartic Acid Racemisation in Dentine as a Measure of Ageing. *Nature* **1976,** *262*, 279–281.

[20] Masuda, W.; Nouso, C.; Kitamura, C.; Terashita, M.; Noguchi, T. D-Aspartic Acid in Bovine Dentine Non-collagenous Phosphoprotein. *Arch. Oral Biol.* **2002,** *47*, 757–762.

[21] Ritz, S.; Turzynski, A.; Schutz, H. W.; Hollmann, A.; Rochholz, G. Identification of Osteocalcin as a Permanent Aging Constituent of the Bone Matrix: Basis for an Accurate Age at Death Determination. *Forensic Sci. Int.* **1996,** *77*, 13–26.

[22] Ohtani, S.; Yamamoto, T.; Matsushima, Y.; Kobayashi, Y. Changes in the Amount of D-aspartic Acid in the Human Femur with Age. *Growth. Dev. Aging* **1998,** *62*, 141–148.

[23] Cloos, P. A.; Fledelius, C. Collagen Fragments in Urine Derived from Bone Resorption are Highly Racemized and Isomerized: A Biological Clock of Protein Aging with Clinical Potential. *Biochem. J.* **2000,** *345* (Pt 3), 473–480.

[24] Powell, J. T.; Vine, N.; Crossman, M. On the Accumulation of D-aspartate in Elastin and Other Proteins of the Ageing Aorta. *Atherosclerosis* **1992,** *97*, 201–208.
[25] Ritz-Timme, S.; Laumeier, I.; Collins, M. J. Aspartic Acid Racemization: Evidence for Marked Longevity of Elastin in Human Skin. *Br. J. Dermatol.* **2003,** *149*, 951–959.
[26] Fisher, G. H.; Garcia, N. M.; Payan, I. L.; Cadilla-Perezrios, R.; Sheremata, W. A.; Man, E. H. D-aspartic Acid in Purified Myelin and Myelin Masic Protein. *Biochem. Biophys. Res. Commun.* **1986,** *135*, 683–687.
[27] Shapira, R.; Wilkinson, K. D.; Shapira, G. Racemization of Individual Aspartate Residues in Human Myelin Basic Protein. *J. Neurochem.* **1988,** *50*, 649–654.
[28] Fisher, G. H.; D'Aniello, A.; Vetere, A.; Cusano, G. P.; Chavez, M.; Petrucelli, L. Quantification of D-aspartate in Normal and Alzheimer Brains. *Neurosci. Lett.* **1992,** *143*, 215–218.
[29] Roher, A. E.; Lowenson, J. D.; Clarke, S.; Wolkow, C.; Wang, R.; Cotter, R. J., et al. Structural Alterations in the Peptide Backbone of Beta-Amyloid Core Protein May Account for its Deposition and Stability in Alzheimer's Disease. *J. Biol. Chem.* **1993,** *268*, 3072–3083.
[30] Fujii, N.; Tajima, S.; Tanaka, N.; Fujimoto, N.; Takata, T.; Shimo-Oka, T. The Presence of D-beta-aspartic Acid-containing Peptides in Elastic Fibers of Sun-Damaged Skin: A Potent Marker for Ultraviolet-Induced Skin Aging. *Biochem. Biophys. Res. Commun.* **2002,** *294*, 1047–1051.
[31] Mori, Y.; Aki, K.; Kuge, K.; Tajima, S.; Yamanaka, N.; Kaji, Y., et al. UV B-irradiation Enhances the Racemization and Isomerizaiton of Aspartyl Residues and Production of Nepsilon-Carboxymethyl Lysine (CML) in Keratin of Skin. *J. Chromatogr. B Analyt. Technol. Biomed. Life Sci.* **2011,** *879*, 3303–3309.
[32] Ingrosso, D.; Perna, A. F. D-amino Acids in Aging Erythrocytes. *EXS* **1998,** *85*, 119–141.
[33] Ueno, K.; Ueda, T.; Sakai, K.; Abe, Y.; Hamasaki, N.; Okamoto, M., et al. Evidence for a Novel Racemization Process of an Asparaginyl Residue in Mouse Lysozyme Under Physiological Conditions. *Cell. Mol. Life Sci.* **2005,** *62*, 199–205.
[34] Geiger, T.; Clarke, S. Deamidation, Isomerization, and Racemization at Asparaginyl and Aspartyl Residues in Peptides. Succinimide-Linked Reactions that Contribute to Protein Degradation. *J. Biol. Chem.* **1987,** *262*, 785–794.
[35] Fujii, N.; Harada, K.; Momose, Y.; Ishii, N.; Akaboshi, M. D-amino Acid Formation Induced by a Chiral Field within a Human Lens Protein During Aging. *Biochem. Biophys. Res. Commun.* **1999,** *263*, 322–326.
[36] Fujii, N.; Shimmyo, Y.; Sakai, M.; Sadakane, Y.; Nakamura, T.; Morimoto, Y., et al. Age-Related Changes of Alpha-crystallin Aggregate in Human Lens. *Amino Acids* **2007,** *32*, 87–94.
[37] Sakaue, H.; Takata, T.; Fujii, N.; Sasaki, H. Alpha B- and βA3-crystallins Containing D-aspartic Acids Exist in a Monomeric State. *Biochim. Biophys. Acta* **2015,** *1854*, 1–9.
[38] Takata, T.; Fujii, N. Isomerization of Asp Residues Plays an Important Role in αA-crystallin Dissociation. *FEBS J.* **2016,** *283*, 850–859.
[39] Maeda, H.; Takata, T.; Fujii, N.; Sakaue, H.; Nirasawa, S.; Takahashi, S., et al. Rapid Survey of Four Asp Isomers in Disease-Related Proteins by LC-MS Combined with Commercial Enzymes. *Anal. Chem.* **2015,** *87*, 561–568.
[40] Aki, K.; Fujii, N.; Fujii, N. Kinetics of Isomerization and Inversion of Aspartate 58 of αA-crystallin Peptide Mimics Under Physiological Conditions. *PLoS One* **2013,** *8*, e58515.

CHAPTER 7

Graph Theoretical Tools in Two- and Three-Dimensional Chirality Problems*

Ottilia Fülöp[1] and Béla Barabás[2]
[1]Institute of Mathematics, Department of Differential Equations, Budapest University of Technology and Economics, Budapest, Hungary
[2]Institute of Mathematics, Department of Stochastics, Budapest University of Technology and Economics, Budapest, Hungary

Abstract

Two-dimensional chirality problems in mathematics can be studied by so-called lattice animals or 2D animals. Generalizing this idea for 3D problems we consider in the three-dimensional space R^3 a Cartesian grid of the first octant consisting of n^3 small unit cubic cells and define there solid animals or 3D animals.

The simplest approach is to say that a 2D (or 3D) animal is chiral or achiral. One could assume that there is no other possibility. On the other hand a few scientists recognized that it is possible to measure the degree of chirality. For example F. Harary, R. W. Robinson, P. G. Mezey, A. I. Kitaigorodski, K. Mislow, J. Siegel, G. Gilat, D. Avnir, and A. Y. Meyer have made important contributions to this field. In this paper we provide a solution for quantification of the problem of chirality for 2D and 3D objects using graphs.

Keywords: Chirality; Jordan curve; 3D Jordan property; graph representations

7.1 INTRODUCTION

Chirality in the 2D and 3D spaces plays an important role in many fields of natural science. In mathematics the chirality of an object A in the ordinary 2D or 3D space (i.e., in R^2 or R^3) means that it cannot produce a perfect overlap with its mirror image $A^\diamond$ within the specified space. Otherwise A is said to be *achiral*.

An object A that is chiral if embedded in R^2 is achiral if embedded in R^3. For example, the left planar object from Fig. 7.1 is chiral in R^2, since no rotation or translation (in R^2) can result in a perfect overlap with

* Some results of the present paper were described in preliminary publications [1,2] and reported at Congress [3] and Symposium [4].

Advances in Asymmetric Autocatalysis and Related Topics
DOI: http://dx.doi.org/10.1016/B978-0-12-812824-4.00007-1

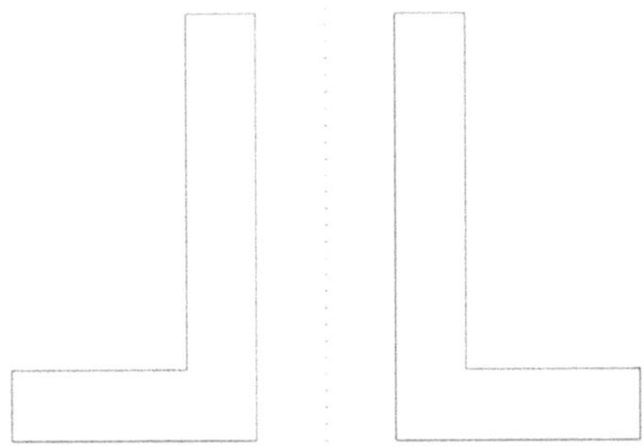

Figure 7.1 Planar chiral object and its mirror image (the dotted line is the axis of reflection).

its mirror image (right object from Fig. 7.1). In case we embed it in the 3D space R^3, it becomes achiral because it can be mapped to its mirror image by a 3D rotation about the dotted line of Fig. 7.1. Moreover, the dimension-dependency of chirality is generalized in Refs. [5] and [6]:

Theorem 1

An object A that is chiral if embedded in R^n is achiral if embedded in R^m where $m \geq n + 1$.

Thus chirality of any object may occur only in the lowest dimension where the specified object is embeddable.

Following the idea introduced and developed by the authors listed in the abstract we consider the 2D case discussed in our paper [1] and the 3D case [2–4]. We are going to show that there always exist special star graph representations for 2D and 3D chiral objects.

7.2 CHEMISTRY AND GRAPH THEORY: A BEAUTIFUL CONNECTION

... in writing organic chemical formulas and figuring out all constitutional isomers, chemists made unconscious use of graph theory. Realizing that a chemist needs to use consciously graph theory, I tried to become familiar with the "chemical versus graph-theoretical vocabulary".

(A. T. Balaban) [7]

The earliest known graph theoretical problem (the Problem of Seven Bridges of Königsberg) was solved and presented to the St. Petersburg Academy on August 26, 1735 and published 1 year later by Leonhard Euler [8]. The city of Königsberg (today Kaliningrad) was set on both sides of the Pregel River and included two islands that were connected to each other and the mainland by seven bridges. Euler proved that there is no walk through the city that crosses each bridge once and only once, i.e., there is no Eulerian path in the graph corresponding to historical Königsberg. Graph theory may be said to have its famous beginning here. Since then it has developed into an extensive branch of mathematics with many applications. Nowadays graphs are often used in chemistry, biochemistry, electrical engineering, social and informational systems, operational research, and combinatorial optimization, since a lot of practical problems can be represented by them. The first known chemical application of graphs was more than one century after Euler's work, and before the introduction of the term "graph": in 1857 Arthur Cayley published a paper [9] where he counted trees with certain properties. Later, for the intention of enumeration of isomers, Cayley simply considered "diagrams" where some of the atoms were linked, these are considered to be the first molecular graphs in the history of chemical graph theory. Using graph theoretical tools, any structural formula of a chemical compound has a molecular graph representation. The vertex set of this molecular graph is formed by the atoms of the compound and the edges represent the chemical bonds. Molecular graphs can distinguish between structural isomers, but they do not contain any information about the 3D arrangement of the bonds, and therefore cannot distinguish between stereoisomers [10]. In the case of chirality, instead of molecular graphs we have another application for graphs: in this paper we provide a quantification of chirality using graph representations of objects.

Other types of chemical graphs are the reaction graphs (finite directed graphs that describe the topological structures of the communicabilities of processes, with each of their directed edges labeled by either + or − [11]) or synthon graphs in which vertices symbolize molecules from multistep synthetic reactions, starting with a target molecule followed by retrosynthetic steps till simple starting materials are reached [7]. All these chemical graphs raise a lot of questions that stimulate graph theoretical research.

7.3 BASIC NOTIONS

Let us consider in the plane a square grid of the first quadrant consisting of n^2 small unit squares (we take the nth regular subdivision $[0, 1], [1, 2], \ldots, [n-1, n]$ of both coordinate axes and the Cartesian product $\{0, 1, 2, \ldots, n\} \times \{0, 1, 2, \ldots, n\}$ together with the corresponding grid edges). Let $G = (V, E)$ be the graph with node set V formed by the $(n+1)^2$ grid points and the edge set E containing all sides of n^2 squares (i.e., all grid edges).

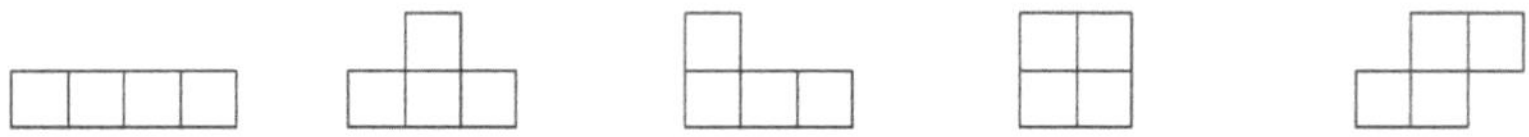

Figure 7.2 Tetrominoes in R^2.

Definition 1

A *Jordan cycle* C of the graph G is a connected subgraph of G having only two-degree nodes.

Definition 2

A subgraph A of G is called *animal* if it contains all the nodes and edges of that fall on a Jordan cycle C of G or within the interior of C.

Definition 3

A *cell* C of an animal A is a 4-cycle contained in A.

Note that the perimeter of an arbitrary animal A is a *Jordan curve*, i.e., a nonself-intersecting continuous closed curve. Let us denote this curve by $J(A)$. In this paper we deal only with Jordan curves resulting from taking the perimeter of a given animal A.

Definition 4

Animals with n cells are called *n-polyominoes* or *n-ominoes*.

For example, the 4-ominoes (tetrominoes) are the "straight", "T", "L", square and skew tetrominoes (see Fig. 7.2).

In case of any arbitrary fixed Jordan curve J of the square grid we determine a unique labeling of the vertices. Labels are ordered pairs of not necessarily different elements from $\{-1, 1, 0, \emptyset\}$ as follows. We label considering only grid vertices that belong to J. The first entry of the label of any vertex v indicates the horizontal grid edges (of curve J) that are incident to v as follows: two horizontal grid edges (of J) incident to v equals 0, a single grid edge to the right equals 1, and a single edge from the left equals -1. If there is no horizontal edge (of J) incident to v, then the label is $\emptyset$. Similarly, the second entry of the label informs us about the vertical grid edges (of curve J) incident to v: two vertical grid edges (of J) incident to v is 0, no incident vertical edge is $\emptyset$, one edge upwards is 1, and a single downwards grid edge is -1.

For example the sequence of labels of vertices from the perimeter of the first tetromino of Fig. 7.2 starting with the closest vertex to the origin and moving counterclockwise along the curve is (1, 1), (0, $\emptyset$), (0, $\emptyset$),

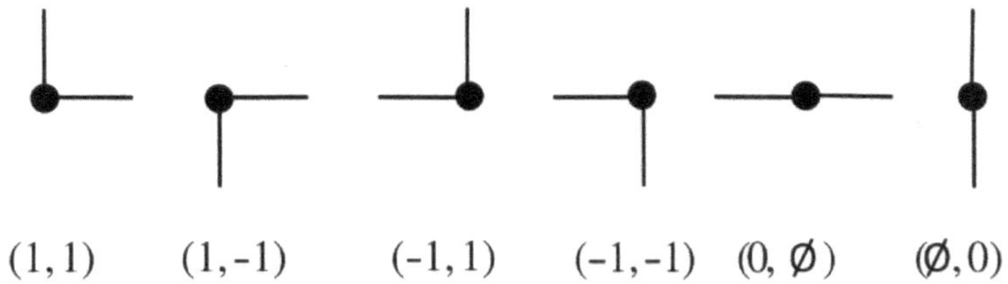

Figure 7.3 Possibilities of labeling a vertex of *J*.

(0, Ø), (−1, 1), (−1, −1), (0, Ø), (0, Ø), (0, Ø), (1, −1). The labels of the vertices from the perimeter of the fifth tetromino of Fig. 7.2 are (1, 1), (0, Ø), (−1, 1), (1, −1), (−1, 1), (−1, −1), (0, Ø), (1, −1), (−1, 1), (1, −1). Observe that any Jordan curve *J* is closed, thus the sum of all first (or all second) entries of labels along *J* is 0 and in the case of any vertex of *J*, if an entry of the label is 0, then the other is Ø (Fig. 7.3).

Now if we consider the 3D space, a Cartesian grid of the first octant consisting of n^3 small unit cubes (*cells*), the graph $G = (V, E)$ here similarly has a vertex set V formed by the $(n+1)^3$ grid points and the edge set E contains all grid edges.

Definition 5
Two grid faces (belonging to the same cell or to different cells) are called *adjacent* if they have exactly one common edge.

Definition 6
Two cells are called *adjacent* if they have exactly one common face.

Definition 7
A union of (finite number of) adjacent cells is called a 3D *solid animal* if it is connected and it does not contain a "hole." In this paper in R^3 we deal with solid animals only. Thus, we will omit the attribute "solid" further on.

Definition 8
Given an arbitrary *3D* animal *A*, we say that he *boundary* of *A* is the union all faces that belong to exactly one cell of *A*.

The boundaries of these animals are nonself-intersecting, continuous, simple, closed surfaces in the *3D* space. They divide the space into an "interior" region and an "exterior" region. This property is often called a *3D Jordan property*.

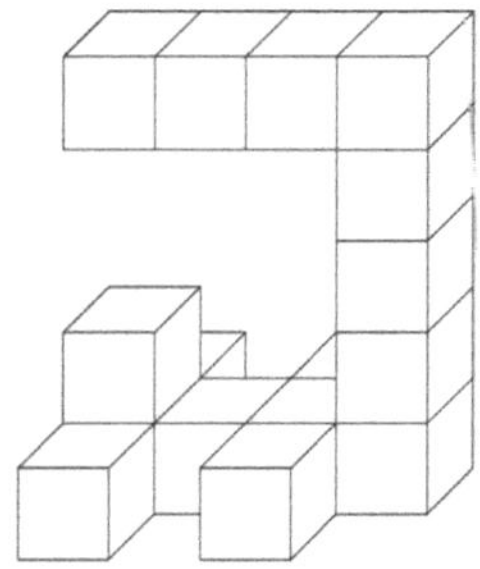

Figure 7.4 Example of a 3D animal *A*.

Definition 9
The *interior* of A is defined as the interior region determined by the boundary B of A.

An example of a 3D animal is shown in Fig. 7.4.

Here we label the grid edges e of the boundary of the *3D* animal A by ordered pairs as follows. If e is parallel to x-axis then the first entry of the label is 0 in the case of two horizontal faces (of boundary of A) incident to e, 1 in the case of a single face (of boundary of A) incident to e from the right and -1 if the incident face is from the left, Ø in the case of no incident horizontal face of boundary of A. The second entry of the label indicates the existence of two incident vertical faces of boundary A (0), only one incident vertical face upwards (1), downwards (-1) or no incident vertical face of the boundary (Ø). Note that in the case of grid edges e parallel to y-axis the positive direction of coordinate axes x and z determine the sign of the entries of labels, in case of edges parallel to z-axis signs are determined by the positive direction of axes x and y. The 3D Jordan property of the boundary of the 3D animal A implies that if an entry of the label of any edge of the boundary is 0 then the other one is Ø.

7.4 2D RESULTS

If A is an arbitrary 2D animal [1,5,6,12] with perimeter $J(A)$, we denote shortly the Jordan curve $J(A)$ by J and its mirror image by $J^{\diamond}$. One can easily observe that the convexity of $J \cup Int(J)$ implies achirality of A (we have only horizontal and vertical grid edges). The converse is not true, see the second tetromino of Fig. 7.2. Assume that J and $J^{\diamond}$ are positioned so that the intersection of their interior $Int(J)$ and $Int(J^{\diamond})$ has the maximum possible area, i.e., *area* $[Int(J) \cap Int(J^{\diamond})] = maximum$

(idea originating from Kitaigorodski [13]). Let us denote the union of J and $J^{\diamond}$, satisfying the condition of maximum intersection by $J \oplus J^{\diamond}$.

This object $J \oplus J^{\diamond}$ partitions the plane into $k+1$ disjoint subsets, namely $S_0, S_1, \ldots, S_k$, where S_0 is the unbounded exterior part of the plane lying on the outside of both Jordan curves J and $J^{\diamond}$. Let us consider $S_1 = Int(J) \cap Int(J^{\diamond})$ and for $i = 2, 3, \ldots, k$ let S_i be the maximum connected subset of the partition, which belongs to the interior of precisely one of J or $J^{\diamond}$ having no common points with any of $S_0, S_1, \ldots, S_{i-1}$. ($S_1$ may be the union of more disjoint components, and $S_2, S_3, \ldots, S_k$ are connected components.) If $J \oplus J^{\diamond}$ is not unique, one with the smallest k is chosen ("minimum k condition").

In Ref. [6] it is shown that the conditions of maximum intersection and minimum k condition for $J \oplus J^{\diamond}$ are essential.

The integer $k-1$ denoted by $gtk(J)$ in Ref. [6] is an important attribute of chirality of Jordan curve J. In the above-mentioned paper the author defines the graph representation $g(J \oplus J^{\diamond})$ of a Jordan curve considering adjacency of nodes i and j of a graph with node set $\{1, 2, \ldots, k\}$ in the case when the corresponding subsets S_i and S_j of the partition are separated by a line segment (i.e., sequence of grid edges) of $J \oplus J^{\diamond}$ of positive length.

In this paper the definition of adjacency of nodes (subsets) is more strict.

Definition 10

Nodes i and j are *adjacent* if the corresponding subsets S_i and S_j of the partition are separated by a *simple* line segment of $J \oplus J^{\diamond}$ of positive length (so "parallel" grid edges from $J \cap J^{\diamond}$ are excluded). In this case partition sets S_i and S_j are also called *adjacent*.

As an example let us consider the chiral Jordan curve J and its mirror image $J^{\diamond}$ from Fig. 7.5.

Notice that in this paper we consider graph representation of a Jordan curve $J(A)$ based on adjacency from Definition 10. Assume that J and $J^{\diamond}$ from Fig. 7.5 are positioned so that the intersection of their interiors $Int(J) \cap Int(J^{\diamond})$ has the maximum possible area.

$J \oplus J^{\diamond}$ in this case is not unique because both objects from Fig. 7.6 satisfy condition of maximum intersection and the minimum k condition.

The graph representation of J (defined using adjacency from Definition 10) denoted by $g(J \oplus J^{\diamond})$ in both cases is a "cherry graph" as shown in Fig. 7.7. (Here we follow the notations of Ref. [6].)

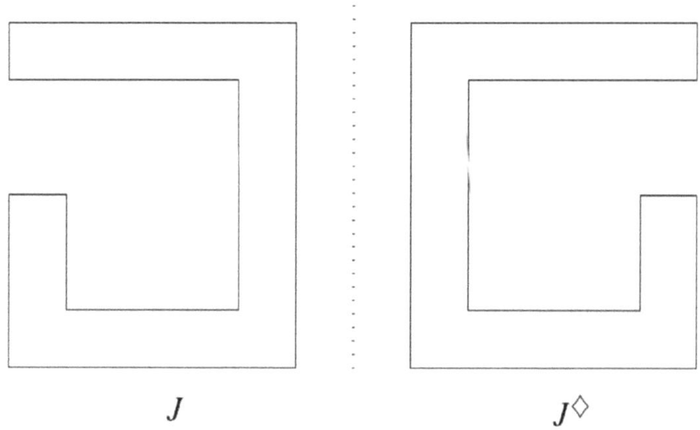

Figure 7.5 J and its mirror image $J^\diamond$.

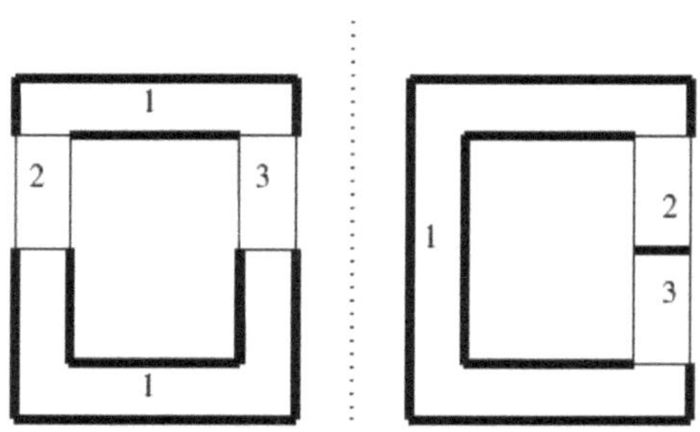

Figure 7.6 On the right figure J is rotated by 180 degrees, on the left one S_1 is not connected.

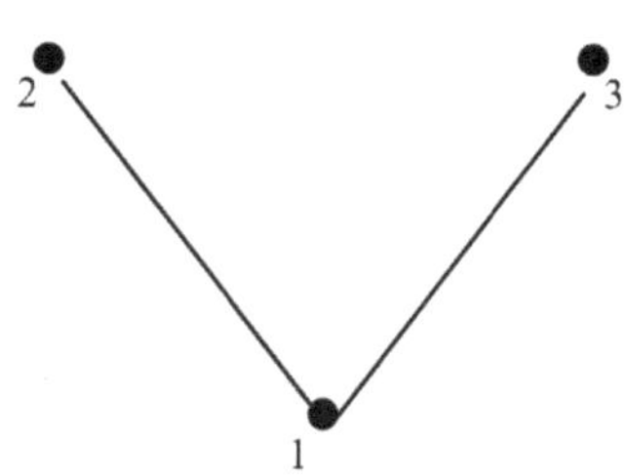

Figure 7.7 Graph representation of J from Fig. 7.5.

Nodes 2 and 3 are not adjacent for the second object of Fig. 7.6, since S_2 and S_3 are separated by a "parallel" line segment (grid edges from $J \cap J^\diamond$ are denoted by bold lines).

For simplicity, let us call the grid edges of Jordan curve J red ones and the edges from $J^\diamond$ blue ones. The other edges of the square grid are considered uncolored. Hence each vertex has a blue label and a red one. If a vertex v belongs to the intersection $J \cap J^{\diamond}$ (i.e., is an endpoint of a double grid edge), then it has two labels: a red one, determined by the edges of J that are incident to v and a blue one considering incident edges of $J^{\diamond}$ to vertex v. If $v \notin J \cup J^{\diamond}$ then both colored labels are $(\emptyset, \emptyset)$.

Lemma 1

For any $i \in \{1, 2, \ldots, k\}$ $Int(S_i)$ does not contain colored edges.

Proof

If $i = 1$, $S_1 = Int(J) \cap Int(J^{\diamond})$. $Int(J)$ doesn't contain red edges and $Int(J^{\diamond})$ doesn't contain blue grid edges because J and $J^{\diamond}$ are Jordan curves. Thus S_1 can contain only uncolored grid edges.

Let us suppose indirectly that there exists $i \in \{2, \ldots, k\}$ such that $Int(S_i)$ contains a colored edge e. If e is red then $S_i \subseteq Int(J^{\diamond})$. J is a Jordan curve that crosses S_i, thus $F = Int(J) \cap Int(S_i) \neq \emptyset$.

We have $\emptyset \neq F \subseteq Int(J) \cap Int(J^{\diamond})$, that is a contradiction with the condition of maximum intersection. If e is a blue edge (i.e., belongs to the Jordan curve $J^{\diamond}$) we get similarly a contradiction with the condition of maximum intersection.

Corollary 1

Any sequence of consecutive simple grid edges (simple line segment) that separates two adjacent partition sets from $\{S_i: i \in \{0, \ldots, k\}\}$ is monochromatic. (Contains only red edges or only blue ones.)

Proof

$J \oplus J^{\diamond}$ partitions the plane into sets $S_0, S_1, \ldots, S_k$, thus the simple line segment that separates two arbitrary adjacent partition sets S_i and S_j cannot contain uncolored grid edges. If we suppose indirectly that this line segment contains both red and blue grid edges (i.e., edges from both Jordan curves J and $J^{\diamond}$) let us consider two different colored grid edges with a common endpoint P.

S_i and S_j are adjacent, J and $J^{\diamond}$ are Jordan curves, thus the red label of P can be only $(1, 1)$, $(1, -1)$, $(-1, 1)$, $(-1, -1)$, contradiction with Lemma 1.

Corollary 2

Any two adjacent partition sets S_i and S_j different from S_0 and S_1 belong to the interior of different Jordan curves (one of them belongs to $Int(J) \setminus Int(J^{\diamond})$ and the other to $Int(J^{\diamond}) \setminus Int(J)$).

Proof

Suppose indirectly that there exist adjacent partition sets S_i and S_j different from S_0 and S_1 that belong to $Int(J)\backslash Int(J^{\diamond})$. (If they belong to $Int(J^{\diamond})\backslash Int(J)$ the proof is similar.) S_i and S_j are separated by a blue line segment (because J is a Jordan curve). Then $J^{\diamond}$ crosses J such that either S_i or S_j belongs to $Int(J) \cap Int(J^{\diamond})$, that is a contradiction with the condition of maximum intersection.

Theorem 2

If S_i and S_j $(i, j \in \{1, 2, \ldots, k\})$ are adjacent, then one of them is S_1.

Proof

This theorem says that two partition sets S_i and S_j $(i \neq j)$ different from S_0 and S_1 cannot be adjacent.

Suppose indirectly that there exist $i \neq j$ $(i, j \in \{2, 3, \ldots, k\})$ such that S_i and S_j are adjacent. Assume that $S_i \in Int(J)\backslash Int(J^{\diamond})$. Corollary 2 implies that $S_j \in Int(J^{\diamond})\backslash Int(J)$.

Corollary 1 guarantees that the simple line segment that separates S_i and S_j is monochromatic. Let us consider an endpoint E of this monochromatic separating line segment. J and $J^{\diamond}$ are Jordan curves thus all grid edges incident to E are colored (red or blue). Any coloration of them leads us to a contradiction with $S_i \in Int(J)$ or with the condition of maximum intersection.

Theorem 3

For each $i \in \{2, \ldots, k\}$ S_i and S_1 are adjacent partition sets.

Proof

Suppose indirectly that there exists an $i \in \{2, \ldots, k\}$ such that S_1 and S_i are not adjacent sets. Theorem 2 implies that there is no adjacent partition set for S_i.

J and $J^{\diamond}$ are Jordan curves, thus Lemma 1 implies that parallel line segments separates S_i from the other partition sets situated in its neighborhood, that is a superposition of J with its mirror image $J^{\diamond}$, contradiction with the fact that J is chiral.

We can summarize our results for $g(J \oplus J^{\diamond})$ in the following theorem:

Theorem 4

The graph representation $g(J \oplus J^{\diamond})$ of a chiral Jordan curve J is a star graph, i.e., it contains a node that is adjacent to all other nodes and there are no other adjacent nodes in the graph.

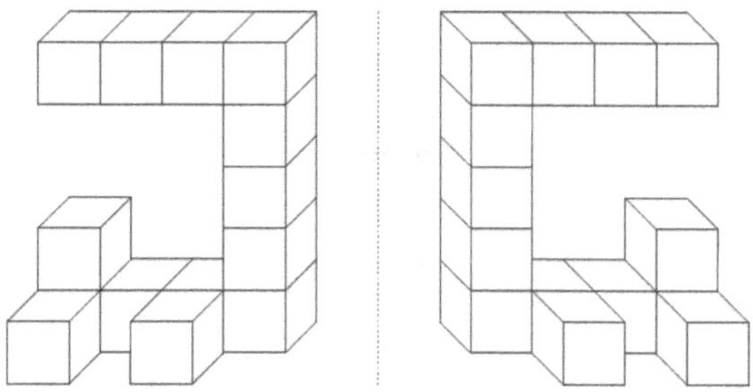

Figure 7.8 *A* and its mirror image $A^{\diamond}$.

7.5 CHIRALITY IN R^3

In this section we consider the scenario when A is an arbitrary 3D animal with boundary B and its mirror image $A^{\diamond}$ with boundary $B^{\diamond}$ (see Fig. 7.8 where the dotted line represents the mirror-plane).

The 3D case follows a natural analogy with the 2D solution, so assume that A and $A^{\diamond}$ are positioned in the three-dimensional space so that the intersection of their interior $Int(A) \cap Int(A^{\diamond})$ has maximum possible volume, i.e.,

$$\text{Volume}\,[Int(A) \cap Int(A^{\diamond})] = \text{maximum}$$

We denote the union of B and $B^{\diamond}$ satisfying the condition of maximum intersection by $B \oplus B^{\diamond}$. Note that $B \oplus B^{\diamond}$ contains only the faces of the boundaries B and $B^{\diamond}$. Grid faces of $B \cap B^{\diamond}$ are considered *double faces* and in our figures double grid edges (of $B \cap B^{\diamond}$) are denoted by bold lines.

If we consider the condition of maximum intersection we obtain $B \oplus B^{\diamond}$, as shown in Fig. 7.9.

Object $B \oplus B^{\diamond}$ partitions $R^3 \backslash (B \oplus B^{\diamond})$ into $k+1$ disjoint subsets, namely $S_0, S_1, \ldots, S_k$ where S_0 is the unbounded exterior part of the space lying outside of both 3D Jordan-property surfaces B and $B^{\diamond}$. Let us keep the notations of Refs. [2,4] and [6], considering $S_1 = Int(A) \cap Int(A^{\diamond})$ and for $i = 2, 3, \ldots, k$ let S_i be the maximum connected subset of the partition, which belongs to the interior of precisely one of A or $A^{\diamond}$ having no common points with any of $S_0, S_1, \ldots, S_{i-1}$. Observe that S_1 may be the union of more disjoint components, but $S_2, S_3, \ldots, S_k$ are all connected components.

If $B \oplus B^{\diamond}$ is not unique, one with the smallest k must be chosen ("minimum k condition").

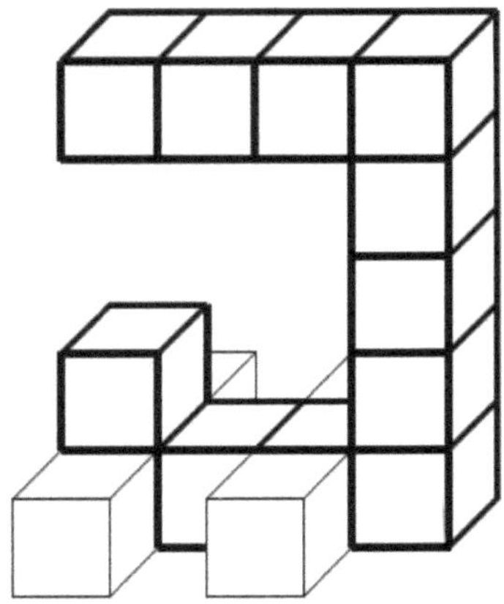

Figure 7.9 $B \oplus B^{\diamond}$ of our example.

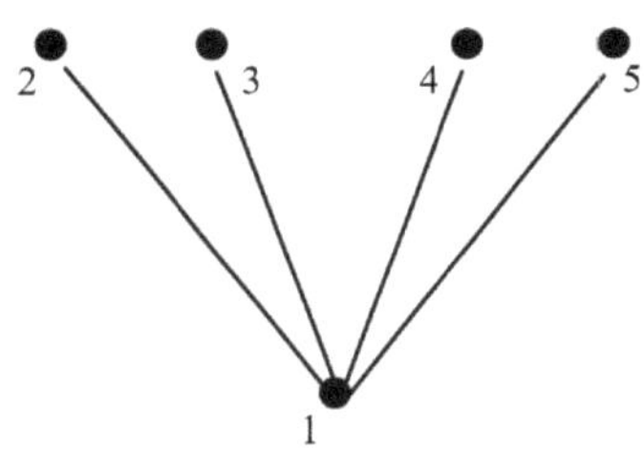

Figure 7.10 The graph associated to *A* in our example.

Now we are going to associate to the partition sets $S_1, S_2, \ldots, S_k$ a graph G with node set $V(G) = \{1, 2, \ldots, k\}$. Each node i of the graph G corresponds to a subset S_i, $i = 1, \ldots, k$. Adjacency of nodes in G will be defined as follows.

Definition 11

Nodes i and j are *adjacent* in the graph G if the corresponding subsets S_i and S_j of the partition are separated by a set of connected adjacent simple faces, i.e., double grid faces from $B \oplus B^{\diamond}$ are excluded. Such a separating (hole-free) union of connected adjacent simple grid faces is called *wall* further on.

We can associate a graph to A as shown in Fig. 7.10.

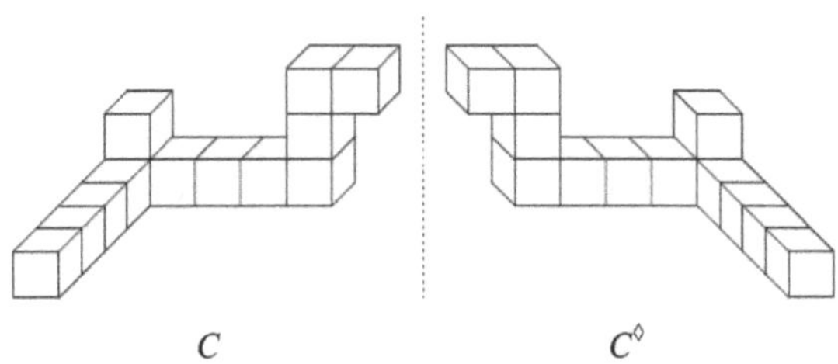

Figure 7.11 A chiral object C and its mirror image $C^{\diamond}$.

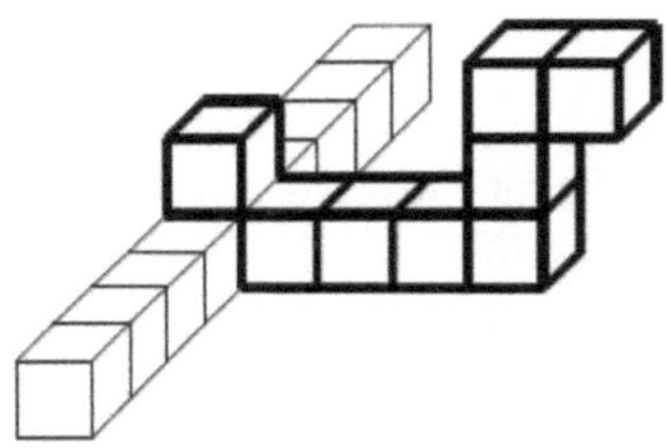

Figure 7.12 Case 1: Maximum intersection of C and $C^{\diamond}$ when $k = 3$.

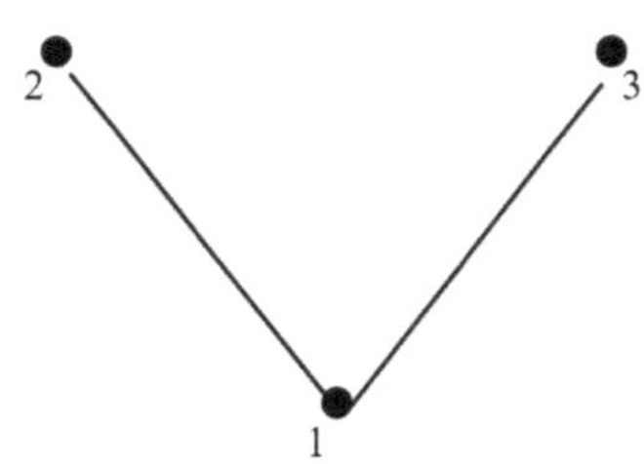

Figure 7.13 Case 1: Graph representation of C.

7.5.1 Minimum *k* Condition

We are going to illustrate the importance of "minimum k condition" by the following example. Consider now the animal C shown in Fig. 7.11.

This chiral object C from Fig. 7.11 consists of 13 unit cubes. One can reach the maximum intersection with its mirror image in two different ways. There are nine common cubes in both cases. Fig. 7.12 shows one possibility. In this case $k = 3$.

The graph associated to the object considered in Case 1 is shown in Fig. 7.13.

Another way to get the maximum intersection is shown in Fig. 7.14.

Notice that Case 2 from Fig. 7.14 cannot be considered because of the minimum k condition.

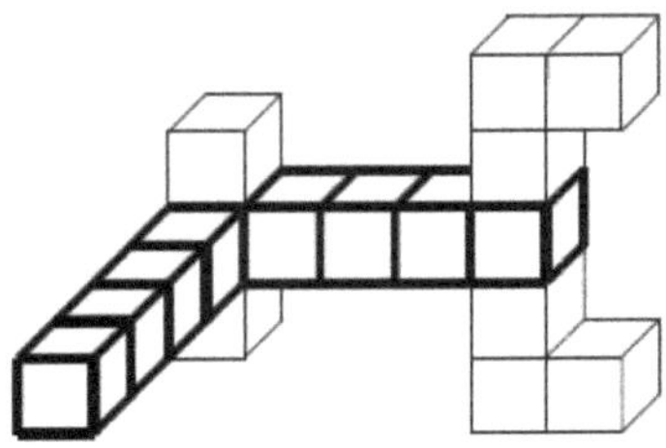

Figure 7.14 Case 2: Maximum intersection of C and $C^{\diamond}$ when $k = 5$.

7.5.2 3D Theorems

Let us consider now an arbitrary animal A in the 3D space with boundary B and its mirror image $A^{\diamond}$ with boundary $B^{\diamond}$.

Imagine we color the grid faces of B red and those of $B^{\diamond}$ blue.

All other faces of the grid are considered uncolored.

Lemma 2

Neither A nor $A^{\diamond}$ crosses any S_i for $i = 1, 2, \ldots, k$.

Proof

Neither A nor $A^{\diamond}$ crosses S_1, because $S_1 = Int(A) \cap Int(A^{\diamond})$, so

Lemma 2 holds for $i = 1$. In the case of $i \in \{2, 3, \ldots, k\}$ suppose indirectly that A or $A^{\diamond}$ crosses S_i. By definition S_i belongs to the interior of precisely one of A or $A^{\diamond}$. If the other one crosses S_i then we obtain a contradiction with the condition of maximum intersection.

Lemma 2 implies that $Int(S_i)$ can contain only "uncolored" grid faces for $i = 1, 2, \ldots, k$. Boundaries B and $B^{\diamond}$ are surfaces with 3D Jordan properties, so Lemma 2 also implies that two adjacent partition sets S_i and S_j can be separated only by a monochromatic wall.

Theorem 5

If S_i and S_j are adjacent, then either $i = 1$ or $j = 1$.

Proof

Suppose indirectly that there exist i and j, ($i \neq j \in \{2, 3, \ldots, k\}$) such that S_i and S_j are adjacent partition sets. One of them (let us suppose it to be S_i) belongs to $Int(A) \backslash Int(A^{\diamond})$ and the other (S_j) to $Int(A^{\diamond}) \backslash Int(A)$. (Otherwise we have a monochromatic wall in $Int(A) \backslash Int(A^{\diamond})$ or $Int(A^{\diamond}) \backslash Int(A)$, which is a contradiction of the condition of maximum intersection.)

The simple wall that separates S_i and S_j is monochromatic. Jordan property of B and $B^{\diamond}$ implies that any coloration of this wall leads us to a contradiction with $S_i \in Int(A)$ or with the condition of maximum intersection.

Theorem 6
For each $i \in \{2, \ldots, k\}$ S_i and S_1 are adjacent partition sets.

Proof
Suppose indirectly that there exists an $i \in \{2, \ldots, k\}$ such that S_i and S_1 are not adjacent sets. Theorem 5 implies that there is no adjacent partition set for S_i.

B and $B^{\diamond}$ have 3D Jordan property, thus Lemma 2 implies that a wall of double faces separates S_i from the other partition sets situated in its neighborhood, which is a superposition of B with its mirror image $B^{\diamond}$, so we obtain a contradiction with the fact that B is chiral.

We can summarize our results for the graph representation of A in the following theorem:

Theorem 7
The graph representation of a chiral surface with 3D Jordan property is a star graph, i.e., it contains a node that is adjacent to all other nodes and there are no other adjacent nodes in the graph.

7.6 CONCLUSION

In this paper we show that a chiral object in the 2D space can be represented with a star graph. The number of nodes of the graph measures the degree of chirality of an object. This idea can be extended to the 3D objects, i.e., a 3D chiral object can be represented by a star graph, which provides solution for the quantification problem of chirality in R^3.

REFERENCES

[1] Barabás, B.; Fülöp, O. Star Graph Representations of Chiral Objects in Graph Theory. *J. Math. Chem.* **2012,** *50*, 1514–1520.
[2] Fülöp, O.; Barabás, B. Three-Dimensional Chiral Objects and Their Star Graph Representations. *J. Math. Chem.* **2013,** *51*, 2354–2360.

[3] Fülöp, O.; Barabás, B. Graph Representations of Three-Dimensional Chiral Objects. Lecture Notes in Engineering and Computer Science. In *Proceedings of The World Congress on Engineering*, London, UK, July 3–5, 2013; pp 101–102.
[4] Fülöp, O.; Barabás, B. Two- and Three-Dimensional Chiral Objects and Their Graph Representations. In *Third International Symposium on the Soai Reaction and Related Topic*, Felsőmocsolád, Hungary, September 2–5, 2015, Abstracts; pp 10–11.
[5] Mezey, P. G. Similarity Analysis in Two and Three Dimensions Using Lattice Animals and Polycubes. *J. Math. Chem.* **1992,** *11*, 27–45.
[6] Mezey, P. G. Chirality Measure and Graph Representations. *Comput. Math. Appl.* **1997,** *34* (11), 105–112.
[7] Balaban, A. T. Chemical Graph Theory and the Sherlock Holmes Principle. *HYLE* **2013,** *19* (1), 107–134.
[8] Euler, L. Solutio Problematis ad Geometriam Situs Pertinentis. *Comment. Acad. Sci. Imp. Petrop.* **1736,** *8*, 128–140.
[9] Cayley, A. On the Theory of the Analytical Forms Called Trees. *Philos. Mag.* **1857,** *4* (13), 172–176.
[10] https://en.wikipedia.org/wiki/Molecular_graph.
[11] Fu, Y. Reaction Graph. *J. Comput. Sci. Technol.* **1998,** *3* (13), 510–530.
[12] Mezey, P. G. *Shape in Chemistry: An Introduction to Molecular Shape and Topology;* VCH Publishers: New York, 1993.
[13] Kitaigorodski, A. I. Chapter 4. Application of Close - Packing Theory to Organic Crystals. In *Organic Chemical Crystallography;* Consultants Bureau: New York, 1961; pp 75–100.

FURTHER READING

Avnir, D.; Meyer, A. Y. Quantifying the Degree of Molecular Shape Distortion: A Chirality Measure. *J. Mol. Struct. (Theochem)* **1991,** *226*, 211–222.
Cahn, R. S.; Ingold, C. K.; Prelog, V. Specification of Molecular Chirality. *Angew. Chem., Int. Ed. Engl.* **1966,** *5*, 385–415.
Gilat, G. On the Quantification of Asymetry in Nature. *Found. Phys. Lett.* **1990,** *3*, 189–196.
Gilat, G. On Quantifying Chirality: Obstacles and Problems Towards Unification. *J. Math. Chem.* **1994,** *15*, 197–205.
Harary, F.; Lewinter, M. Spanning Subgraphs of a Hypercube. *Int. J. Compu. Math.* **1988,** *25* (1), 1–4.
Harary, F.; Mezey, P. G. Chiral and Achiral Square-Cell Configurations and the Degree of Chirality. In *New Develoments in Molecular Chirality;* Kluver Academic Publisher: Dordrecht, 1991; pp 241–256.
Harary, F.; Robinson, R. W. The Number of Achiral Trees. *J. Reine Angew. Math.* **1975,** *278*, 322–332.
http://www.thalidomide.ca/.
Mislow, K.; Siegel, J. Stereoisomerism and Local Chirality. *J. Am. Chem. Soc.* **1984,** *106*, 3319–3328.

CHAPTER 8

Synthesis of Pyrimidine-Terminated Chiral Large Molecular Architectures with Functions of Self-Replication and Self-Improvement by Asymmetric Autocatalysis

Tsuneomi Kawasaki[1,2], Arimasa Matsumoto[2,3], Itaru Sato[4] and Kenso Soai[2,3]
[1]Department of Materials Science and Engineering, University of Fukui, Bunkyo, Fukui, Japan
[2]Research Center for Chirality, Research Institute for Science and Technology (RIST), Tokyo University of Science, Kagurazaka, Shinjuku-ku, Tokyo, Japan
[3]Department of Applied Chemistry, Tokyo University of Science, Kagurazaka, Shinjuku-ku, Tokyo, Japan
[4]Graduate School of Science and Engineering, Ibaraki University, Bunkyo, Ibaraki, Japan

Abstract

The synthesis of huge chiral molecular architectures with functions of self-replication and self-improvement is one of the great topics of research in synthetic, systems and prebiotic chemistry. Here we describe a new chemical system in which large chiral multifunctionalized molecules achieve asymmetric autocatalytic multiplication and amplification of their enantiomeric and diastereomeric purities. The large chiral multifunctionalized molecules, that is, hexakis(5-pyrimidyl alkanol)hexaalkylsilane, replicatively catalyze the production of themselves with the same structure, including information on the chirality at newly formed asymmetric carbon atoms, in the reaction of corresponding achiral aldehydes and reagents. The chirality of the large molecules efficiently controlled the enantioselectivity of the reaction to construct multiasymmetric stereogenic centers in a single reaction step.

Keywords: Asymmetric catalytic amplification; multiasymmetric stereogenic centers; multifunctional chiral molecules; self-improvement; self-replication

8.1 INTRODUCTION

The origin and amplification of biologically related organic compounds such as L-amino acids and D-sugars have attracted much attention [1–7]; therefore self-replication and self-improvement of huge chiral molecular

Advances in Asymmetric Autocatalysis and Related Topics
DOI: http://dx.doi.org/10.1016/B978-0-12-812824-4.00008-3

architectures are among the significant challenges of great interest in synthetic, systems and prebiotic chemistry [8–11]. Self-replications of oligonucleotides [12–16], oligopeptides [17,18] and complementary artificial molecules [19,20] in synthetic chemical systems have been developed to date [21]. In these systems, replicating compounds were formed through a nonenzymatic template-directed synthesis; thus, these chemical approaches are considered good models for prebiotic evolution of replicating molecules. Therefore it has been considered that such research would be closely related to the origin of life.

On the other hand, the chirality in the replicated products in previously reported chemical systems is based upon the preexisting chirality of each building block. To our knowledge, there has been no report of an efficient chemical process in which large chiral molecules are autocatalytically formed from completely achiral substrates and reactants. We describe here the synthesis of large chiral molecules with autocatalytic self-replicating ability; and in addition, evaluate their function of self-improvement, that is, amplification of their own enantiomeric and diastereomeric imbalances (Fig. 8.1) [22].

Previously, we have reported the asymmetric autocatalysis [23–29] of 5-pyrimidyl alkanols [30] with a significant amplification of the enantiomeric excess (*ee*) [31,32], which enables the amplification of *ee* from extremely low to near enantiopure values (>99.5% *ee*) through consecutive reactions (Fig. 8.2) [33–39]. When diisopropylzinc (i-Pr_2Zn) is added to pyrimidine-5-carbaldehyde with an ethynyl substituent at the 2-position [40–42] in the presence of a catalytic amount of (*S*)-5-pyrimidyl alkanol with a low *ee* (*ca.* 0.00005% *ee*), asymmetric autocatalytic amplification of the *ee* produces (*S*)-5-pyrimidyl alkanol with a high *ee* as the final product

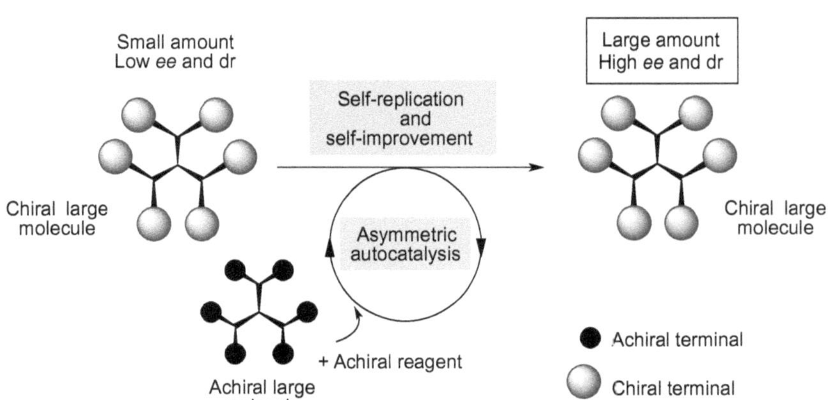

Figure 8.1 The concept of this work.

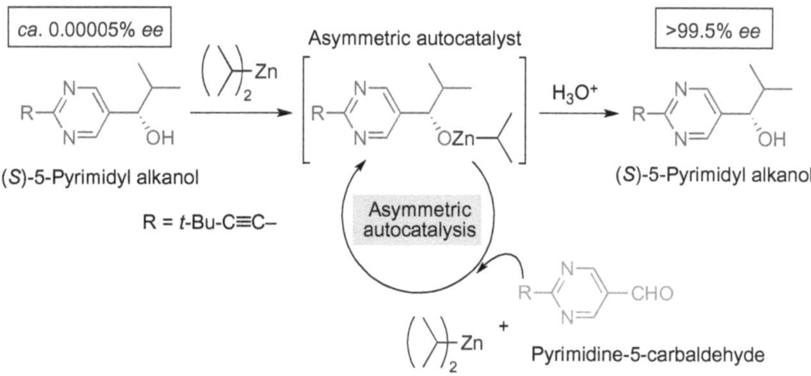

Figure 8.2 Asymmetric autocatalysis of 5-pyrimidyl alkanol with amplification of enantiomeric excess.

(>99.5% *ee*). Without adding any chiral substances, enantiomerically enriched alkanol can be obtained in conjunction with the amplification of a stochastic imbalance of enantiomer [43–49]; therefore a spontaneous absolute asymmetric synthesis has been achieved [50]. In addition, chiral substances [51–56], such as chiral compounds arising from carbon isotope substitution [57], and a chiral physical factor [58], such as circularly polarized light can trigger this asymmetric autocatalytic cycle to afford the enantioenriched alkanol with corresponding absolute configurations. Therefore, asymmetric autocatalysis with amplification of *ee* can connect the proposed origins of chirality with almost enantiomerically pure organic compounds and has been considered one of the reaction models for generation and evolution of biological homochirality [59].

8.2 SELF-REPLICATION OF HEXAKIS(5-PYRIMIDYL ALKANOL) HEXAALKYLSILANE WITH SELF-IMPROVEMENT OF ITS ENANTIOMERIC AND DIASTEREOMERIC PURITY

We have found here that chiral macromolecules possessing a multifunctionalized structure with as many as six asymmetric carbon centers automultiply in the manner of asymmetric autocatalysis [22]. Even when large molecules with low diastereomeric and enantiomeric purities were used as the initial catalyst, the consecutive asymmetric autocatalysis multiplied the number of large molecules and amplified the isomeric ratio. The enantioenrichment based upon the homoisomer of the newly formed large molecules achieved >99.5% *ee*.

The large chiral molecules have been designed based upon 5-pyrimidyl alkanols, which act as highly efficient asymmetric autocatalysts in the addition reaction of i-Pr_2Zn to the corresponding pyrimidine-5-carbaldehyde. The large molecule is formed from a branched alkylsilane backbone and the periphery of the structure has a pyrimidine moiety with an asymmetric carbon atom. In the initial replication system, we have achieved the asymmetric autocatalysis of hexakis(5-pyrimidyl alkanol)hexaalkylsilane. Such types of chiral molecules catalyze the production of molecules with the same structure, including the chirality of newly formed asymmetric carbon atoms, in the reaction of corresponding achiral hexaaldehyde and achiral i-Pr_2Zn. It is noteworthy that the ratio of the diastereomers including enantiomers continuously improved, finally forming a nearly enantiopure homochiral product; thus, the present large molecules have the ability to both self-replicate and self-improve.

8.2.1 Synthesis of Pyrimidine-Terminated Hexakis(alkanol) hexaalkyl-silane 1 and Hexakis(carbaldehyde)hexaalkylsilane 2

The synthetic strategy of hexakis(2-ethynyl-5-pyrimidyl alkanol)hexaalkylsilane **1** and corresponding hexaaldehyde **2** is the coupling reaction between pyrimidines and an alkylsilane backbone with terminal acetylene. Pyrimidine compounds such as aldehyde (equivalent) and alkanol were synthesized first (Fig. 8.3). Lithiation and formylation of 5-bromo-2-chloropyrimidine **3** by treatment with n-BuLi and ethyl formate afforded 2-chloropyrimidine-5-carbaldehyde in 53% yield, and its formyl group was protected as the diethyl acetal to give **4** in excellent yield (Fig. 8.3A). Enantioenriched (*S*)-2-chloro-5-pyrimidyl alkanol **5** was synthesized by the asymmetric addition of i-Pr_2Zn to 2-chloropyrimidine-5-carbaldehyde using (1*S*,2*R*)-MPP (2-morpholino-1-phenylpropan-1-ol) as a chiral ligand. After recrystallization, (*S*)-**5** with >99.5% *ee* was obtained. 5-Bromo-2-chloropyrimidine **3** could be directly converted to *rac*-**5** in 39% yield by treatment with n-BuLi followed by isobutylaldehyde.

The branched hexaalkylsilane core **9** with terminal acetylene was constructed starting from the reaction between methyl trichlorosilane and allylmagnesium bromide (Fig. 8.3B). The obtained triallylmethylsilane **6** was submitted to the hydrosilylation of a terminal olefin using platinum catalyst **7** to afford the terminal silylchloride, which was reacted with alkenyl lithium reagent to introduce a long alkyl chain, affording **8** with terminal olefin. A chlorodimethylsilyl group was introduced by

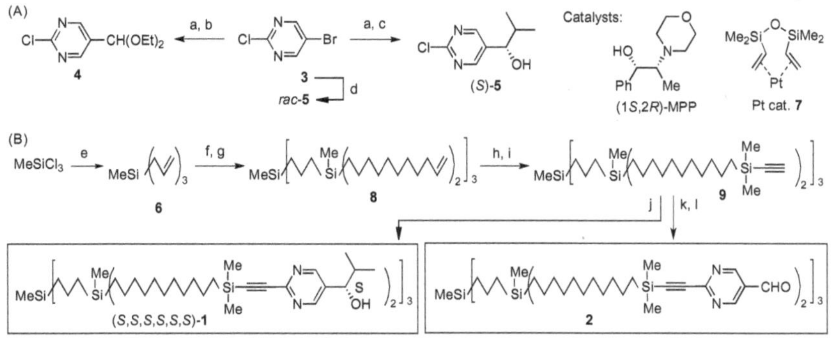

Figure 8.3 Synthesis of (A) terminal pyrimidines and (B) large molecules with the structure of pyrimidine-terminated hexaalkylsilane core. Reaction conditions: a, *n*-BuLi, HCO_2Et, THF (53%); b, $TsOH{\cdot}H_2O$, $CH(OEt)_3$, EtOH (97%); c, (1*S*,2*R*)-MPP, *i*-Pr_2Zn, toluene, 0°C (63%, 93% *ee*) then recrystallization (>99.5% *ee*); d, *n*-BuLi, Me_2CHCHO, THF, −100 to −78°C (39%); e, $CH_2=CHCH_2MgBr$, Et_2O, reflux (84%), f, Pt cat. **7**, $HSiMeCl_2$ (neat), reflux; g, Li, $Br(CH_2)_9CH=CH_2$, Et_2O-hexane, 0°C (two steps in 70%); h, Pt cat. **7**, $HSiMe_2Cl$ (neat), reflux; i, HC≡CMgBr, THF (two steps in 33%); j, (*S*)-**5**, $(Ph_3P)_4Pd$, CuI, TBAI (tetra-*n*-butylammonium iodide), *i*-Pr_2NH (51%). k, **4**, $(Ph_3P)_4Pd$, CuI, TBAI, *i*-Pr_2NH, 50–60°C; l, $TsOH{\cdot}H_2O$, THF-H_2O (two steps in 44%).

hydrosilylation of the terminal alkenyl group, and treatment of the resulting chlorosilane with ethynylmagnesium bromide afforded hexabranched alkynylsilane **9** with a terminal acetylene. Sonogashira coupling between synthesized **9** and highly enantioenriched (*S*)-**5** afforded the desired homochiral hexakis((*S*)-alkanol)hexaalkylsilane **1** in 51% yield in almost enantiopure form. When *rac*-**5** was used as the coupling partner, racemic hexamer **1** was obtained in almost the same chemical yield. The corresponding achiral hexakis(pyrimidine-5-aldehyde)hexaalkylsilane **2** could also be prepared by a similar synthetic method using diethyl acetal **4** as the monomeric coupling partner followed by acidic hydrolysis of the diethyl acetal in two steps (44% yield).

8.2.2 Chiral HPLC Analysis of Hexakis(5-pyrimidyl alkanol) hexaalkylsilane 1

Analysis of the isomers containing racemic hexakis(5-pyrimidyl alkanol) hexaalkylsilane **1** was conducted using HPLC employing a chiral stationary phase (Fig. 8.4). Six equivalents of asymmetric carbon atoms at the terminal pyrimidine produce seven diastereomeric and enantiomeric isomers **1**. The seven isomers of hexamer **1** are described by the number of each *S*- and *R*-asymmetric carbon atoms as follows: (*S*,*S*,*S*,*S*,*S*,*S*)-**1**

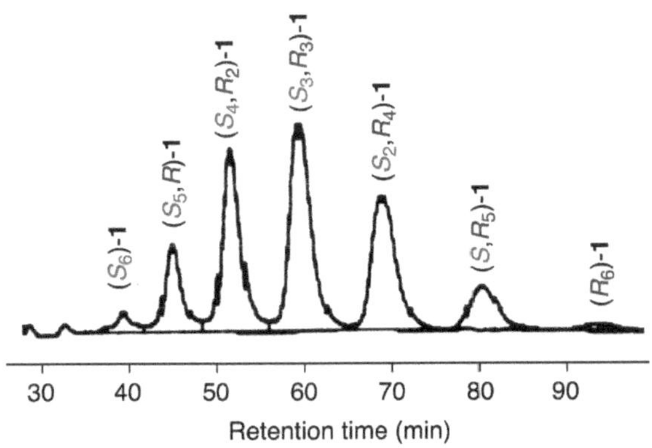

Figure 8.4 The chromatogram of racemic hexamer **1** by HPLC employing a chiral stationary phase. Conditions: Daicel Chiralpak IA, eluent THF/hexane = 23/77 (v/v), flow rate 1.0 mL/min, 254 nm UV detector.

(abbreviated as (S_6)-**1**), (*S*,*S*,*S*,*S*,*S*,*R*)-**1** (abbreviated as (S_5,*R*)-**1**), (*S*,*S*,*S*,*S*,*R*,*R*)-**1** (abbreviated as (S_4,R_2)-**1**), (*S*,*S*,*S*,*R*,*R*,*R*)-**1** (abbreviated as (S_3, R_3)-**1**), (*S*,*S*,*R*,*R*,*R*,*R*)-**1** (abbreviated as (S_2,R_4)-**1**), (*S*,*R*,*R*,*R*,*R*,*R*)-**1** (abbreviated as (*S*,R_5)-**1**) and (*R*,*R*,*R*,*R*,*R*,*R*)-**1** (abbreviated as (R_6)-**1**). The isomers of **1** could be separated using chiral HPLC.

The analyzed isomer ratio of the racemic mixture of hexamer **1** was 2.6/10.7/24.2/30.5/22.1/8.4/1.5 from chiral HPLC. This ratio shows an approximately theoretical binomial distribution; therefore *S* and *R* alkanols were introduced randomly to the periphery of the skeleton; that is, seven isomers of **1** were formed in the stochastic racemic ratio, therefore each signal could be assigned as shown in Fig. 8.4 and Table 8.1. The isomer with the shortest retention time is homochiral (S_6)-**1**, the next one is (S_5,*R*)-**1**, then (S_4,R_2)-**1**. The largest signal is assigned as achiral (S_3,R_3)-**1** and other signals also determined in this order including homochiral (R_6)-**1** with longest retention time.

Hexakis(5-pyrimidyl alkanol)hexaalkylsilane **1** has 10 stereoisomers, including the substitutional isomers of (S_4,R_2)/(S_2,R_4)-**1** and (S_3,R_3)-**1** (Table 8.1). Considering asymmetric silicon atoms, there are more stereoisomers; however, chiral HPLC can only recognize the isomers based on the number of asymmetric carbon centers of pyrimidine moieties. Thus, the ratio of seven isomers **1**, ignoring the isomers caused by the asymmetric silicon atoms and substitution pattern of (*S*)- and (*R*)-pyrimidyl alkanols on the branches, can be analyzed using HPLC on a chiral stationary phase. We evaluated the selectivity of the replicating reaction by analyzing the amount of these seven isomers **1**.

Table 8.1 Chiral HPLC analysis of *rac*-**1** and its substitutional isomers

Isomer	Retention time (min)	Area (%)		Number of substitutional isomer
		Analytical	Theoretical[a]	
(S_6)	39.9	2.6	1.6	1
(S_5,R)	45.6	10.7	9.4	1
(S_4,R_2)	52.2	24.2	23.4	2[b]
(S_3,R_3)	60.1	30.5	31.2	2[b]
(S_2,R_4)	69.4	22.1	23.4	2[b]
(S,R_5)	80.5	8.4	9.4	1
(R_6)	93.5	1.5	1.6	1
Total		100	100	10

[a]The values are based on the binomial distribution.
[b]The substitution pattern of the isomers are shown below.

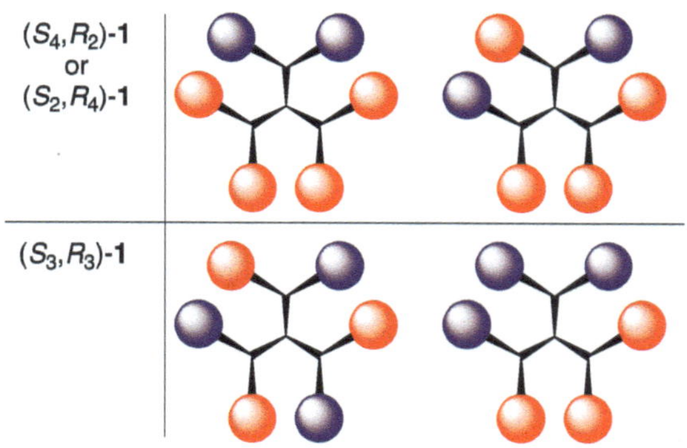

8.2.3 Asymmetric Autocatalysis of Chiral Hexakis(pyrimidyl alkanol)hexaalkylsilane 1 in the Addition of *i*-Pr_2Zn to the Corresponding Hexaaldehyde With Amplification of Isomer Ratio

We have found here that chiral macromolecules possessing a hexameric multifunctionalized structure with as many as six asymmetric carbon centers automultiply in the manner of asymmetric autocatalysis (Fig. 8.5). Even when large molecules with low diastereomeric and enantiomeric purities were used as the initial catalyst, five consecutive asymmetric autocatalyses multiplied the amount of large molecules (homoisomer: by a factor of *ca.* 92,000 times) and amplified the diastereomeric and enantiomeric ratio; thus homoisomer (S_6)-**1** of the newly formed chiral molecules was achieved as >99.5% *ee*.

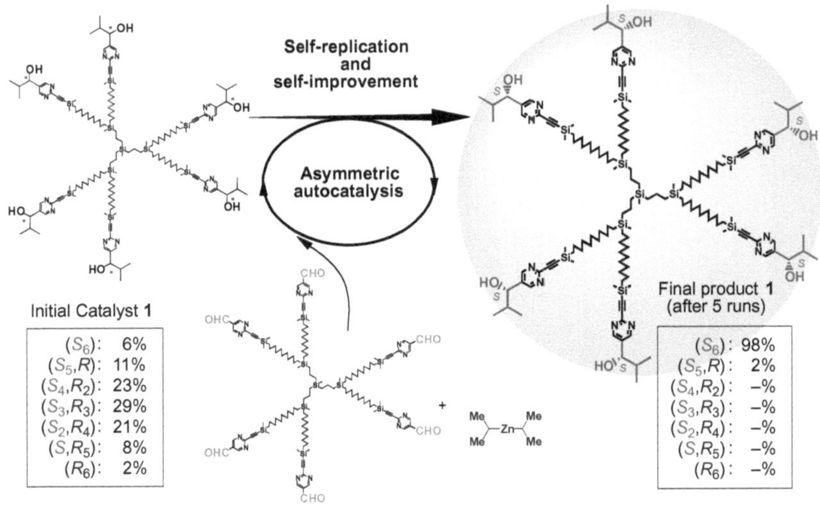

Figure 8.5 Self-replication and self-improvement of *ee* and diastereomeric ratio of chiral molecules (S_6)-**1**.

Table 8.2 The self-replication and self-improvement of pyrimidine-terminated large chiral molecules **1**

Run	Catalyst 1		Product and catalyst 1		Yield (%) (newly)
	Ratio[a] (%)	mol (%)	Isomer ratio[a] (%)	*ee* (%)[b]	
1	6/10/23/29/21/8/1	20	15/14/20/23/18/8/2	75	55
2	15/14/20/23/18/8/2	20	29/16/17/17/12/7/2	84	65
3	29/16/17/17/12/7/2	10	71/18/5/3/1/1/1	98	60
4	71/18/5/3/1/1/1	10	93/6/1/–/–/–/–	>99.5	59
5	93/6/1/–/–/–/–	10	98/2/–/–/–/–/–	>99.5	60

[a]The isomer ratio of (S_6)/(S_5,R)/(S_4,R_2)/(S_3,R_3)/(S_2,R_4)/(S,R_5)/(R_6)-**1** is described.
[b]*ee* (%) = {[(S_6)-**1**−(R_6)-**1**]/[(S_6)-**1** + (R_6)-**1**]} × 100.

The detailed experimental outcomes of the asymmetric autocatalysis using hexamer **1** are shown in Table 8.2. Using these large synthetic molecules, we performed the asymmetric autocatalytic reaction and determined the diastereomeric and enantiomeric ratio of the formed large molecules **1**. The hexaisopropylzinc alkoxide of **1** has a molecular weight of 3258, and was assumed to be the actual catalytic species. Initially, the addition of *i*-Pr_2Zn to hexaaldehyde **2** was conducted in the presence of 20 mol% of the initial catalyst **1** (*ee* based on the homoisomers (S_6)- and (R_6)-**1** was 59%), isomer ratios:

$$(S_6)/(S_5,R)/(S_4,R_2)/(S_3,R_3)/(S_2,R_4)/(S,R_5)/(R_6)-\mathbf{1} = 6/10/23/29/21/8/1.$$

The i-Pr_2Zn addition to each aldehyde moiety proceeded readily to afford hexaalkylated product **1** as a mixture with the initial catalyst **1**. The *ee* of (S_6)-**1** was found to be enhanced to 75% *ee*, including the initial catalyst (Run 1). Then, product **1** (including the initial catalyst) was submitted to the next autocatalytic reaction as the catalyst. After three additional cycles of reaction, we observed further amplification of the *ee* of (S_6)-**1** and diastereomeric ratio (Runs 2–4). The ratio of (S_6)-**1** was significantly increased to 93% and the multiplication factor of homoisomer **1** from the initial catalyst was achieved to >12,500 times and the *ee* of (S_6)-**1** was significantly amplified to >99.5% *ee* (Run 4). One more cycle of the asymmetric autocatalytic reaction using 10 mol% of catalyst **1** afforded the large chiral molecule **1** as the final product with a yield of 60% as the newly formed product. The ratio of (S_6)-**1** has been achieved as 98%, the *ee* of homoisomer (S_6)-**1** amplified to >99.5% *ee* and the isomer ratio also improved to (S_6)/(S_5,R)/(S_4,R_2)/(S_3,R_3)/(S_2,R_4)/(S,R_5)/(R_6)-**1** = 98/2/–/–/–/–/– (Run 5 and Fig. 8.6).

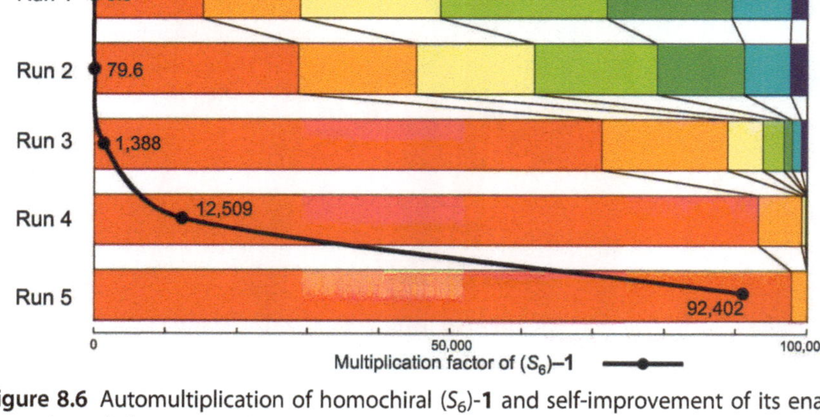

Figure 8.6 Automultiplication of homochiral (S_6)-**1** and self-improvement of its enantiomeric and diastereomeric purities.

Thus, homochiral (S_6)-**1** has exclusively increased by a factor of more than 92,000 times during these steps, and the ratio of other isomers containing the *R*-configuration almost disappeared, except for 2% of (S_5,*R*)-**1** (Fig. 8.6). That is, self-replication of large chiral molecule (S_6)-**1** has been obtained in this autocatalytic system and in addition, the improvement of the isomeric ratio and amplification of enantioenrichment, that is, self-improvement, was observed during the present chemical replication cycle. We have established, for the first time, the asymmetric autocatalysis of (S_6)-**1** with amplification of chirality in which six asymmetric carbon centers have been constructed in a highly enantioselective manner in a single reaction step.

8.3 SELF-REPLICATION AND SELF-IMPROVEMENT OF ENANTIOMERIC AND DIASTEREOMERIC RATIO OF TETRAKIS(PYRIMIDYL ALKANOL)TETRAALKYLSILANE

Next, we demonstrate the creation of large chiral tetrameric molecules with self-replicating and self-improving functionality. It was found that the chiral artificial molecules with four asymmetric carbon centers automultiplied by asymmetric autocatalysis and the multiplication factor of homochiral large molecules achieved *ca.* 2.8 × 10^6 times during five consecutive asymmetric autocatalyses (Fig. 8.7). In addition, the initial low enantiomeric and diastereomeric imbalances could be enhanced to an almost enantiopure state (> 99.5% *ee*).

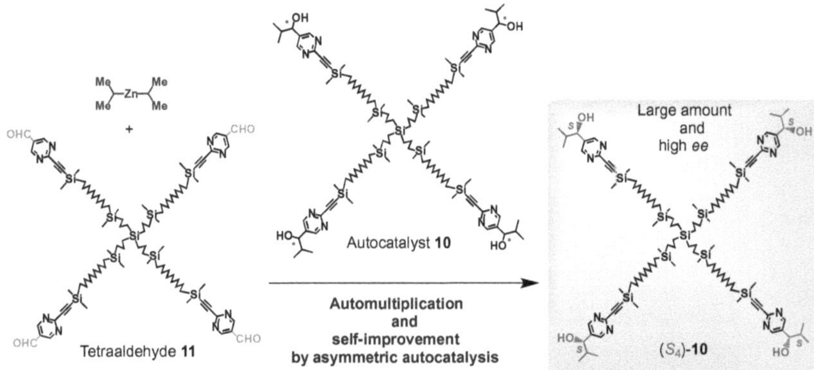

Figure 8.7 Replication and self-improvement of *ee* and diastereomeric ratio of chiral molecules (S_4)-**10**.

8.3.1 Synthesis and Chiral Analysis of Pyrimidine-Terminated Tetrakis(aldehyde) and Tetrakis(alkanol)

Tetraalkanol **10** and tetraaldehyde **11** were designed and synthesized in a similar manner to that of hexamers **1** and **2** using the same pyrimidine compounds as the coupling partner (Fig. 8.8). The tetrabranched alkylsilane backbone was synthesized starting from tetraallylsilane (Fig. 8.8A). After platinum-catalyzed hydrosilylation of terminal olefin using chlorodimethylsilane, the resulting silylchloride was reacted with alkenyl lithium reagent to afford **12** with terminal olefin in two steps (93% yield). Again, the hydrosilylation of **12** followed by alkynylation of the resulting terminal chlorosilane by the treatment of ethynylmagnesium bromide afforded tetrabranched core **13** with terminal acetylene in two steps (72% yield). The Sonogashira coupling reaction with (*S*)-**5** afforded homochiral tetrakis (5-pyrimidyl alkanol)tetraalkylsilane (S_4)-**10** in almost enantiopure form in 51% yield. Racemic **10** can also be prepared under the same conditions using *rac*-**5** as a coupling partner. The corresponding achiral tetrakis(pyrimidine-5-carbaldehyde)tetraalkylsilane **11** can also be formed by the coupling reaction using acetal **4** followed by deprotection of the diethyl acetal.

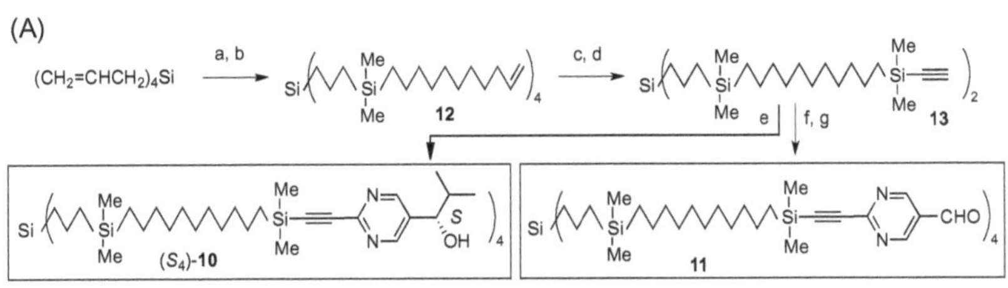

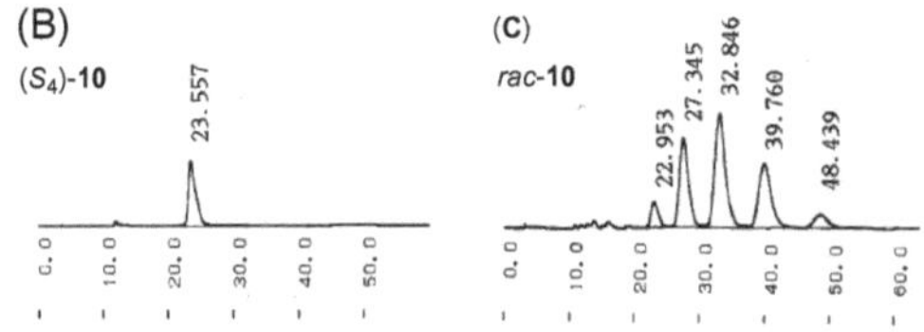

(D)

Isomer	Time (min)	Area (%) Analytical	Area (%) Theoretical
(S_4)-**10**	23.0	6.0	6.25
(S_3,*R*)-**10**	27.3	25.4	25.00
(S_2,R_2)-**10**	32.8	38.2	37.50
(*S*,R_3)-**10**	39.8	24.8	25.00
(R_4)-**10**	48.4	5.6	6.25
	Total	100.0	100.00

Figure 8.8 (A) Synthesis of large pyrimidine-terminated molecules. Conditions: a, Pt cat. **7**, $HSiMe_2Cl$ (neat), reflux; b, Li, $Br(CH_2)_9CH=CH_2$, Et_2O-hexane, 0°C (two steps with 93% yield); c, Pt cat. **7**, $HSiMeCl_2$ (neat), reflux; d, HC≡CMgBr, THF (two steps with 72% yield); e, (*S*)-**5**, $(Ph_3P)_4Pd$, CuI, TBAI, *i*-Pr_2NH (51%); f, **4**, $(Ph_3P)_4Pd$, CuI, TBAI, *i*-Pr_2NH (67%); g, TsOH•H_2O, THF-H_2O (72%). (B) Chiral HPLC chromatogram of (S_4)-**10** and (C) *rac*-**10**. Conditions: Daicel Chiralpak IA, eluent: THF/hexane = 20/80 (v/v), flow rate: 1.0 mL/min, 254 nm UV detector. (D) Analytical and theoretical ratio of five diastereomeric and enantiomeric isomers of tetrakis(5-pyrimidyl alkanol)tetraalkylsilane **10**.

The retention time of homochiral (S_4)-**10** is 23.6 min (Fig. 8.8B) and HPLC employing a chiral stationary phase can separate five isomers of **10** (Fig. 8.8C). Thus, it appeared that the ratio of the racemic tetraalkanol **10** is (S_4)/(S_3,*R*)/(S_2,R_2)/(*S*,R_3)/(R_4)-**10** = 6.0/25.4/38.2/24.8/5.6 from its chromatogram. The analyzed ratio shows an approximately theoretical tbinomial distribution; therefore each signal of the isomer was assigned as shown in Fig. 8.8D. The isomer with shortest retention time is homochiral (S_4)-**10**, the next one is (S_3,*R*)-**10**, then achiral (S_2,R_2)-**10** was assigned to he largest signal. Of course, there are no isomers arising from the substitution pattern of (*S*)- and (*R*)-pyrimidine terminals in these large tetrameric molecules.

8.3.2 Asymmetric Autocatalysis of Hexakis(pyrimidyl alkanol) hexaalkylsilane in the Addition of *i*-Pr_2Zn to the Corresponding Hexaaldehyde

We performed the automultiplication and self-improvement of large chiral tetrafunctionalized molecules **10** with four asymmetric carbon centers by asymmetric autocatalysis and the results are summarized in Table 8.3. Initial compound **10** with a small imbalance of enantiomers and diastereomers was prepared by adding a small amount of (S_4)-**10** to *rac*-**10** with a stochastic racemic mixture of its isomers. The *ee* of (S_4)-**10** was calculated to be 0.25% from the amount of its enantiomer (R_4)-**10**. In Run 1, the above-indicated **10** with a small excess of enantiomer was utilized initially to perform the addition of *i*-Pr_2Zn to tetrakis(pyrimidine-5-carbaldehyde)tetraalkylsilane **11**. After tetraalkylation of the tetraaldehyde **11**, compound **10** with the isomeric ratio of (S_4)/(S_3,*R*)/(S_2,R_2)/(*S*,R_3)/(R_4)-**10** = 28/26/21/16/9 was isolated in 77% yield. The *ee* of homoisomer (S_4)-**10** amplified significantly, reaching 52% and the homoisomer (S_4)-**10** was multiplied by a factor of 23 times. The obtained mixture of the initial compound and product **10** was subjected to the next replication and improvement cycle as the starting compound of Run 2. After an additional four runs of replication cycles (Runs 2–5), the final compound (S_4)-**10** was obtained with an *ee* of >99.5% and the isomeric ratio of (S_4)/(S_3,*R*)/(S_2,R_2)/(*S*,R_3)/(R_4)-**10** = 98/1/1/–/– by the continuing self-replicating and self-improving processes.

The homoisomer (S_4)-**10** was automultiplied *ca.* 2.8×10^6 times to reach an almost enantiomerically and diastereomerically pure form after the fifth cycle of asymmetric autocatalysis (Fig. 8.9). The amounts of isomers of (*S*,R_3) and (R_4)-**10** in the final product were below the detection level in chiral HPLC analysis. In this chemical system, homoisomer (S_4)-**10** replicated significantly with improvement of its diastereomeric and enantiomeric

Table 8.3 Asymmetric autocatalysis of (S_4)-tetrakis(5-pyrimidyl alkanol)tetraalkylsilane **10** with amplification of enantiomeric and diastereomeric purities

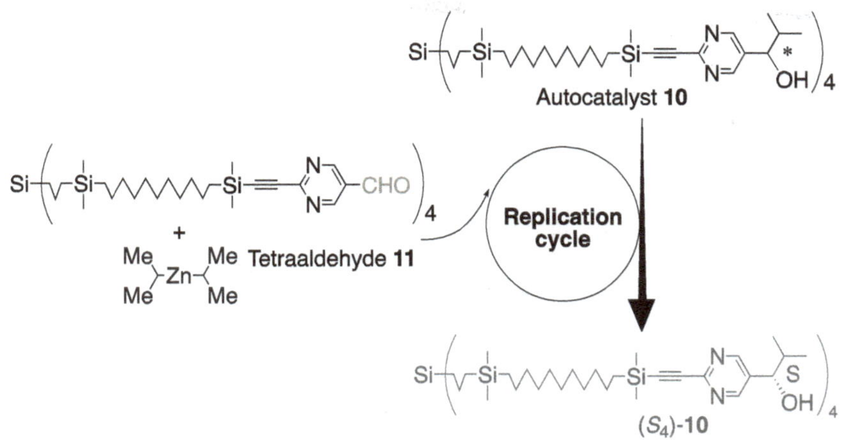

Run	Catalyst 10			Product and catalyst 10	Newly Yield (%)	Multiplication factor of (S_4)-10
	Ratio[a]	*ee*[b]	mol (%)	Ratio[a]		
1	6.03/25/38/25/6	0.25	20	28/26/21/16/9	77	×23
2	28/26/21/16/9	52	20	74/14/7/4/1	86	×14
3	74/14/7/4/1	96	5	96/3/1/–/–	89	×24
4	96/3/1/–/–	>99.5	5	97/2/1/–/–	86	×18
5	97/2/1/–/–	>99.5	5	98/1/1/–/–	91	×19
				Total: × *ca.* 2,800,000		

[a]The isomeric ratios of (S_4)/(S_3,R)/(S_2,R_2)/(S,R_3)/(R_4)-**10** are described. The ratio was determined by HPLC using a chiral stationary phase.
[b]The *ee* based on the homoisomers (S_4)- and (R_4)-**10**.

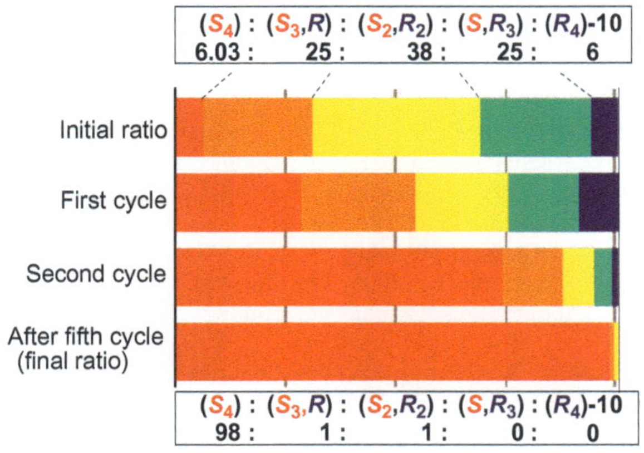

Figure 8.9 Self-replication of tetrakis(pyrimidyl alkanol)tetraalkylsilane with self-improvement of enantiomeric and diastereomeric purities of chiral large molecules (S_4)-**10**.

ratio during the continuing reaction cycles. Compound **10** containing the moiety with the *R*-configuration had its ratio greatly reduced.

8.4 SUMMARY

As described, we have synthesized large chiral molecular architectures, that is, hexakis(5-pyrimidyl alkanol)hexaalkylsilane **1** and tetrakis(5-pyrimidyl alkanol)tetraalkylsilane **10** with functions of self-replication and self-improvement. These reactions are asymmetric autocatalysis of large artificial chiral molecules with amplification of enantiomeric and diastereomeric purities. Thus, the asymmetric autocatalysis of 5-pyrimidyl alkanol with amplification of *ee* can work on large branched molecules with many reaction points. In addition, it is the first example of a real chemical reaction, that is, large chiral molecules catalyze the production of chiral molecules with the same structure and same configuration from completely achiral substrate and reagent. The enantioselectivity of this reaction was controlled by the chirality of the large multifunctionalized molecules. The synthetically important aspect of this reaction is that in a single reaction, multiasymmetric carbon centers are constructed with high selectivity.

ACKNOWLEDGMENT

The authors thank Ms. Mai Nakaoda, Mr. Yutaro Takahashi, Mr. Yusuke Kanto, Ms. Nanako Kuruhara and Mr. Kenji Hosoi for participating in the experiments. This work was supported by a Grant-in-Aid for Scientific Research from Japan Society for the Promotion of Science (JSPS), MEXT-Supported Program for the Strategic Research Foundation at Private Universities, 2012–16. This research was partially supported by JSPS KAKENHI (Grant No. 17750044 and 20750037) and the Iketani Science and Technology Foundation.

REFERENCES

[1] Bolli, M.; Micura, R.; Eschenmoser, A. Pyranosyl-RNA: Chiroselective Self-Assembly of Base Sequences by Ligative Oligomerization of Tetranucleotide-2′,3′-Cyclophosphates (with a Commentary Concerning the Origin of Biomolecular Homochirality). *Chem. Biol.* **1997,** *4*, 309–320.

[2] Satyanarayana, T.; Abraham, S.; Kagan, H. B. Nonlinear Effects in Asymmetric Catalysis. *Angew. Chem., Int. Ed.* **2009,** *48*, 456–494.

[3] Frank, F. C. On Spontaneous Asymmetric Synthesis. *Biochim. Biophys. Acta* **1953,** *11*, 459–463.

[4] Soai, K.; Kawasaki, T.; Matsumoto, A. Asymmetric Autocatalysis of Pyrimidyl Alkanol and its Application to the Study on the Origin of Homochirality. *Acc. Chem. Res.* **2014,** *47*, 3643–3654.

[5] Mislow, K. Absolute Asymmetric Synthesis: A Commentary. *Collect. Czech. Chem. Commun.* **2003,** *68*, 849–864.
[6] Soai, K.; Kawasaki, T. Asymmetric Autocatalysis – Discovery and State of the Art. In *The Soai Reaction and Related Topic;* Palyi, G., Zicchi, C., Caglioti, L., Eds.; Academia Nationale di Scienze Lettere e Arti Modena: Modena, 2012; pp 8–34.
[7] Feringa, B. L.; van Delden, R. A. Absolute Asymmetric Synthesis: The Origin, Control, and Amplification of Chirality. *Angew. Chem., Int. Ed.* **1999,** *38*, 3418–3438.
[8] Orgel, L. E. Molecular Replication. *Nature* **1992,** *358*, 203–209.
[9] Ruiz-Mirazo, K.; Briones, C.; de la Escosura, A. Prebiotic Systems Chemistry: New Perspectives for the Origins of Life. *Chem. Rev.* **2014,** *114*, 285–366.
[10] Nitschke, J. R. Systems Chemistry: Molecular Networks Come of Age. *Nature* **2009,** *462*, 736–738.
[11] von Kiedrowski, G.; Otto, S.; Herdewijn, P. Welcome Home, Systems Chemists! *J. Syst. Chem.* **2010,** *1*, 1.
[12] von Kiedrowski, G. A Self-Replicating Hexadeoxynucleotide. *Angew. Chem., Int. Ed. Engl.* **1986,** *25*, 932–935.
[13] Sievers, D.; von Kiedrowski, G. Self-Replication of Complementary Nucleotide-Based Oligomers. *Nature* **1994,** *369*, 221–224.
[14] Paul, N.; Joyce, G. F. A Self-Replicating Ligase Ribozyme. *Proc. Natl. Acad. Sci. USA* **2002,** *99*, 12733–12740.
[15] Li, T.; Nicolaou, K. C. Chemical Self-Replication of Palindromic Duplex DNA. *Nature* **1994,** *369*, 218–221.
[16] Pitsch, S.; Krishnamurthy, R.; Bolli, M.; Wendeborn, S.; Holzner, A.; Minton, M., et al. Pyranosyl-RNA ("p-RNA"): Base-Pairing Selectivity and Potential to Replicate. Preliminary Communication. *Helv. Chim. Acta* **1995,** *78*, 1621–1635.
[17] Lee, D. H.; Granja, J. R.; Martinez, J. A.; Severin, K.; Ghadiri, M. R. Self-Replicating Peptide. *Nature* **1996,** *382*, 525–528.
[18] Yao, S.; Ghosh, I.; Zutshi, R.; Chmielewski, J. A pH-Modulated, Self-Replicating Peptide. *J. Am. Chem. Soc.* **1997,** *119*, 10559–11560.
[19] Tjivikua, T.; Ballester, P.; Rebek, J., Jr. Self-Replicating System. *J. Am. Chem. Soc.* **1990,** *112*, 1249–1250.
[20] Hong, J.-I.; Feng, Q.; Rotello, V.; Rebek, J., Jr. Competition, Cooperation, and Mutation: Improving a Synthetic Replicator by Light Irradiation. *Science* **1992,** *255*, 848–850.
[21] Bissette, A. J.; Fletcher, S. P. Mechanisms of Autocatalysis. *Angew. Chem., Int. Ed.* **2013,** *52*, 12800–12826.
[22] Kawasaki, T.; Nakaoda, M.; Takahashi, Y.; Kanto, Y.; Kuruhara, N.; Hosoi, K., et al. Self-Replication and Amplification of Enantiomeric Excess of Chiral Multifunctionalized Large Molecules by Asymmetric Autocatalysis. *Angew. Chem., Int. Ed.* **2014,** *53*, 11199–11202.
[23] Soai, K.; Shibata, T.; Sato, I. Enantioselective Automultiplication of Chiral Molecules by Asymmetric Autocatalysis. *Acc. Chem. Res.* **2000,** *33*, 382–390.
[24] Soai, K.; Kawasaki, T. Asymmetric Autocatalysis with Amplification of Chirality. *Top. Curr. Chem.* **2008,** *284*, 1–33.
[25] Soai, K.; Kawasaki, T.; Matsumoto, A. The Origins of Homochirality Examined by Using Asymmetric Autocatalysis. *Chem. Rec.* **2014,** *14*, 70–83.
[26] Kawasaki, T.; Soai, K. Asymmetric Induction Arising from Enantiomerically Enriched Carbon-13 Isotopomers and Highly Sensitive Chiral Discrimination by Asymmetric Autocatalysis. *Bull. Chem. Soc. Jpn.* **2011,** *84*, 879–892.
[27] Blackmond, D. G. Asymmetric Autocatalysis and its Implications for the Origin of Homochirality. *Proc. Natl. Acad. Sci. USA* **2004,** *101*, 5732–5736.

[28] Avalos, M.; Babiano, R.; Cintas, P.; Jimenez, J. L.; Palacios, J. C. Chiral Autocatalysis: Where Stereochemistry Meets the Origin of Life. *Chem. Commun.* **2000,** 887–892.
[29] Podlech, J.; Gehring, T. New Aspects of Soai's Asymmetric Autocatalysis. *Angew. Chem., Int. Ed.* **2005,** *44*, 5776–5777.
[30] Matsumoto, A.; Abe, T.; Hara, A.; Tobita, T.; Sasagawa, T.; Kawasaki, T., et al. Crystal Structure of the Isopropylzinc Alkoxide of Pyrimidyl Alkanol: Mechanistic Insights for Asymmetric Autocatalysis with Amplification of Enantiomeric Excess. *Angew. Chem., Int. Ed.* **2015,** *54*, 15218–15221.
[31] Soai, K.; Shibata, T.; Morioka, H.; Choji, K. Asymmetric Autocatalysis and Amplification of Enantiomeric Excess of a Chiral Molecule. *Nature* **1995,** *378*, 767–768.
[32] Sato, I.; Urabe, H.; Ishiguro, S.; Shibata, T.; Soai, K. Amplification of Chirality from Extremely Low to Greater than 99.5% ee by Asymmetric Autocatalysis. *Angew. Chem., Int. Ed.* **2003,** *42*, 315–317.
[33] Held, F. E.; Fingerhut, A.; Tsogoeva, S. B. Insights into the Spontaneous Emergence of Enantioselectivity in an Asymmetric Mannich Reaction Carried out without External Catalyst. *Tetrahedron: Asymmetry* **2012,** *23*, 1663–1669.
[34] Gridnev, I. D.; Serafimov, J. M.; Brown, J. M. Solution Structure and Reagent Binding of the Zinc Alkoxide Catalyst in the Soai Asymmetric Autocatalytic Reaction. *Angew. Chem., Int. Ed.* **2004,** *43*, 4884–4887.
[35] Gehring, T.; Quaranta, M.; Odell, B.; Blackmond, D. G.; Brown, J. M. Observation of a Transient Intermediate in Soai's Asymmetric Autocatalysis: Insights from ^{1}H-NMR Turnover in Real Time. *Angew. Chem., Int. Ed.* **2012,** *51*, 9539–9542.
[36] Schiaffino, L.; Ercolani, G. Unraveling the Mechanism of the Soai Asymmetric Autocatalytic Reaction by First-Principles Calculations: Induction and Amplification of Chirality by Self-Assembly of Hexamolecular Complexes. *Angew. Chem., Int. Ed.* **2008,** *47*, 6832–6835.
[37] Ercolani, G.; Schiaffino, L. Putting the Mechanism of the Soai Reaction to the Test: DFT Study of the Role of Aldehyde and Dialkylzinc Structure. *J. Org. Chem.* **2011,** *76*, 2619–2626.
[38] Gridnev, I. D.; Vorobiev, A. K. Quantification of Sophisticated Equilibria in the Reaction Pool and Amplifying Catalytic Cycle of the Soai Reaction. *ACS Catal* **2012,** *2*, 2137–2149.
[39] Micheau, J.-C.; Coudret, C.; Cruz, J.-M.; Buhse, T. Amplification of Enantiomeric Excess, Mirror-Image Symmetry Breaking and Kinetic Proofreading in Soai Reaction Models with Different Oligomeric Orders. *Phys. Chem. Chem. Phys.* **2012,** *14*, 13239–13248.
[40] Shibata, T.; Yonekubo, S.; Soai, K. Practically Perfect Asymmetric Autocatalysis with (2-Alkynyl-5-pyrimidyl)alkanols. *Angew. Chem., Int. Ed.* **1999,** *38*, 659–661.
[41] Lutz, F.; Kawasaki, T.; Soai, K. Asymmetric Autocatalysis of a Ferrocene-Containing Chiral Compound with Amplification of Chirality. *Tetrahedron: Asymmetry* **2006,** *17*, 486–490.
[42] Kawasaki, T.; Kamimura, S.; Amihara, A.; Suzuki, K.; Soai, K. Enantioselective C–C Bond Formation as a Result of the Oriented Prochirality of an Achiral Aldehyde at the Single-Crystal Face upon Treatment with a Dialkyl Zinc Vapor. *Angew. Chem., Int. Ed.* **2011,** *50*, 6796–6798.
[43] Dóka, É.; Lente, G. Stochastic Modeling of the Soai Reaction. In *The Soai Reaction and Related Topic;* Palyi, G., Zucchi, C., Caglioti, L., Eds.; Academia Nationale di Scienze Lettere e Arti Modena: Modena, 2012; pp 123–147.
[44] Dóka, É.; Lente, G. Mechanism-Based Chemical Understanding of Chiral Symmetry Breaking in the Soai Reaction. A Combined Probabilistic and

Deterministic Description of Chemical Reactions. *J. Am. Chem. Soc.* **2011,** *133*, 17878–17881.

[45] Barabás, B.; Tóth, J.; Pályi, G. Stochastic Aspects of Asymmetric Autocatalysis and Absolute Asymmetric Synthesis. *J. Math. Chem.* **2010,** *48*, 457–489.

[46] Micskei, K.; Póta, G.; Caglioti, L.; Palyi, G. Empirical Description of Chiral Autocatalysis. *J. Phys. Chem. A* **2006,** *110*, 5982–5984.

[47] Barabas, B.; Caglioti, L.; Zucchi, C.; Maioli, M.; Gál, E.; Micskei, K., et al. Violation of Distribution Symmetry in Statistical Evaluation of Absolute Enantioselective Synthesis. *J. Phys. Chem. B* **2007,** *111*, 11506–11510.

[48] Caglioti, L.; Zucchi, C.; Pályi, G. Single-Molecule Chirality. *Chimica Oggi* **2005,** *23* (38–39), 42–43.

[49] Saito, Y.; Hyuga, H. Colloquium: Homochirality: Symmetry Breaking in Systems Driven far from Equilibrium. *Rev. Mod. Phys.* **2013,** *85*, 603–621.

[50] Soai, K.; Sato, I.; Shibata, T.; Komiya, S.; Hayashi, M.; Matsueda, Y., et al. Asymmetric Synthesis of Pyrimidyl Alkanol without Adding Chiral Substances by the Addition of Diisopropylzinc to Pyrimidine-5-carbaldehyde in Conjunction with Asymmetric Autocatalysis. *Tetrahedron: Asymmetry* **2003,** *14*, 185–188.

[51] Soai, K.; Osanai, S.; Kadowaki, K.; Yonekubo, S.; Shibata, T.; Sato, I. *d*- and *l*-Quartz-Promoted Highly Enantioselective Synthesis of a Chiral Organic Compound. *J. Am. Chem. Soc.* **1999,** *121*, 11235–11236.

[52] Shindo, H.; Shirota, Y.; Niki, K.; Kawasaki, T.; Suzuki, K.; Araki, Y., et al. Asymmetric Autocatalysis Induced by Cinnabar: Observation of the Enantioselective Adsorption of a 5-Pyrimidyl Alkanol on the Crystal Surface. *Angew. Chem., Int. Ed.* **2013,** *52*, 9135–9138.

[53] Kawasaki, T.; Suzuki, K.; Hakoda, Y.; Soai, K. Achiral Nucleobase Cytosine Acts as an Origin of Homochirith Asymmetric Autocatalysis. *Angew. Chem., Int. Ed.* **2008,** *47*, 496–499.

[54] Shibata, T.; Yamamoto, J.; Matsumoto, N.; Yonekubo, S.; Osanai, S.; Soai, K. Amplification of a Slight Enantiomeric Imbalance in Molecules Based on Asymmetric Autocatalysis: The First Correlation between High Enantiomeric Enrichment in a Chiral Molecule and Circularly Polarized Light. *J. Am. Chem. Soc.* **1998,** *120*, 12157–12158.

[55] Lutz, F.; Igarashi, T.; Kinoshita, T.; Asahina, M.; Tsukiyama, K.; Kawasaki, T., et al. Mechanistic Insights in the Reversal of Enantioselectivity of Chiral Catalysts by Achiral Catalysts in Asymmetric Autocatalysis. *J. Am. Chem. Soc.* **2008,** *130*, 2956–2958.

[56] Kawasaki, T.; Harada, Y.; Suzuki, K.; Tobita, T.; Florini, N.; Palyi, G., et al. Enantioselective Synthesis Utilizing Enantiomorphous Organic Crystal of Achiral Benzils as a Source of Chirality in Asymmetric Autocatalysis. *Org. Lett.* **2008,** *10*, 4085–4088.

[57] Kawasaki, T.; Matsumura, Y.; Tsutsumi, T.; Suzuki, K.; Ito, M.; Soai, K. Asymmetric Autocatalysis Triggered by Carbon Isotope ($^{13}C/^{12}C$) Chirality. *Science* **2009,** *324*, 492–495.

[58] Kawasaki, T.; Sato, M.; Ishiguro, S.; Saito, T.; Morishita, Y.; Sato, I., et al. Enantioselective Synthesis of Near Enantiopure Compound by Asymmetric Autocatalysis Triggered by Asymmetric Photolysis with Circularly Polarized Light. *J. Am. Chem. Soc.* **2005,** *127*, 3274–3275.

[59] Soai, K.; Kawasaki, T. Asymmetric Autocatalysis of Pyrimidyl Alkanol Using an Organozinc Reagent. Chapter 6 In *Organometallic Chirality;* Palyi, G., Zucchi, C., Caglioti, L., Eds.; Mucchi Editore: Modena, 2008; pp 107–125.

CHAPTER 9

Model Equivalency in Interpreting the Stochastic Enantiomer Distribution in the Soai Reaction

Gábor Lente
Department of Inorganic and Analytical Chemistry, University of Debrecen, Debrecen, Hungary

Abstract

Two previous models proposed to interpret the stochastic distribution of enantiomers observed in the Soai reaction are investigated in detail. The first model comprises only four steps with two different rate constants and implies simple second order autocatalysis. The second model features 18 steps and 7 variable parameters. Based on the observation that these two models give practically identical predictions, it is shown that the more complicated model can be essentially reduced to the simpler one by assuming that all six intermediates in it can be described by the steady state assumption. It is concluded that even simple models can be quite useful in interpreting absolute asymmetric synthesis if the experimental observations are limited to the values of enantiomeric excess formed.

Keywords: Stochastic kinetics; statistical analysis; steady state treatment; enantiomeric excess distribution; equivalent kinetic models; absolute asymmetric synthesis

9.1 INTRODUCTION

By now, the Soai reaction has become by far the best characterized example of enantioselective autocatalysis. The overall chemical process is fundamentally a carbon–carbon bond formation between an organozinc reagent and a prochiral aromatic aldehyde, which produces a chiral secondary alcohol. The Soai group has published numerous papers on these highly notable results about enantioselective autocatalysis since 1995 [1–14]. In a recent scientific paper, the impact of the initial discovery was shown to be extremely significant and analyzed by the tools of mathematical graph theory [15].

Advances in Asymmetric Autocatalysis and Related Topics
DOI: http://dx.doi.org/10.1016/B978-0-12-812824-4.00009-5

It is arguable that the most interesting phenomenon observed in this system is the appearance of sizable enantiomeric excesses (*ee*) in the product secondary alcohol without an external chiral influence [2–4]. It is also well documented that the reaction system is very sensitive to the initial presence of any chiral impurities [1–15]. Recent studies show that this sensitivity also includes isotope chirality, which is present even in seemingly nonchiral material [5–6,16–17]. However, the studies done with the exclusion of external chiral influences are viewed as being very careful from an experimental point of view, and the observed phenomena are generally considered examples of absolute asymmetric synthesis [2–4,18–21], which is of great interest in terms of the efforts to interpret the origins of biological (homo)chirality [22–23].

Naturally, these results have inspired a large group of interested scientists and the mechanism of the Soai reaction, with special attention being paid to the details of enantioselective autocatalysis, has been also studied from many different aspects using many different techniques [18–19,24–27]. Yet, a number of open questions still remain in this field.

Some of the theoretical lines of thought pursued in the attempt to understand the essential characteristics of the Soai reaction have involved some quantum mechanical studies on structures suspected to be important and mechanism-based kinetic models [28–30]. A number of empirical approaches toward the interpretation of experimental observations have also been reported [31–32] and numerous modeling studies gave information about the conditions necessary for chiral amplification to occur [33–42]. A significant novel element was added to the pool of theoretical tools when it was pointed out that a stochastic approach to kinetics and statistical evaluations are useful in this system [43–55]. In essence, these techniques use a strictly molecule-based (discrete) approach to chemical kinetics instead of a continuous one described by concentrations [56–57]. These techniques show that the very early part of the reaction under absolute asymmetric conditions has a large influence on the final product distribution and also emphasize the need to view chirality on a single-molecule level rather than as a bulk property [45–46,48]. In a different approach along the same basic principles, reaction noise has been shown to be an important factor in determining the final outcomes of processes in various model calculations [58–61].

An important way of characterizing absolute asymmetric reactions is to determine the statistical distribution of enantiomeric excesses formed in a batch of strictly identical experiments. The Soai group has published two such notable distributions: the first with 37 [3], and the second with 84 repetitions [4]. In theory, it is clearly expected that these distributions should be symmetric, which means that a given *ee* must be formed with identical probability for the two possible enantiomers [44]. Indeed, later analysis [44,62] showed that there is no significant asymmetry in the smaller distribution [3]. The larger one with 84 repetitions, on the other hand, which was measured in the presence of achiral silica gel [4], shows a slight but significant preference for the formation of the S enantiomer [50,52,62,63]. It was shown that this slight asymmetry could even be caused by the presence of less than 10 molecules of chiral impurity in the initial mixture [52].

Some efforts have also been made to interpret not only the symmetry, but also the shape of this distribution curve. For this purpose, the symmetric experimentally measured distribution is preferred because no assumptions about initial impurities are needed in this case. Simple models showing first-order [43] and second-order autocatalysis [44] were considered for this purpose, and, as shown in Fig. 9.1 (*orange* (gray in print versions) curve), the second-order case gave good agreement with experimental data. A later study developed a method to predict these distributions based on a much more complicated network of reactions [51]. It was shown that a chemical mechanism proposed earlier by Buhse [33]

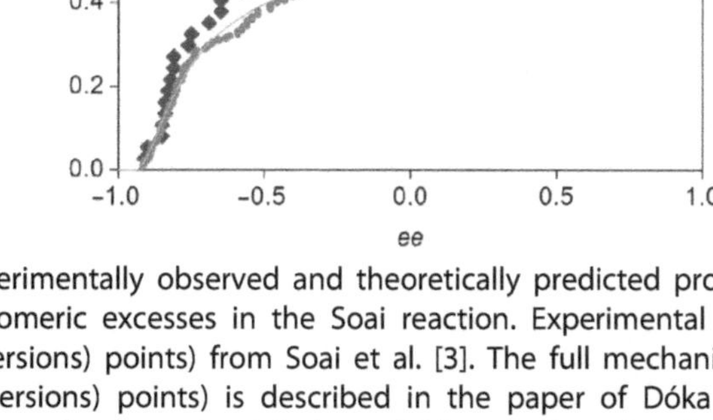

Figure 9.1 Experimentally observed and theoretically predicted probability distributions of enantiomeric excesses in the Soai reaction. Experimental data (*blue* (dark gray in print versions) points) from Soai et al. [3]. The full mechanism fit (*red* (light gray in print versions) points) is described in the paper of Dóka and Lente [51], details of the second order autocatalysis model (*orange* (gray in print versions) curve) can be found in an article by Lente [44]. The graph was constructed by unifying the relevant graphs published in these two previous papers [44,51].

gives a good interpretation of the experimentally found fluctuations with a proper selection of rate constants [51]. This is shown by the *red* (light gray in print versions) points in Fig. 9.1.

Fig. 9.1 shows that both the detailed Buhse model of the Soai reaction and a simple second-order autocatalytic process give a reasonable fit to the experimental data with some (not unexpected) deviation. However, from a purely theoretical point of view, it is even more remarkable that the two model calculations gave very similar results.

In fact, taking into account the fact that the predictions of the more complicated model had to be calculated based on Monte Carlo simulations with only 100 repetitions and symmetrization (i.e., the reason why the *red* (light gray in print versions) points do not define a very smooth line and are shown as a collection of points rather than a continuous curve) [51], the two models give practically identical final results. The present book chapter aims to find the reasons for this coincidence, which seems highly unlikely from a statistical point of view and calls for the pursuit of a causal relation between the two cases.

9.2 RESULTS AND DISCUSSION

First, both earlier model calculations [44,51] were carefully reproduced. Table 9.1 shows the simple model containing second-order autocatalysis [44]. Here, the only parameter value that influences the final distribution is the dimensionless ratio of the catalytic and uncatalyic rate constants, $\beta = k_c/\left(V^2 N_A^2 k_u\right)$, where N_A is Avogadro's constant and V is the volume of the sample. This parameter had to be re-optimized compared to the original work because $n = 3.0 \times 10^{20}$ was used there as the initial number of particles, whereas this quantity was $n = 6.0 \times 10^{19}$ in the actual

Table 9.1 Chemical reactions and rate constants in the simple second-order autocatalytic scheme proposed earlier [44]

Reaction	Rate constant
$A \rightarrow B_R$	k_u
$A \rightarrow B_S$	k_u
$A + 2B_R \rightarrow 3B_R$	k_c
$A + 2B_S \rightarrow 3B_S$	k_c

The final distribution is only sensitive to the dimensionless ratio of the two rate constants: $\beta = k_c/\left(V^2 N_A^2 k_u\right) = 1.6 \times 10^{-21}$ was used in this study. N_A is Avogadro's constant and V is the volume of the sample.

experiment [3]. The new optimized value of β is 1.6×10^{-21}, the one used in the original paper [44] is 3.8×10^{-22}. This difference is not unreasonable for a fivefold change in the initial particle number. Essentially, the same algorithm was used as in the previous work: stochastic calculations were done iteratively until a sufficiently high number of particles was reached for deterministic continuation. However, the advances in computation power made it possible to calculate the first, stochastic part up to 10,000 molecules (instead of 2000 in the previous report [44]), which gave a better starting point for deterministic continuation.

Table 9.2 shows the full mechanism proposed by Buhse [33] and is used in the detailed calculations reported earlier [51]. Again, the same algorithm was used in this work. The higher computation speed was not

Table 9.2 Elementary reactions, rate constants and parameter values in the mechanism originally proposed by Buhse [33]

Reaction	Rate constant
c + z → czr	k_1
c + z → czs	k_1
czr + czr → dr2	k_2
dr2 → 2 czr	k_{-2}
czs + czs → ds2	k_2
ds2 → 2 czs	k_{-2}
czr + czs → drs	$\alpha \times k_2$
drs → czr + czs	k_{-2}
dr2 + c → tr2	k_3
tr2 → dr2 + c	k_{-3}
ds2 + c → ts2	k_3
ts2 → ds2 + c	k_{-3}
drs + c → trs	k_3
trs → drs + c	k_{-3}
tr2 + z → dr2 + czr	k_4
ts2 + z → ds2 + czs	k_4
trs + z → drs + czr	k_4
trs + z → drs + czs	k_4

Parameter values: n_0 (CHO) = 0.10 mmol; n_0 (Zn) = 0.20 mmol; $V = 0.90$ cm^3; $\alpha = 1.0 \times 10^{-5}$; $k_1 = 7.0 \times 10^{-18}$ M^{-1}s^{-1}; $k_2 = 800$ M^{-1}s^{-1}; $k_{-2} = 110$ s^{-1}; $k_3 = 100$ M^{-1}s^{-1}; $k_{-3} = 100$ s^{-1}; $k_4 = 8.0 \times 10^5$ M^{-1}s^{-1} [51]. Substance notations: c, reactant carbaldehyde (CHO); czr and czs, enantiomers of the product zinc alkoxide (COZn); dr2 and ds2, homochiral enantiomers of the dimer of the zinc alkoxide ($(COZn)_2$); drs, heterochiral isomer of the dimer of the zinc alkoxide ($(COZn)_2$); tr2 and ts2, homochiral enantiomers of the association product of the dimer and the carbaldehyde ($(COZn)_2$–CHO); trs, heterochiral isomer of the association product of the dimer and the carbaldehyde ($(COZn)_2$–CHO); z, reactant dialkyl zinc compound (Zn).

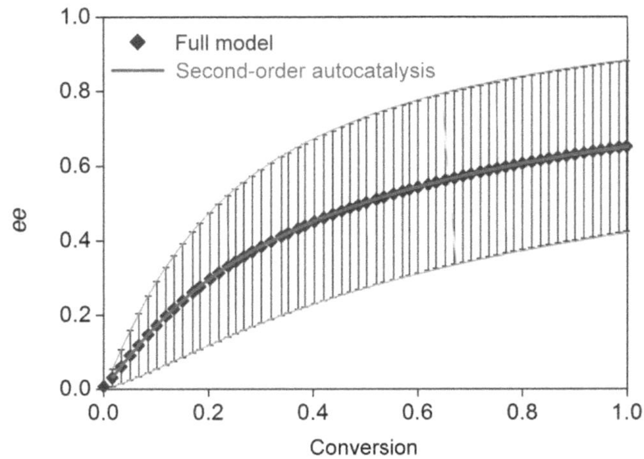

Figure 9.2 The expectation for the enantiomeric excess of chiral material formed and its standard deviation as a function of overall conversion in the two models described in this work. Model details are given in Table 9.1 (second-order autocatalysis) and Table 9.2 (full model).

used here to extend the calculations to larger molecule numbers in the initial stochastic simulations (10,000 in both the original [51] and the present work). However, instead of 100 repetitions in the Monte Carlo simulations [51], it was possible to carry out 1000 repetitions. This gave a much better defined final distribution, which was much smoother, as expected. All the parameters were the same as in the original study [51].

The predictions from the two models were compared in a number of different representations. Fig. 9.2 shows one such comparison. Here, the expectation for the enantiomeric excess was calculated from both models as a function of the conversion of the reactant aldehyde. It is very remarkable that the two models gave identical predictions for both the expectation and its standard error.

A second method of demonstrating the striking similarity of the two models used the distributions of *ee*. Fig. 9.3 shows such comparisons at three different conversion values. The red curve is a comparison after the formation of the initial 10,000 product molecules. Here the distribution is very narrow and is in fact very close to a binomial distribution, which was already remarked on in one of the previous works [51]. A second comparison (red curve and points) is shown at the conversion value of about 15%. This is a much wider distribution, which already implies very sizable stochastic fluctuations in the results of individual experiments.

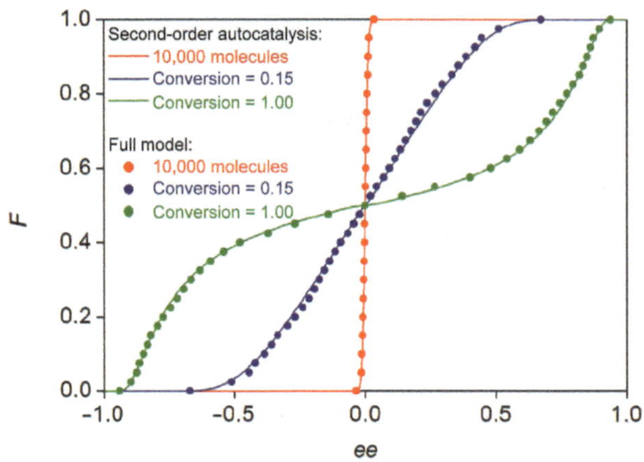

Figure 9.3 Theoretically predicted probability distributions of enantiomeric excesses in the Soai reaction at three different conversions. The full mechanism fit is given as points (only a selection of the 1000 calculated points shown for clarity) by the method described in the paper Dóka and Lente [51]. Continuous curves show results from the second-order autocatalysis model originally proposed by Lente [44].

Finally, the prediction at 100% conversion is shown by the green points and curve. It is again highly remarkable that the two models give identical predictions for all cases.

To gain some further insight into the kinetic details of the system, classical deterministic model calculations [64] were carried out in the full model shown in Table 9.2. All initial values (concentrations, volume) were the same as in the actual experimental study [3], and the parameters were the ones shown in Table 9.2. Because of the deterministic nature of these calculations, no formation of *ee* was predicted from symmetric initial conditions [57]. However, this simulation study gives an approximate idea about the typical concentration levels of the intermediates formed in the mechanism shown in Table 9.2, which is very important information in the assessment of relative reaction rates and the possible kinetic simplifications that can be used to describe the system. Fig. 9.4 shows the results of such a calculation. Here, the concentrations are not given as a function of time, but as a function of conversion (now measured in absolute concentration rather than relative units).

When plotted as a function of time, the results showed a rather long initial inhibition period followed by a very sudden autocatalytic reaction portion. Therefore, plotting the time dependence would not be very

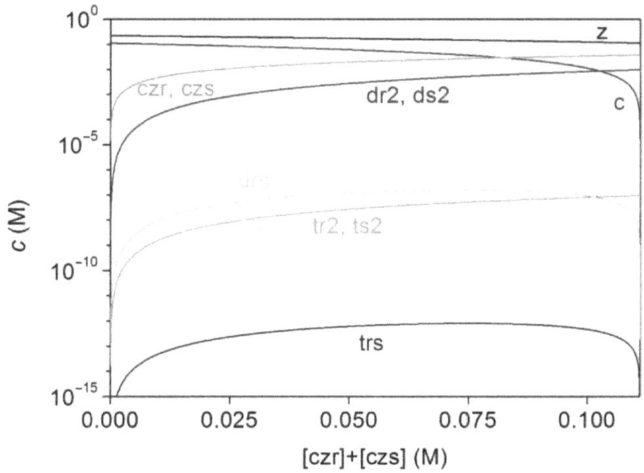

Figure 9.4 The results of the deterministic simulation based on the full mechanism shown in Table 9.2. Concentrations of various species (reagents, intermediates, products) are plotted on a logarithmic scale as a function of conversion.

informative about typical concentration levels under conditions at which the overall reaction is actually proceeding.

With the aid of the observations shown in Figs. 9.2, 9.3, and 9.4, it is possible to make some simplifications in the mathematical treatment of the full model shown in Table 9.2. These simplifications, combined with other calculations, illuminate the fundamental reasons behind the striking similarity of the predictions of the two models, which otherwise show very different levels of complexity.

First, Fig. 9.4. suggests that all the intermediates in the mechanism (ds2, dr2, drs, ts2, tr2, trs) reach only low concentrations compared to the concentrations of the products, czr and czs. This observation is valid for any reaction time. Therefore it is reasonable to attempt to handle the six mentioned species as steady-state intermediates [64] under the conditions of the present calculations.

The steady-state assumption sets the rate of concentration change for a reactive intermediate at zero, and calculates the steady-state concentration of this species as a function of the other concentrations in the system. This actually gives time dependence for this concentration as well, but its rate of change is much lower in absolute terms than the rate of concentration change for any major sepcies [64]. The steady-state assumption can be used for multiple species in a single system simultaneously. In the system shown in Table 9.2, six intermediates (ds2, dr2, drs, ts2, tr2, trs) can be assumed to be in steady state, which gives six independent equations. These are as follows:

For species dr2:

$$0 = k_2[\mathrm{czr}]^2 - k_{-2}[\mathrm{dr2}] - k_3[\mathrm{dr2}][\mathrm{c}] + k_{-3}[\mathrm{tr2}] + k_4[\mathrm{tr2}][\mathrm{z}] \tag{9.1}$$

For species ds2:

$$0 = k_2[\mathrm{czs}]^2 - k_{-2}[\mathrm{ds2}] - k_3[\mathrm{ds2}][\mathrm{c}] + k_{-3}[\mathrm{ts2}] + k_4[\mathrm{ts2}][\mathrm{z}] \tag{9.2}$$

For species drs:

$$0 = \alpha k_2[\mathrm{czs}][\mathrm{czr}] - k_{-2}[\mathrm{drs}] - k_3[\mathrm{drs}][\mathrm{c}] + k_{-3}[\mathrm{trs}] + 2k_4[\mathrm{trs}][\mathrm{z}] \tag{9.3}$$

For species tr2:

$$0 = k_3[\mathrm{dr2}][\mathrm{c}] - k_{-3}[\mathrm{tr2}] - k_4[\mathrm{tr2}][\mathrm{z}] \tag{9.4}$$

For species ts2:

$$0 = k_3[\mathrm{ds2}][\mathrm{c}] - k_{-3}[\mathrm{ts2}] - k_4[\mathrm{ts2}][\mathrm{z}] \tag{9.5}$$

Finally for species trs:

$$0 = k_3[\mathrm{drs}][\mathrm{c}] - k_{-3}[\mathrm{trs}] - 2k_4[\mathrm{trs}][\mathrm{z}] \tag{9.6}$$

Eqs. (9.1)–(9.6) constitute a system of simultaneous linear equations for the unknowns [ds2], [dr2], [drs], [ts2], [tr2], and [trs]. After some straightforward mathematical transformations, all six steady-state concentrations can be obtained as follows:

$$[\mathrm{drs}]_{\mathrm{ss}} = \frac{\alpha k_2[\mathrm{czs}][\mathrm{czr}]}{k_{-2}} \tag{9.7}$$

$$[\mathrm{trs}]_{\mathrm{ss}} = \frac{\alpha k_2 k_3[\mathrm{c}][\mathrm{czs}][\mathrm{czr}]}{(k_{-3} + 2k_4[\mathrm{z}])k_{-2}} \tag{9.8}$$

$$[\mathrm{dr2}]_{\mathrm{ss}} = \frac{k_2[\mathrm{czr}]^2}{k_{-2}} \tag{9.9}$$

$$[\mathrm{ds2}]_{\mathrm{ss}} = \frac{k_2[\mathrm{czs}]^2}{k_{-2}} \tag{9.10}$$

$$[\mathrm{tr2}]_{\mathrm{ss}} = \frac{k_2 k_3[\mathrm{c}][\mathrm{czr}]^2}{(k_{-3} + k_4[\mathrm{z}])k_{-2}} \tag{9.11}$$

$$[\mathrm{ts2}]_{\mathrm{ss}} = \frac{k_2 k_3[\mathrm{c}][\mathrm{czs}]^2}{(k_{-3} + k_4[\mathrm{z}])k_{-2}} \tag{9.12}$$

A further step in using the steady-state assumption is to write the rates of concentration change for the remaining species by substituting the steady-state concentrations obtained above [64]. In this special case, the concentration of czr and czs are of interest.

For czr, the following formula can be derived:

$$\begin{aligned}\frac{\mathrm{d[czr]}}{\mathrm{d}t} &= k_1[\mathrm{c}][\mathrm{z}] - 2k_2[\mathrm{czr}]^2 + 2k_{-2}[\mathrm{dr2}]_{\mathrm{ss}} - \alpha k_2[\mathrm{czs}][\mathrm{czr}] \\ &\quad + k_{-2}[\mathrm{drs}]_{\mathrm{ss}} + k_4[\mathrm{z}][\mathrm{tr2}]_{\mathrm{ss}} + k_4[\mathrm{z}][\mathrm{trs}]_{\mathrm{ss}} \\ &= k_1[\mathrm{c}][\mathrm{z}] + \frac{k_2k_3k_4[\mathrm{c}][\mathrm{z}][\mathrm{czr}]^2}{(k_{-3}+k_4[\mathrm{z}])k_{-2}} + \frac{\alpha k_2k_3k_4[\mathrm{c}][\mathrm{z}][\mathrm{czs}][\mathrm{czr}]}{(k_{-3}+2k_4[\mathrm{z}])k_{-2}}\end{aligned} \tag{9.13}$$

A very similar derivation can be used to find the concentration change for czs:

$$\frac{\mathrm{d[czs]}}{\mathrm{d}t} = k_1[\mathrm{c}][\mathrm{z}] + \frac{k_2k_3k_4[\mathrm{c}][\mathrm{z}][\mathrm{czs}]^2}{(k_{-3}+k_4[\mathrm{z}])k_{-2}} + \frac{\alpha k_2k_3k_4[\mathrm{c}][\mathrm{z}][\mathrm{czs}][\mathrm{czr}]}{(k_{-3}+2k_4[\mathrm{z}])k_{-2}} \tag{9.14}$$

A good way to characterize the relative rates of formation of the two molecules is to divide Eq. (9.13) by Eq. (9.14):

$$\frac{\dfrac{\mathrm{d[czr]}}{\mathrm{d}t}}{\dfrac{\mathrm{d[czs]}}{\mathrm{d}t}} = \frac{k_1 + \dfrac{k_2k_3k_4[\mathrm{czr}]^2}{(k_{-3}+k_4[\mathrm{z}])k_{-2}} + \dfrac{\alpha k_2k_3k_4[\mathrm{czs}][\mathrm{czr}]}{(k_{-3}+2k_4[\mathrm{z}])k_{-2}}}{k_1 + \dfrac{k_3k_2k_4[\mathrm{czs}]^2}{(k_{-3}+k_4[\mathrm{z}])k_{-2}} + \dfrac{\alpha k_2k_3k_4[\mathrm{czs}][\mathrm{czr}]}{(k_{-3}+2k_4[\mathrm{z}])k_{-2}}} \tag{9.15}$$

From the predicted final distributions, it is also clear that the concentration of czr and czs are never different by more than a factor 20. The very low value of α ($=1.0\times10^{-5}$) in this case implies that the last term both in the denominator and enumerator are negligible compared to the second one under any conditions because:

$$[\mathrm{czr}] \gg \alpha[\mathrm{czs}] \quad \text{and} \quad [\mathrm{czs}] \gg \alpha[\mathrm{czr}] \tag{9.16}$$

This leaves the following expression from Eq. (9.15):

$$\frac{\dfrac{d[\mathrm{czr}]}{\mathrm{d}t}}{\dfrac{d[\mathrm{czs}]}{\mathrm{d}t}} \cong \frac{1 + \dfrac{k_2k_3k_4[\mathrm{czr}]^2}{k_1k_{-2}(k_{-3}+k_4[z])}}{1 + \dfrac{k_2k_3k_4[\mathrm{czs}]^2}{k_1k_{-2}(k_{-3}+k_4[z])}} \tag{9.17}$$

From the simple scheme in Table 9.1, a straightforward derivation gives the following formula for the rates of formation of the two enantiomers:

$$\frac{\frac{d[B_R]}{dt}}{\frac{d[B_S]}{dt}} = \frac{1 + \frac{k_c[B_R]^2}{k_u}}{1 + \frac{k_c[B_S]^2}{k_u}} \tag{9.18}$$

In a parametric sense, Eq. (9.18) is very similar to Eq. (9.17). The only difference is the appearance of the additive $k_4[z]$ term in Eq. (9.17). However, the influence of this change can only be minor during the course of a single reaction run, as z is the reagent that is used in the highest initial amount, so its concentration is less than a factor of 2. The notable analogy between the two expressions suggests that the following parameter combination in the model shown in Table 9.2 should have a value that is very close to parameter β in Table 9.1:

$$\beta = \frac{k_c}{k_u V^2 N_A^2} \cong \frac{k_2 k_3 k_4}{k_1 k_{-2}(k_{-3} + k_4[z]) V^2 N_A^2} \tag{9.19}$$

The parameter combination on the right of Eq. (9.19) can be calculated as 1.8×10^{-21} for a value of [z] close to the initial concentration (0.200 mol/dm^3). The value optimized for β in Table 9.1 is 1.6×10^{-21}. Indeed, these two values are very close to each other, which proves the consistency of the line of thought shown in this book chapter.

9.3 CONCLUSIONS

Model calculations carried in this contribution have shown that the expectation values and their standard errors of the *ee* predicted by the very different models of the Soai reaction are strikingly similar. A detailed explanation of this apparent coincidence was presented based on steady-state handling of the full model, which facilitated mathematical simplifications without losing the predictive power of the model. The final results imply that the two models, which are very different in terms of complexity, are kinetically equivalent when the *ee* of the product formed is the only variable of interest. Although the model showing simple second-order autocatalysis has no reasonable chemical background, its predictive power is still the same as that of the full model. This observation points out that simple kinetic models can be very useful in interpreting the observed phenomena in enantioselective autocatalysis if the experimental information that needs to be interpreted is highly limited.

REFERENCES

[1] Soai, K.; Shibata, T.; Morioka, H.; Choji, K. Asymmetric Autocatalysis and Amplification of Enantiomeric Excess of a Chiral Molecule. *Nature* **1995,** *378*, 767–768.

[2] Soai, K.; Shibata, T.; Kowata, Y. Production of Optically Active Pyrimidylalkyl Alcohol by Spontaneous Asymmetric Synthesis. Japan Kokai Tokkyo Koho JP 1997 9-268179. Application date: February 1 and April 18, 1996 (in Japanese).

[3] Soai, K.; Sato, I.; Shibata, T.; Komiya, S.; Hayashi, M.; Matsueda, Y., et al. Asymmetric Synthesis of Pyrimidyl Alkanol without Adding Chiral Substances by the Addition of Diisopropylzinc to Pyrimidine-5-carbaldehyde in Conjunction with Asymmetric Autocatalysis. *Tetrahedron: Asymmetry* **2003,** *14*, 185–188.

[4] Kawasaki, T.; Suzuki, K.; Shimizu, M.; Ishikawa, K.; Soai, K. Spontaneous Absolute Asymmetric Synthesis in the Presence of Achiral Silica Gel in Conjunction with Asymmetric Autocatalysis. *Chirality* **2006,** *18*, 479–482.

[5] Kawasaki, T.; Shimizu, M.; Nashiyama, D.; Ito, M.; Ozawa, H.; Soai, K. Asymmetric Autocatalysis Induced by Meteoritic Amino Acids with Hydrogen Isotope Chirality. *Chem. Commun.* **2009,** 4396–4398.

[6] Kawasaki, T.; Matsamura, Y.; Tsutsumi, T.; Suzuki, K.; Ito, M.; Soai, K. Asymmetric Autocatalysis Triggered by Carbon Isotope ($^{13}C/^{12}C$) Chirality. *Science* **2009,** *324*, 492–495.

[7] Soai, K.; Kawasaki, T. Asymmetric Autocatalysis: Automultiplication of Chiral Molecules. *Chim. Oggi.* **2009,** *27* (6, Supp.), 3–7.

[8] Soai, K.; Kawasaki, T. Asymmetric Autocatalysis: Discovery and State of the Art. In *The Soai Reaction and Related Topic;* Pályi, G., Zucchi, C., Caglioti, L., Eds.; Artestampa - Accademia Nazionale di Scienze, Lettere ed Arti: Modena, 2012; pp 9–34.

[9] Mineki, H.; Hanasaki, T.; Matsumoto, A.; Kawasaki, T.; Soai, K. Asymmetric Autocatalysis Initiated by Achiral Nucleic Acid Base Adenine: Implications on the Origin of Homochirality of Biomolecules. *Chem. Commun.* **2012,** *48*, 10538–10540.

[10] Shindo, H.; Shirota, Y.; Niki, K.; Kawasaki, T.; Suzuki, K.; Araki, Y., et al. Asymmetric Autocatalysis Induced by Cinnabar: Observation of the Enantioselective Adsorption of a 5-Pyrimidyl Alkanol on the Crystal Surface. *Angew. Chem., Int. Ed.* **2013,** *52*, 9135–9138.

[11] Soai, K.; Kawasaki, T.; Matsumoto, A. The Origins of Homochirality Examined by Using Asymmetric Autocatalysis. *Chem. Rec.* **2014,** *14*, 70–83.

[12] Soai, K.; Kawasaki, T.; Matsumoto, A. Asymmetric Autocatalysis of Pyrimidyl Alkanol and Its Application to the Study on the Origin of Homochirality. *Acc. Chem. Res.* **2014,** *47*, 3643–3654.

[13] Matsumoto, A.; Abe, T.; Hara, A.; Tobita, T.; Sasagawa, T.; Kawasaki, T., et al. Crystal Structure of Isopropylzinc Alkoxide of Pyrimidyl Alkanol: Mechanistic Insights for Asymmetric Autocatalysis with Amplification of Enantiomeric Excess. *Angew. Chem., Int. Ed.* **2015,** *54*, 15218–15221.

[14] Matsumoto, A.; Ozawa, H.; Inumaru, A.; Soai, K. Asymmetric Induction by Retgersite, Nickel Sulfate Hexahydrate, in Conjunction with Asymmetric Autocatalysis. *New J. Chem.* **2015,** *39*, 6742–6745.

[15] Fülöp, O.; Barabás, B. Impact of the Soai-Autocatalysis on Natural Sciences. *J. Math. Chem.* **2016,** *54*, 10–17.

[16] Barabás, B.; Caglioti, L.; Micskei, K.; Zucchi, C.; Pályi, G. Isotope Chirality and Asymmetric Autocatalysis: A Possible Entry to Biological Chirality. *Orig. Life Evol. Biosph.* **2008,** *38*, 317–327.

[17] Barabás, B.; Kurdi, R.; Pályi, G. Natural Abundance Isotopic Chirality in the Reagents of the Soai Reaction. *Symmetry* **2016,** *8*, 2.
[18] Gridnev, I. D.; Serafimov, J. M.; Quiney, H.; Brown, J. M. Reflections on Spontaneous Asymmetric Synthesis by Amplifying Autocatalysis. *Org. Biomol. Chem.* **2003,** *1*, 3811–3819.
[19] Gridnev, I. D. Chiral Symmetry Breaking in Chiral Crystallization and Soai Autocatalytic Reaction. *Chem. Lett.* **2006,** *35*, 148–153.
[20] Caglioti, L.; Barabás, B.; Faglioni, F.; Florini, N.; Lazzeretti, P.; Maioli, M., et al. On the Track of Absolute Enantioselective Catalysis. *Chim. Oggi.* **2008,** *26*, 30–32.
[21] Barabás, B.; Caglioti, L.; Faglioni, F.; Florini, N.; Lazzaretti, P.; Maioli, M., et al. On the Traces of Absolute Enantioselective Synthesis. *AIP Conf. Proc.* **2008,** *963B*, 1150–1152.
[22] Pályi, G.; Zucchi, C.; Caglioti, L., Eds. *Progress in Biological Chirality;* Elsevier: Oxford, 2004.
[23] Wu, M.; Walker, S. I.; Higgs, P. G. Autocatalytic Replication and Homochirality in Biopolymers: Is Homochirality a Requirement of Life or a Result of it? *Astrobiology* **2012,** *12*, 818–829.
[24] Viedma, C. Chiral Symmetry Breaking During Crystallization: Complete Chiral Purity Induced by Nonlinear Autocatalysis and Recycling. *Phys. Rev. Lett.* **2005,** *94*, 065504.
[25] Gridnev, I. D.; Vorobiev, A. K. Quantification of Sophisticated Equilibria in the Reaction Pool and Amplifying Catalytic Cycle of the Soai Reaction. *ACS Catal.* **2012,** *2*, 2137–2149.
[26] Gridnev, I. D.; Vorobiev, A. K. On the Origin and Structure of the Recently Observed Acetal in Soai Reaction. *Bull. Chem. Soc. Jpn.* **2015,** *88*, 333–340.
[27] Romagnoli, C.; Sieng, B.; Amedjkouh, M. Asymmetric Amplification Coupling Enantioselective Autocatalysis and Asymmetric Induction for Alkylation of Azaaryl Aldehydes. *Eur. J. Org. Chem.* **2015,** *19*, 4087–4092.
[28] Faglioni, F.; Lazzeretti, P.; Pályi, G. Parity Violation Energy of 5-Pyrimidyl Alkanol, a Chiral Autocatalytic Molecule. *Chem. Phys. Lett.* **2007,** *435*, 346–349.
[29] Schiaffino, L.; Ercolani, G. Unraveling the Mechanism of the Soai Asymmetric Autocatalytic Reaction by First-Principles Calculations: Induction and Amplification of Chirality by Self-assembly of Hexanuclear Complexes. *Angew. Chem., Int. Ed.* **2008,** *47*, 6832–6835.
[30] Ercolani, G.; Schiaffino, L. Mechanistic Insights into the Soai Reaction from Formal Kinetics and Density Functional Theory Calculations. In *The Soai Reaction and Related Topic;* Pályi, G., Zucchi, C., Caglioti, L., Eds.; Artestampa - Accademia Nazionale di Scienze, Lettere ed Arti: Modena, 2012; pp 331–346.
[31] Micskei, K.; Póta, G.; Caglioti, L.; Pályi, G. Empirical Description of Chiral Autocatalysis. *J. Phys. Chem. A* **2006,** *110*, 5982–5984.
[32] Barabás, B.; Zucchi, C.; Maioli, M.; Micskei, K.; Pályi, G. Stochastic and Empirical Models of the Absolute Asymmetric Synthesis by the Soai-Autocatalysis. *J. Mol. Model.* **2015,** *21*, 33.
[33] Buhse, T. A Tentative Kinetic Model for Chiral Amplification in Autocatalytic Alkylzinc Additions. *Tetrahedron: Asymmetry* **2003,** *14*, 1055–1061.
[34] Rivera Islas, J.; Lavabre, D.; Jean-Michel Grevy, J. M.; Hernández Lamoneda, R.; Rojas Cabrera, H.; Micheau, J. C., et al. Mirror-Symmetry Breaking in the Soai Reaction: A Kinetic Understanding. *Proc. Natl. Acad. Sci. USA* **2005,** *101*, 5732–5736.
[35] Hochberg, D.; Zorzano, M. P. Mirror Symmetry Breaking as a Problem in Dynamic Critical Phenomena. *Phys. Rev. E* **2007,** *76*, 021109.
[36] Lavabre, D.; Micheau, J. C.; Rivera Islas, J.; Buhse, T. Enantioselectivity Reversal by Achiral Additives in the Soai Reaction: A Kinetic Understanding. *J. Phys. Chem. A* **2007,** *111*, 281–286.

[37] Hochberg, D. Effective Potential and Chiral Symmetry Breaking. *Phys. Rev. E* **2010,** *81*, 016106.

[38] Micheau, J. C.; Coudret, C.; Buhse, T. Systems Chemistry of the Soai Reaction. In *The Soai Reaction and Related Topic;* Pályi, G., Zucchi, C., Caglioti, L., Eds.; Artestampa - Accademia Nazionale di Scienze, Lettere ed Arti: Modena, 2012; pp 169–196.

[39] Plasson, R. Mirror Symmetry Breaking in High Order Autocatalytic Systems. In *The Soai Reaction and Related Topic;* Pályi, G., Zucchi, C., Caglioti, L., Eds.; Artestampa - Accademia Nazionale di Scienze, Lettere ed Arti: Modena, 2012; pp 275–296.

[40] Ribó, J. M.; Crusats, J.; Moyano, A.; Veintemillas-Verdaguer, S. Actual Chemical Scenarios for Absolute Asymmetric Synthesis. In *The Soai Reaction and Related Topic;* Pályi, G., Zucchi, C., Caglioti, L., Eds.; Artestampa - Accademia Nazionale di Scienze, Lettere ed Arti: Modena, 2012; pp 297–318.

[41] Micheau, J. C.; Coudret, C.; Cruz, J. M.; Buhse, T. Amplification of Enantiomeric Excess, Mirror-Image Symmetry Breaking and Kinetic Proofreading in Soai Reaction Models with Different Oligomeric Orders. *Phys. Chem. Chem. Phys.* **2012,** *14*, 13239–13248.

[42] Ribó, J. M.; Blanco, C.; Crusats, J.; El-Hachemi, Z.; Hochberg, D.; Moyano, A. Absolute Asymmetric Synthesis in Enantioselective Autocatalytic Reaction Networks: Theoretical Games, Speculations on Chemical Evolution and Perhaps a Synthetic Option. *Chem. Eur. J.* **2014,** *20*, 17250–17271.

[43] Lente, G. Homogeneous Chiral Autocatalysis: A Simple, Purely Stochastic Kinetic Model. *J. Phys. Chem. A* **2004,** *108*, 9475–9478.

[44] Lente, G. Stochastic Kinetic Models of Chiral Autocatalysis: A General Tool for the Quantitative Interpretation of Total Asymmetric Synthesis. *J. Phys. Chem. A* **2005,** *109*, 11058–11063.

[45] Caglioti, L.; Zucchi, C.; Pályi, G. Single-Molecule Chirality. *Chim. Oggi.* **2005,** *23*, 38–43.

[46] Caglioti, L.; Hajdu, C.; Holczknecht, O.; Zékány, L.; Zucchi, C.; Micskei, K., et al. The Concept of Racemates and the Soai-Reaction. *Viva Origino* **2006,** *34*, 62–80.

[47] Saito, Y.; Sugimori, T.; Hyuga, H. Stochastic Approach to Enantiomeric Excess Amplification and Chiral Symmetry Breaking. *J. Phys. Soc. Jpn.* **2007,** *76*, 044802.

[48] Caglioti, L.; Pályi, G. Chiral Chemistry of Single Molecules. *Chim. Oggi.* **2008,** *26*, 41–42.

[49] Sugimori, T.; Hyuga, H.; Saito, Y. Fluctuation Induced Homochirality. *J. Phys. Soc. Jpn.* **2008,** 77, 064606.

[50] Barabás, B.; Caglioti, L.; Micskei, K.; Pályi, G. Data-Based Stochastic Approach to Absolute Asymmetric Synthesis by Autocatalysis. *Bull. Chem. Soc. Jpn.* **2009,** *82*, 1372–1376.

[51] Dóka, É.; Lente, G. Mechanism-Based Chemical Understanding of Chiral Symmetry Breaking in the Soai Reaction. A Combined Probabilistic and Deterministic Description of Chemical Reactions. *J. Am. Chem. Soc.* **2011,** *133*, 17878–17881.

[52] Lente, G. Stochastic Interpretation of the Asymmetry of Enantiomeric Distribution Observed in the Absolute Asymmetric Soai Reaction. *Tetrahedron: Asymmetry* **2011,** *22*, 1595–1599.

[53] Lente, G. Stochastic Modeling of the Soai Reaction. In *The Soai Reaction and Related Topic;* Pályi, G., Zucchi, C., Caglioti, L., Eds.; Artestampa - Accademia Nazionale di Scienze, Lettere ed Arti: Modena, 2012; pp 123–147.

[54] Saito, Y.; Hyuga, H. ee: Amplification and Homochirality in Chemical Reaction. In *The Soai Reaction and Related Topic;* Pályi, G., Zucchi, C., Caglioti, L., Eds.; Artestampa - Accademia Nazionale di Scienze, Lettere ed Arti: Modena, 2012; pp 319–330.

[55] Saito, Y.; Hyuga, H. Colloquium: Homochirality: Symmetry Breaking in Systems Driven far from Equilibrium. *Rev. Mod. Phys.* **2013,** *85*, 603–621.
[56] Érdi, P.; Tóth, J. *Mathematical Models of Chemical Reactions;* Manchester University Press: Manchester, 1989; pp 91–161.
[57] Érdi, P.; Lente, G. *Stochastic Chemical Kinetics: Theory and (Mostly) Systems Biological Applications;* Springer: New York, 2014.
[58] Hochberg, D.; Zorzano, M. P. Reaction-Noise Induced Homochirality. *Chem. Phys. Lett.* **2006,** *431*, 185–189.
[59] Cruz, J. M.; Parmananda, P.; Buhse, T. Noise-Induced Enantioselection in Chiral Autocatalysis. *J. Phys. Chem. A* **2008,** *112*, 1673–1676.
[60] Micskei, K.; Rábai, G.; Gál, E.; Caglioti, L.; Pályi, G. Oscillatory Symmetry Breaking in the Soai Reaction. *J. Phys. Chem. B* **2008,** *112*, 9196–9200.
[61] Hochberg, D. Mirror Symmetry Breaking and Restoration: The Role of Noise and Chiral Bias. *Phys. Rev. Lett.* **2009,** *102*, 248101.
[62] Barabás, B.; Caglioti, L.; Zucchi, C.; Maioli, M.; Gál, E.; Micskei, K., et al. Violation of Distribution Symmetry in Statistical Evaluation of Absolute Enantioselective Synthesis. *J. Phys. Chem. B* **2007,** *111*, 11506–11510.
[63] Lente, G. The Effect of Parity Violation on Kinetic Models of Enantioselective Autocatalyis. *Phys. Chem. Chem. Phys.* **2007,** *9*, 6134–6141.
[64] Lente, G. *Deterministic Kinetics in Chemistry and Systems Biology;* Springer: New York, 2014.

CHAPTER 10

Structural Study of Asymmetric Autocatalysis by X-Ray Crystallography

Arimasa Matsumoto, Tsuneomi Kawasaki and Kenso Soai
Department of Applied Chemistry, Tokyo University of Science, Kagurazaka, Shinjuku, Tokyo, Japan

Abstract

Asymmetric amplification during self-replication is a key feature that is used to explain the origin of homochirality. Asymmetric autocatalysis of 5-pyrimidyl alkanol in the asymmetric addition of diisopropylzinc to pyrimidine-5-carbaldehyde is a unique example of this phenomenon. To obtain mechanistic insight into this reaction, the crystallization of possible intermediates, specifically zinc alkoxides of this 5-pyrimidyl alkanol, was investigated. Single-crystal X-ray diffraction analysis of the obtained alkoxide crystals reveals the existence of tetramer or higher oligomer structures in this asymmetric autocatalytic system.

Keywords: Asymmetric autocatalysis; amplification of *ee*; zinc alkoxide; X-ray crystallography

10.1 INTRODUCTION

Biological homochirality is an unresolved question, and has consequently attracted much attention in a variety of research fields [1]. There are several possible origins of the induction of a small imbalance of enantiomeric excess (*ee*). Propagation and amplification of chirality generated from the initial breaking of symmetry are important topics in explaining the evolution of homochirality. One possible scenario for the emergence of homochirality from the trace imbalance of *ee* is asymmetric autocatalysis with amplification of *ee*. In asymmetric autocatalysis, a chiral product serves as an asymmetric catalyst for its own production with the same chirality. Thus, asymmetric autocatalysis is an automultiplication of the chiral compound. When asymmetric autocatalysis takes place with enantiomeric amplification during the reaction, homochirality from the small imbalance of *ee* is achieved. Although the theory of asymmetric amplification during

Advances in Asymmetric Autocatalysis and Related Topics
DOI: http://dx.doi.org/10.1016/B978-0-12-812824-4.00010-1

self-replication was first proposed in 1953 [2], the actual chemical reaction involving asymmetric autocatalysis that leads to high levels of amplification of *ee* was not developed until long after.

We have found asymmetric autocatalysis with amplification of *ee* in a real chemical reaction (Scheme 10.1) [3–5]. In the addition reaction of diisopropylzinc (i-Pr_2Zn) to the pyrimidine-5-carbaldehyde **1**, (*S*)-pyrimidyl alkanol **2** acts as an asymmetric catalyst to form (*S*)-**2**. During this reaction, asymmetric amplification occurs, and (*S*)-**2** with high *ee* is obtained as the final product, even when the initial *ee* of the catalyst is low. Initially, this asymmetric autocatalysis was found in the reaction with pyridine-3-carbaldehyde; asymmetric amplification during the reaction was achieved in the reaction with pyrimidine-5-carbaldehyde and 2-methylpyrimidine-5-carbaldehyde. Finally, almost perfect amplification by 2-alkynyl analogues was found; excess amplification from about 5×10^{-5}% *ee* to >99.5% *ee* (product) was achieved by consecutive asymmetric autocatalysis reactions.

This unique high amplification efficiency enabled the detection of trace imbalances of *ee*, induced by various chiral initiators, such as circularly polarized light [6], crystal chirality of organic [7] and inorganic [8] compounds, various chiral materials [9], and chirality due to isotope substitution [10]. The alkanol with high *ee* was obtained with corresponding absolute configurations to the chiral initiator. The high efficiency also enabled spontaneous absolute asymmetric synthesis [11].

To understand the mechanism of this unique reaction, approaches based on kinetic experiments [12], reaction modeling [13], nuclear magnetic resonance (NMR) [14], and density functional theory (DFT) calculation [15]

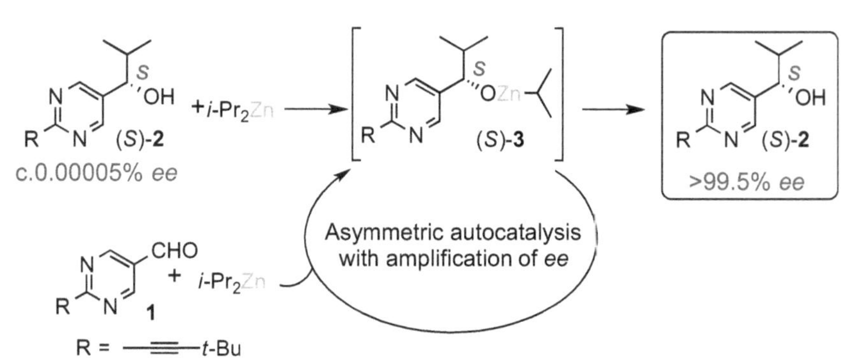

Scheme 10.1 Asymmetric autocatalysis of 5-pyrimidyl alkanol.

have been considered by various groups. Sometimes, in nonautocatalytic asymmetric reactions a product with higher *ee* than the catalyst was observed, as a nonlinear effect [16]. This could be explained by catalyst dimerization. For example, the preferable formation of an unreactive racemic dimer results in the amplification of the *ee* of the actual reactive monomer in the reaction system, and thus amplification of the *ee* of the product. Another possible explanation is that the dimer is the actual catalyst and that an enantiopure dimer has a higher reactivity than a racemic dimer.

Early studies of the kinetics and amplification rate of *ee* during the reaction revealed that the active monomer model from the analogue of the nonautocatalytic alkylzinc addition reaction [16c] is inadequate to explain the amplification and to support the dimer as a catalyst [12b]. Detailed kinetic studies revealed almost second-order kinetics in aldehydes [12d]. The reaction pass that a dimeric catalyst reacts with two aldehydes via a tetrameric transition state was suggested based on these kinetics. A DFT-calculation study based on this mechanism suggests that the amplification is based on the difference of reactivity of the enantiopure and the racemic dimers [14b]. NMR observations support the existence of dimeric species [15d].

However, further reaction-model studies indicated that this model is still inadequate for explaining the significant amplification observed in the actual reaction [12a,c]. It was proposed that tetramers or higher oligomers be included in the reaction models [15c,e]. A recent NMR study that included diffusion-ordered spectroscopy (DOSY) analysis suggested the existence of tetrameric or higher oligomeric species [14c,d]. Consideration of higher oligomeric species in the reaction system also provides a good explanation for the characteristic temperature dependence of the reaction rate [14c]. It can be concluded from previous results that the oligomer structure plays a very important role in understanding the mechanism of asymmetric autocatalysis. Although a structure study of various types of alkoxide aggregates by DFT calculation has been performed [15h], no direct observation of an asymmetric autocatalyst, i.e., the isopropylzinc alkoxide of pyrimidyl alkanol **2**, has yet been undertaken by X-ray diffraction analysis; even crystallographic analysis of the structure of the *i*-Pr_2Zn complex has rarely been studied [17].

We therefore studied the structure of asymmetric autocatalysis by single-crystal X-ray diffraction analysis. We found that the various alkoxide structures depend on a variety of reaction conditions (Fig. 10.1) [18].

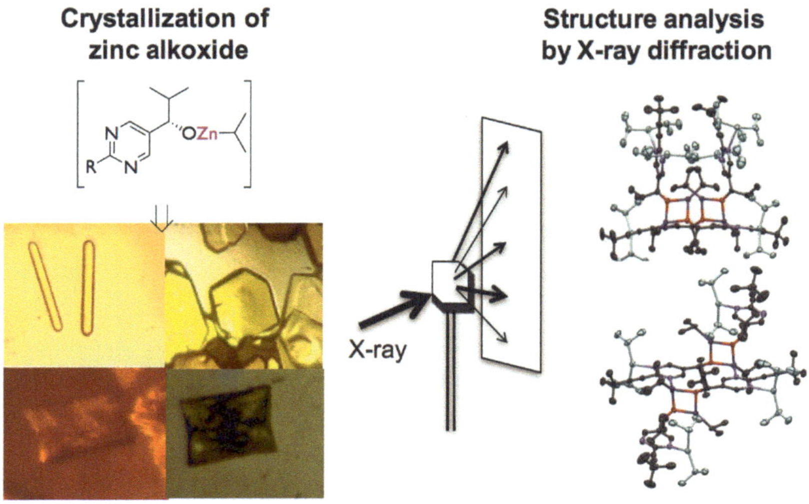

Figure 10.1 X-ray structure analysis of asymmetric autocatalysis.

10.2 RESULTS AND DISCUSSION

10.2.1 Crystallization and Single-Crystal Analysis of an Asymmetric Autocatalyst

Crystallization conditions of a zinc alkoxide are shown in Scheme 10.2. A single crystal of the enantiopure zinc alkoxide of (*S*)-5-pyrimidyl alkanol **2** was obtained by the addition of excess neat *i*-Pr_2Zn (8.4 equiv.) at room temperature to the alkanol **2**. All alkoxides were dissolved by heating at c.80°C. After being left to stand at room temperature for several days, the crystal of alkoxide was obtained as a slightly yellow rod (crystal **I**). Racemic zinc alkoxide was obtained by using racemic alkanol **2** instead of the enantiopure alkanol. Plate-like crystals were obtained (crystal **II**). Reducing the amount of diisopropylzinc in the crystallization conditions afforded other types of crystals, **III** and **IV**, for enantiopure and racemic alkoxides, respectively. In the presence of tetrahydrofuran (THF), the enantiopure alkoxide afforded the THF-containing crystal **V**. These obtained crystals were covered with Nujol to prevent decomposition due to air and moisture.

The crystals were immediately mounted on the single-crystal X-ray diffractometer and cooled to 100 K by nitrogen gas flow. X-ray diffraction analysis results are summarized in Table 10.1. The enantiopure alkoxide crystallized in the orthorhombic space group $P2_12_12_1$ as a tetramer with 6 *i*-Pr_2Zn (crystal **I**, Scheme 10.2). The racemic alkoxide crystal **II** also has

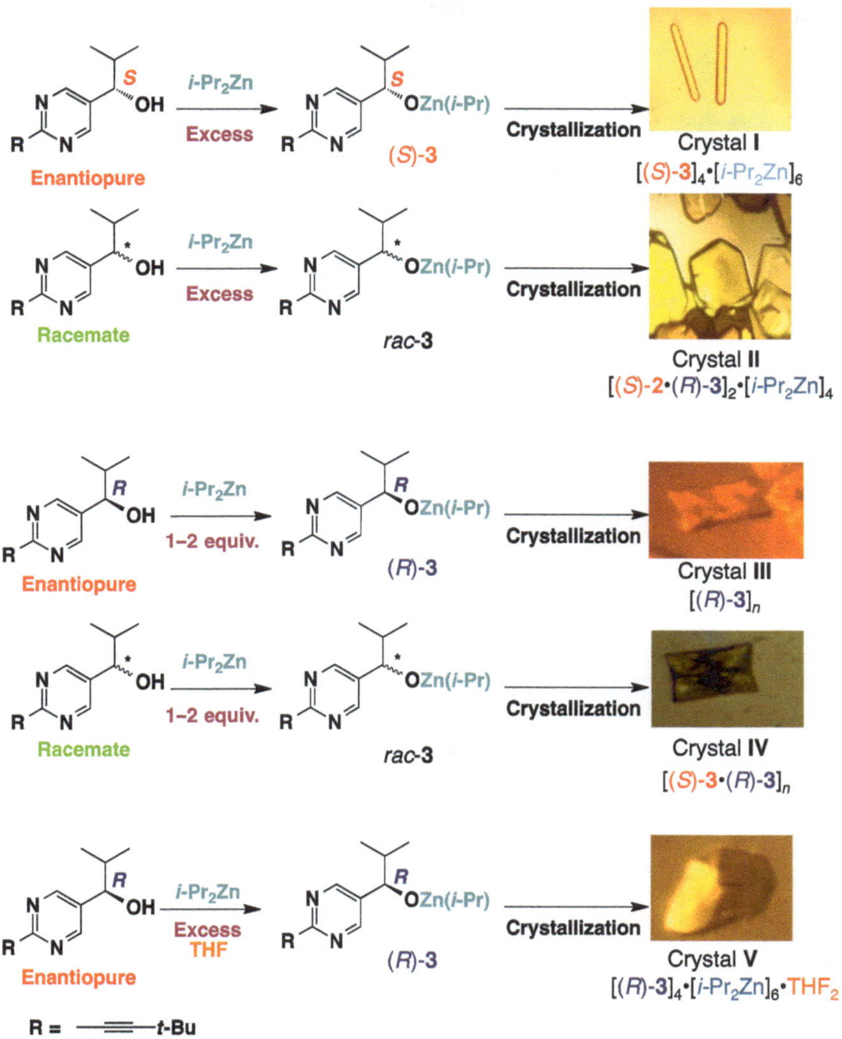

Scheme 10.2 Crystallization of zinc alkoxide of pyrimidyl alkanol **2**.

a tetramer structure with 4 i-Pr_2Zn in the monoclinic space group $C2/c$. Crystals **III** and **IV** have an oligomeric structure in the space groups $P2_1$ and $P2_1/c$, respectively. Details of the structures are discussed in the following sections.

10.2.2 Tetramer Structure

In the crystal **I** prepared from enantiopure alkoxide with an excess amount of diisopropylzinc, the zinc alkoxide forms a tetramer structure

Table 10.1 Summary of single-crystal diffraction analysis of alkoxide crystals **I–V**

	Crystal I **CCDC 1420664**	**Crystal II** **CCDC 1420666**	**Crystal III** **CCDC 1420663**	**Crystal IV** **CCDC 1420665**	**Crystal V** **CCDC 1420667**
Formula	$C_{104}H_{188}N_8O_4Zn_{10}$	$C_{92}H_{160}N_8O_4Zn_8$	$C_{34}H_{52}N_4O_2Zn_2$	$C_{34}H_{52}N_4O_2Zn_2$	$C_{112}H_{204}N_8O_6Zn_{10}$
Configuration of alkanol	Enantiopure Tetramer only *S*	Racemic (*meso*) Tetramer	Enantiopure Oligomer only *R*	Racemic (*meso*) Oligomer	Enantiopure Tetramer with THF only *R*
Z	4	4	2	4	2
Space group	$P2_12_12_1$	$C2/c$	$P2_1$	$P2_1/c$	$P2_1$
a (Å)	16.8592 (3)	29.93 (2)	6.2846 (4)	6.3721 (9)	11.4499 (7)
b (Å)	21.3576 (4)	10.151 (8)	10.9782 (8)	10.7370 (15)	25.1924 (15)
c (Å)	33.1066 (6)	34.38 (3)	25.3989 (19)	25.601 (4)	22.7023 (14)
α (deg.)	90	90	90	90	90
β (deg.)	90	104.154 (10)	95.355 (3)	96.730 (2)	103.4220 (10)
γ (deg.)	90	90	90	90	90
Radiation	Cu Kα	Mo Kα	Mo Kα	Mo Kα	Mo Kα
Temp. (K)	100	100	100	100	100
R_1	0.0416	0.0533	0.0655	0.0527	0.0340
$wR2$	0.1110	0.1469	0.2147	0.1409	0.0810
Flack parameter	0.022 (19)	–	0.12 (4)	–	0.005 (4)

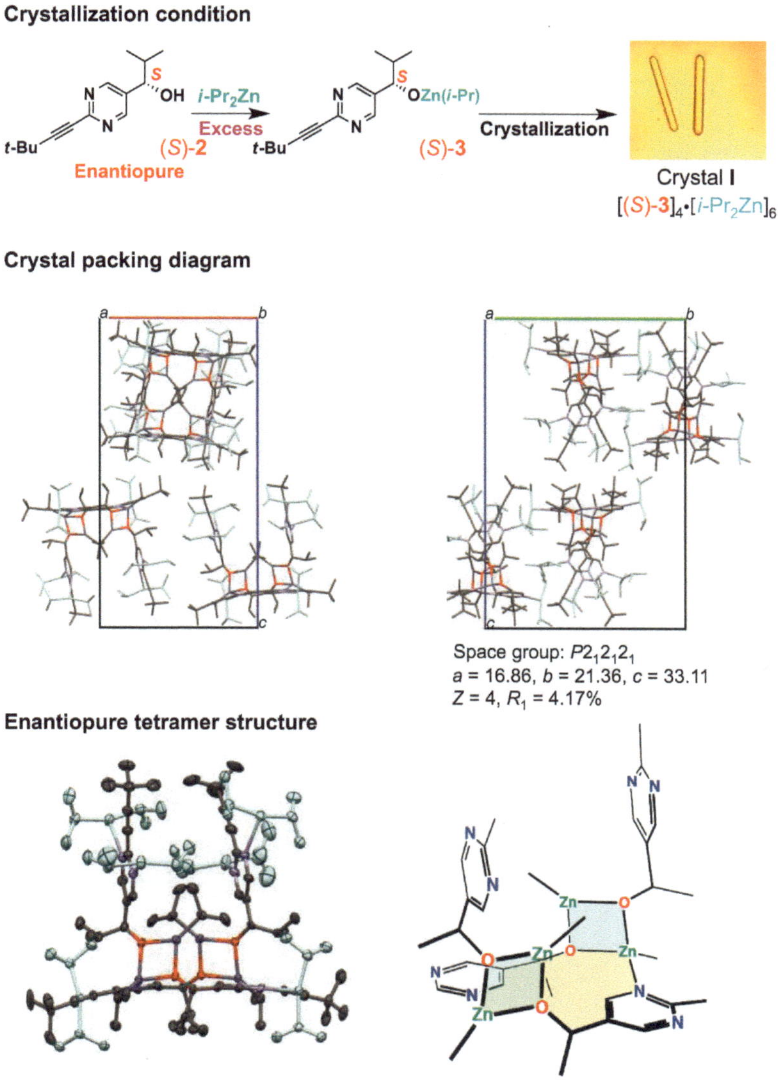

Figure 10.2 Crystal structure of enantiopure tetramer crystal **I**.

(Fig. 10.2). Two alkoxides form a square Zn_2O_2 dimer. One of the alkoxides in this Zn_2O_2 dimer is bridged to another dimer by two Zn–N coordination bonds to form a 12-membered macrocyclic structure. The two Zn atoms that do not form part of the macrocyclic structure remain in a coordinatively unsaturated 3-coordination state. The unsaturated zinc should have higher catalytic activity. When one ignores the subtle difference of the isopropyl group direction, this tetramer has C_2 symmetry.

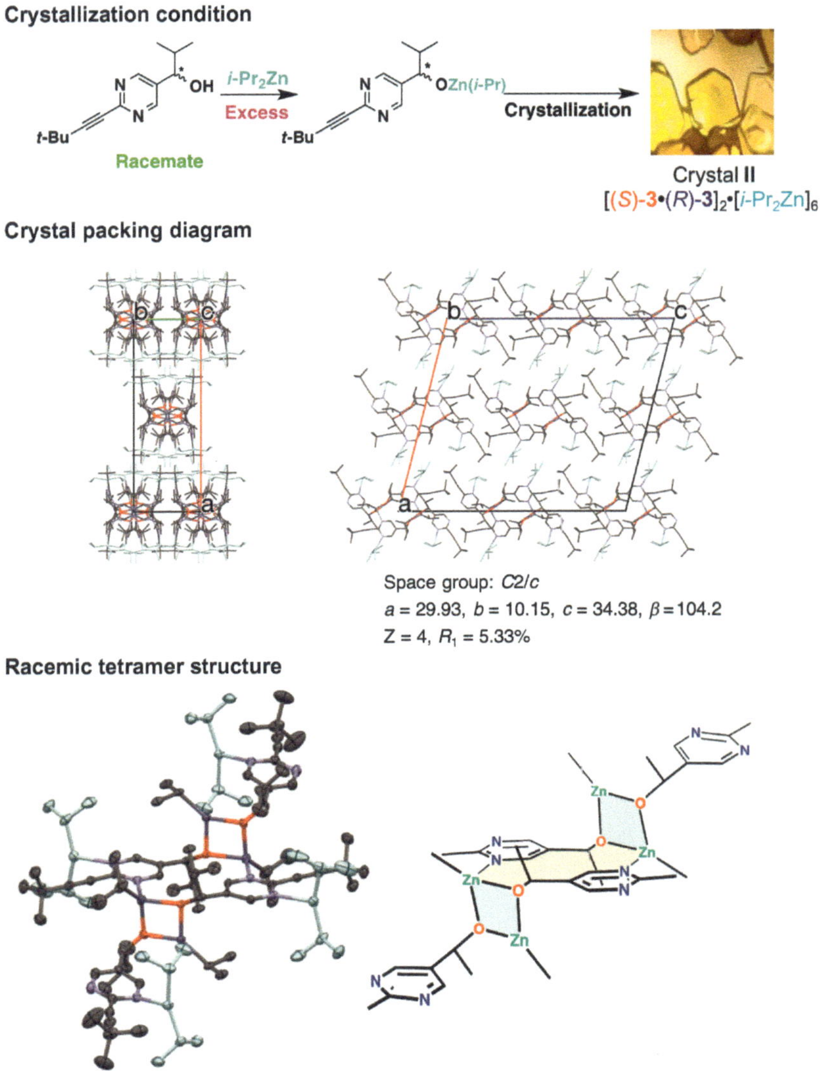

Figure 10.3 Crystal structure of racemic tetramer crystal **II**.

The racemic alkoxide crystal **II** also has a tetramer structure (Fig. 10.3). Again, two alkoxides form square Zn_2O_2 dimers, and Zn–N coordination bonds afford a 12-membered macrocyclic structure. In this structure, the Zn_2O_2 dimers are constructed by *R*- and *S*-alkoxides. The 12-membered macrocycle also contains the *R*- and *S*-alkoxides pair. The total tetramer structure has C_i symmetry. It should be noted that DFT calculation studies predict that these types of tetramer coordination structures are preferred for the isopropylzinc alkoxide, whereas the less hindered dimethylzinc alkoxide prefers other types of structures [15c].

In the chiral tetramer **I**, six *i*-Pr_2Zn moieties coordinate to the nitrogen atoms that do not coordinate to the Zn_2O_2 square. In the racemic alkoxide crystal **II**, four *i*-Pr_2Zn moieties coordinate. These coordinated isopropyl zinc moieties have a slightly bent C—Zn—C bond, c.156—164°, and a slightly longer C—Zn bond, c.1.99—2.01 Å, compared to a normal linear structure, due to nitrogen coordination [17,19].

Both the enantiopure and racemic tetramers have the same coordination structure. However, the most remarkable difference between the structures of the two tetramers is the conformational difference around the macrocycles (Fig. 10.4). In the enantiopure tetramer, the Zn_2O_2 square is positioned on the same side of the 12-membered macrocycle and the pyrimidyl rings form a face-to-face structure. On the contrary, in the racemic tetramer structure, the Zn_2O_2 square is positioned on different sides of the macrocycle. This enantiopure tetramer structure is quite different from the predicted structure from previous calculation studies [15]. Due to the steric hindrance of this face-to-face structure, the Zn—N distance in the macrocycle is slightly extended, to 2.26 Å, for the enantiopure tetramer. Thus, the enantiopure tetramer seems less stable than the racemic one. The stability difference was also confirmed by DFT calculation. Of course, the crystal structure cannot be directly compared with a solution structure. This DFT calculation is preliminary; it ignores the solvent effect. The characteristic conformation of the enantiopure tetramer structure resulted in 4.8 kcal/mol higher energy than for the racemic tetramer.

10.2.3 Structure of Oligomers

Reducing the amount of diisopropylzinc in the crystallization of an enantiopure alkoxide afforded crystal **III** (Fig. 10.5). In this crystal, the alkoxide also formed a Zn_2O_2 dimer and the dimers connected by Zn—N coordination to form 12-membered macrocycles. When the amount of diisopropylzinc is reduced, there is no coordinative diisopropylzinc in the structure, and the remaining nitrogen atom in the pyrimidine ring forms another 12-membered macrocycle with another dimer, to form the one-dimensional (1D) oligomer structure by the alternating reputation of Zn_2O_2 squares and 12-membered macrocycles. This 1D oligomeric structure is also possible in the racemic zinc alkoxide in crystal **IV** (Fig. 10.6), crystallized under similar conditions. In contrast to the tetramer structure,

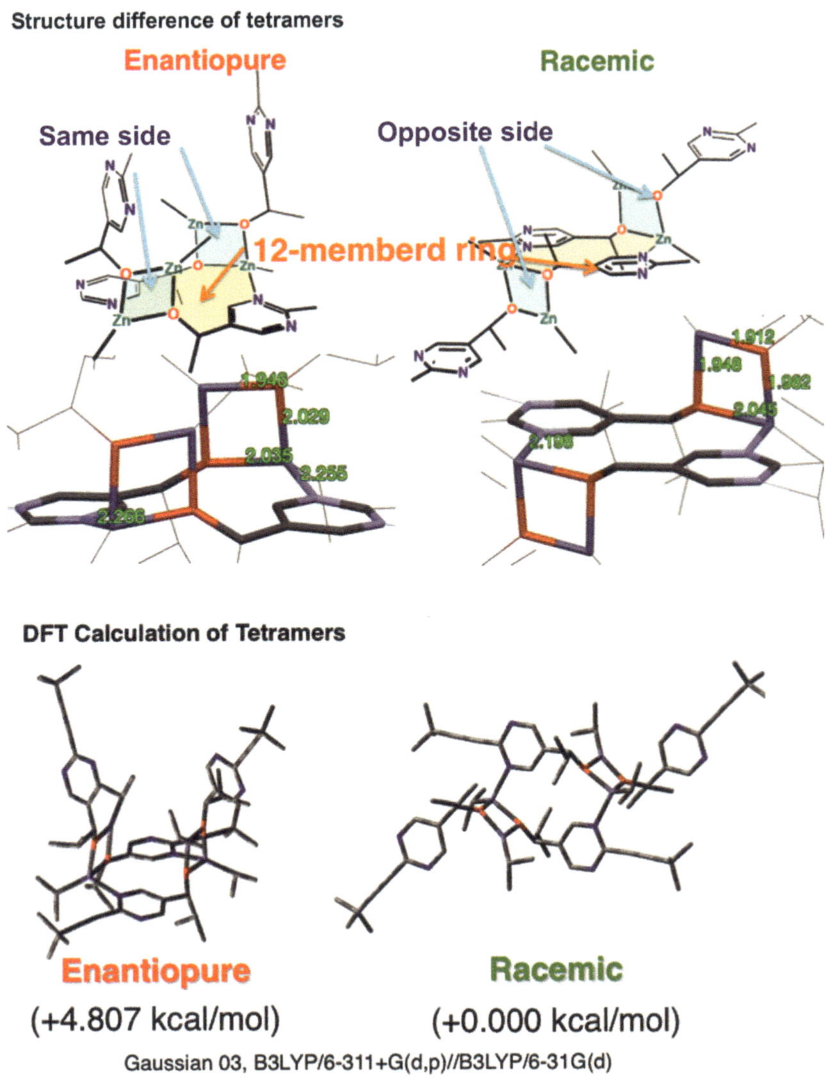

Figure 10.4 Structure difference between enantiopure and racemic tetramers.

Zn_2O_2 squares are always positioned on the opposite side of the macrocycle in both the enantiopure and racemic oligomers, and both oligomer structures have similar conformations. Indeed, when ignoring the direction of the stereochemistry of the isopropyl group, the racemic and enantiopure oligomer structures are almost superimposable. These results clearly suggest that a higher oligomeric structure than a tetramer can exist in the reaction mixture, depending on the concentration of the alkanol and *i*-Pr_2Zn.

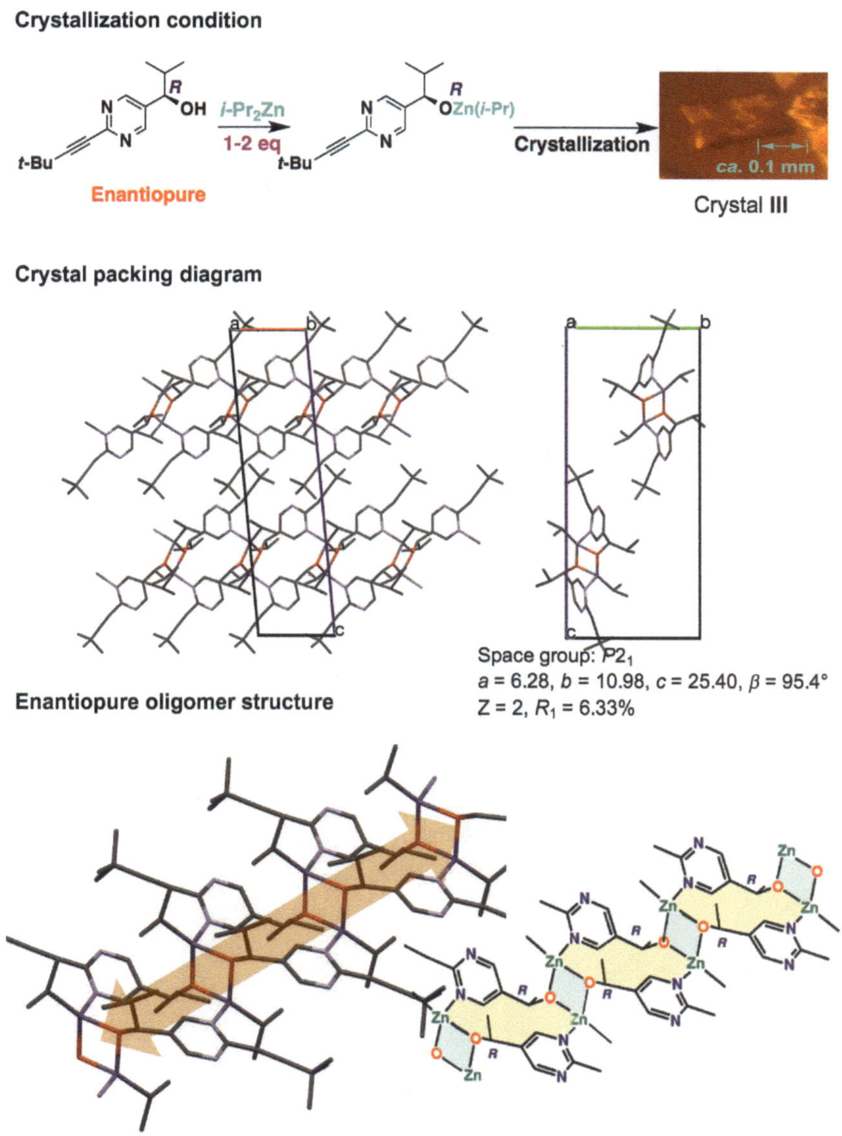

Figure 10.5 Enantiopure oligomer structure of crystal **III**.

10.2.4 Structural Change Due to Solvent

Another interesting crystal structure change is due to the solvent effect. In the crystallization of enantiopure alkoxides in THF, tetramer crystal **V** containing THF was obtained. Once again, the tetramer of zinc alkoxide of 5-pyrimydyl alkanol form the Zn_2O_2 square and 12-membered

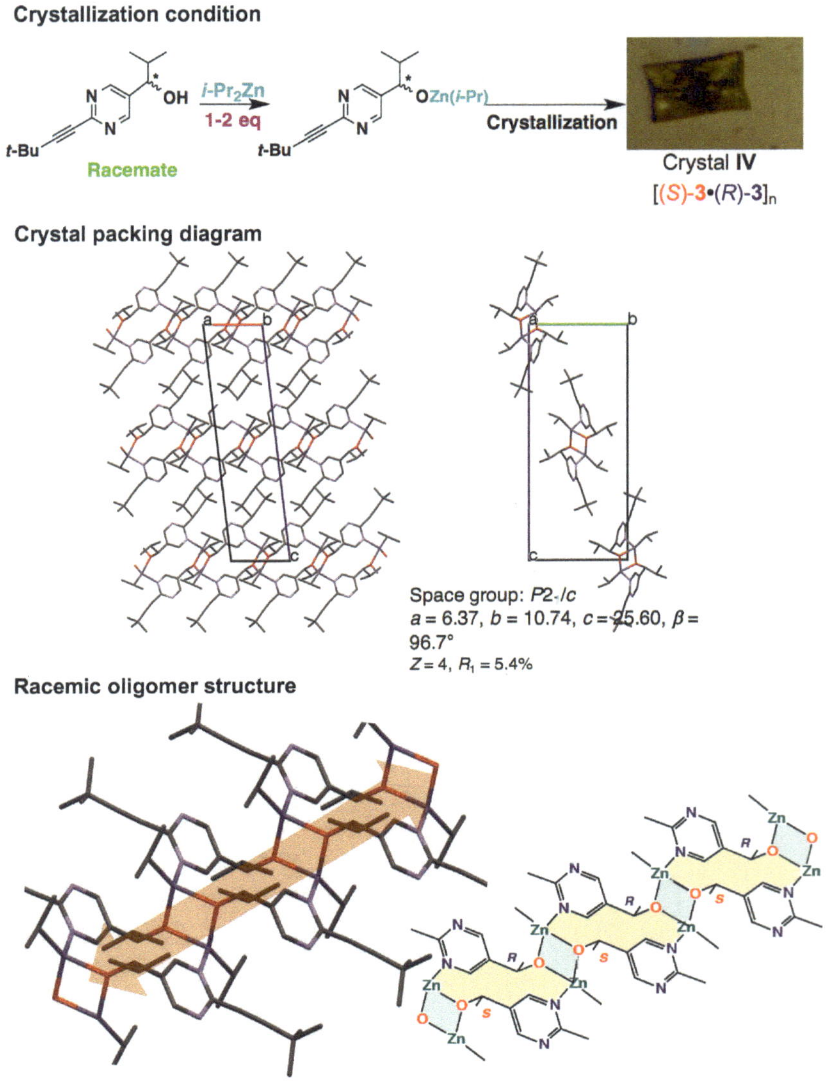

Figure 10.6 Racemic oligomer structure of crystal **IV**.

macrocycle, with the square dimer positioned on the same side of the 12-membered ring. However, THF molecules are coordinated to one of the Zn atoms of Zn_2O_2. Thus, there are no coordinatively unsaturated zinc atoms, which are considered to have high reactivity. This feature offers a good reason why the asymmetric autocatalysis reaction of pyrimidyl-5-alkanol is very sluggish in the presence of THF (Fig. 10.7).

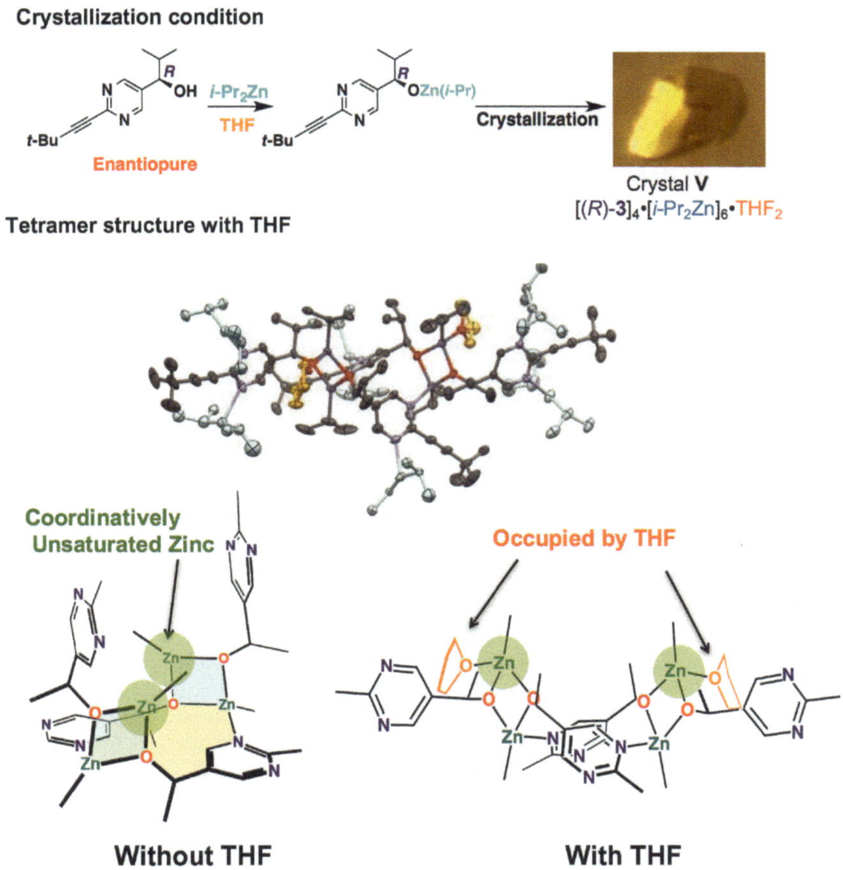

Figure 10.7 THF coordinated structure of crystal **V** and its difference from crystal **I**.

10.3 CONCLUSION

In this chapter, various alkoxide structures of pyrimidine-5-carbaldehyde were analyzed by single-crystal X-ray diffraction analysis. In this study, various types of alkoxide crystals were obtained by changing the crystallization conditions, such as enantiopure or racemic alkoxides, the concentration of diisopropylzinc, and the solvent. X-ray crystal structure analysis directly revealed that these crystals have different tetrameric or oligomeric structures. These coordinative structure variations of pyrimidyl alkanols should play an important role in significant amplification of *ee* and spontaneous symmetry breaking in the asymmetric autocatalysis.

We believe that this study of alkoxide structures, using X-ray diffraction analysis, represents a milestone for understanding the mechanism of asymmetric autocatalysis and will shed light on further mechanistic studies by NMR and DFT calculations.

ACKNOWLEDGMENTS

This work has been financially supported by Grant-in-Aid for Scientific Research from Japan Society for the Promotion of Science (JSPS KAKENHI Grant Numbers 23685012, 26810026, and 15H03781) and MEXT-Supported Program for the Strategic Research Foundation at Private Universities, 2012–16.

We thank enthusiastic graduate students whose names appear in references.

REFERENCES

[1] (a) Weissbuch, I.; Addadi, L.; Berkovitch-Yellin, Z.; Gati, E.; Lahav, M.; Leiserowitz, L. Spontaneous Generation and Amplification of Optical Activity in α-Amino Acids by Enantioselective Occlusion into Centrosymmetric Crystals of Glycine. *Nature* **1984,** *310*, 161–164.
(b) Kondepudi, D. K.; Kaufman, R. J.; Singh, N. Chiral Symmetry Breaking in Sodium Chlorate Crystallizaton. *Science* **1990,** *250*, 975–976.
(c) Bolli, M.; Micura, R.; Eschenmoser, A. Pyranosyl-RNA: Chiroselective Self-Assembly of Base Sequences by Ligative Oligomerization of Tetranucleotide-2′,3′-Cyclophosphates (with a Commentary Concerning the Origin of Biomolecular Homochirality). *Chem. Biol.* **1997,** *4*, 309–320.
(d) Siegel, J. S. Homochiral Imperative of Molecular Evolution. *Chirality* **1998,** *10*, 24–27.
(e) Feringa, B. L.; van Delden, R. A. Absolute Asymmetric Synthesis: The Origin, Control, and Amplification of Chirality. *Angew. Chem., Int. Ed.* **1999,** *38*, 3418–3438.
(f) Green, M. M.; Park, J.-W.; Sato, T.; Teramoto, A.; Lifson, S.; Selinger, R. L. B., et al. The Macromolecular Route to Chiral Amplification. *Angew. Chem., Int. Ed.* **1999,** *38*, 3138–3154.
(g) Ribó, J. M.; Crusats, J.; Sagués, F.; Claret, J.; Rubires, R. Chiral Sign Induction by Vortices During the Formation of Mesophases in Stirred Solutions. *Science* **2001,** *292*, 2063–2066.
(h) Zepik, H.; Shavit, E.; Tang, M.; Jensen, T. R.; Kjaer, K.; Bolbach, G., et al. Chiral Amplification of Oligopeptides in Two-Dimensional Crystalline Self-Assemblies on Water. *Science* **2002,** *295*, 1266–1269.
(i) Mislow, K. Absolute Asymmetric Synthesis: A Commentary. *Collect. Czechoslov. Chem. Commun.* **2003,** *68*, 849–864.
(j) Pizzarello, S.; Weber, A. L. Prebiotic Amino Acids as Asymmetric Catalysts. *Science* **2004,** *303*, 1151.
(k) Viedma, C. Chiral Symmetry Breaking During Crystallization: Complete Chiral Purity Induced by Nonlinear Autocatalysis and Recycling. *Phys. Rev. Lett.* **2005,** *94*, 065504.
(l) Soloshonok, V. A.; Ueki, H.; Yasumoto, M.; Mekala, S.; Hirschi, J. S.; Singleton, D. A. Phenomenon of Optical Self-Purification of Chiral Non-racemic Compounds. *J. Am. Chem. Soc.* **2007,** *129*, 12112–12113.

(m) Breslow, R.; Cheng, Z.-L. L-Amino Acids Catalyze the Formation of an Excess of D-Glyceraldehyde, and Thus of Other D Sugars, under Credible Prebiotic Conditions. *Proc. Natl. Acad. Sci. U. S. A.* **2010,** *107*, 5723–5725.
(n) Saito, Y.; Hyuga, H. Colloquium: Homochirality: Symmetry Breaking in Systems Driven Far from Equilibrium. *Rev. Mod. Phys.* **2013,** *85*, 603–621.
(o) Ernst, K.-H. Molecular Chirality at Surfaces. *Phys. Status Solidi* **2012,** *249*, 2057–2088.
(p) Saito, Y.; Hyuga, H. Colloquium: Homochirality: Symmetry Breaking in Systems Driven Far from Equilibrium. *Rev. Mod. Phys.* **2013,** *85*, 603–621.
(q) Valero, G.; Ribó, J. M.; Moyano, A. A Closer Look at Spontaneous Mirror Symmetry Breaking in Aldol Reactions. *Chem. Eur. J.* **2014,** *20*, 17395–17408.
(r) Olsson, S.; Björemark, P. M.; Kokoli, T.; Sundberg, J.; Lennartson, A.; McKenzie, C. J., et al. Absolute Asymmetric Synthesis: Protected Substrate Oxidation. *Chem. Eur. J.* **2015,** *21*, 5211–5219.

[2] Frank, F. C. On Spontaneous Asymmetric Synthesis. *Biochim. Biophys. Acta* **1953,** *11*, 459–463.

[3] (a) Soai, K.; Shibata, T.; Morioka, H.; Choji, K. Asymmetric Autocatalysis and Amplification of Enantiomeric Excess of a Chiral Molecule. *Nature* **1995,** *378*, 767–768.
(b) Shibata, T.; Yonekubo, S.; Soai, K. Practically Perfect Asymmetric Autocatalysis with (2-Alkynyl-5-pyrimidyl)alkanols. *Angew. Chem., Int. Ed.* **1999,** *38*, 659–661.
(c) Sato, I.; Urabe, H.; Ishiguro, S.; Shibata, T.; Soai, K. Amplification of Chirality from Extremely Low to Greater than 99.5% ee by Asymmetric Autocatalysis. *Angew. Chem., Int. Ed.* **2003,** *42*, 315–317.

[4] (a) Soai, K.; Kawasaki, T. Asymmetric Autocatalysis with Amplification of Chirality. In *Amplification of Chirality. Topics in Current Chemistry;* Soai, K., Ed.; Springer Berlin Heidelberg: Berlin, Heidelberg, 2008; Vol. 284, pp 1–31.
(b) Kawasaki, T.; Soai, K. Asymmetric Induction Arising from Enantiomerically Enriched Carbon-13 Isotopomers and Highly Sensitive Chiral Discrimination by Asymmetric Autocatalysis. *Bull. Chem. Soc. Jpn.* **2011,** *84*, 879–892.
(c) Soai, K.; Kawasaki, T. Asymmetric Autocatalysis: Discovery and State of the Art. In *The Soai Reaction and Related Topic;* Palyi, G., Zicchi, C., Caglioti, C., Eds.; Edizioni Artestampa: Modena, 2012; pp 9–34.
(d) Kawasaki, T.; Soai, K. Asymmetric Autocatalysis Triggered by Chiral Crystals Formed from Achiral Compounds and Chiral Isotopomers. *Isr. J. Chem.* **2012,** *52*, 582–590.
(e) Kawasaki, T.; Sato, I.; Mineki, H.; Matsumoto, A.; Soai, K. Approaches Toward the Origin of Homochirality Using the Synthetic Organic Chemistry—; Asymmetric Autocatalysis with Amplification of Enantiomeric Excess—. *J. Synth. Org. Chem. Jpn.* **2013,** *71*, 109–123.
(f) Soai, K.; Kawasaki, T. Asymmetric Autocatalysis of Pyrimidyl Alkanol. In *Inventing Reactions;* Goossen, L. J., Ed.; Springer: Berlin and New York, 2013; pp 261–279.
(g) Soai, K.; Kawasaki, T.; Matsumoto, A. *The Origins of Homochirality Examined by Using Asymmetric Autocatalysis. Chem. Rec.* **2014,** 70–83.
(h) Soai, K.; Kawasaki, T.; Matsumoto, A. Asymmetric Autocatalysis of Pyrimidyl Alkanol and Its Application to the Study on the Origin of Homochirality. *Acc. Chem. Res.* **2014,** *47*, 3643–3654.

[5] (a) Avalos, M.; Babiano, R.; Cintas, P.; Jiménez, J. L.; Palacios, J. C. Chiral Autocatalysis: Where Stereochemistry Meets the Origin of Life. *Chem. Commun.* **2000,** 887–892.
(b) Blackmond, D. G. Asymmetric Autocatalysis and Its Implications for the Origin of Homochirality. *Proc. Natl. Acad. Sci. U. S. A.* **2004,** *101*, 5732–5736.

(c) Podlech, J.; Gehring, T. New Aspects of Soai's Asymmetric Autocatalysis. *Angew. Chem., Int. Ed.* **2005,** *44*, 5776–5777.
(d) Brown, J. M.; Gridnev, I. D.; Klankermayer, J. Asymmetric Autocatalysis with Organozinc Complexes; Elucidation of the Reaction Pathway. In *Amplification of Chirality. Topics in Current Chemistry;* Soai, K., Ed.; Springer Berlin Heidelberg: Berlin, Heidelberg, 2008; Vol. 284, pp 35–65.
(e) Gehring, T.; Busch, M.; Schlageter, M.; Weingand, D.; Concise, A. Summary of Experimental Facts About the Soai Reaction. *Chirality* **2010,** *22* (Suppl 1), E173–E182.
(f) Barabás, B.; Tóth, J.; Pályi, G. Stochastic Aspects of Asymmetric Autocatalysis and Absolute Asymmetric Synthesis. *J. Math. Chem.* **2010,** *48*, 457–489.
(g) Lente, G. The Role of Stochastic Models in Interpreting the Origins of Biological Chirality. *Symmetry (Basel)* **2010,** *2*, 767–798.
(h) Micheau, J.-C.; Coudret, C.; Buhse, T. Systems Chemistry of the Soai Reaction. In *The Soai Reaction and Related Topic;* Palyi, G., Zucchi, C., Caglioti, C., Eds.; Edizioni Artestampa: Modena, 2012; pp 169–196.
(i) Bissette, A. J.; Fletcher, S. P. Mechanisms of Autocatalysis. *Angew. Chem., Int. Ed.* **2013,** *52*, 12800–12826.
(j) Fülöp, O.; Barabás, B. Impact of the Soai-Autocatalysis on Natural Sciences. *J. Math. Chem.* **2016,** *54*, 10–17.

[6] (a) Sato, I.; Sugie, R.; Matsueda, Y.; Furumura, Y.; Soai, K. Asymmetric Synthesis Utilizing Circularly Polarized Light Mediated by the Photoequilibrium of Chiral Olefins in Conjunction with Asymmetric Autocatalysis. *Angew. Chem., Int. Ed.* **2004,** *43*, 4490–4492.
(b) Kawasaki, T.; Sato, M.; Ishiguro, S.; Saito, T.; Morishita, Y.; Sato, I., et al. Enantioselective Synthesis of Near Enantiopure Compound by Asymmetric Autocatalysis Triggered by Asymmetric Photolysis with Circularly Polarized Light. *J. Am. Chem. Soc.* **2005,** *127*, 3274–3275.

[7] (a) Kawasaki, T.; Jo, K.; Igarashi, H.; Sato, I.; Nagano, M.; Koshima, H., et al. Asymmetric Amplification Using Chiral Cocrystals Formed from Achiral Organic Molecules by Asymmetric Autocatalysis. *Angew. Chem., Int. Ed.* **2005,** *44*, 2774–2777.
(b) Kawasaki, T.; Suzuki, K.; Hatase, K.; Otsuka, M.; Koshima, H.; Soai, K. Enantioselective Synthesis Mediated by Chiral Crystal of Achiral Hippuric Acid in Conjunction with Asymmetric Autocatalysis. *Chem. Commun.* **2006,** *2*, 1869–1871.
(c) Kawasaki, T.; Suzuki, K.; Hakoda, Y.; Soai, K. Achiral Nucleobase Cytosine Acts as an Origin of Homochirality of Biomolecules in Conjunction with Asymmetric Autocatalysis. *Angew. Chem., Int. Ed.* **2008,** *47*, 496–499.
(d) Kawasaki, T.; Harada, Y.; Suzuki, K.; Tobita, T.; Florini, N.; Pályi, G., et al. Enantioselective Synthesis Utilizing Enantiomorphous Organic Crystal of Achiral Benzils as a Source of Chirality in Asymmetric Autocatalysis. *Org. Lett.* **2008,** *10*, 4085–4088.
(e) Kawasaki, T.; Hakoda, Y.; Mineki, H.; Suzuki, K.; Soai, K. Generation of Absolute Controlled Crystal Chirality by the Removal of Crystal Water from Achiral Crystal of Nucleobase Cytosine. *J. Am. Chem. Soc.* **2010,** *132*, 2874–2875.
(f) Kawasaki, T.; Nakaoda, M.; Kaito, N.; Sasagawa, T.; Soai, K. Asymmetric Autocatalysis Induced by Chiral Crystals of Achiral Tetraphenylethylenes. *Orig. Life Evol. Biosph.* **2010,** *40*, 65–78.
(g) Kawasaki, T.; Sasagawa, T.; Shiozawa, K.; Uchida, M.; Suzuki, K.; Soai, K. Enantioselective Synthesis Induced by Chiral Crystal Composed of DL-Serine in Conjunction with Asymmetric Autocatalysis. *Org. Lett.* **2011,** *13*, 2361–2363.

(h) Carter, D. J.; Rohl, A. L.; Shtukenberg, A.; Bian, S.; Hu, C.; Baylon, L., et al. Prediction of Soai Reaction Enantioselectivity Induced by Crystals of N-(2-Thienylcarbonyl)glycine. *Cryst. Growth Des.* **2012,** *12*, 2138–2145.
(i) Mineki, H.; Hanasaki, T.; Matsumoto, A.; Kawasaki, T.; Soai, K. Asymmetric Autocatalysis Initiated by Achiral Nucleic Acid Base Adenine: Implications on the Origin of Homochirality of Biomolecules. *Chem. Commun.* **2012,** *48*, 10538–10540.
(j) Kawasaki, T.; Uchida, M.; Kaimori, Y.; Sasagawa, T.; Matsumoto, A.; Soai, K. Enantioselective Synthesis Induced by the Helical Molecular Arrangement in the Chiral Crystal of Achiral Tris(2-Hydroxyethyl) 1,3,5-Benzenetricarboxylate in Conjunction with Asymmetric Autocatalysis. *Chem. Lett.* **2013,** *42*, 711–713.
(k) Mineki, H.; Kaimori, Y.; Kawasaki, T.; Matsumoto, A.; Soai, K. Enantiodivergent Formation of a Chiral Cytosine Crystal by Removal of Crystal Water from an Achiral Monohydrate Crystal under Reduced Pressure. *Tetrahedron: Asymmetry* **2013,** *24*, 1365–1367.
(l) Matsumoto, A.; Ide, T.; Kaimori, Y.; Fujiwara, S.; Soai, K. Asymmetric Autocatalysis Triggered by Chiral Crystal of Achiral Ethylenediamine Sulfate. *Chem. Lett.* **2015,** *44*, 688–690.
[8] (a) Soai, K.; Osanai, S.; Kadowaki, K.; Yonekubo, S.; Shibata, T.; Sato, I. *d*- and *l*-Quartz-Promoted Highly Enantioselective Synthesis of a Chiral Organic Compound. *J. Am. Chem. Soc.* **1999,** *121*, 11235–11236.
(b) Sato, I.; Kadowaki, K.; Soai, K. Asymmetric Synthesis of an Organic Compound with High Enantiomeric Excess Induced by Inorganic Ionic Sodium Chlorate. *Angew. Chem., Int. Ed.* **2000,** *39*, 1510–1512.
(c) Sato, I.; Kadowaki, K.; Ohgo, Y.; Soai, K. Highly Enantioselective Asymmetric Autocatalysis Induced by Chiral Ionic Crystals of Sodium Chlorate and Sodium Bromate. *J. Mol. Catal. A Chem.* **2004,** *216*, 209–214.
(d) Shindo, H.; Shirota, Y.; Niki, K.; Kawasaki, T.; Suzuki, K.; Araki, Y., et al. Asymmetric Autocatalysis Induced by Cinnabar: Observation of the Enantioselective Adsorption of a 5-Pyrimidyl Alkanol on the Crystal Surface. *Angew. Chem., Int. Ed.* **2013,** *52*, 9135–9138.
(e) Matsumoto, A.; Ozawa, H.; Inumaru, A.; Soai, K. Asymmetric Induction by Retgersite, Nickel Sulfate Hexahydrate, in Conjunction with Asymmetric Autocatalysis. *New J. Chem.* **2015,** *39*, 6742–6745.
[9] (a) Sato, I.; Kadowaki, K.; Urabe, H.; Jung, J. H.; Ono, Y.; Shinkai, S., et al. Highly Enantioselective Synthesis of Organic Compound Using Right- and Left-Handed Helical Silica. *Tetrahedron Lett.* **2003,** *44*, 721–724.
(b) Hitosugi, S.; Matsumoto, A.; Kaimori, Y.; Iizuka, R.; Soai, K.; Isobe, H. Asymmetric Autocatalysis Initiated by Finite Single-Wall Carbon Nanotube Molecules with Helical Chirality. *Org. Lett.* **2014,** *16*, 645–647.
(c) Kawasaki, T.; Araki, Y.; Hatase, K.; Suzuki, K.; Matsumoto, A.; Yokoi, T., et al. Helical Mesoporous Silica as an Inorganic Heterogeneous Chiral Trigger for Asymmetric Autocatalysis with Amplification of Enantiomeric Excess. *Chem. Commun.* **2015,** *51*, 8742–8744.
[10] (a) Sato, I.; Omiya, D.; Saito, T.; Soai, K. Highly Enantioselective Synthesis Induced by Chiral Primary Alcohols Due to Deuterium Substitution. *J. Am. Chem. Soc.* **2000,** *122*, 11739–11740.
(b) Kawasaki, T.; Matsumura, Y.; Tsutsumi, T.; Suzuki, K.; Ito, M.; Soai, K. Asymmetric Autocatalysis Triggered by Carbon Isotope ($^{13}C/^{12}C$) Chirality. *Science* **2009,** *324*, 492–495.
(c) Kawasaki, T.; Shimizu, M.; Nishiyama, D.; Ito, M.; Ozawa, H.; Soai, K. Asymmetric Autocatalysis Induced by Meteoritic Amino Acids with Hydrogen Isotope Chirality. *Chem. Commun.* **2009,** 4396–4398.

(d) Kawasaki, T.; Okano, Y.; Suzuki, E.; Takano, S.; Oji, S.; Soai, K. Asymmetric Autocatalysis: Triggered by Chiral Isotopomer Arising from Oxygen Isotope Substitution. *Angew. Chem., Int. Ed.* **2011,** *50*, 8131–8133.
(e) Kawasaki, T.; Ozawa, H.; Ito, M.; Soai, K. Enantioselective Synthesis Induced by Compounds with Chirality Arising from Partially Deuterated Methyl Groups in Conjunction with Asymmetric Autocatalysis. *Chem. Lett.* **2011,** *40*, 320–321.
(f) Matsumoto, A.; Oji, S.; Takano, S.; Tada, K.; Kawasaki, T.; Soai, K. Asymmetric Autocatalysis Triggered by Oxygen Isotopically Chiral Glycerin. *Org. Biomol. Chem.* **2013,** *11*, 2928–2931.
(g) Barabás, B.; Kurdi, R.; Pályi, G. Natural Abundance Isotopic Chirality in the Reagents of the Soai Reaction. *Symmetry (Basel)* **2016,** *8*, 2.

[11] (a) Soai, K.; Sato, I.; Shibata, T.; Komiya, S.; Hayashi, M.; Matsueda, Y., et al. Asymmetric Synthesis of Pyrimidyl Alkanol without Adding Chiral Substances by the Addition of Diisopropylzinc to Pyrimidine-5-carbaldehyde in Conjunction with Asymmetric Autocatalysis. *Tetrahedron: Asymmetry* **2003,** *14*, 185–188.
(b) Singleton, D. A.; Vo, L. K. A Few Molecules Can Control the Enantiomeric Outcome. Evidence Supporting Absolute Asymmetric Synthesis Using the Soai Asymmetric Autocatalysis. *Org. Lett.* **2003,** *5*, 4337–4339.
(c) Kawasaki, T.; Suzuki, K.; Shimizu, M.; Ishikawa, K.; Soai, K. Spontaneous Absolute Asymmetric Synthesis in the Presence of Achiral Silica Gel in Conjunction with Asymmetric Autocatalysis. *Chirality* **2006,** *18*, 479–482.
(d) Suzuki, K.; Hatase, K.; Nishiyama, D.; Kawasaki, T.; Soai, K. Spontaneous Absolute Asymmetric Synthesis Promoted by Achiral Amines in Conjunction with Asymmetric Autocatalysis. *J. Syst. Chem.* **2010,** *1*, 5.
(e) Soi, K.; Shibata, T.; Kowata, Y. Asymmetric Synthesis of Enantioenriched Alkanol by Spontaneous Asymmetric Synthesis. *Jpn. Kokai Tokkyo Koho* **1997,** *9* (268), 179.

[12] (a) Sato, I.; Omiya, D.; Tsukiyama, K.; Ogi, Y.; Soai, K. Evidence of Asymmetric Autocatalysis in the Enantioselective Addition of Diisopropylzinc to Pyrimidine-5-carbaldehyde Using Chiral Pyrimidyl Alkanol. *Tetrahedron: Asymmetry* **2001,** *12*, 1965–1969.
(b) Blackmond, D. G.; McMillan, C. R.; Ramdeehul, S.; Schorm, A.; Brown, J. M. Origins of Asymmetric Amplification in Autocatalytic Alkylzinc Additions. *J. Am. Chem. Soc.* **2001,** *123*, 10103–10104.
(c) Sato, I.; Omiya, D.; Igarashi, H.; Kato, K.; Ogi, Y.; Tsukiyama, K., et al. Relationship Between the Time, Yield, and Enantiomeric Excess of Asymmetric Autocatalysis of Chiral 2-Alkynyl-5-pyrimidyl Alkanol with Amplification of Enantiomeric Excess. *Tetrahedron: Asymmetry* **2003,** *14*, 975–979.
(d) Buono, F. G.; Blackmond, D. G. Kinetic Evidence for a Tetrameric Transition State in the Asymmetric Autocatalytic Alkylation of Pyrimidyl Aldehydes. *J. Am. Chem. Soc.* **2003,** *125*, 8978–8979.

[13] (a) Buhse, T.; Tentative, A. Kinetic Model for Chiral Amplification in Autocatalytic Alkylzinc Additions. *Tetrahedron: Asymmetry* **2003,** *14*, 1055–1061.
(b) Islas, J. R.; Lavabre, D.; Grevy, J.-M.; Lamoneda, R. H.; Cabrera, H. R.; Micheau, J.-C., et al. Mirror-Symmetry Breaking in the Soai Reaction: A Kinetic Understanding. *Proc. Natl. Acad. Sci. U. S. A.* **2005,** *102*, 13743–13748.
(c) Lente, G. Stochastic Kinetic Models of Chiral Autocatalysis: A General Tool for the Quantitative Interpretation of Total Asymmetric Synthesis. *J. Phys. Chem. A* **2005,** *109*, 11058–11063.
(d) Micskei, K.; Póta, G.; Caglioti, L.; Palyi, G. Empirical Description of Chiral Autocatalysis. *J. Phys. Chem. A* **2006,** *110*, 5982–5984.

(e) Barabas, B.; Caglioti, L.; Zucchi, C.; Maioli, M.; Gál, E.; Micskei, K., et al. Violation of Distribution Symmetry in Statistical Evaluation of Absolute Enantioselective Synthesis. *J. Phys. Chem. B* **2007,** *111*, 11506–11510.
(f) Micskei, K.; Rábai, G.; Gál, E.; Caglioti, L.; Pályi, G. Oscillatory Symmetry Breaking in the Soai Reaction. *J. Phys. Chem. B* **2008,** *112*, 9196–9200.
(g) Busch, M.; Schlageter, M.; Weingand, D.; Gehring, T. Systematic Studies Using 2-(1-Adamantylethynyl)pyrimidine-5-carbaldehyde as a Starting Material in Soai's Asymmetric Autocatalysis. *Chem. Eur. J.* **2009,** *15*, 8251–8258.
(h) Crusats, J.; Hochberg, D.; Moyano, A.; Ribó, J. M. Frank Model and Spontaneous Emergence of Chirality in Closed Systems. *ChemPhysChem* **2009,** *10*, 2123–2131.
(i) Dóka, É.; Lente, G. Mechanism-Based Chemical Understanding of Chiral Symmetry Breaking in the Soai Reaction. A Combined Probabilistic and Deterministic Description of Chemical Reactions. *J. Am. Chem. Soc.* **2011,** *133*, 17878–17881.
(j) Micheau, J.-C.; Coudret, C.; Cruz, J.-M.; Buhse, T.: Amplification of Enantiomeric Excess, Mirror-Image Symmetry Breaking and Kinetic Proofreading in Soai Reaction Models with Different Oligomeric Orders. Phys. Chem. Chem. Phys. **2012,** *14*, 13239–13248.
(k) Ercolani, G. Principles for Designing an Achiral Receptor Promoting Asymmetric Autocatalysis with Amplification of Chirality. *Tetrahedron: Asymmetry* **2014,** *25*, 405–410.
(l) Barabás, B.; Zucchi, C.; Maioli, M.; Micskei, K.; Pályi, G. Stochastic and Empirical Models of the Absolute Asymmetric Synthesis by the Soai-Autocatalysis. *J. Mol. Model.* **2015,** *21*, 33.

[14] (a) Gridnev, I. D.; Serafimov, J. M.; Quiney, H.; Brown, J. M. Reflections on Spontaneous Asymmetric Synthesis by Amplifying Autocatalysis. *Org. Biomol. Chem.* **2003,** *1*, 3811–3819.
(b) Gridnev, I. D.; Serafimov, J. M.; Brown, J. M. Solution Structure and Reagent Binding of the Zinc Alkoxide Catalyst in the Soai Asymmetric Autocatalytic Reaction. *Angew. Chem., Int. Ed.* **2004,** *43*, 4884–4887.
(c) Quaranta, M.; Gehring, T.; Odell, B.; Brown, J. M.; Blackmond, D. G. Unusual Inverse Temperature Dependence on Reaction Rate in the Asymmetric Autocatalytic Alkylation of Pyrimidyl Aldehydes. *J. Am. Chem. Soc.* **2010,** *132*, 15104–15107.
(d) Gehring, T.; Quaranta, M.; Odell, B.; Blackmond, D. G.; Brown, J. M. Observation of a Transient Intermediate in Soai's Asymmetric Autocatalysis: Insights from ^{1}H NMR Turnover in Real Time. *Angew. Chem., Int. Ed.* **2012,** *51*, 9539–9542.

[15] (a) Gridnev, I. D.; Brown, J. M. Asymmetric Autocatalysis: Novel Structures, Novel Mechanism? *Proc. Natl. Acad. Sci. U. S. A.* **2004,** *101*, 5727–5731.
(b) Gridnev, I. D. Chiral Symmetry Breaking in Chiral Crystallization and Soai Autocatalytic Reaction. *Chem. Lett.* **2006,** *35*, 148–153.
(c) Klankermayer, J. J.; Gridnev, I. D.; Brown, J. M. Role of the Isopropyl Group in Asymmetric Autocatalytic Zinc Alkylations. *Chem. Commun.* **2007,** 3151–3153.
(d) Schiaffino, L.; Ercolani, G. Unraveling the Mechanism of the Soai Asymmetric Autocatalytic Reaction by First-Principles Calculations: Induction and Amplification of Chirality by Self-Assembly of Hexamolecular Complexes. *Angew. Chem., Int. Ed.* **2008,** *47*, 6832–6835.
(e) Schiaffino, L.; Ercolani, G. Amplification of Chirality and Enantioselectivity in the Asymmetric Autocatalytic Soai Reaction. *ChemPhysChem* **2009,** *10*, 2508–2515.

(f) Schiaffino, L.; Ercolani, G. Mechanism of the Asymmetric Autocatalytic Soai Reaction Studied by Density Functional Theory. *Chem. Eur. J* **2010,** *16*, 3147–3156.

(g) Ercolani, G.; Schiaffino, L. Putting the Mechanism of the Soai Reaction to the Test: DFT Study of the Role of Aldehyde and Dialkylzinc Structure. *J. Org. Chem.* **2011,** *76*, 2619–2626.

(h) Gridnev, I. D.; Vorobiev, A. K. Quantification of Sophisticated Equilibria in the Reaction Pool and Amplifying Catalytic Cycle of the Soai Reaction. *ACS Catal.* **2012,** *2*, 2137–2149.

[16] (a) Puchot, C.; Samuel, O.; Dunach, E.; Zhao, S.; Agami, C.; Kagan, H. B. Nonlinear Effects in Asymmetric Synthesis. Examples in Asymmetric Oxidations and Aldolization Reactions. *J. Am. Chem. Soc.* **1986,** *108*, 2353–2357.

(b) Oguni, N.; Matsuda, Y.; Kaneko, T. Asymmetric Amplifying Phenomena in Enantioselective Addition of Diethylzinc to Benzaldehyde. *J. Am. Chem. Soc.* **1988,** *110*, 7877–7878.

(c) Kitamura, M.; Okada, S.; Suga, S.; Noyori, R. Enantioselective Addition of Dialkylzincs to Aldehydes Promoted by Chiral Amino Alcohols. Mechanism and Nonlinear Effect. *J. Am. Chem. Soc.* **1989,** *111*, 4028–4036.

(d) Satyanarayana, T.; Abraham, S.; Kagan, H. B. Nonlinear Effects in Asymmetric Catalysis. *Angew. Chem., Int. Ed.* **2009,** *48*, 456–494.

[17] Lennartson, A.; Hedström, A.; Håkansson, M. Diisopropyl(*N,N,N′,N′*-tetramethylethylene-diamine)zinc(II), the First Crystal Structure of a Diisopropylzinc Complex. *Acta Crystallogr.* **2007,** *63*, 123–125.

[18] Matsumoto, A.; Abe, T.; Hara, A.; Tobita, T.; Sasagawa, T.; Kawasaki, T., et al. Crystal Structure of the Isopropylzinc Alkoxide of Pyrimidyl Alkanol: Mechanistic Insights for Asymmetric Autocatalysis with Amplification of Enantiomeric Excess. *Angew. Chem., Int. Ed.* **2015,** *54*, 15218–15221.

[19] (a) Bacsa, J.; Hanke, F.; Hindley, S.; Odedra, R.; Darling, G. R.; Jones, A. C., et al. The Solid-State Structures of Dimethylzinc and Diethylzinc. *Angew. Chem., Int. Ed.* **2011,** *50*, 11685–11687.

(b) Hursthouse, M. B.; Motevalli, M.; O'Brien, P.; Walsh, J. R.; Jones, A. C. X-Ray Crystal Structure of a Triazine Adduct of Dimethylzinc: an Important Precursor for the Deposition of II/VI Materials. *J. Mater. Chem.* **1991,** *1*, 139.

CHAPTER 11

On the Relative Measures of the Degrees of Chirality and General Asymmetry

Paul G. Mezey
Canada Research Chair in Scientific Modelling and Simulation, Department of Chemistry and Department of Physics and Physical Oceanography, Memorial University of Newfoundland, St. John's, NL, Canada

Abstract

Presented here are comparisons of the numerical measures for the degrees of chirality and analogous in addition to the less often used measures for some more general asymmetry types, and some general trends are investigated. The relations between these two types of measures reveal a virtual hierarchy of symmetry deficiencies, which appear informative from a chemical, but especially, from a biochemical perspective. The relative roles of various types of imperfect symmetries can be characterized by numerical measures, and these measures are suggested for correlations with some of the physically measurable properties of biochemically important molecules.

Keywords: Biochemically important molecules; chirality, degrees of; chirality, measure of; imperfect symmetry; symmetry deficiency

11.1 INTRODUCTION

In nature, one of the most important aspects of the presence or lack of symmetry is related to chirality: the nonsuperimposability of an object on its mirror image. All objects, including molecules, if they have mirror plane symmetries, as represented by reflection plane symmetry elements usually denoted by σ, or rotation–reflection symmetries associated with S_{2m}-type symmetry elements, are achiral, that is, they are superimposable on their mirror images. Of course, most actual objects are not rigid, and in practice, achirality is dependent on deformability; if high enough energy is available, deformations, ultimately, decompositions, in the extreme case atomization, and eventual reassembly, are always sufficient to turn any object into its mirror image. Most molecules, however, as long as they preserve their chemical identity, do not possess enough energy for deformations sufficiently extensive to turn them into their mirror images.

Advances in Asymmetric Autocatalysis and Related Topics
DOI: http://dx.doi.org/10.1016/B978-0-12-812824-4.00011-3

Hence, most actual molecules are chiral, not only in some of their instantaneous "geometrical" arrangements, but even in a practical sense, allowing for some limited deformations in a "range of geometries" available to them under the usual energy conditions.

However, chirality is only one manifestation of the lack of certain types of symmetries, and similar considerations, although with no direct consequence for possible superimposability of the object on its mirror image, do apply to symmetry elements different from reflection planes σ, and rotation–reflection symmetry elements of the S_{2m} type. Such considerations of approximate symmetries are in fact more important than previously regarded during the early developments of group theoretical descriptions of symmetry. At this time the great, simplifying advantages of existing symmetries received most of the attention, for example, in reducing the mathematical difficulties of applying some physical models, such as those used in spectroscopy and quantum chemistry. Symmetry can be treated with mathematical rigor, and it often provides some advantages of simplification that are recognized on various levels of modeling molecules. Symmetry can be used on the geometrical level and also on some abstract levels where it is not the geometrical object itself, but the properties of the physical model used that possess some inherent symmetry. In fact, efforts have even been made to apply such simplifications, in an approximate sense, when it is known that the actual molecule does not have the symmetry needed for the simplification, and the corrections to the model have been attempted only after a molecule without symmetry has been used.

An entirely different approach has been introduced whereby the actual deviation from a specific symmetry has become the object of study. Some of the early approaches can be traced back to the works of Mislow, Bolstad, Kitaigorodski, Dubois, Alliot, and Panaye [1–3]. Whereas the description of exact symmetries can be based on the full algebraic elegance of symmetry groups, by contrast, these early realizations of the need to provide some intervention for approximate symmetries [1,2] used some, perhaps less elegant but practical, approaches, which have formed the basis of much of the later developments. A similar philosophy has been the motivation when expanding the "prochirality" approach [3]. Of course, the recognition of the significance of having or not having perfect symmetry has already played a major role in the more theoretical aspects of symmetries of model formulations, such as those in some fundamental studies of early quantum chemical approaches [4–11], even without a formal proposal for a "measure" of approximate symmetries.

Motivated by these earlier studies, several approaches have been proposed for the characterization and numerical evaluation of the degree of chirality and the local as well as global approximate symmetry properties of molecules [12–50].

Since most actual objects actually lack perfect symmetry, these types of problems can also be viewed as reckoning with reality, through a characterization of the imperfectness of approximate symmetries and, specifically, the characterization of the "degree of chirality". These imperfections have been approached from many perspectives [1–3,13–16,19–35,39–43,48–50].

Here we shall be concerned primarily with two families of general symmetry–deficiency measures and related chirality measures.

One approach is based on a method originally designed for bounded 3D objects; a method that can also be applied, for example, to molecular isodensity contours, as well as the convex hulls and more advanced "hulls" of the 3D electron density shapes, or the point sets of the nuclear arrangements of molecules.

The other approach is more specifically chemical, since it refers to the (3N-6)-dimensional nuclear configuration space M, a metric space, of all the conformational changes and all the chemical reactions associated with a given stoichiometry.

In spite of the rather fundamental differences between these two families of approaches, one is 3D and the other is multidimensional, there are enough common features to allow us to investigate some of treatments applicable to both, and to discuss the benefits they may provide.

11.2 A GENERAL 3D SYMMETRY DEFICIENCY AND CHIRALITY MEASURE APPROACH

We shall consider more generally the first approach, making use of the so-called internal and external symmetry deficiency measures, as applied in 3D (see, e.g., Refs. [34,49]), as briefly reviewed below.

11.2.1 General Internal and External Symmetry Deficiency Measures

These two types of measures are based on a rather general and, in fact, natural approach to investigating and evaluating symmetry deficiency. These measures can be applied to a rather wide range of objects, including molecules, and also in fields that can be very different from chemistry.

The generality of the approach is related to the fact that it is formulated using the general concepts of set theory. Here only a short review of the main definitions is given, for the relevant theorems and their proofs one may consult the Refs. [31,32,34]. First we shall review some of the basic concepts:

A set X is an R-*set* if X has the symmetry element R.

A set Y is an R-*subset* of a set X if Y is a subset of X, and Y is an R-set.

Set Y is a *maximum volume* R-*subset* of a set X if Y is an R-subset of X and if volume V(Y) is maximum among all R-subsets of X. Note that, while Y is not necessarily unique, the value of this volume V(Y) is a unique value.

A set Z is an R-*superset* of a set X if Z is a superset of X (X is a subset of Z), and Z is an R-set.

Set Z is a *minimum volume* R-*superset* of a set X if Z is an R-superset of X and if volume V(Z) is minimum among the volumes of all R-supersets of X. Note that, while Z is not necessarily unique, the value V(Z) is unique.

The internal R-symmetry deficiency measure, ISD(X,R) of a set X is defined as:

$$\mathrm{ISD(X,R)} = 1 - \mathrm{V(Y)/V(X)}, \tag{11.1}$$

where Y is a maximum volume R-subset of X.

The external R-symmetry deficiency measure, ESD(X,R) of a set X is defined as:

$$\mathrm{ESD(X,R)} = 1 - \mathrm{V(X)/V(Z)}, \tag{11.2}$$

where Z is a minimum volume R-superset of X.

At this point it is useful to distinguish two types of symmetry elements, as well as the associated two types of symmetry deficiencies: those relevant to chirality and those with no impact on chirality.

Following the notations of Ref. [50], let us denote the family of all achirality-inducing symmetry elements by $\boldsymbol{R}_a$, where:

$$\boldsymbol{R}_a = \{\sigma, S_{2m}, m = 1, 2, \ldots\} \tag{11.3}$$

Clearly, in chemical applications, the m value must be limited, since, e.g., $m = 20$ is impossible for any molecule, as this would imply impossible crowdedness around some central atom.

Nevertheless, for some general object, such symmetry elements are possible; consequently, for the sake of generality, we do not set limits for index m.

Evidently, if a molecule is deficient in each "potentially accessible" symmetry linked to a symmetry element in the above set, then the molecule must be chiral.

It is also natural to define another family of symmetry elements, those not associated to achirality.

Again, for generality, we do not limit the number of such symmetry elements, as for some general object, for example, rotation axes C_m of very high m values are possible, even if for actual molecules, again, this value m must be limited as a consequence of the impossibility of overcrowding.

Take then a collection of "no-chirality-inducing" symmetry elements, denoted by $\boldsymbol{R}_g$:

$$\boldsymbol{R}_g = \left\{R_1, R_2, R_3, \ldots, R_p\right\}. \tag{11.4}$$

Of course, the union of these two sets:

$$\boldsymbol{R} = \boldsymbol{R}_a \cup \boldsymbol{R}_g \tag{11.5}$$

provides the full range of the relevant symmetry elements.

For all three of these families, $\boldsymbol{R}_a$, $\boldsymbol{R}_g$, and $\boldsymbol{R}$, if the subject of study, for example, a molecular model, has a well-defined 3D body X associated with it, then the volume-based internal and external symmetry deficiencies can be obtained for each of these symmetry elements, such as:

$$\mathrm{ISD}(\mathrm{X}, S_{2m}),\ \text{ and }\ \mathrm{ESD}(\mathrm{X}, S_{2m}),\ \text{ or }\ \mathrm{ISD}(\mathrm{X}, R_p),\ \text{ and }\ \mathrm{ESD}(\mathrm{X}, R_p).$$

In the specific case of chirality, it is also of importance to consider all symmetry elements and all the symmetry deficiency measures associated with set $\boldsymbol{R}_a$, both individually and collectively, since zero symmetry deficiency just for a single symmetry element in set $\boldsymbol{R}_a$ may already imply achirality, even if the symmetry deficiency values are high for all of the other symmetry elements from the same set $\boldsymbol{R}_a$.

That is, we may indeed consider two chirality-inducing symmetry deficiency vectors, one for the internal, another for the external symmetry deficiency measures:

$$\boldsymbol{d}(\boldsymbol{X}, \boldsymbol{R}_a, \mathrm{ISD}) = (\mathrm{ISD}(\mathrm{X}, \sigma),\ \mathrm{ISD}(\mathrm{X}, S_2),\ \mathrm{ISD}(\mathrm{X}, S_4),\ \mathrm{ISD}(\mathrm{X}, S_6),\ \ldots) \tag{11.6}$$

$$\boldsymbol{d}(\boldsymbol{X}, \boldsymbol{R}_a, \mathrm{ESD}) = (\mathrm{ESD}(\mathrm{X}, \sigma),\ \mathrm{ESD}(\mathrm{X}, S_2),\ \mathrm{ESD}(\mathrm{X}, S_4),\ \mathrm{ESD}(\mathrm{X}, S_6),\ \ldots) \tag{11.7}$$

If the maximum value of m is n, then these are $(n+1)$ dimensional vectors, providing a detailed description of the chirality-inducing symmetry deficiencies.

Two overall chirality measures can be obtained from these vectors, defined as "achirality deficiency measures":

$$\mathrm{IACD}(\mathrm{X}, \boldsymbol{R}_a) = \min(\boldsymbol{d}_i(\boldsymbol{X}, \boldsymbol{R}_a, \mathrm{ISD}),\ i = 1\ldots n+1) \tag{11.8}$$

and

$$\mathrm{EACD}(\mathrm{X}, \boldsymbol{R}_a) = \min(\boldsymbol{d}_i(\boldsymbol{X}, \boldsymbol{R}_a, \mathrm{ESD}),\ i = 1\ldots n+1) \tag{11.9}$$

where IACD(X,$\boldsymbol{R}_a$) is referred to as "internal achirality deficiency measure", and where EACD(X,$\boldsymbol{R}_a$) is referred to as "external achirality deficiency measure".

Similar measures can also be obtained for a collection of "no-achirality-inducing" symmetry elements of a set $\boldsymbol{R}_g = \{R_1,\ R_2,\ R_3,\ \ldots,\ R_p,\ \ldots\}$, which has its focus on "chirality—independent" symmetry deficiencies, or in short, "ci-symmetry deficiencies". Here the possibility of zero symmetry deficiency for a particular symmetry element has a less dramatic effect on the overall molecular properties than in the case of set $\boldsymbol{R}_a$, where a single value of zero deficiency might imply achirality; nevertheless, a single zero value of a ci-symmetry deficiency here also has a general effect: no longer can it be said that the molecule is completely asymmetric.

Accordingly we may consider two additional symmetry deficiency vectors, the ci-symmetry deficiency vectors, one for the internal, another for the external ci-symmetry deficiency measures:

$$\boldsymbol{d}(\boldsymbol{X}, \boldsymbol{R}_g, \mathrm{ISD}) = (\mathrm{ISD}(\mathrm{X}, R_1), \mathrm{ISD}(\mathrm{X}, R_2), \mathrm{ISD}(\mathrm{X}, R_3), \ldots, \mathrm{ISD}(\mathrm{X}, R_p), \ldots) \tag{11.10}$$

$$\boldsymbol{d}(\boldsymbol{X}, \boldsymbol{R}_g, \mathrm{ESD}) = (\mathrm{ESD}(\mathrm{X}, R_1), \mathrm{ESD}(\mathrm{X}, R_2), \mathrm{ESD}(\mathrm{X}, R_3), \ldots, \mathrm{ESD}(\mathrm{X}, R_p), \ldots) \tag{11.11}$$

Here again, if the maximum value of p is n, then these are $(n+1)$ dimensional vectors, providing a detailed description of the no-chirality-inducing, that is, chirality-independent "ci" symmetry deficiencies.

By analogy with the case of chirality, two overall ci-symmetry deficiency measures can be obtained from these vectors:

$$\mathrm{ICISD}(\mathrm{X}, \boldsymbol{R}_g) = \min(\boldsymbol{d}_i(\boldsymbol{X}, \boldsymbol{R}_a, \mathrm{ISD}),\ i = 1\ldots n+1) \tag{11.12}$$

and

$$\mathrm{ECISD}(\mathrm{X}, \boldsymbol{R}_g) = \min(\boldsymbol{d}_i(\boldsymbol{X}, \boldsymbol{R}_a, \mathrm{ESD}),\ i = 1\ldots n+1) \tag{11.13}$$

where the quantity ICISD(X,$\boldsymbol{R}_g$) is referred to as the "internal ci-symmetry deficiency measure", and where the quantity ECISD(X,$\boldsymbol{R}_g$) is referred to as the "external ci-symmetry deficiency measure", with the understanding that the reference now excludes achirality-inducing symmetry elements.

If one makes no distinctions between chirality generating and chirality-independent symmetry deficiencies, then it is natural to consider the full set $\boldsymbol{R} = \boldsymbol{R}_a \cup \boldsymbol{R}_g$. Accordingly, one may repeat the procedure for the full set $\boldsymbol{R}$.

In this case, we may consider two additional, overall symmetry deficiency vectors, two general symmetry deficiency vectors or, in short, two symmetry deficiency vectors, one for the internal, another one for the external symmetry deficiency measures:

$$\begin{aligned}\boldsymbol{d}(\boldsymbol{X}, \boldsymbol{R}, \mathrm{ISD}) = (&\mathrm{ISD}(\mathrm{X}, \sigma), \mathrm{ISD}(\mathrm{X}, S_2), \mathrm{ISD}(\mathrm{X}, S_4), \mathrm{ISD}(\mathrm{X}, S_6), \ldots\\ &\mathrm{ISD}(\mathrm{X}, R_1), \mathrm{ISD}(\mathrm{X}, R_2), \mathrm{ISD}(\mathrm{X}, R_3), \ldots, \mathrm{ISD}(\mathrm{X}, R_p), \ldots)\end{aligned} \tag{11.14}$$

and

$$\begin{aligned}\boldsymbol{d}(\boldsymbol{X}, \boldsymbol{R}, \mathrm{ESD}) = (&\mathrm{ESD}(\mathrm{X}, \sigma), \mathrm{ESD}(\mathrm{X}, S_2), \mathrm{ESD}(\mathrm{X}, S_4), \mathrm{ESD}(\mathrm{X}, S_6), \ldots\\ &\mathrm{ESD}(\mathrm{X}, R_1), \mathrm{ESD}(\mathrm{X}, R_2), \mathrm{ESD}(\mathrm{X}, R_3), \ldots, \mathrm{ESD}(\mathrm{X}, R_p), \ldots)\end{aligned} \tag{11.15}$$

These vectors provide a detailed description of the symmetry deficiencies of all symmetry elements, including both achirality-inducing an no-achirality-inducing symmetries of the studied object X.

As it has been seen in the two sub-cases, two overall, total symmetry deficiency measures can be obtained from these vectors:

$$\mathrm{IOSD}(\mathrm{X}, \boldsymbol{R}) = \min(\boldsymbol{d}_j(\boldsymbol{X}, \boldsymbol{R}, \mathrm{ISD}), j = 1, 2, \ldots) \tag{11.16}$$

and

$$\mathrm{EOSD}(\mathrm{X}, \boldsymbol{R}) = \min(\boldsymbol{d}_j(\boldsymbol{X}, \boldsymbol{R}, \mathrm{ESD}), j = 1,\ 2, \ldots) \tag{11.17}$$

where IOSD(X,$\boldsymbol{R}$) is the internal overall symmetry deficiency measure, EOSD(X,$\boldsymbol{R}$) is the external overall symmetry deficiency measure, where the vector component index j runs through all the symmetry elements, and where the achirality inducing symmetry elements are assumed to be found in the front of the list, that is, they have the lower j index values.

11.3 SOME NATURAL CHOICES FOR THE MOLECULAR REPRESENTATIONS X, SUITABLE FOR THE APPLICATION OF THE R-SET–BASED 3D SYMMETRY DEFICIENCY AND CHIRALITY MEASURES

For these R-set–based 3D symmetry deficiency measures, the choice of X offers some interesting options. One natural choice involves the actual "material" making up the most directly relevant substance, chemically, of molecules: the electron density cloud. These clouds, actually fuzzy, boundaryless objects, are the most direct contributors to chemical bonding; however, they are not closed and bounded sets to which the R-set–based 3D internal and external symmetry deficiency measures are directly applicable.

However, this problem can be circumvented, using the spatial regions enclosed by isodensity contours of such density clouds, called Density Domains [28–30], and denoted as DD(a,K) for some nuclear configuration K and density threshold value a, are defined as the set of points $\mathbf{r}$ in 3D where the electron density $\rho(\mathbf{r}$,K) greater than some threshold a:

$$\mathrm{DD}(a, \mathrm{K}) = \{\mathbf{r}\text{: } \rho(\mathbf{r}, \mathrm{K}) \geq a\} \tag{11.18}$$

Such density domains DD(a,K) are already suitable choices for set X in the definitions of internal and external symmetry deficiency measures, although the electron density value chosen for the isocontour can be selected in an infinite number of ways, and some selection must be made to ensure uniformity, for example, one may select the:

$$a = 0.01 \ \text{a.u. (atomic unit)} \tag{11.19}$$

contour that is both bulky enough for appropriate volume representation, and detailed enough to represent specific shape features for molecules.

With this choice for the threshold a, and applying the definitions (11.1) and (11.2) to the choice of X as the density domain, we obtain, for example, the following chirality and symmetry deficiency measures:

$$\mathrm{ISD}(\mathrm{DD}(0.01, \mathrm{K}), S_{2m}), \ \text{and} \ \mathrm{ESD}(\mathrm{DD}(0.01, \mathrm{K}), \ S_{2m}),$$

or

$$\mathrm{ISD}(\mathrm{DD}(0.01, \mathrm{K}), R_p), \ \text{and} \ \mathrm{ESD}(\mathrm{DD}(0.01, \mathrm{K}), R_p).$$

Of course, another threshold value *a* may be useful for specific purposes, and the treatment is general for any electron density threshold value *a*.

In some cases a simpler and computationally easier approach is suitable for the chemical problem at hand, and in some cases instead of electron density clouds, the much simpler representation of the nuclear arrangements K, specified as a 3D geometrical arrangement G of the nuclei is sufficient.

Consider a nuclear arrangement, given by the formal internal geometry G of the nuclei of the molecule in the 3D space. One of the simplest "body" representations of this nuclear arrangement is provided by the convex hull H(G) of the point set G of 3D nuclear locations, where any one of the equivalent definitions of convex hull may be used [51], for example:

$$\mathrm{H(G)} = \text{smallest convex set containing G}, \tag{11.20}$$

although the definition of the intersection of all half spaces containing G is the most directly connected to the further developments we follow.

Finding good approximations to the various R-subsets of H(G) is a computationally less demanding task than doing the same for a density domain DD(*a*,K).

Only marginally more demanding of a computational task is the determination of the α-hull [52] and the more detailed α,β,γ-hull [53] of a nuclear arrangement G. A similar computational time increase is expected for the replacement of the convex hull H(DD(*a*,K)) by the α-hull and the more detailed α,β,γ-hull of the given density domain DD(*a*,K) of some specified nuclear configuration K.

The convex hull H(X) of set X, as the smallest convex set containing X, lends itself to useful generalizations.

11.3.1 The the α-Hull

One such generalization, the α-hull introduced by Edelsbrunner, Kirkpatrick, and Seidel [52], takes advantage of the fact that half planes can be easily modified to curved surfaces.

These concept are the simplest if applied to finite point sets, but generalizations to continua are rather trivial.

For a set X of *n* points of the plane, the *convex hull* of X, denoted by H(X) can be defined as the intersection of all closed halfplanes that contain X.

The α-hull of X, H(α,X) is obtained by replacing the halfplanes with discs.

Following Ref. [52], one may define a *generalized disc of radius* $1/\alpha$ as:

i. a disc of radius $1/\alpha$ if $\alpha > 0$,
ii. the complement of a disc of radius $-1/\alpha$ if $\alpha < 0$, and
iii. a halfplane

$$\text{if} \quad \alpha = 0. \tag{11.21}$$

The α-hull H(α,X) of X is the intersection of all closed generalized discs of radius $1/\alpha$ that contain X. Clearly, if $\alpha = 0$, then the α-hull H(α,X) of X becomes the ordinary convex hull H(X) of X:

$$H(0,\ X) = H(X). \tag{11.22}$$

If one takes the convention that the empty intersection is the entire space, then for any set X, the α-hull exists for any α value.

For a sufficiently small value of α, the α-hull is the finite point set X itself.

The definitions for higher dimensions follow in a straightforward way.

The following, natural generalization to 3D finite point sets has been briefly outlined in Ref. [52]. A complete definition is as follows.

A generalized ball of radius $1/\alpha$ is:

i. a ball of radius $1/\alpha$ if $\alpha > 0$,
ii. the complement of a ball of radius $-1/\alpha$ if $\alpha < 0$, and
iii. a half space

$$\text{if } \alpha = 0. \tag{11.23}$$

If X is a finite point set in a 3D Euclidean space, then the α-hull (X) of X, denoted by H(α,X), is the intersection of all closed generalized balls of radius $1/\alpha$ which contain X.

The definition of α-hull can be easily generalized to continua, such as electron density clouds, and this is one of the features that we need; however, if one considers a continuum X of points instead of a finite point set, then one must typically deal with the intersection of infinitely many closed disks (or balls in 3D).

11.3.2 The α,β,γ-Hull and Oriented α,β,γ-Hull

The α,β,γ-hull of a point set X is useful when one needs some simple shape biases or shape restrictions, in terms of some distortions of a perfectly symmetric sphere into some ellipsoid.

The oriented α,β,γ-hull of a point set X is useful when one needs some orientational constraints, with respect to external references, such as a set of three orthogonal coordinate axes.

In order to follow or take advantage of such restrictions, we may consider oriented, solid ellipsoids of half axes of lengths $1/\alpha$, $1/\beta$, and $1/\gamma$, which may, or may not be restricted to be aligned with three, mutually orthogonal coordinate directions of the 3D Euclidean space.

A generalized solid ellipsoid $E(\alpha,\beta,\gamma)$ of half axes of lengths $1/\alpha$, $1/\beta$, and $1/\gamma$, is:

i. the solid ellipsoid of half axes of lengths of $1/\alpha$, $1/\beta$, and $1/\gamma$, if for these quantities $\alpha > 0$, $\beta > 0$, and $\gamma > 0$;

ii. the complement of the solid ellipsoid of half axes of lengths $-1/\alpha$, $-1/\beta$, and $-1/\gamma$ if for these quantities $\alpha < 0$, $\beta < 0$, and $\gamma < 0$ and;

iii. a half space

$$\text{if } \alpha = 0,\ \beta = 0, \text{ and } \gamma = 0. \tag{11.24}$$

One may first consider such an ellipsoid without orientation constraints.

11.3.3 The α,β,γ-Hull H(α,β,γ X) of a Point Set X

The α,β,γ-hull of H(α,β,γ X) of a point set X is the intersection of all closed generalized ellipsoids $E(\alpha,\beta,\gamma)$ that contain X.

One may introduce *orientation constraints* if one considers only ellipsoids $E(\alpha,\beta,\gamma)$ of a fixed orientation of their principal axes relative to the three external directions, such as coordinate axes. In the latter, orientationally constrained case, the ellipsoids may undergo arbitrary translations but no rotations, when testing whether the set X fits within the ellipsoids.

11.3.4 The Oriented α,β,γ-Hull H(α,β,γ,o, X) of a Point Set X

The oriented α,β,γ-hull H(α,β,γ,o, X) of a point set X is the intersection of all closed generalized ellipsoids $E(\alpha,\beta,\gamma)$ *of orientation o* that contain X.

If a reference coordinate system is given, then the orientation

$$o = o(a, b, c) \tag{11.25}$$

can be given by the a, b, c angles between the α, β, and γ axes of the ellipsoids $E(\alpha,\beta,\gamma)$ and the x, y, and z Cartesian axes, respectively.

The most advanced among the "generalized hulls" discussed here is the T-hull of a set X, where T is some other, possibly quite irregular set.

11.3.5 The *T*-Hull of a Point Set X

It is possible to generalize even further the concept of convex hull H(X) by replacing ellipsoids of the α,β,γ-hull H(α,β,γ X) with more general test objects, denoted by *T*, leading to the concept of the *T*-hull H(*T*,X) of a point set X.

One may consider a closed, but otherwise arbitrary 3D set *T*. As in earlier studies, for the sake of simplicity in the notation, we write T' for the closure of the complement of *T*, i.e.:

$$T' = \mathrm{clos}(E^3 \backslash T). \tag{11.26}$$

Then the set T' can also be chosen as a reference object.

When applying *T*-halls, we should keep in mind the following:

Any set generated by translation and rotation of *T* is called a *version* of *T*.

If set *T* is subject to orientation constraints, then a version of *T* is a set obtained from *T* only by translation.

The first task is to test the containment of set X in all possible rotated and translated versions of the reference set *T*.

The *T*-hull H(*T*,X) of a point set X is the intersection of all rotated and translated versions of *T* that contain X.

If one considers a reference set *T* of some fixed orientation with respect to some Cartesian axes of the 3D Euclidean space, then the associated *oriented T-hull* H(*T*,*o*,X) of a point set X is the intersection of all translated versions of the oriented set *T* that contain X.

Note that a *T*-hull H(*T*,Y) or an oriented *T*-hull Y = H(*T*,*o*,X) of a point set X is not necessarily connected, even for a connected reference set *T*.

Each of the above outlined "hull" operations is suitable to bring about some simplification, for example, the convex hulls, α-hulls, α,β,γ-hulls, and *T*-hulls, the latter two with or without orientation, can be treated in a simpler way than, for example, the X = DD(*a*,K) density domains. Each such hull generation is a single step, with no need for any further adjustment to accomplish the given information reduction, that is also manifested by the idempotency of these hull operations:

$$\mathrm{H(H(X))} = \mathrm{H(X)} \tag{11.27}$$

$$\mathrm{H}(\alpha, \mathrm{H}(\alpha, \mathrm{X})) = \mathrm{H}(\alpha, \mathrm{X}) \tag{11.28}$$

$$\mathrm{H}(\alpha, \beta, \gamma\ \mathrm{H}(\alpha, \beta, \gamma\ \mathrm{X})) = \mathrm{H}(\alpha, \beta, \gamma\ \mathrm{X}) \tag{11.29}$$

$$H(\alpha,\beta,\gamma,o,H(\alpha,\beta,\gamma,o,X)) = H(\alpha,\beta,\gamma,o,X) \tag{11.30}$$

$$H(T,H(T,X)) = H(T,X) \tag{11.31}$$

$$H(T,o,H(T,o,X)) = H(T,o,X) \tag{11.32}$$

That is, no further simplification; in fact, no change can be achieved by any attempted repeated or iterative application of any of these "hall-generation" steps, unless some of the conditions are also changed, and a different type of hall is generated.

11.4 A CHEMICALLY MOTIVATED, CONFIGURATION SPACE APPROACH APPLIED TO SOME SYMMETRY DEFICIENCY AND CHIRALITY MEASURE PROBLEMS

Since some of the aspects of using the configuration space approach for the treatment of some symmetry deficiencies have been already described in Ref. [50], here only a brief introduction will be given, and the emphasis will be placed on those aspects that provide useful links to the approaches described in the previous sections of this report.

The starting point of the nuclear configuration space model is a description based on the 3D positions of the nuclei of the molecule. The description is conceptually the simplest if the nuclear positions are considered within a common laboratory frame, that is, with respect to some origin and a choice of orthogonal x, y, and z coordinate axes. In this model, each nuclear location is described by three Cartesian coordinates. Essential aspects of conformational changes, including symmetry changes, as well as chemical reactions can then be represented by some changes in these coordinate values. If there are N nuclei, these coordinates span a 3N-dimensional Euclidean space, with proper distance function (e.g., by the Pythagorean rule), that describes the changes of nuclear arrangements. Of course, if a single molecule is considered, then rigid translations and rigid rotations within such a 3N-dimensional coordinate system do not correspond to any actual chemical process, and in the case of $N>2$ nuclei, there are, in fact, six coordinates unnecessary. To take advantage of this chance for dimension-reduction, various "internal coordinates," such as bond stretching, bond angle, and dihedral angle "internal coordinates" are often used, reducing the dimension to 3N-6. However, the choice of internal coordinates is far from unique, and it has also been shown [54,55], that such a (3N-6)-dimensional "internal configuration space" can never be a vector space, and some of the convenience of the Pythagorean distance function is lost. However, it has also been shown [54,55], that a proper "metric", that is, a well-defined distance function always can be introduced even to such a "reduced" space.

It has also been shown that if some physically impossible "molecular" structures with pathologically short internuclear distances are excluded, than a representation of the internal coordinates within a metric space M of 3N-6 dimensions can also be considered as a manifold with boundary [54,55].

In one, rather general representation [54,55], the M-space distance d(K,K′) between any two internal configurations, a proper metric, is based on the Euclidean distance between sets of equivalent nuclear arrangements within the 3N-dimensional space of Cartesian coordinates of the nuclei. As required, this metric fulfills the following four required conditions for any three internal nuclear configurations K, K′, and K″:

$$d(K, K') \geq 0, \tag{11.33}$$

$$d(K, K') = 0 \text{ if and only if } \quad K = K', \tag{11.34}$$

$$d(K, K') = d(K', K), \tag{11.35}$$

$$d(K, K') \leq d(K, K') + d(K, K'). \tag{11.36}$$

Utilizing this distance function among internal configurations, various symmetry deficiency measures can be obtained within M. For each nuclear configuration K and with respect to any symmetry element *R*, a measure of symmetry deficiency SD(K,*R*) has been defined as the minimum of the M-space distances of the actual nuclear configuration K from all those K′ configurations that do possess this very symmetry element *R*:

$$SD(K, R) = \min\{d(K, K'){:}K' \text{ has } R\}. \tag{11.37}$$

The SD(K,*R*) quantity is the internal configuration space distance, that is, the M-space distance of the configuration K to the set of nuclear configurations possessing this symmetry element, that distance, by definition, is the distance to the nearest nuclear configuration K′ that has the symmetry element *R*.

One evident consequence of this definition is the following: if the nuclear configuration K itself has the symmetry element *R*, that is, if it is not at all deficient (has zero deficiency) in this symmetry R, then the choice of $\mathrm{K}' = \mathrm{K}$ realizes the minimum distance, where this distance is obviously zero, consequently, as expected, the measure of symmetry deficiency is zero:

$$\mathrm{SD}(\mathrm{K}, R) = 0. \tag{11.38}$$

Although several configurations may actually realize this minimum distance value, that is, nuclear configuration K′ needs not be unique for a given choice of nuclear configuration K and symmetry element R; nevertheless, the value SD(K,R) of this distance is unique, hence, the symmetry deficiency value is well-defined.

Now we are in the position of considering for a given nuclear configuration K the three sets, $\boldsymbol{R}_a$, $\boldsymbol{R}_g$, and $\boldsymbol{R} = \boldsymbol{R}_a \cup \boldsymbol{R}_g$ of symmetry elements, and consider all the associated symmetry deficiencies, now expressed in terms of then SD(K,R) M-space symmetry deficiency measures, instead of the internal and external 3D symmetry deficiencies.

Simply by replacing ISD(X,R) with SD(K,R) in each of the vectors and overall measures, one obtains the corresponding descriptors:

One obtains for the chirality generating $\boldsymbol{R}_a$ set and for the chirality independent $\boldsymbol{R}_g$ set the M-space symmetry deficiency vectors

$$\boldsymbol{d}(K, \boldsymbol{R}_a, \mathrm{SD}) = (\mathrm{SD}(\mathrm{K}, \sigma),\ \mathrm{SD}(\mathrm{K}, S_2),\ \mathrm{SD}(\mathrm{K}, S_4),\ \mathrm{SD}(\mathrm{K}, S_6),\ \ldots) \tag{11.39}$$

and

$$\boldsymbol{d}(\mathrm{K}, \boldsymbol{R}_g, \mathrm{SD}) = (\mathrm{SD}(\mathrm{K}, R_1),\ \mathrm{SD}(\mathrm{K}, R_2),\ \mathrm{SD}(\mathrm{K}, R_3),\ \ldots,\ \mathrm{SD}(\mathrm{K}, R_p),\ \ldots), \tag{11.40}$$

respectively, and for the full set $\boldsymbol{R} = \boldsymbol{R}_a \cup \boldsymbol{R}_g$,

$$\begin{aligned}\boldsymbol{d}(\mathrm{K}, \boldsymbol{R}, \mathrm{SD}) = {} & (\mathrm{SD}(\mathrm{K}, \sigma),\ \mathrm{SD}(\mathrm{K}, S_2),\ \mathrm{SD}(\mathrm{K}, S_4),\ \mathrm{SD}(\mathrm{K}, S_6),\ \ldots \\ & \mathrm{SD}(\mathrm{K}, R_1),\ \mathrm{SD}(\mathrm{K}, R_2),\ \mathrm{SD}(\mathrm{K}, R_3),\ \ldots,\ \mathrm{SD}(\mathrm{K}, R_p),\ \ldots)\end{aligned} \tag{11.41}$$

Based on the components of these vectors, an overall M-space achirality deficiency measure is obtained as:

$$\mathrm{MACD}(\mathrm{K}, \boldsymbol{R}_a) = \min(\boldsymbol{d}_i(\mathrm{K}, \boldsymbol{R}_a, \mathrm{SD}),\ i = 1\ldots n + 1) \tag{11.42}$$

referred to as "M-space achirality deficiency measure of nuclear configuration K".

Similarly, an overall M-space ci-symmetry deficiency measure can be obtained as:

$$\mathrm{MCISD}(\mathrm{K}, \boldsymbol{R}_g) = \min\,(\boldsymbol{d}_i(K, \boldsymbol{R}_g, \mathrm{SD}),\ i = 1\ldots m), \tag{11.43}$$

where is referred to as "M-space chirality-independent symmetry deficiency measure of nuclear configuration K", and where index i runs through the serial numbers of all symmetry elements in set $\boldsymbol{R}_g$.

For the overall, total symmetry deficiency measure of K in M, one obtains

$$\mathrm{MOSD}(\mathrm{K}, \boldsymbol{R}) = \min(\boldsymbol{d}_j(\mathrm{K}, \boldsymbol{R}, \mathrm{SD}),\ j = 1, 2, \ldots), \tag{11.44}$$

referred to as the M-space overall symmetry deficiency measure of nuclear configuration K, where index j runs through the serial numbers of all symmetry elements in set $\boldsymbol{R}$.

11.5 CHOICES FOR CHIRALITY MEASURES AND DESCRIPTORS IN TERMS OF A "SYMMETRY DEFICIENCY SPECTRUM"

Whereas the quantities IOSD(X,$\boldsymbol{R}$), EOSD(X,$\boldsymbol{R}$), and MOSD(K,$\boldsymbol{R}$) serve as overall measures of symmetry efficiency of general "chemical objects" (electron density domains, nuclear convex hulls, α-halls, α,β,γ-halls, T-halls, or even M-space nuclear configurations K), there is often a need for case-to-case comparisons involving specific symmetry elements. In such cases, it is useful to consider the individual components of the various symmetry deficiency vectors, such as:

$$\begin{aligned}\boldsymbol{d}(\mathbf{X}, \boldsymbol{R}, \mathrm{ISD}) = (&\mathrm{ISD}(\mathrm{X}, \sigma),\ \mathrm{ISD}(\mathrm{X}, S_2),\ \mathrm{ISD}(\mathrm{X}, S_4),\ \mathrm{ISD}(\mathrm{X}, S_6),\ \ldots\\ &\mathrm{ISD}(\mathrm{X}, R_1),\ \mathrm{ISD}(\mathrm{X}, R_2),\ \mathrm{ISD}(\mathrm{X}, R_3),\ \ldots,\ \mathrm{ISD}(\mathrm{X}, R_p),\ \ldots),\end{aligned}$$

and

$$\begin{aligned}\boldsymbol{d}(\mathbf{X}, \boldsymbol{R}, \mathrm{ESD}) = (&\mathrm{ESD}(\mathrm{X}, \sigma),\ \mathrm{ESD}(\mathrm{X}, S_2),\ \mathrm{ESD}(\mathrm{X}, S_4),\ \mathrm{ESD}(\mathrm{X}, S_6),\ \ldots\\ &\mathrm{ESD}(\mathrm{X}, R_1),\ \mathrm{ESD}(\mathrm{X}, R_2),\ \mathrm{ESD}(\mathrm{X}, R_3),\ \ldots,\ \mathrm{ESD}(\mathrm{X}, R_p),\ \ldots),\end{aligned}$$

and

$$\begin{aligned}\boldsymbol{d}(\mathbf{X}, \boldsymbol{R}, \mathrm{SD}) = (&\mathrm{SD}(\mathrm{X}, \sigma),\ \mathrm{SD}(\mathrm{X}, S_2),\ \mathrm{SD}(\mathrm{X}, S_4),\ \mathrm{SD}(\mathrm{X}, S_6),\ \ldots\\ &\mathrm{SD}(\mathrm{X}, R_1),\ \mathrm{SD}(\mathrm{X}, R_2),\ \mathrm{SD}(\mathrm{X}, R_3),\ \ldots,\ \mathrm{SD}(\mathrm{X}, R_p),\ \ldots),\end{aligned}$$

In fact, by considering the sequence of vector components for the vectors involving all chemically symmetry elements, the list of these components may serve as a "symmetry deficiency spectrum" for the given "chemical objects." Direct comparisons of individual symmetry deficiencies obtained for specific symmetry elements are not unlike comparisons of spectra, and the fact that in these lists the chirality-inducing symmetry deficiencies are given first underlines the relative chemical importance of these symmetry deficiencies, providing, numerical chirality measures.

ACKNOWLEDGMENT

The author acknowledges stimulating conversations with Professors G. Palyi, P. Surjan, E. Szathmary, and B. Gulyas, the hospitality of the Hungarian Academy of Sciences, Budapest, Hungary, as well as the continuing financial support of the Canada Research Chair program.

REFERENCES

[1] Mislow, K.; Bolstad, R. Molecular Dissymetry and Optical Activity. *J. Am. Chem. Soc.* **1955,** *77*, 6712–6713.
[2] Kitaigorodski, A. I. Chapter 4. In *Organic Chemical Crystallography;* Consultants Bureau: New York, 1961; pp 75–100.
[3] Dubois, J.-E.; Alliot, M.-J.; Panaye, A. Système DARC. Présentation Unitaire de Chiralité et Prochiralité à L'aide du Concept d'Environnement Orienté. *Compt. Rend. Acad. Sci. Paris (Série C)* **1971,** *273*, 224–227.
[4] Klein, D. J.; Kramling, R. W. The Symmetry of Reduced Density Matrices. I. General Algebraic Formulation. *Int. J. Quant. Chem. S* **1970,** *4*, 661–674.
[5] Klein, D. J. The Symmetry of Reduced Density Matrices. II. Spin-Free Quantum Chemistry and Point Group Symmetry. *Int. J. Quant. Chem. S* **1970,** *3*, 675–688.
[6] Klein, D. J.; Carlisle, C. H.; Matsen, F. A. Symmetry Adaptation to Sequences of Finite Groups. In *Advances in Quantum Chemistry;* Lowdin, P. O., Ed.; Academic Press: Amsterdam, 1970; Vol. 5, pp 219–260.
[7] Klein, D. J. Symmetry Adapted Perturbation Theory for Interatomic and Intermolecular Exchange Energies. *Int. J. Quant. Chem. S* **1971,** *4*, 271–283.
[8] Klein, D. J. Comments on Kets not Strongly Symmetry Adapted. *Int. J. Quant. Chem.* **1974,** *8*, 234–245.
[9] Klein, D. J. Finite Groups and Semisimple Algebras in Quantum Mechanics. In *Group Theory and Its Applications;* Loebl, E. M., Ed.; Academic Press: New York, 1975; Vol. III, pp 1–93.
[10] Hasselbarth, W.; Ruch, E.; Klein, D. J.; Seligman, T. H. Double Classes: A New Classification Scheme for Group Elements. In *Proceedings of the V International Colloquium on Group Theory and Its Applications;* Sharpe, R. T., Ed.; Academic Press: New York, 1977; pp 617–622.
[11] Klein, D. J.; Seitz, W. A. Symmetric-Group Algebraic Variational Solutions for Heisenberg Models at Finite Temperature. *Int. J. Quant. Chem.* **1992,** *41*, 43–52.

[12] Mislow, K.; Siegel, J. Stereoisomerism and Local Chirality. *J. Am. Chem. Soc.* **1984,** *106*, 3319–3322.
[13] Rassat, A. Un Critere de Classement des Sytemes Chiraux de Points a Partir de la Distance au Sens. de Haussdorf. *Compt. Rend. Acad. Sci. (Paris) II* **1984,** *299*, 53–55.
[14] Gilat, G. Chiral Interactions in Biomolecules. *Chem. Phys. Lett.* **1985,** *121*, 9–12.
[15] Gilat, G.; Schulman, L. S. Chiral Interaction, Magnitude of the Effects and Application to Natural Selection of L-enantiomer. *Chem. Phys. Lett.* **1985,** *121*, 13–16.
[16] Barron, L. D. Symmetry and Molecular Chirality. *Chem. Soc. Rev.* **1986,** *15*, 189–223.
[17] Mezey, P. G. Tying Knots Around Chiral Centres: Chirality Polynomials and Conformational Invariants for Molecules. *J. Am. Chem. Soc.* **1986,** *108*, 3976–3984.
[18] Mezey, P. G. Symmetry and Periodicity of Potential Surfaces. *Theor. Chim. Acta* **1986,** *73*, 221–228.
[19] Gilat, G. Chiral Coefficient: A Measure of the Amount of Structural Chrality. *J. Phys. A, Math. Gen.* **1989,** *22*, L545–L550.
[20] Gilat, G. On the Quantification of Symmetry in Nature. *Found. Phys. Lett.* **1990,** *3*, 189–196.
[21] Hel-Or, Y.; Peleg, S.; Avnir, D. Two-Dimensional Rotational Dynamic Chirality and a Chirality Scale. *Langmuir* **1990,** *6*, 1691–1695.
[22] Mezey, P. G., Ed. *New Developments in Molecular Chirality;* Kluwer: Dordrecht, 1991.
[23] Auf der Heyde, T. P. E.; Buda, A. B.; Mislow, K. Desymmetrization and the Degree of Chirality. *J. Math. Chem.* **1991,** *6*, 255–265.
[24] Buda, A. B.; Mislow, K. On Geometric Measures of Chirality. *J. Mol. Struct. (Theochem)* **1991,** *232*, 1–12.
[25] Meyer, A. Y.; Avnir, D. The Relation Between Molecular Shape and Molecular Rotations in Chiral Halogenated Alkanes. *Struct. Chem.* **1991,** *2*, 475–478.
[26] Mezey, P. G. The Degree of Similarity of Three-Dimensional Bodies; Applications to Molecular Shapes. *J. Math. Chem.* **1991,** 7, 39–49.
[27] Mezey, P. G. On the Allowed Symmetries of All Distorted Forms of Conformers, Molecules, and Transition Structures. *Can. J. Chem.* **1992,** *70*, 343–347.
[28] Mezey, P. G. *Shape in Chemistry: An Introduction to Molecular Shape and Topology;* VCH Publishers: New York, 1993.
[29] Mezey, P. G. Quantum Chemistry of Macromolecular Shape,. *Int. Rev. Phys. Chem.* **1997,** *16*, 361–388.
[30] Mezey, P. G. The Proof of the Metric Properties of a Fuzzy Chirality Measure of Molecular Electron Density Clouds. *J. Mol. Struct. Theochem* **1998,** *455*, 183–190.
[31] Mezey, P. G. Generalized Chirality and Symmetry Deficiency. *J. Math. Chem.* **1998,** *23*, 65–84.
[32] Mezey, P. G. Mislow's Label Paradox, Chirality-Preserving Conformational Changes, and Related Chirality Measures. *Chirality* **1998,** *10*, 173–179.
[33] Mezey, P. G. Theory of Biological Homochirality: Chirality, Symmetry Deficiency, and Electron Cloud Holography in the Shape Analysis of Biomolecules. In *Advances in BioChirality;* Palyi, G., Zucchi, C., Caglioti, L., Eds.; Elsevier Science Publishing: Amsterdam, 1999; pp 35–46.
[34] Mezey, P. G. The Role of Imperfect Symmetry in Nature, Art, Mathematics, and Chemistry: Approximate Symmetry and Symmetry Deficiency Measures. *J. Int. Soc. Interdiscip. Study Symmetry* **2004,** *2004*, 166–169.
[35] Mezey, P. G. The Theory of Chirality Induction and Chirality Reduction in Biomolecules. In *Progress in Biological Chirality;* Palyi, G., Zucchi, C., Cagliotti, L., Eds.; Elsevier: Oxford, 2004; pp 209–219.

[36] Mezey, P. G. Molecular Similarity, Quantum Topology, and Shape. In *Computational Medicinal Chemistry and Drug Discovery;* Bultinck, P., Tollenaere, J. P., De Winter, H., Langenaeker, W., Eds.; Marcel Decker Inc.: New York, 2004; pp 345–364.
[37] Mezey, P. G. The Topological Model of Non-rigid Molecules and Reaction Mechanisms. In *Symmetries and Properties of Non-Rigid Molecules: A Comprehensive Survey;* Maruani, J., Serre, J., Eds.; Elsevier Science Publishing: Amsterdam, 1983; pp 335–353.
[38] Mezey, P. G. Topological Theory of Molecular Conformations. In *Structure and Dynamics of Molecular Systems;* Daudel, R., Korb, J.-P., Lemaistre, J.-P., Maruani, J., Eds.; Reidel: Dordrecht, 1985; pp 57–70.
[39] Mezey, P. G.; Maruani, J. The Concept of "Syntopy": A Continuous Extension of the Symmetry Concept for Quasi-Symmetric Structures Using Fuzzy-Set Theory. *Mol. Phys.* **1990,** *69*, 97–113.
[40] Mezey, P. G.; Maruani, J. The Fundamental Syntopy of Quasi-Symmetric Systems: Geometric Criteria and the Underlying Syntopy of a Nuclear Configuration Space. *Int. J. Quant. Chem.* **1993,** *45*, 177–187.
[41] Zabrodsky, H.; Avnir, D. Continuous Symmetry Measures. 4. Chirality. *J. Am. Chem. Soc.* **1995,** *117*, 462–473.
[42] Mezey, P. G. A Proof of the Metric Properties of the Symmetric Scaling-Nesting Dissimilarity Measure and Related Symmetry Deficiency Measures. *Int. J. Quant. Chem.* **1997,** *63*, 105–109.
[43] Mezey, P. G. Generalized Chirality and Symmetry Deficiency. *J. Math. Chem.* **1998,** *23*, 65–84.
[44] Mezey, P. G.; Fukui, K.; Arimoto, S.; Taylor, K. Polyhedral Shapes of Functional Group Distributions in Biomolecules and Related Similarity Measures. *Int. J. Quant. Chem.* **1998,** *66*, 99–105.
[45] Maggiora, G. M.; Mezey, P. G. A Fuzzy Set Approach to Functional Group Comparisons Based on an Asymmetric Similarity Measure. *Int. J. Quant. Chem.* **1999,** *74*, 503–514.
[46] Mezey, P. G.; Fukui, K.; Arimoto, S. A Treatment of Small Deformations of Polyhedral Shapes of Functional Group Distributions in Biomolecules. *Int. J. Quant. Chem.* **2000,** *76*, 756–761.
[47] Mezey, P. G. Theory and Detailed Computer Modelling of Biomolecules. In *Fundamentals of Life;* Palyi, G., Zucchi, C., Cagliotti, L., Eds.; Elsevier: Paris, 2002; pp 401–416.
[48] Mezey, P. G. Rules on the Changes of Approximate Symmetry Measures Along Reaction Paths. *Mol. Phys.* **2006,** *104*, 723–729 Corrigendum, *Mol. Phys.*, **104**, 2575 (**2006**).
[49] Wang, L.; Wang, L.; Arimoto, S.; Mezey, P. G. Large-Scale Chirality Measures and General Symmetry Deficiency Measures for Functional Group Polyhedra of Proteins. *J. Math. Chem.* **2006,** *40*, 145–153.
[50] Mezey, P. G. Theoretical Aspects of Organometallic Chirality. In *Organometallic Chirality;* Palyi, G., Zucchi, C., Caglioti, L., Eds.; Accademia Nazionale di Scienze, Lettere e Arti – Mucchi Editore: Modena, 2008; pp 29–46.
[51] Grunbaum, B. *Convex Polytopes;* Wiley: New York, 1967.
[52] Edelsbrunner, H.; Kirkpatrick, D. G.; Seidel, R. On the Shape of a Set of Points in the Plane. *IEEE Trans. Inf. Theory* **1983,** *29*, 551–559.
[53] Mezey, P. G. The α,β,γ- Hull and the T-Hull of a Point Set: Tools for the Analysis of Shapes and Relative Orientations of Objects in 3D - Space. *J. Math. Chem.* **1991,** *8*, 91–102.
[54] Mezey, P. G. Manifold Theory of Multidimensional Potential Surfaces. *Int. J. Quant. Chem., Quant. Biol. Symp.* **1981,** *8*, 185–196.
[55] Mezey, P. G. *Potential Energy Hypersurfaces;* Elsevier: Amsterdam, 1987.

FURTHER READING

Mezey, P. G. The T-Hull Approach to Shape Analysis. *Theor. Chim. Acta* **1995,** *91*, 67–71.

Mezey, P. G. Two Symmetry Constraints on the Identity and Deformations of Chemical Species. *J. Phys. Chem.* **1995,** *99*, 4947–4954.

Mezey, P. G. Molecular Similarity Measures of Conformational Changes and Electron Density Deformations. *Adv. Mol. Similarity* **1996,** *1*, 89–120.

CHAPTER 12

Asymmetric Reduction of Ketones to Chiral Platform Molecules

Bálint Fridrich, József M. Tukacs and László T. Mika
Department of Chemical and Environmental Process Engineering, Budapest University of Technology and Economics, Budapest, Hungary

Abstract

The enantioselective reduction of prochiral substrates including C=O double bonds is fundamentally important in the synthesis of biologically active compounds. Due to the depletion of fossil-based resources, the production of these species from renewable resources is highly desired. The enantiopure γ-valerolactone was proposed as a biomass-based chiral platform molecule that can be easily obtained by the asymmetric reduction of levulinic acid and its esters. Synthetic schemes of the production of optically active γ-valerolactone are presented. The effect of catalyst composition and reaction parameters on the activity and selectivity will be discussed in detail.

Keywords: Hydrogenation; optically active; γ-valerolactone; levulinic acid; biomass conversion; enantioselective synthesis

12.1 INTRODUCTION

The sustainability of our civilization exclusively depends on whether we can supply the increasing population with enough energy, food, drinking water, and carbon-based chemicals without threatening the Earth's long-term health. There is no doubt that the chemical industry, which continuously provides value-added chemicals, including biologically active compounds such as pharmaceuticals and pesticides, is also exclusively based on fossil resources. Since it is difficult to estimate the depletion of crude oil and natural gas, the utilization of renewable feedstocks for production of carbon-based end-products should be accelerated. Although, the conversion of carbon dioxide could be an elegant approach and several studies were published on its activation, no effective processes to produce organic chemicals on a large scale have been developed yet. Accordingly, selective biomass conversion as one of the most preferred alternative solutions has become a focus of interest and several novel

Advances in Asymmetric Autocatalysis and Related Topics
DOI: http://dx.doi.org/10.1016/B978-0-12-812824-4.00012-5

building blocks, so-called platform chemicals, have been identified. They could either replace the currently used fossil-based chemicals or serve as renewable feedstock for their production [1]. While these bulk chemicals can be produced by biorefinery concept, several synthesis concerning biologically active compounds could utilize chiral starting materials. The optically active building blocks could easily be obtained from naturally available carbohydrates and amino acids *via* selective transformation, which do not affect their chiral center(s). However, most of the chemical reactions performed on natural resources led to the loosening of chirality and resulted in the formation of either racemic mixture or nonchiral products. Consequently, to obtain enantiopure platform molecules the chirality has to be incorporated by applying an asymmetric synthesis stage.

The main part of a biomass is present in the form of sugars, starch, cellulose, lignocellulose, and chitin. Since the main agricultural products, i.e., wheat, rice and corn, are supplied for continuously increasing food and feed production, the selection and consumption of appropriate resources have become controversial issues for sustainability [2]. Recent trends move towards the valorization of various biomass-based waste stream having very similar composition to biomass.

The acid-catalyzed dehydration and subsequent hydration reactions of C_6 sugars such as glucose and fructose lead to the formation of 5-hydroxymethyl furfural (5-HMF) and levulinic acid (LA). 5-HMF has been considered as one of the C_6-platforms and can be used for the synthesis of a wide-range of carbon-based chemicals and materials [3]. The conversion of 5-HMF resulted in an equimolar amount of LA and formic acid. The latter can also be used as a hydrogen source for the transfer hydrogenation reaction including reduction of LA. Due to its reactive functional groups, LA has been identified as a C_5-platform molecule [4]. Its various ester derivatives may be used as gasoline and biodiesel additives. While D-aminolevulinate is a known biodegradable herbicide, its calcium salt is intravenously injected as a calcium carrier [5]. Its bisphenol derivative could be a substitute of bisphenol-A in polymer manufacture [4]. The selective hydrogenation of LA to 4-hydroxyvaleric acid (4-HVA) opens the way to the facile synthesis of γ-valerolactone (GVL), which was proposed by Horváth as a sustainable liquid and the key molecule in GVL economy [6]. The simplified conversion route of carbohydrates to GVL as valorization of biomass-based waste is depicted in Scheme 12.1.

γ-valerolactone has already been used for the production of alkanes and alkenes [7], transportation fuels [8], ionic liquids [9], 1,4-pentanediol,

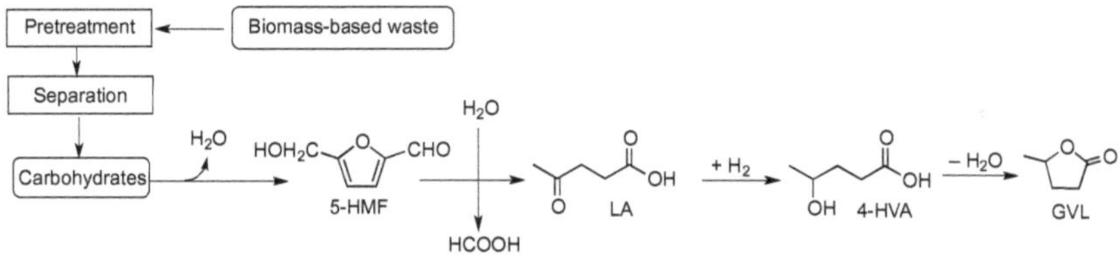

Scheme 12.1 Production of GVL from biomass-based waste.

adipic acid [10], polymers [11], illuminating liquid and lighter fluid [12], etc. Recently, it was demonstrated that GVL can be applied as an environmentally benign reaction medium for the catalytic reactions for, e.g., cross-coupling [13], hydroformylation [14], acid catalyzed dehydration of carbohydrates [16]. Due to its outstanding physical and chemical properties, GVL can be globally used as a carbon and energy source; so far no environmental concerns have been reported. Since GVL occurs naturally in fruits and fermented products such as wines and beers [15], it is nontoxic. GVL is a liquid between −31°C and +207°C, has a high open cup flash point (96°C), a definite smell for easy recognition in case of need, and is miscible with water so facilitating biodegradation. It is stable in a wide range of temperatures up to 150°C [16]. Importantly, GVL does not form any measurable amount of peroxides under air, in sharp contrast to 2-methyltetrahydrofuran [17] or dialkyl ethers (diethyl ether, diisopropyl ether, etc.) making it a relatively safe molecule. Compared to commercially used industrial solvents, its vapor pressure is very low even at high temperature [18] resulting in low VOC emission.

The optically active γ-lactones occur naturally [19] and can be used as chiral building blocks for the synthesis of biologically active compounds [20]. Interestingly, the optically active GVL can be found in the reaction scheme of some agricultural and pharmaceutical agents [20]. The selected examples are presented below. The aggregation pheromone Sulcatol [21] for ambrosia beetles can be synthesized in two steps from optically active γ-valerolactone (Scheme 12.2). A series of aroma compounds can also be obtained from chiral GVL via a short synthesis (Scheme 12.3).

Although, the synthesis of the antihypertensive WS75624B Scheme 12.4 [22], and insecticide Geodiamolide (Scheme 12.5) [23] seems to be extremely complex, application of the optically active GVL in their synthesis scheme could reduce the number of corresponding synthesis steps.

Obviously, the most efficient protocol for the manufacture of GVL is the acid-catalyzed conversion of the carbohydrate content of biomass or

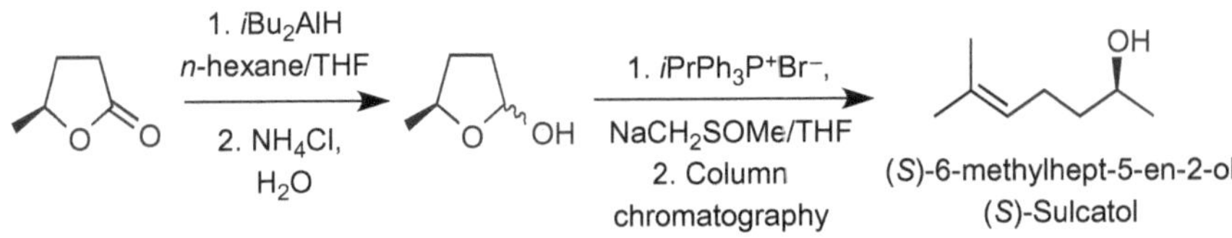

Scheme 12.2 Synthesis of (*S*)-Sulcatol from (*S*)-GVL.

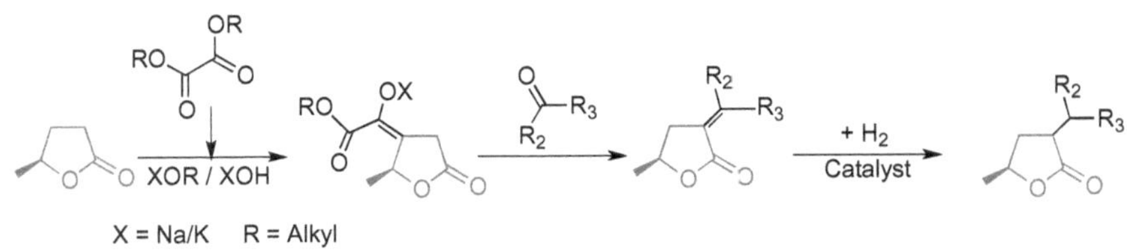

Scheme 12.3 Synthesis of alkylated (*S*)-GVL.

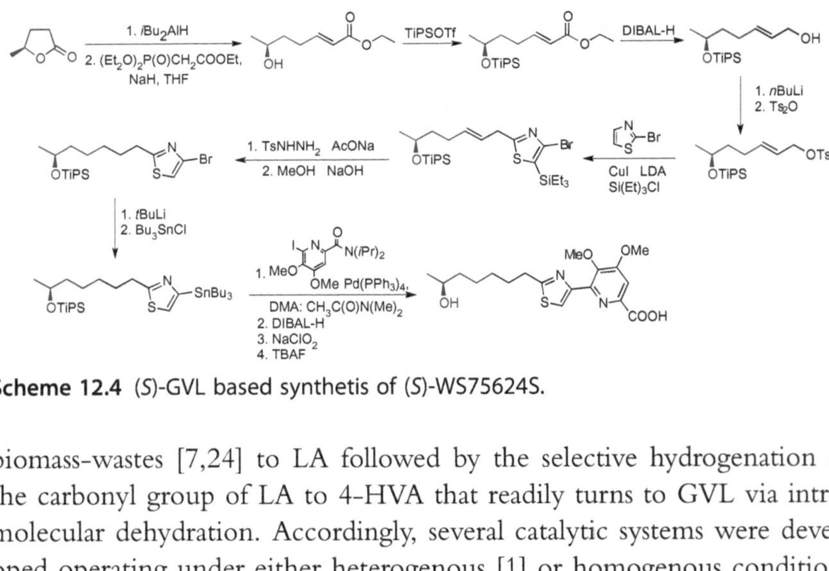

Scheme 12.4 (*S*)-GVL based synthetis of (*S*)-WS75624S.

biomass-wastes [7,24] to LA followed by the selective hydrogenation of the carbonyl group of LA to 4-HVA that readily turns to GVL via intramolecular dehydration. Accordingly, several catalytic systems were developed operating under either heterogenous [1] or homogenous conditions [25,26]. The enantioselective hydrogenation of LA could open a new way to produce optically active 4-HVA that readily turns to optically active GVL [27] (Scheme 12.6).

The enantioselective reduction of ketones for the synthesis of optically active alcohols is of utmost importance and can be an atom-efficient protocol both on laboratory and industrial scales [28]. The most studied model substrates are different aromatic ketones, e.g., acetophenone in the enantioselective reduction of the carbonyl group. That is why novel

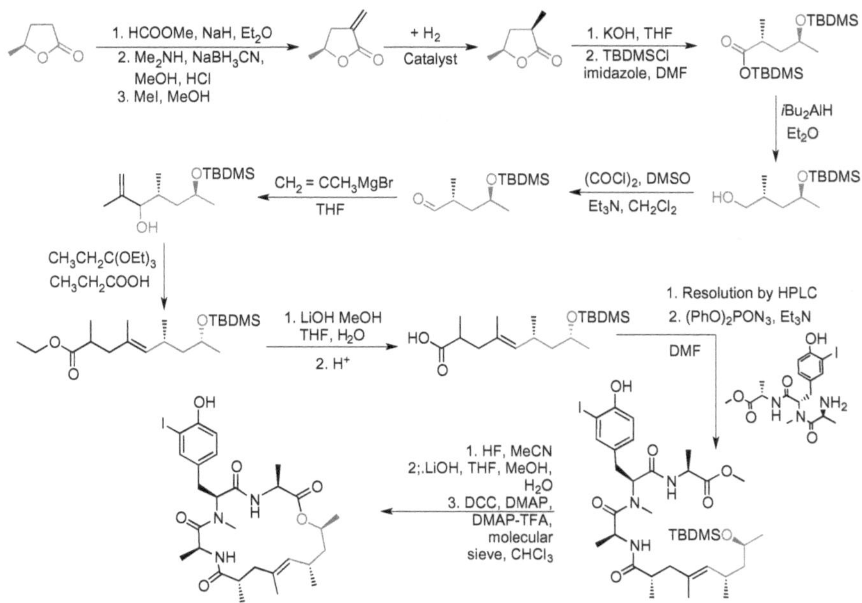

Scheme 12.5 Synthesis of Geodiamolide, an insectide of marine algae origin, from (*S*)-GVL.

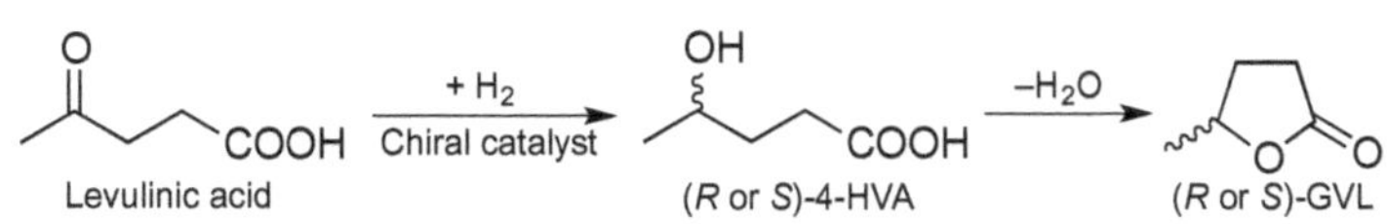

Scheme 12.6 Enantioselective hydrogenation of LA.

catalyst systems have been tested on them. It was shown in several cases that enantioselectivity was dramatically influenced by the nature of the functional groups in the vicinity of the carbonyl group [28d]. Contrary to aromatic ketones, the reduction of aliphatic compounds resulted in lower yield and enantiomeric excess (*ee*) [29,30]. Concerning the production of optically active GVL, it can be obtained via enantioselective hydrogenation of LA, by using alternative synthetic routes, i.e., enzyme-catalyzed transformations, enantioselective reduction of levulinate esters. It should be noted that optical isomers of GVL were separated by resolution, as well.

To isolate two isomers of GVL, Toshima et al. proposed a process in which a L-phenylalanine derivate was used for the resolution of γ-valerolactone via its diastereomeric salt (Scheme 12.7) [31] with sufficient enantiomeric excess. It is noteworthy that the stoichiometric synthesis of enantipure GVL was developed by Vigneron. A facile

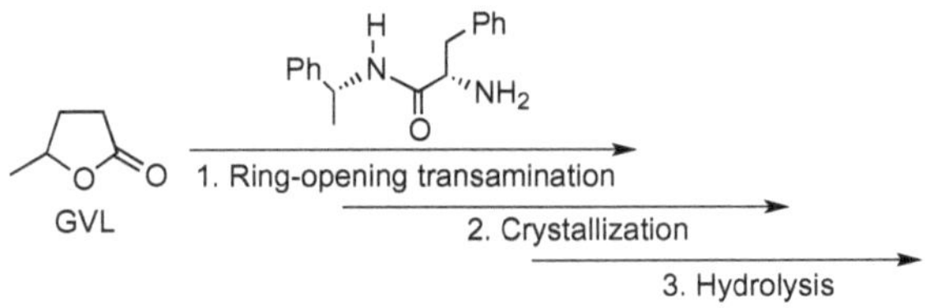

Scheme 12.7 Resolution of GVL.

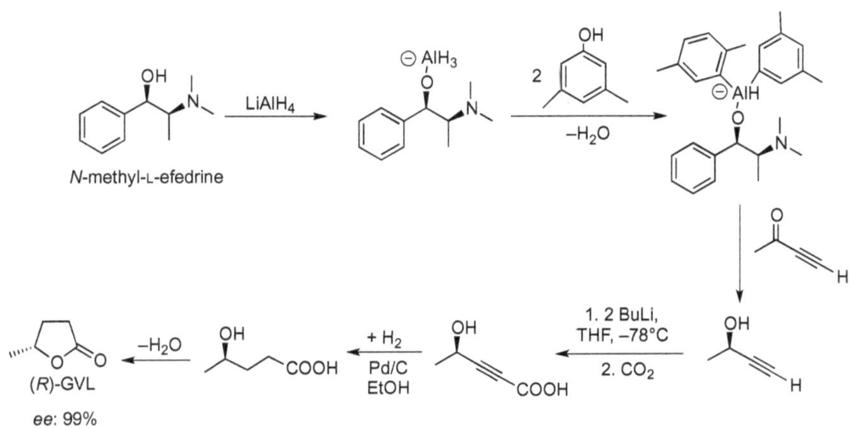

Scheme 12.8 Proposed synthesis of (*R*)-GVL by α-ketyl-acetylene.

enantioselective reduction of α-ketyl-acetylene was carried out by using *N*-methyl-L-efedrin for the activation of $LiAlH_4$ forming an optically active hydrogenation reagent. Its subsequent conversion led to the formation of the corresponding carboxylic acid. The triple bond of the latter was saturated to form (*R*)-4-HVA (Scheme 12.8).

Figadère and Cavé proposed the synthesis of (*R*)-GVL from D-glutamic acid [32]. Expectedly, due to the chirality of the natural starting material, the *ee* of the final product was remarkable (Scheme 12.9). Karnik et al. reported the synthesis of optically active γ-lactones including GVL by using (*S*)-menthyl- or (*S*)-bornyl esters of 4-carboxylates by using $NaBH_4$. In the case of LA, the yield of (*S*)-GVL was moderate (61%) (Scheme 12.10) [33]. So far, a few studies were reported [34,35] on the reduction of alkyl levulinates in the presence of Ru-based catalysts to optically active GVL. Noyori reduced methyl, ethyl, and octyl levulinates to the corresponding γ-lactones in the presence of Ru/BINAP complexes with excellent (>90%) *ee*. The enantioselective hydrogenation was

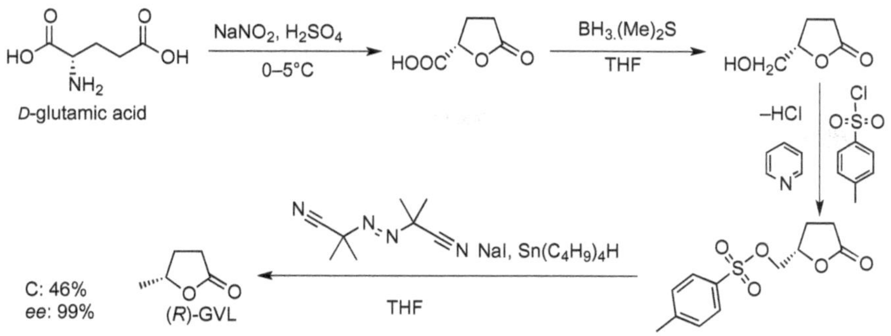

Scheme 12.9 Proposed synthesis of (*R*)-GVL.

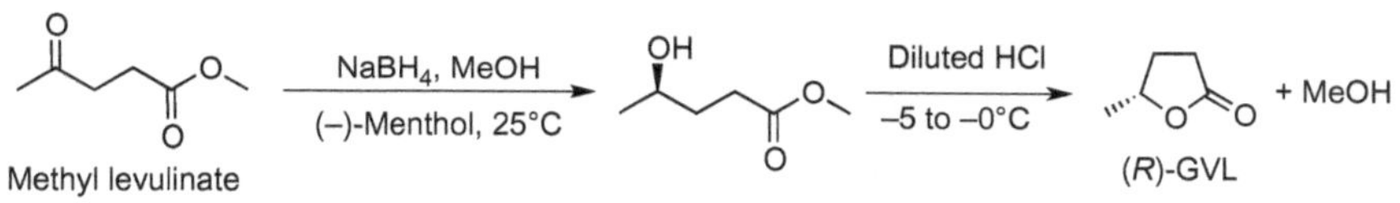

Scheme 12.10 Reduction by $NaBH_4$ in the presence of (–)-menthol.

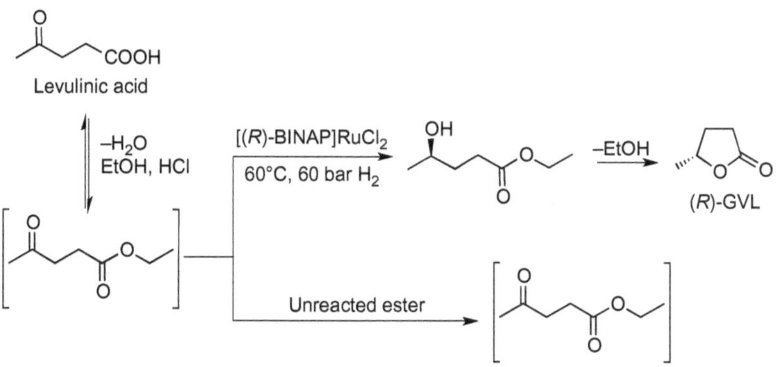

Scheme 12.11 One-pot reduction of levulinc acid to a mixture of optically active lactone and by-products.

extended to 2-acylbenzoic esters for the synthesis of optically active phtalides [33]. Vinogradov et al. synthesized (*R*)-GVL from LA via ethyl levulinate (EL), which in situ formed in the reaction mixture in the presence of ethanol and HCl, by the use of Ru/BINAP catalyst in EtOH under 60 bar H_2 at 60°C [34]. The final reaction mixture contained (*R*)-GVL and (*R*)-4-hydroxyethyl levulinate and unreacted EL (Scheme 12.11).

Compared to transition metal catalyzed transformation, the selectivity of the reduction of C=O bonds in the presence of enzymes was outstanding. Thus, several attempts were made to reduce γ-keto acid esters to

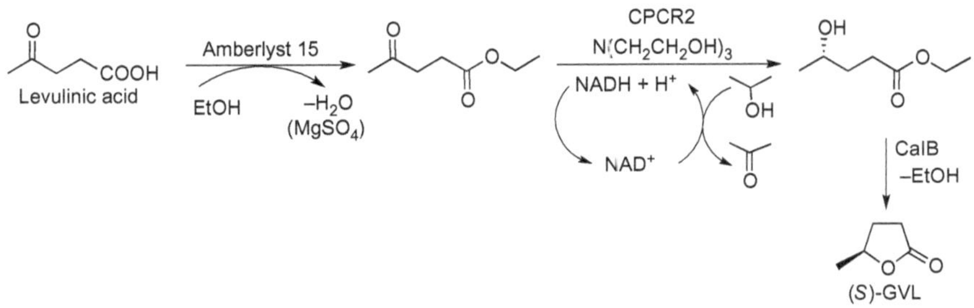

Scheme 12.12 Enzymatic synthesis of (*S*)-GVL via ethyl levulinate.

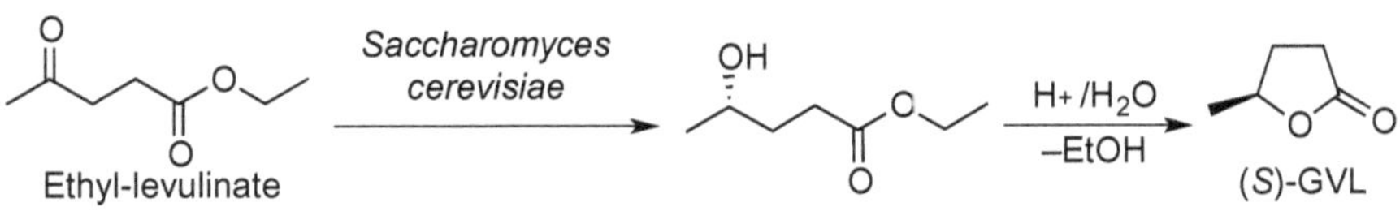

Scheme 12.13 Bakers' yeast assisted reduction of ethyl levulinate.

form the corresponding optically active γ-lactones via chemoenzymatic routes. Hilterhaus et al. developed a reaction sequence to produce (*S*)-GVL via EL. Due to the enzymatic conversion, the *ee* and overall yields were high (~90%), as expected (Scheme 12.12) [36]. Jacobs et al. showed that bakers' yeast assisted the reduction of alkyl levulinates and subsequent HCl-catalyzed hydrolysis of the corresponding hydroxy esters resulted in (*S*)-GVL with a yield of 73% [37] (Scheme 12.13).

The promising aspect of the application of enantipure GVL initiated the investigation of the possibility for the asymmetric reduction of LA to optically active 4-HVA. To take into consideration the ring opening and/or closing, the stability of the chiral center of GVL was crucial for further synthetic strategies. The ring opening of GVL under acidic conditions resulted in 4-HVA, which forms GVL *via* ring closure under neutral conditions. Thus, the stability of the chiral center of GVL was investigated by using the ^{18}O isotope-labeling technique [27]. When (*S*)-GVL having 98.5% *ee* was reacted with 97 atom % ^{18}O water in the presence of HCl at room temperature, the in-situ ^{1}H-NMR verified the equilibrium reaction between GVL and 4-HVA. In order to neutralize the reaction mixture, the equimolar amount of sodium hydroxide was added, then the GVL was extracted with ethyl acetate. Based on chiral GC analysis, no change in the *ee* was observed and the GC-MS confirmed the presence of ^{18}O-GVL as well that can only be formed over ^{18}O-4-HVA (Fig. 12.1) [27].

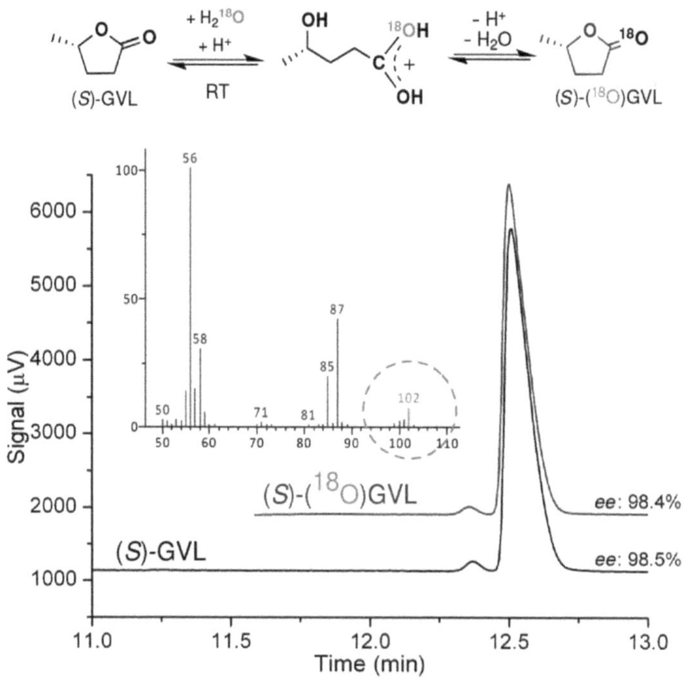

Figure 12.1 ^{18}O incorporation into the optically active GVL.

We demonstrated that GVL can easily be prepared by selective hydrogenation of LA in the presence of bidentate phosphine-modified Ru catalysts without any added solvent with TOF of 100–21,000 h^{-1}. When racemic BINAP was used as a ligand, 98.6% conversion was obtained under 100 bar H_2 at 140°C (TOF = 6978 h^{-1}) [26c]. Consequently, the reduction of LA was achieved by using a catalyst in situ formed from Ru $(acac)_3$ and a 10-fold excess of (*R*)-BINAP under identical conditions. While full conversion was detected after 12 h, the *ee* was moderate (26%). When the reaction was repeated in a 120-mL Parr high-pressure reactor, in which the reaction mixture can be intensively stirred by a propeller, a significantly higher reaction rate was observed, but the *ee* remained the same. Furthermore, it was revealed that the conversion rate had no influence on enantioselectivity (Fig. 12.2). Unexpectedly, the addition of (*R*,*R*)-1,2-diphenyl-ethylenediamine (*R*,*R*-DPEN) did not have a significant effect on the final *ee* values.

It was demonstrated that both in-situ generated and preformed transition metal complexes can be applied for selective hydrogenation; however, the latter usually showed higher activity and selectivity [28d].

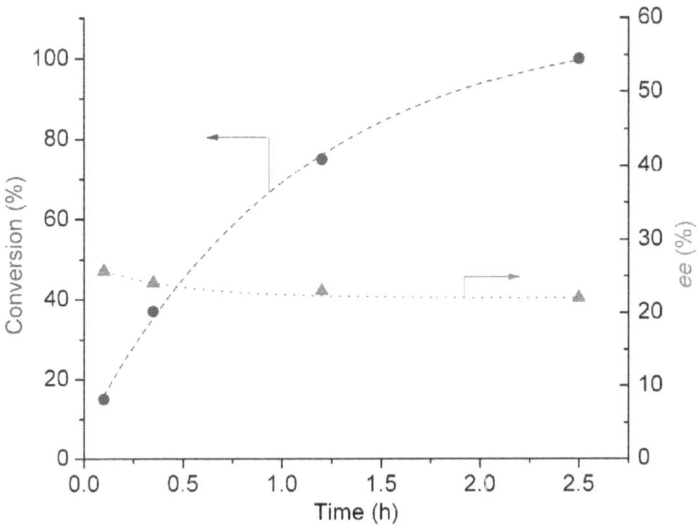

Figure 12.2 Solvent-free asymmetric reduction of LA to optically active GVL. Conditions: [Ru] = 7.6×10^{-4} mol/dm^3, [(R)-BINAP] = 7.6×10^{-3} mol/dm^3, $T = 140°C$, $p = 60$ bar, RPM = 600.

Noyori et al. showed that 4-phenyl-2-butanone could be reduced with *ee* = 51% [28d] by applying a modified BINAP-based Ru/XylBINAP/(*S*,*S*)-DPEN catalyst. The *ee* was further increased by introducing $RuCl_2$[(*S*)-XylBINAP][(*R*,*R*)-DAIPEN] as a catalyst. For cyclopropyl ketones, higher selectivity with opposite sense of asymmetric induction was achieved representing an outstanding *ee* (>95%) [38]. Thus, the reduction of LA in the presence of XylBINAP-modified Ru-based catalysts was performed; however, the *ee* was modest (*ee* = 11%). Similarly, *ee* = 13% was achieved when methanol was used as a co-solvent. By using bidentate phopshine-modified Ru catalyst, both the electronic and steric effects of the bis(diarylphosphine) backbone significantly enhanced the reactivity, and improved the stereorecognition [39]. Saito et al. synthesized a new ligand family (SEGPHOS) with smaller dihedral angle representing an outstanding catalytic performance for the reduction of carbonyl compounds including aliphatic ones [40]. By using a substrate/catalyst ratio of 1000, EL was hydrogenated at 50°C for 20 h, and ethyl-4-(*R*)-hydroxypentanoate was obtained with a yield of 99% [39]. The conversion of LA to GVL in methanol, as the proposed solvent [39], was monitored by using various SEGPHOS-based Ru catalysts (Table 12.1) [27].

Table 12.1 Conversion rates and *ee* values obtained for the hydrogenation of LA to GVL in the presence of various SEGPHOS-based Ru catalysts

Catalyst		**Conv. (%)**	***ee* (%)**
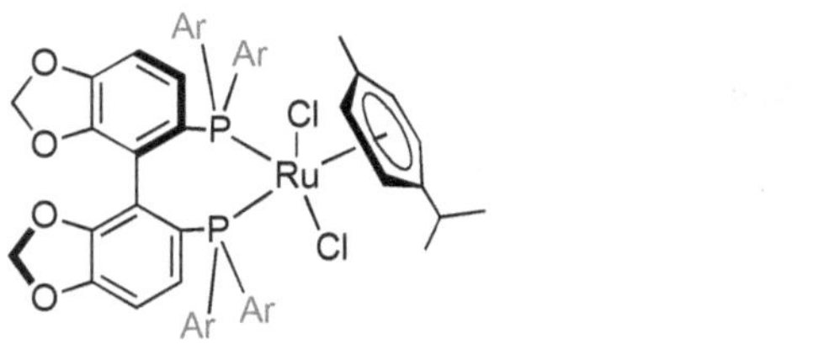 (*R*)-RuCl(*p*-cymene)(SEGPHOS)Cl types	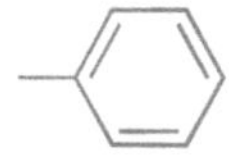	100	18
	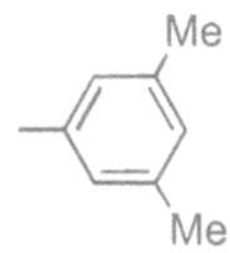	100	9
	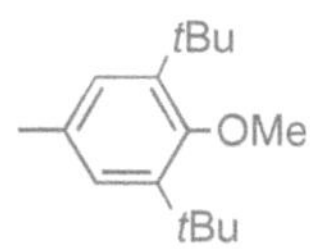	95	13
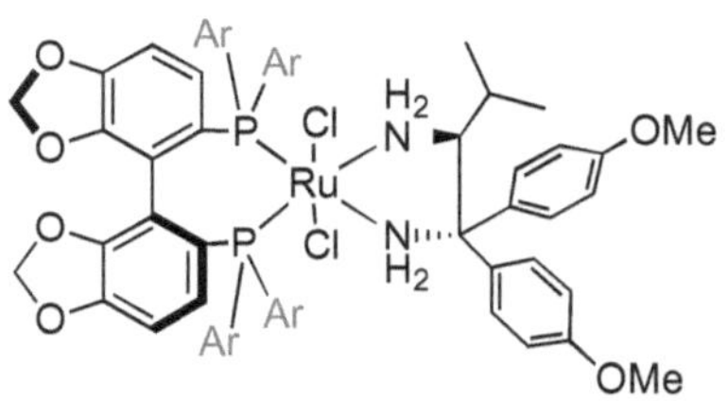 $RuCl_2$[(*S*)-(DM-SEGPHOS)][(*S*)-DAIPEN]	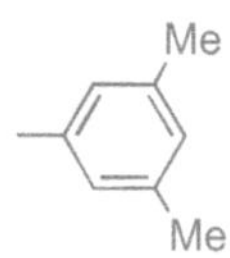	100	13
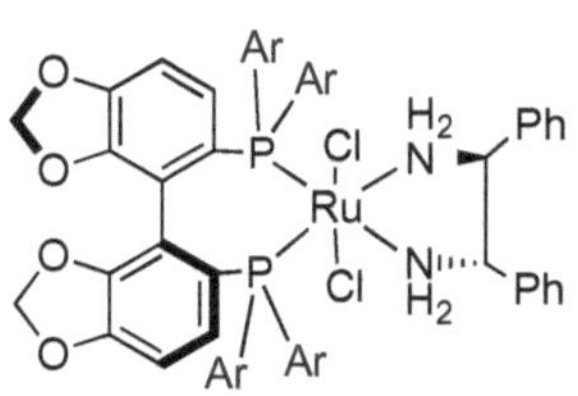 $RuCl_2$[(*S*)-(DM-SEGPHOS)][(*S,S*)-DPEN]	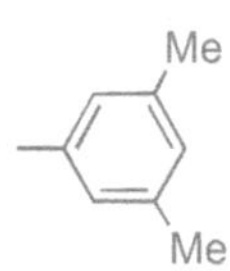	100	16
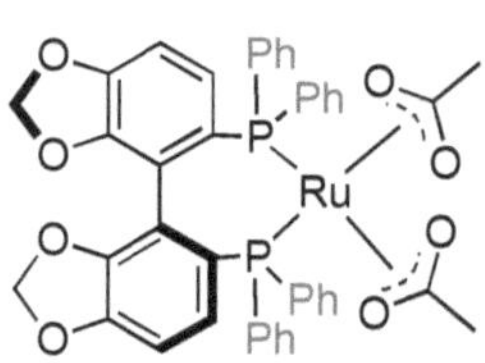 (*S*)-Ru(OAc)$_2$(SEGPHOS)		100	12
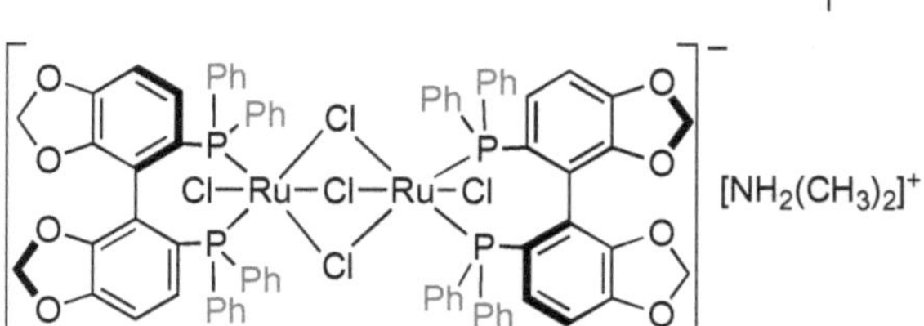 (*S*)-[(RuCl(SEGPHOS))$_2$(μ-Cl)$_3$]$^-$[NH_2Me_2]$^+$		100	56

Reaction conditions: 1 mL (9.8 mmol) LA in 1.4 mL MeOH; $T = 140°C$; $t = 20$ h; $p = 60$ bar, catalyst: 0.004 mmol, S/C = 2400.

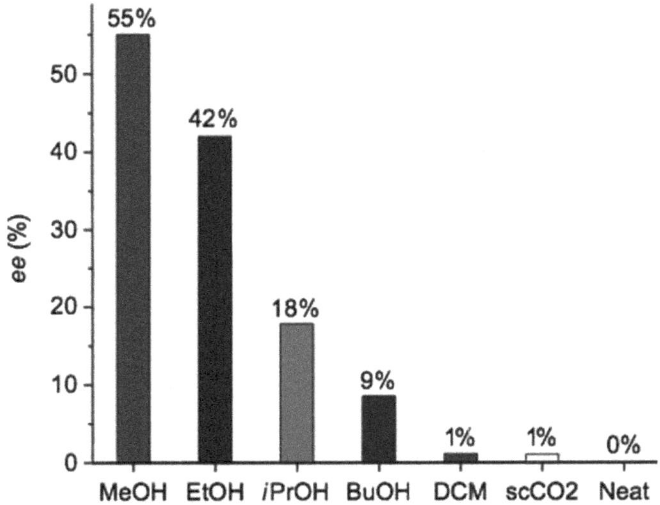

Figure 12.3 Asymmetric hydrogenation of LA to GVL in different solvents. Conditions: [Ru] = 0.004 mol/dm^3, T = 140°C, p = 60 bar, RPM = 400. Catalyst precursor: (*S*)-[(RuCl(SEGPHOS))$_2$(μ-Cl)$_3$][NH_2Me_2].

When mononuclear (*R*)-$RuCl_2$ [(*p*-cymene) (SEGPHOS)] precursor was applied complete conversion was achieved with *ee* = 18% (similar yields were obtained under solvent-free conditions where BINAP-based catalysts were used). When substituents on the SEGPHOS ligand were varied, a small decrease on the *ee* values was observed. Comparing DPEN and DAIPEN analogues of Noyori's catalyst, no significant change in the *ee* was detected. Similar results were observed when (*S*)-Ru(OAc)$_2$(SEGPHOS) precursor was applied. However, the use of (*S*)-[(RuCl(SEGPHOS))$_2$(μ-Cl)$_3$][$NH_2(CH_3)_2$] precursor resulted in a dramatic increase in the enantioselectivity. The reduction of LA in the presence of (*S*)-[(RuCl(SEGPHOS))$_2$(μ-Cl)$_3$][NH_2Me_2] catalyst leads to the complete formation of (*S*)-GVL with *ee* = 56%, that was 4.5 and 2.5 times higher than that obtained by using Noyori's catalyst in methanol or Ru/BINAP catalyst under solvent-free conditions.

Since asymmetric induction was strongly affected by the reaction medium, alcohols, methylene chloride and supercritical CO_2 were evaluated for reduction of LA at 60 bar H_2 and 140°C. Although, complete conversion was obtained in all cases, the enantiomeric excess was in a wide range depending on the solvent (Fig. 12.3). Without any added solvent, no chiral induction occurred, and negligible values were obtained when methylene chloride or supercritical CO_2 was used. However, significantly higher *ee* values were obtained in alcohols, a protic reaction media, which

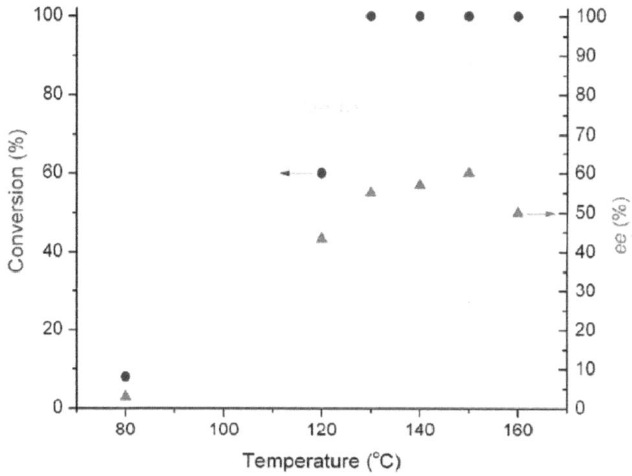

Figure 12.4 Asymmetric hydrogenation of LA to GVL at different temperatures.

is in accordance with the data published on the reduction of nonaromatic ketones by Saito et al. [39]. Supposedly, methanol acts as a co-solvent and may have an effect on the formation and stabilization of the catalytically active species. Importantly, the ring closure of 4-hydroxypentanoic acid resulted in the formation of an equimolar amount of water and GVL. Accordingly, the application of a water-free solvent is inefficient.

The Ru-catalyzed asymmetric reductions are usually performed in the range of 20–40°C, although SEGPHOS-based catalysts operate at higher (65°C) temperatures. The investigation of the effect of the reaction temperature on the activity and selectivity of LA reduction was investigated in the range of 80–160°C. It was shown that negligible or moderate conversions were achived below 120°C (Fig. 12.4).

Interestingly, alkyl levulinates were converted to the corresponding alcohols at 30°C [33] and 60°C [34]. When, 9.8 mmol of LA was dissolved in 1.4 mL MeOH was reduced at 80°C and 60 bar H_2 for 20 h, negligible conversion (8%) and *ee* (3%) was observed; and only a slight increase was found at 120°C. Although, full conversion was obtained over 130°C, the maximum of *ee* value (56%) was obtained at 140°C [27].

It is supposed that the reduction of LA to GVL in the presence of HCl with Ru-BINAP-HCl system undergoes via in-situ formation of EL in ethanol. When LA was reduced by catalyst (*S*)-[(RuCl(SEGPHOS))$_2$(μ-Cl)$_3$][$NH_2(CH_3)_2$] the concentration of methyl levulinate in the final reaction mixture was below detection limit. For comparison, methyl levulinate in methanol was reduced under identical conditions resulting

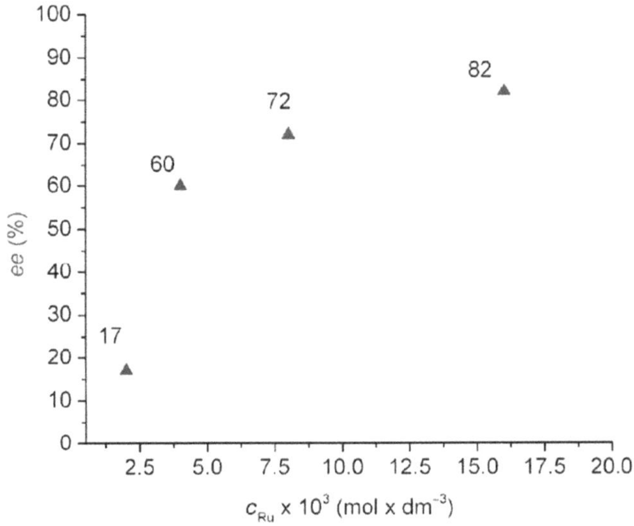

Figure 12.5 Observation of change in the enantiomeric excess with the variation of catalyst concentration.

in complete conversion and interestingly with *ee* = 27%. It is significantly lower than the value obtained for LA (56%). In addition, the methyl levulinate was reduced at lower temperature. Although, the in-situ equilibrium formation of methyl levulinate from methanol and LA cannot be completely excluded, if LA was hydrogenated to (*S*)-4-hydroxyvaleric acid it was spontaneously dehydrated to (*S*)-GVL.

Unexpectedly, for the reduction of LA, significant improvement in enantioselectivity was detected by varying the Ru concentration. For example, when LA was hydrogenated in MeOH as co-solvent at [Ru] = 0.016 mol/dm^3 at 60 bar and 150°C, complete conversion was obtained with *ee* = 82% for (*S*)-GVL. This selectivity fits to the *ee* values obtained for dialkyl ketones; however, a further increase of the catalyst concentration had no effect on the selectivity (Fig. 12.5).

In order to demonstrate the conversion of "real" biomass-based wastes a mixture containing wheat straw, rice husk, corn straw, nut and pea-pod was treated in 2 M H_2SO_4 at 170°C for 8 h. After our published workup procedure was performed, c.1 mL of deep dark brown solution was obtained, which was subsequently purified by vacuum distillation. The hydrogenation of the colorless LA under optimized conditions resulted in complete conversion with enantioselectivity for (*S*)-GVL 78% [27]. Indeed, we have shown that GVL with high *ee* can be produced from biomass-based LA (Scheme 12.14).

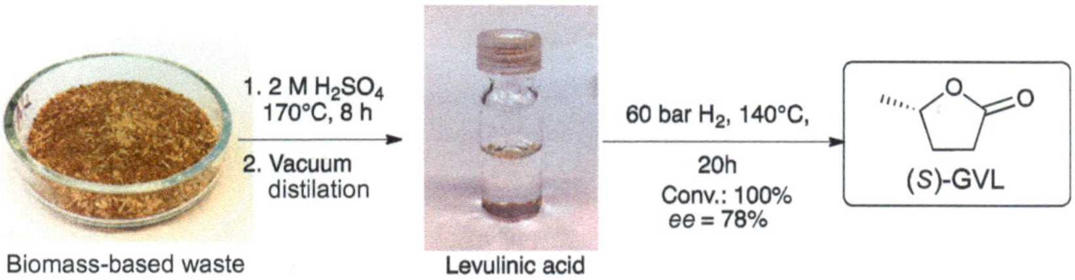

Scheme 12.14 Conversion of biomass-based waste to (*S*)-GVL.

12.2 CONCLUSION

We demonstrated that LA could be directly converted to optically active (*S*)-GVL, a proposed chiral platform molecule. It was revealed that no alkyl levulinate was necessary to obtain optically active GVL. We have shown that in the presence of a catalyst in situ generated from Ru(III)-acetylacetonate and (*S*)-BINAP, LA was converted to (*S*)-GVL with *ee* = 26% without adding any solvent and/or additive. By applying (*S*)-$\{[RuCl(SEGPHOS)]_2(\mu\text{-}Cl)_3\}^-[NH_2(CH_3)_2]^+$ catalyst precursor in methanol, the *ee* was increased to 82%. The conversion of "real" biomass waste to optically active GVL was also demonstrated.

ACKNOWLEDGMENT

This work was funded by National Research, Development and Innovation Office—NKFIH (PD 116559). L. T. Mika is grateful to the support of János Bolyai Research Scholarship of the Hungarian Academy of Sciences.

REFERENCES

[1] Werpy, T.; Petersen, G.; Aden, A.; Bozell, J.; Holladay, J.; White, J., et al. *Top Value Added Chemicals from Biomass*, Vol. 1. U.S. Department of Energy: Washington, DC, 2004.

[2] Horváth, I. T.; Anastas, P. *Chem. Rev.* **2007,** *107*, 2169–2173.

[3] (a) Gallezot, P. *Chem. Soc. Rev.* **2012,** *41*, 1538–1558.
(b) van Putten, R. J.; van der Waal, J. C.; de Jong, E.; Rasrendra, C. B.; Heeres, H. J.; de Vries, J. G. *Chem. Rev.* **2013,** *133*, 1499–1597.

[4] (a) Bozell, J. J.; Moens, L.; Elliot, D. C.; Wang, Y.; Neuenscwander, G. G.; Fitzpatrik, S. W., et al. *Resour. Conserv. Recycl.* **2000,** *28*, 227–239.
(b) Girisuta, B.; Janssen, L. P. B. M.; Heeres, H. J. *Chem. Eng. Res. Des.* **2006,** *84*, 339–349.

[5] Leonard, R. H. *Ind. Eng. Chem.* **1956,** *48*, 1331–1341.

[6] Horváth, I. T.; Mehdi, H.; Fábos, V.; Boda, L.; Mika, L. T. *Green Chem.* **2008,** *10*, 238–242.

[7] Mehdi, H.; Fábos, V.; Tuba, R.; Bodor, A.; Mika, L. T.; Horváth, I. T. *Top. Catal.* **2008,** *48*, 49–54.

[8] Bond, J. Q.; Alonso, D. M.; West, R. M.; Dumesic, J. A. *Science* **2010,** *327*, 1110–1114.

[9] (a) Fegyverneki, D.; Orha, L.; Láng, L.; Horváth, I. T. *Tetrahedron* **2010,** *66*, 1078–1081.
(b) Strádi, A.; Molnár, M.; Óvári, M.; Dibó, G.; Richter, F. U.; Mika, L. T. *Green Chem.* **2013,** *15*, 1857–1862.
(c) Strádi, A.; Molnár, M.; Szakál, P.; Gáspár, D.; Dibó, G.; Mika, L. T. *RSC Adv.* **2015,** *5*, 72529–72535.

[10] Wong, P. K.; Li, C.; Stubbs, L.; van Menurs, M.; Kumband, D. G. A.; Lim, C. Y., et al. *Pat. Appl.* **2012**. PTC/SG2012/000107.

[11] Chalid, M.; Heeres, H. J.; Broekhuis, A. A. *J. Appl. Polym. Sci.* **2012,** *123*, 3556–3564.

[12] Fábos, V.; Lui, M. Y.; Mui, Y. F.; Wong, Y. Y.; Mika, L. T.; Qi, L., et al. *ACS Sustainable Chem. Eng.* **2015,** *3*, 1899–1904.

[13] (a) Strappaveccia, G.; Luciani, L.; Bartollini, E.; Marrocchi, A.; Pizzo, F.; Vaccaro, L. *Green Chem.* **2015,** *17*, 1071–1076.
(b) Strappaveccia, G.; Ismalaja, E.; Petrucci, C.; Lanari, D.; Marrocchi, A.; Drees, M., et al. *Green Chem.* **2015,** *17*, 365–372.
(c) Ismalaj, E.; Strappaveccia, G.; Ballerini, E.; Elisei, F.; Piermatti, O.; Gelman, D., et al. *ACS Sustainable Chem. Eng.* **2014,** *2*, 2461–2464.

[14] Pongrácz, P.; Kollár, L.; Mika, L. T. *Green Chem.* **2018,** *18*, 842–847.

[15] Izquerdo-Cañas, P. M.; Mena-Morales, E. *García-Romero: LWT Food Sci. Technol.* **2016,** *66*, 468–474.

[16] (a) Qi, L.; Mui, Y. F.; Lo, S. W.; Lui, M. Y.; Akien, G. R.; Horváth, I. T. *ACS Catal.* **2014,** *4*, 1470–1477.
(b) Qi, L.; Horváth, I. T. *ACS Catal.* **2012,** *2*, 2247–2249.

[17] Fábos, V.; Koczó, G.; Mehdi, H.; Boda, L.; Horváth, I. T. *Energy Environ. Sci.* **2009,** *2*, 767–769.

[18] (a) Havasi, D.; Mizsey, P.; Mika, L. T. *J. Chem. Eng. Data* **2016,** *61*, 1502–1508.
(b) Zaitseva, A.; Pokki, J. P.; Le, H. Q.; Alopaeus, V. *J. Chem. Eng. Data* **2016,** *61*, 881–890.

[19] (a) Jennings, W. G.; Sevenants, M. R. *J. Food Sci.* **1964,** *29*, 796–801.
(b) Sevenants, M. R.; Jennings, W. G. *J. Food Sci.* **1966,** *31*, 81–86.
(c) Tang, C. S.; Jennings, W. G. *J. Agric. Food Chem.* **1968,** *16*, 252–254.
(d) Willhalm, B.; Palluy, E.; Winter, M. *Helv. Chim. Acta* **1966,** *49*, 65–67.
(e) Fujimori, T.; Kasuga, R.; Matsushita, H.; Kaneko, H.; Noguchi, M. *Agric. Biol. Chem.* **1976,** *40*, 303.
(f) Ravid, U.; Silverstein, R. M.; Smith, L. R. *Tetrahedron* **1978,** *34*, 1449–1452.

[20] (a) Hilterhaus, L.; Liese, A. Building Blocks. In *White Biotechnology, Advances in Biochemical Engineering/Biotechnology;* Ulber?R., Sell, D., Eds.; Vol. 105; Springer-Verlag: Berlin Heidelberg, 2007; pp 133–173.
(b) Taylor, S. K.; Atkinson, R. F.; Almli, E. P.; Carr, M. D.; van Huis, T. J.; Whittaker, M. R. *Tetrahedron: Asymmetry* **1995,** *6*, 157–164.
(c) Kiyoshi, T.; Tsuneo, I.; Koga, K. *J. Chem. Soc. Chem. Commun.* **1979,** 652–653.
(d) Smith, A. B.; Rano, T. A.; Chida, N.; Sulikowski, G. A.; Wood, J. L. *J. Am. Chem. Soc.* **1992,** *114*, 8008–8022.
(e) Blackwell, C. M.; Davidson, A. H.; Launchbury, S. B.; Lewis, C. N.; Morrice, E. M.; Reeve, M. M., et al. *J. Org. Chem.* **1992,** *57*, 5596–5606.
(f) David, U.; Silverstein, R. M.; Smith, L. R. *Tetrahedron* **1978,** *34*, 1449–1452.

[21] Mori, K. *Tetrahedron* **1975,** *31*, 3011–3012.

[22] Stangeland, E. L.; Sammakia, T. *J. Org. Chem.* **2004,** *69*, 2381–2385.
[23] White, J. D.; Amedio, J. C. *J. Org. Chem.* **1989,** *54*, 736–738.
[24] (a) Novodárszki, G.; Rétfalvi, N.; Dibó, G.; Mizsey, P.; Cséfalvay, E.; Mika, L. T. *RSC Adv.* **2014,** *4*, 2081–2088.
(b) Szabolcs, Á.; Molnár, M.; Dibó, G.; Mika, L. T. *Green Chem.* **2013,** *15*, 439–445.
(c) Kang, M.; Kim, S. W.; Kim, J.-W.; Kim, T. H.; Kim, J. S. *Renewable Energy* **2013,** *54*, 173–179.
(e) Galletti, A. M. R.; Antonetti, C.; Ribechini, E.; Colombini, M. P.; O Di Nasso, N. N.; Bonari, E. *Appl. Energy* **2012,** *102*, 157–162.
(f) Rackemann, D. M.; Doherty, W. O. S. *Biofuels, Bioprod. Biorefin.* **2011,** *4*, 198–214.
(g) Jeong, G.-T.; Park, D.-H. *Appl. Biochem. Biotechnol.* **2010,** *161*, 41–52.
(h) Deng, L.; Li, J.; Lai, D.-M.; Fu, Y.; Guo, Q.-X. *Angew. Chem. Int. Ed.* **2009,** *48*, 6529–6539.
(i) Girisuta, B.; Janssen, L. P. M.; Heeres, H. J. *Chem. Eng. Res. Des.* **2006,** *84*, 339–349.
(j) Tabasso, S.; Montoneri, E.; Carnaroglio, D.; Caporasob, M.; Cravotto, G. *Green Chem.* **2014,** *16*, 73–76.
(k) Li, J.; Jiang, Z.; Hu, L.; Hu, C. *ChemSusChem* **2014,** 7, 2482–2488.
(l) Tukacs, J. M.; Holló, A. T.; Rétfalvi, N.; Cséfalvay, E.; Dibó, G.; Havasi, D., et al. *ChemistrySelect* **2017,** *2*, 1375–1380.
[25] (a) Wright, W. R. H.; Palkovits, R. *ChemSusChem* **2012,** *5*, 1657–1667.
(b) Tukacs, J. M.; Jones, R. V.; Darvas, F.; Dibó, G.; Lezsák, G.; Mika, L. T. *RSC Adv.* **2013,** *3*, 16283–16287.
[26] (a) Tukacs, J. M.; Novák, M.; Dibó, G.; Mika, L. T. *Catal. Sci. Technol.* **2014,** *4*, 2908–2912.
(b) Tukacs, J. M.; Király, D.; Strádi, A.; Novodárszki, G.; Eke, Z.; Dibó, G., et al. *Green Chem.* **2012,** *14*, 2057–2065.
(c) Li, W.; Xie, J.-H.; Lin, H.; Zhou, Q.-L. *Green Chem.* **2012,** *14*, 2388–2390.
(d) Deng, J.; Wang, Y.; Pan, T.; Xu, Q.; Guo, Q.-X.; Fu, Y. *ChemSusChem* **2013,** *6*, 1163–1167.
(e) Fábos, V.; Mika, L. T.; Horváth, I. T. *Organometallics* **2014,** *33*, 181–187.
[27] Tukacs, J. M.; Fridrich, B.; Dibó, G.; Székely, E.; Mika, L. T. *Green Chem.* **2015,** *17*, 5189–5195.
[28] (a) Noyori, R. *Angew. Chem., Int. Ed.* **2002,** *41*, 2008–2022.
(b) Johnson, N. B.; Lennon, I. C.; Moran, P. H.; Ramsden, J. A. *Acc. Chem. Res.* **2007,** *40*, 1291–1299.
(c) Zhang, W.; Chi, Y.; Zhang, X. *Acc. Chem. Res.* **2007,** *40*, 1278–1290.
(d) Noyori, R.; Ohkuma, T. *Angew. Chem., Int. Ed. Engl.* **2001,** *40*, 40–73.
(e) Blaser, H. U. *Adv. Synth. Catal.* **2002,** *344*, 17–31.
[29] Jiang, Q.; Jiang, Y.; Xiao, D.; Cao, P.; Zhang, X. *Angew. Chem., Int. Ed. Engl.* **1998,** *37*, 1100–1103.
[30] Patchett, R.; Magpantay, I.; Saudan, L.; Schotes, C.; Mezzetti, A.; Santoro, F. *Angew. Chem. Int. Ed. Engl.* **2013,** *52*, 10352–10355.
[31] Yumoto, K.; Hasegawa, M.; Toshima, H. *Heterocycles* **2010,** *81*, 421–431.
[32] Figadère, B.; Cavé, A. *Stud. Nat. Prod. Chem.* **1996,** *18*, 193–205.
[33] Karnik, A. V.; Patil, S. T.; Patnekar, S. S.; Semwal, A. *New J. Chem.* **2004,** *28*, 1420–1422.
[34] Ohkuma, T.; Kitamura, M.; Noyori, R. *Tetrahedron Lett.* **1990,** *31*, 5509–5512.
[35] (a) Starodubtseva, E. V.; Turova, O. V.; Vinogradov, M. G.; Gorshkova, L. S.; Ferapontov, V. A. *Russ. Chem. Bull.* **2005,** *54*, 2374–2378.

(b) Turova, O. V.; Starodubtseva, E. V.; Vinogradov, M. G.; Ferapontov, V. A. *J. Mol. Catal. A* **2009,** *311*, 61–65.

[36] Götz, K.; Liese, A.; Ansorge-Schumacher, M.; Hilterhaus, L. *Appl. Microbiol. Biotechnol.* **2013,** *97*, 3865–3873.

[37] Jacobs, H.; Berryman, K.; Jones, J.; Gopalan, A. *Synth. Commun.* **1990,** *20*, 999–1010.

[38] Ohkuma, T.; Koizumi, M.; Doucet, H.; Pham, T.; Kozawa, M.; Murata, K., et al. *J. Am. Chem. Soc.* **1998,** *120*, 13529–13530.

[39] Shimizu, H.; Nagasaki, I.; Saito, T. *Tetrahedron* **2005,** *61*, 5405–5432.

[40] (a) Shimizu, H.; Nagasaki, I.; Matsumura, K.; Sayo, N.; Saito, T. *Acc. Chem. Res.* **2007,** *40*, 1385–1393.

(b) Saito, T.; Yokozawa, T.; Ishizaki, T.; Moroi, T.; Sayo, N.; Muira, T., et al. *Adv. Synth. Catal.* **2001,** *343*, 264–267.

CHAPTER 13

Searching for Spontaneous Mirror-Symmetry Breaking in Organoautocatalytic Reactions

Guillem Valero and Albert Moyano
Department of Inorganic and Organic Chemistry, University of Barcelona, Barcelona, Catalonia, Spain

Abstract

Autocatalytic chemical reactions are assumed to have played a key role in the emergence of life, and asymmetric autocatalytic reactions in particular appear to be the key to understanding the fundamental question of the origin of biological homochirality in prebiotic chemistry. The possibility of chemical reactions that starting from achiral reagents can lead to an essentially homochiral product by amplification of a very tiny enantiomerical imbalance was predicted by Frank in 1953, in a model that postulates mutual antagonism between both enantiomers of the autocatalyst. The experimental realization of this concept appeared to be restricted to the asymmetric autocatalytic dialkylzinc addition to pirimidylcarbaldehydes discovered by Soai and co-workers, but recent reports point to the possibility of spontaneous mirror-symmetry breaking (SMSB) and of asymmetric autocatalysis in organocatalytic reactions. We summarize herein our recent experimental work on the existence of organoautocatalysis and SMSB in aldol and in Mannich reactions.

Keywords: Aldol reaction; autocatalysis; biological homochirality; Mannich reaction; mirror-symmetry breaking; organocatalysis; Soai reaction

13.1 INTRODUCTION

Autocatalytic reactions are those in which the reaction product acts as a catalyst for its own formation. An autocatalytic reaction step has the general form:

$$A + nC = (n + 1)C$$

where A is the reactant, C is the product/catalyst and n is the autocatalysis order. From a mechanistic point of view, elementary reaction steps of order $n = 1$ are the most probable ones. The dynamic signature of an autocatalytic reaction is a sigmoidal evolution of the concentration [C] with time, due to the initial increase of the rate.

Advances in Asymmetric Autocatalysis and Related Topics
DOI: http://dx.doi.org/10.1016/B978-0-12-812824-4.00013-7

Nowadays it is being recognized that autocatalysis is a key concept not only in chemistry but in several branches of knowledge, such as economics, language theory and biology [1,2].

In asymmetric autocatalysis a chiral product (L, D) acts as an enantioselective catalyst for its own formation [3]:

$$A + L = 2L, \quad A + D = 2D$$

Asymmetric autocatalysis leads to the automultiplication of the chiral product without the need of any added chiral catalyst, in an atom-economic fashion and avoiding the separation of the product from the catalyst. However, it should be noted that for a first order autocatalytic reaction, no amplification of the initial enantiomeric purity of the product can take place. The possibility of chemical reactions that starting from nominally achiral reagents lead to chiral products of high enantiomeric purity by amplification of a very tiny initial enantiomeric imbalance (spontaneous mirror-symmetry-breaking, SMSB) was predicted by F. C. Frank in 1953, by coupling asymmetric autocatalysis with mutual inhibition between the product enantiomers in an open flow system [4]. In the steady state, the racemic mixture is metastable and the system evolves to one of the two possible enantiomerically pure configurations (Fig. 13.1).

When both the enantioselective autocatalysis and the inhibition step are reversible and the reactions are run in a system closed to matter flows, thermodynamic equilibrium is reached and the product is of course racemic, but it should be noted that "kinetically controlled" SMSB can still take place before racemization begins to be important, and scalemic products can be obtained during large time windows [5]. In the past few

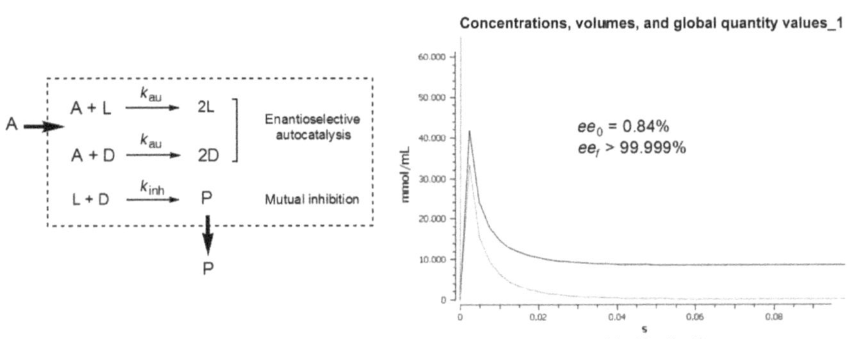

Figure 13.1 Frank's reaction scheme in an open-flow system and results of a numerical simulation with an initial enantiomeric excess of the product of 0.84%, that evolves to a practically enantiopure product.

years, it has become clear that several reaction networks can, in theory, lead to SMSB (either in the steady state or under kinetically controlled conditions), but the existence of enantioselective autocatalysis appears to be a necessary condition [6].

In 1995, Soai and co-workers reported that the addition of diisopropylzinc to pyrimidine carbaldehydes takes place both with highly enantioselective autocatalysis and with amplification of chirality [7], and this reaction has been regarded as the first experimental realization of the Frank model. With 2-alkynylpyrimidine-5-carbaldehydes, asymmetric amplification from $5 \times 10^{-5}\%$ enantiomeric excess (*ee*) to >99.5% is achieved in three batches (Scheme 13.1) [8].

The Soai reaction can be performed in nominally achiral conditions to provide enantioenriched material from the statistical fluctuations of c.$10^{-10}\%$ *ee* around the ideal racemic composition of the initially formed product, as demonstrated by Singleton and Vo in 2002 [9], although a nonstochastic distribution of enantiomers is obtained. After extensive experimentation, Soai developed reproducible experimental conditions that led to approximately stochastic distributions of product enantiomers [10]. There is no single mechanisic proposal that is able to account for all of the characteristics of the Soai reaction, but it is clear that: (a) the rate-limiting step involves enantioselective autocatalysis, probably by a homochiral dimeric zinc alkoxide, and (b) the heterochiral heterodimer is both catalytically inactive and in statistical equilibrium with the homodimers (at room temperature) [11,12].

The fact that only one of the enantiomeric forms of the basic chiral building blocks of biopolymers (i.e., L-amino acids, D-sugars) can be found in living organisms is known as "biological homochirality." Their formation from achiral precursors in the course of prebiotic chemical evolution should have led, however, to racemic (or to nearly racemic) mixtures, even

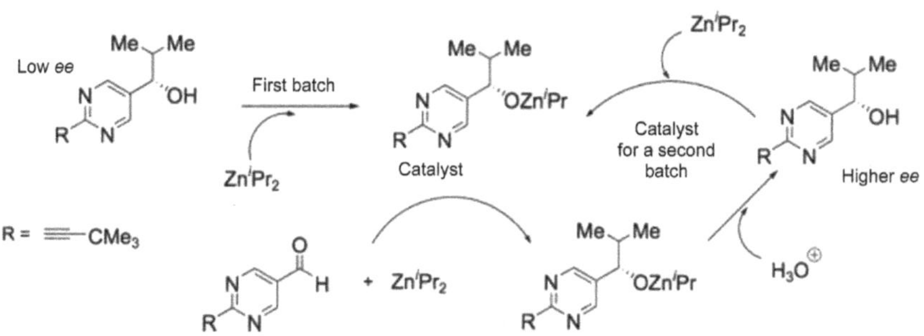

Scheme 13.1 Replicative asymmetric amplification in the Soai reaction.

assuming the possibility of absolute asymmetric synthesis under the influence of chiral physical fields. It has been proposed that chemical reactions taking place with SMSB may be at the origin of biological homochirality (see, Calvin [13]; Avetisov [14]; Kondepudi [15]). The Soai reaction, however, cannot be regarded as a model chemical process in the emergence of biological homochirality, due to the extremely air- and water-sensitive nature of the organometallic reagents involved. The report by Mauksch and Tsogoeva, in 2007, of asymmetric organoautocatalytic processes, such as the aldol and Mannich reactions, taking place with chirality amplification [16] and leading to SMSB [17] was therefore highly relevant in this context. The application of the Frank model and of related reaction networks to these reactions to explain the appearance of SMSB was, however, not straightforward, and we felt therefore that further mechanistic investigations on asymmetric organoautocatalytic reactions were required. We present here a summary of our recent experimental work in this subject.

13.2 RESULTS AND DISCUSSION

13.2.1 Assessment of SMSB and Asymmetric Autocatalysis in the Aldol Reaction [18]

First and following the detailed experimental procedure described in 2010 by Mauksch and Tsogoeva [19], a 1.29 M solution of 4-nitrobenzaldehyde in acetone (containing aq 5 vol% of DMSO) was stirred for 10 days at room temperature, and after aqueous workup, evaporation of the solvent and purification of the crude residue by preparative TLC the aldol addition product was analyzed by chiral HPLC. Out of eleven independent experiments, nonracemic aldol adduct (2.4% and 15% *ee* for the (*R*)-enantiomer) was obtained in two instances (Scheme 13.2). The yield of aldol product was very low (1%–2%, estimated by ^{1}H nuclear magnetic resonance (NMR) in the crude reaction mixture). We decided then to monitor the evolution of the reaction by HPLC, in order to ascertain if the formation of the aldol adduct followed autocatalytic kinetics. After some experimentation, we found out that the use of 2,7-dimethoxynaphthalene as an internal standard and a Chiralpak IC column in an instrument equipped with a UV-VIS diode-array detector, allowed us to follow the evolution both of the yield and of the enantiomeric purity of the aldol product. We noticed that the amount of aldol adduct increased

Scheme 13.2 Spontaneous mirror-symmetry breaking in the aldol reaction between acetone and 4-nitrobenzaldehyde.

linearly with time, at least in the first 500 h of reaction, suggesting pseudo first-order nonautocatalytic kinetics for the observed process.

The enantiomeric composition of the aldol adduct was measured by the area ratios corresponding to each enantiomer, at four different wavelengths (250, 260, 270, and 280 nm). The standard deviations were much lower than the mean values of the measured *ee*'s. We were surprised to find that the highest *ee*'s were always observed (at least when the initial *ee* was higher than 2%) in the first hour of the reaction (c.0.01% yield of aldol), and that they subsequently decreased exponentially with time. This technique allowed us to monitor a great number of reactions in the first 24 h, with the following results:

a. After a set of 39 experiments, we observed that in 22 of them the *ee* of the aldol adduct was less than 2%, both after 4 h and after 24 h from the beginning of the reaction (no detectable SMSB had taken place); in 15 experiments, SMSB is initially observed (*ee* > 2% (*R*), after 4 h). After 24 h, the aldol was essentially racemic in seven instances; in six instances, the *ee* after 24 h had diminished but the aldol was not racemic (*ee* 2.6%–56.4% (*R*)); in one example, both the yield (0.02%) and the *ee* (40% *ee*, (*R*)) were essentially the same after 4 h and after 24 h.

b. In one experiment, run without DMSO, a 6.9% *ee* of the (*S*) aldol was detected after 24 h. In another instance, (*S*) aldol (7% *ee*) was initially present, but the product was racemic (*ee* < 2%) after 24 h.

c. No clear trends were apparent when the reaction was performed in solvent mixtures different from 5% v/v DMSO:acetone. In 1:1 DCM: acetone, 45% *ee* (7% standard deviation) was observed for the (*R*) enantiomer of the aldol in one instance (after 4 h of reaction, 0.02% yield).

d. Independent check of the reliability of the measures: in one experiment, after 95 h, a 7% *ee* (*R*) was measured in situ (0.09% yield). The reaction was quenched, and after chromatographic purification by preparative TLC the isolated aldol adduct showed a 8% *ee* (*R*).

e. Use of other aldehydes: with 2-nitrobenzaldehyde, the reaction took place at a similar rate, but the % *ee* of the aldol could not be measured. With 4-cyanobenzaldehyde, no SMSB was observed: the aldol adduct formed in the first hours of the reaction was always racemic.

In order to find some evidence for the presence of a "local" chiral inducer in trace amounts that could account for the predominance of the (*R*)-enantiomer of the aldol adduct, two sets of four reactions were run in independent laboratories with different glassware, stirrers, magnetic bars, solvents, and reagents. After 6 h, in the first set racemic aldol was present in two reactions, and nonracemic product (13% *ee* and 17% *ee*, always for (*R*)-aldol) was observed in the other two. Very similar results were obtained for the second set of reactions (two racemic, 8% *ee* (*R*), 5% *ee* (*R*)).

In order to ascertain the configurational stability of the aldol product in the reaction conditions, a 1.55×10^{-4} M solution of nonracemic (*R*)-aldol (26.0% *ee*) in acetone containing a 5 vol% of DMSO and a 5 mol% of 4-nitrobenzaldehyde was stirred at room temperature, and the *ee* was monitored by HPLC after 4, 6, 9, and 24 h. Within experimental error, no change in the enantiomeric purity could be detected. Moreover, the same sample was left standing in the laboratory; after 3 months, the enantiomeric purity of the aldol, measured by HPLC, was 24% *ee* (*R*). On the other hand, no rate acceleration was observed when the reaction was run in the presence of variable amounts (0.10–20 mol%) of nonracemic aldol. Thus in an experiment performed in the presence of a 0.15 mol% of (*S*)-aldol (73% *ee*), a 0.35 mol% concentration of (*S*)-aldol of 58% *ee* was measured by in-situ HPLC after 10 days. In a similar way, neither the reaction rate nor the extent of SMSB appeared to be modified when chiral substances such as L- or D-diethyl tartrate were added to the reaction mixture.

With these data in our possession, we concluded that measurable SMSB (>2% *ee* of the aldol after 4 h) takes place in c.50% of the experiments, and in the first stage of the reaction (i.e., yields lower than 0.03%); the process in nonstochastic, because the (*R*)-aldol is formed predominantly. The fast decrease in the enantiomeric purity of the initially formed aldol is due to the fact that after the initial stages of the reaction most of the aldol is produced in racemic form by a slow, nonautocatalytic and nonenantioselective process. In the absence of enantioselective autocatalysis, a Frank-like reaction network could not be at the origin of the observed SMSB.

Upon careful examination of the reaction crudes, we found that a more polar compound than the aldol adduct was also formed in lower amounts. In an experiment stopped after 28 days, we could isolate by preparative thin-layer chromatography (TLC) a product that, according to NMR data, corresponds to that arising from the double aldol addition of 4-nitrobenzaldehyde to acetone, as a 2:1 *syn/anti* mixture of diastereomers, in 0.8% yield. This structural assignment was subsequently confirmed by independent synthesis, using the proline-derived catalyst described by Gryko (Scheme 13.3) [20]. Both enantiomers of the *anti* isomer were available by this procedure.

Since the formation of the bis(aldol) adduct takes place only in very concentrated conditions, even under prolinethioamide catalysis, and it is very unlikely to occur in homogeneous solution in our aldol experiments, we set out to investigate its possible relationship with the SMSB observed in the initial stages of the reaction. We were pleased to find that when the aldol reaction was run in the presence of this bis(aldol) adduct, SMSB takes invariably place. The example shown in Scheme 13.4 is especially instructive: the initial enantiomeric purity of the aldol in the first measure was 50% *ee* for the (*R*)-enantiomer discarding its formation by retro-aldol reaction of the (*S,S*)-bis(aldol). Moreover, when 0.064 M solution of a 1:1 *syn/anti* mixture of the bis(aldol) (78% *ee* of the (*S,S*)-*anti* isomer) was stirred in acetone, after 24 h at room temperature less than 1% yield of the mono-aldol adduct (2% *ee* (*S*)) could be detected, confirming that the retro-aldol reaction was very slow in these conditions.

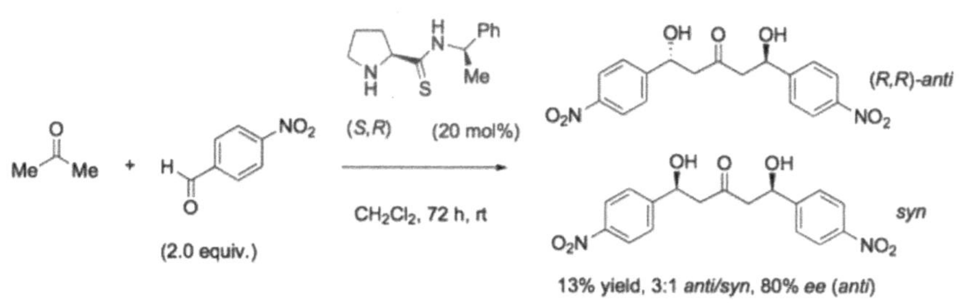

Scheme 13.3 Synthesis of the bis(aldol) adduct formed in the uncatalyzed reaction between acetone and 4-nitrobenzaldehyde.

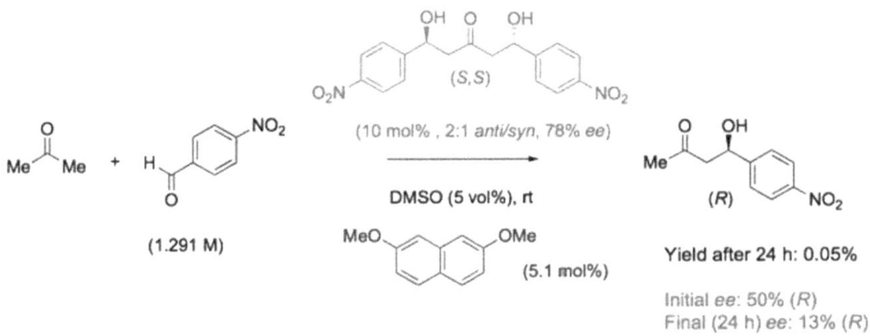

Scheme 13.4 Spontaneous mirror-symmetry breaking in the presence of bis(aldol) adduct.

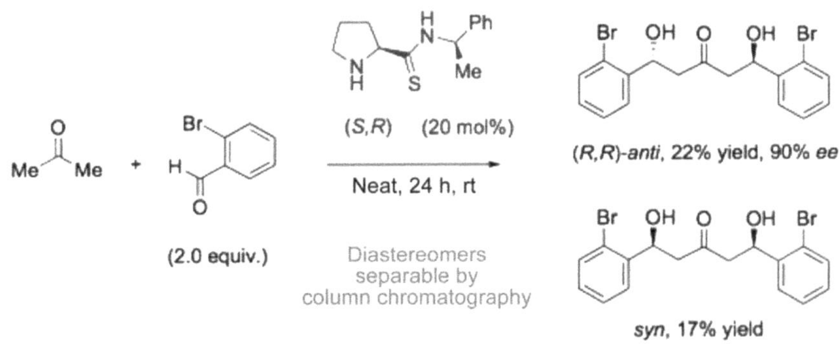

Scheme 13.5 Synthesis of a bis(aldol) mimic.

Due to the impossibility of obtaining the pure *anti* and *syn* diastereomers of this bis(aldol) adduct, we used Gryko's prolinethioamide catalyst to prepare the bis(aldol) adduct of 2-bromobenzaldehyde with acetone, whose *syn* and *anti* diastereomers were separable by standard column chromatography on silicagel (Scheme 13.5), and we studied their role as bis(aldol) mimics in the SMSB reaction.

When we performed the reaction between acetone and 4-nitrobenzaldehyde in the usual conditions, but in the presence of catalytic amounts of this bis(aldol) mimic (either the *anti*-(*S*,*S*), the *anti*-(*R*,*R*), or the *syn* isomer), SMSB was again observed in all instances (Scheme 13.6).

In the search for a reversible reaction network that could lead to SMSB in a closed reaction system, and that does not involve direct autocatalysis by the aldol product, we made the following assumptions:

a. The chiral *anti* bis(aldol) can act as a stereoselective catalyst for the aldol reaction, while the achiral *syn* diastereomer acts as a nonenantioselective catalyst (Scheme 13.7).

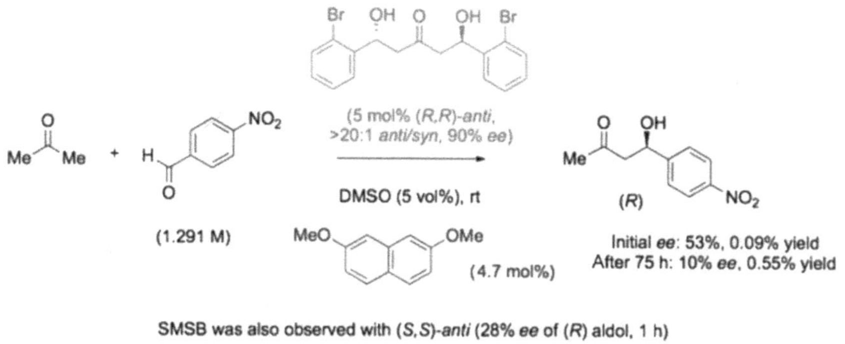

Scheme 13.6 Spontaneous mirror-symmetry breaking (SMSB) in the presence of a mimic of the bis(aldol) adduct.

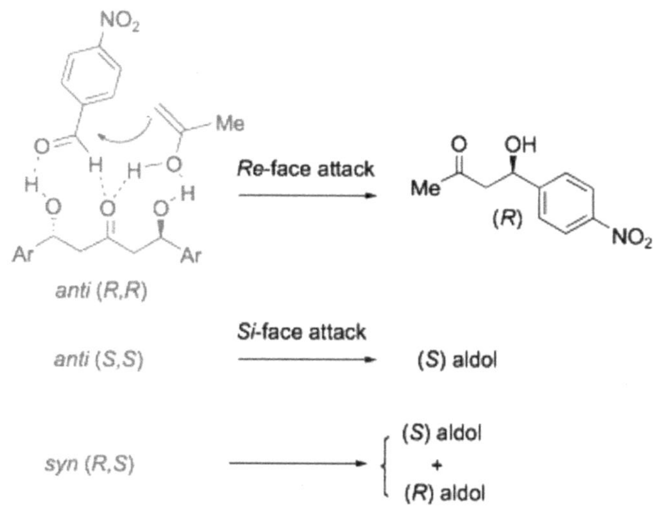

Scheme 13.7 Catalysis of the aldol reaction between acetone and 4-nitrobenzaldehyde by bis(aldol) adducts.

b. The bis(aldol) adducts can only arise by the reversible reaction of initially formed aldol adducts in the initial stages of the reaction, when dissolution of solid 4-nitrobenzaldehyde leads to high concentration gradients on the surface of the crystals of the aldehyde; in this way, catalysis by bis (aldol) is kinetically equivalent to second-order autocatalysis by the aldol. In homogeneous solution, the formation of bis(aldol) becomes extremely slow, due to the very low concentration of aldol.

c. The reversible nonenantioselective catalysis by the *syn* bis(aldol) adduct plays a role analogous to that of mutual inhibition in the Frank reaction network [4], and leads to a strongly nonlinear behavior of the reaction system (Scheme 13.8).

With these assumptions, we devised the reversible reaction network shown in Scheme 13.9, and we undertook a numerical simulation with *Mathematica*, restricting the bis(aldol) adduct formation to the first 7 s of the reaction.

To our satisfaction, we found that this reaction network could lead to a strong amplification of the initial *ee* of the aldol adduct, even starting from an initial enantiomeric purity as low as 10^{-6}% *ee* and a 10^{-15} M concentration of the aldol, an *ee* that corresponds to that expected for the statistical fluctuation at a 1.0 mmol scale [6]. Thus, as shown in Fig. 13.2, for a moderately exoergic aldol reaction ($K_{eq} = 10^2$), and assuming $K_{eq} = 1$ for the bis(aldol) formation by condensation of two aldol adducts

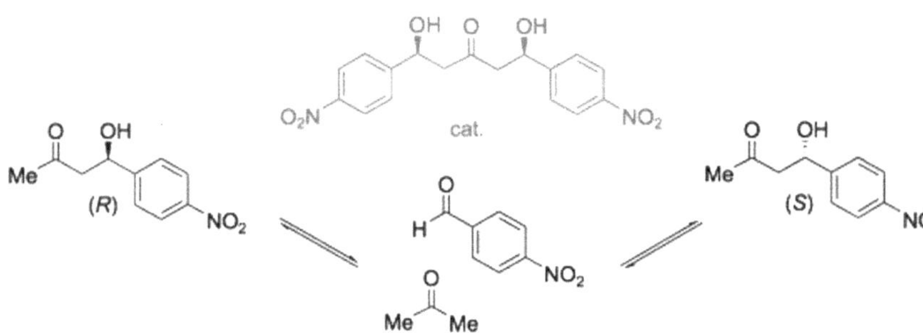

Scheme 13.8 Mutual inhibition between both enantiomers of the aldol product through reversible retro-aldol reaction catalyzed by the *syn* bis(aldol) adduct.

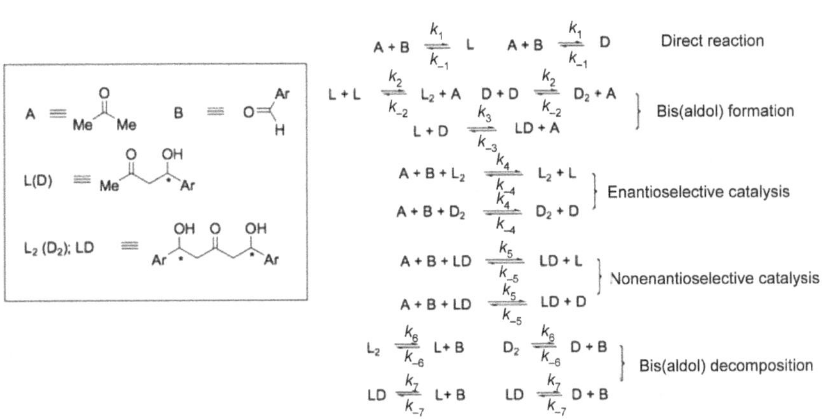

Scheme 13.9 A reversible reaction network involving catalysis by bis(aldol) adducts that leads to spontaneous mirror-symmetry breaking in the aldol formation.

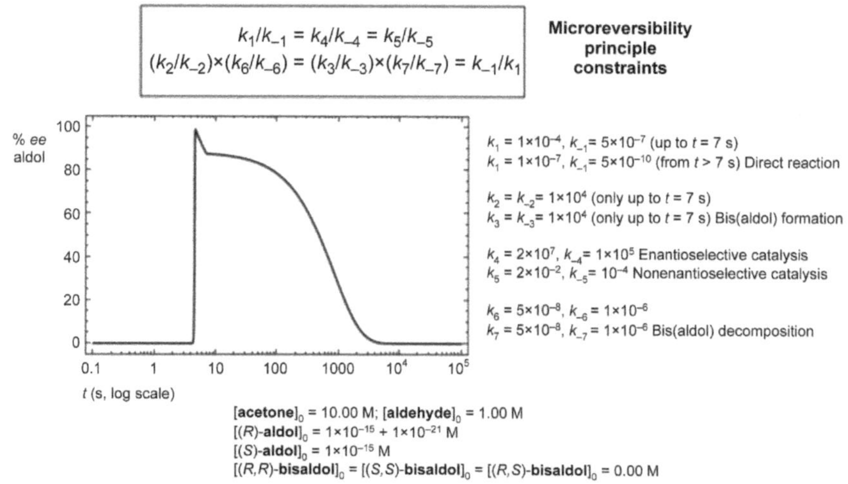

Figure 13.2 Numerical simulation of the reaction network shown in Scheme 13.9, leading to a kinetically controlled spontaneous mirror-symmetry breaking in the initial absence of bis(aldol) adducts.

(which seems reasonable since in this homodesmotic process one aldol C–C bond is broken and another one is formed without significant changes either in conjugation or in strain between reactants and products), it can be seen that at the actual concentrations of our experiments, SMSB takes place between 4 and 5 s from the beginning of the reaction (i.e., presumably when the concentration of bis(aldol) adducts reaches a critical level), and the reaction network is able to maintain a relatively high *ee* for the aldol product during the first minutes of the reaction, provided that the enantioselective catalysis by the bis(aldol) adduct is much more efficient than both the nonenantioselective catalysis and the direct reaction. It is also important to note that the rate constants fulfill the microreversibility principle constraints.

When we performed the numerical simulation with the same concentrations and rate constants, but starting from nonzero initial concentrations of bis(aldol) adducts, similar to those used in the experiment summarized in Scheme 13.4 above (Fig. 13.3), we found that the SMSB took place almost immediately ($t = < 0.002$ s); this finding is in accordance with our observation that SMSB always takes place when the bis(aldol) (or even a bis(aldol) mimic) is present at the beginning of the reaction. Note that in these simulations the stereochemical outcome of the reaction always correlates with the absolute configuration of the enantiomer of aldol that is arbitrarily assumed

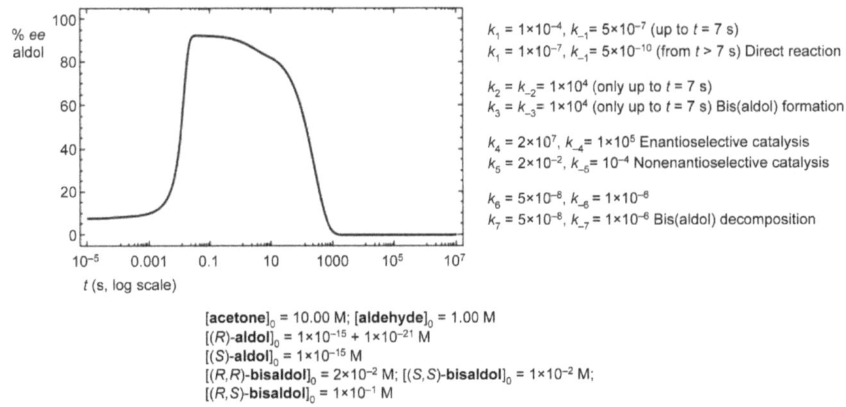

Figure 13.3 Numerical simulation of the reaction network shown in Scheme 13.9, leading to a kinetically controlled spontaneous mirror-symmetry breaking in the initial presence of bis(aldol) adducts.

to be in very small excess; the nonstochastic behavior of SMSB in our experiments (i.e., the predominant formation of (*R*)-configured aldol) should then be explained by assuming that the initial *ee* that is amplified by the reaction network has a sign determined by the unavoidable presence of trace amounts of a chiral nonracemic compound, which may have a very small chiral effect. This is a characteristic of coupled reaction networks that lead to SMSB [5] and it has been experimentally demonstrated in the Soai reaction [9,10,21,22]. Finally, these simulations suggest that the failure to observe SMSB in c.50% of the experiments run in the conditions of Scheme 13.2 could be traced to an inefficient formation of bis(aldol) adducts in the first stages of the reaction, but further experimental work is clearly needed in order to verify this hypothesis.

13.2.2 Reaction Product Kinetic Analysis of an Autocatalytic Mannich Reaction

Asymmetric organoautocatalysis in the Mannich reaction was first described in 2007 by Mauksch, Tsogoeva and co-workers [16], when they found that variable amounts (1−50 mol%) of enantiomerically enriched (29%−99% *ee*) the Mannich adduct of acetone with the imine derived from ethyl glyoxylate and *p*-anisidine catalyzed its own formation, in variable yields (20%−53% yield) and with low to high enantioselectivities (11%−94% *ee*), after 2−6 days at room temperature (Scheme 13.10).

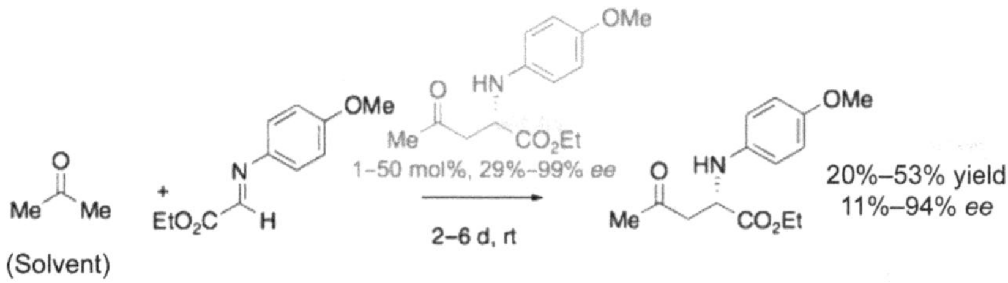

Scheme 13.10 Asymmetric autocatalysis in the Mannich reaction between acetone and an α-iminoethylglyoxylate.

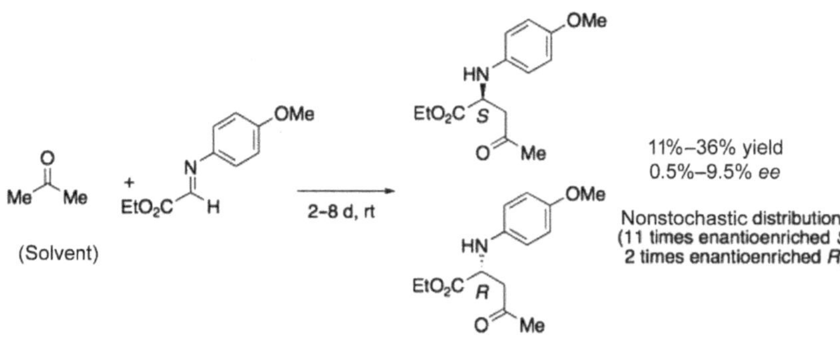

Scheme 13.11 Asymmetric autocatalysis in the Mannich reaction between acetone and an α-iminoethylglyoxylate in the initial absence of any external catalyst.

Subsequently, further examples of asymmetric autocatalysis in Mannich reactions of imine derived from ethyl glyoxylate and *p*-anisidine were reported by Amedjkouh and Brandberg [23] and by Wang and co-workers [24].

In the absence of an external catalyst, the Mannich reaction of Scheme 13.10, a product *ee* of up to 9.5% for the Mannich adduct was reported by Mauksch and Tsogoeva (Scheme 13.11) [17].

The observed SMSB in this reaction was ascribed by the authors to the asymmetric autocatalysis by the Mannich adduct, and a possible structure for the transition state of this step was proposed on the basis of mass spectrometric data and of DFT calculations (Fig. 13.4) [17], although no evidence was provided for the mutual inhibition between both enantiomers of the product.

More recently, Tsogoeva and co-workers found that when hydroxyacetone was used as the donor component instead of acetone, the autocatalytic Mannich reaction took place with much shorter reaction times (1–6 h instead of 2–8 days), and they studied the kinetics of the process

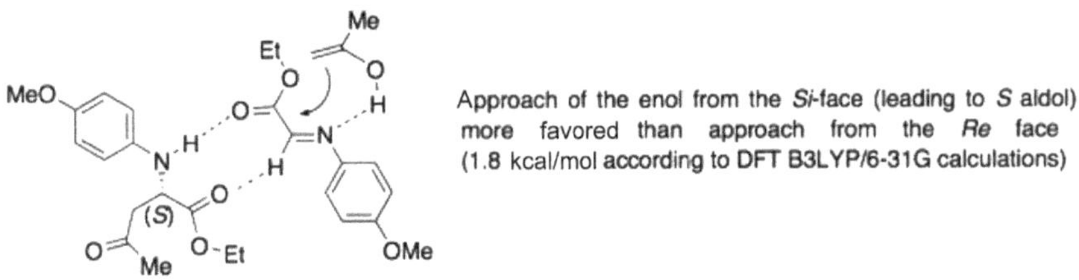

Figure 13.4 Proposed transition state for the asymmetric autocatalysis in the Mannich reaction between acetone and an α-iminoethylglyoxylate.

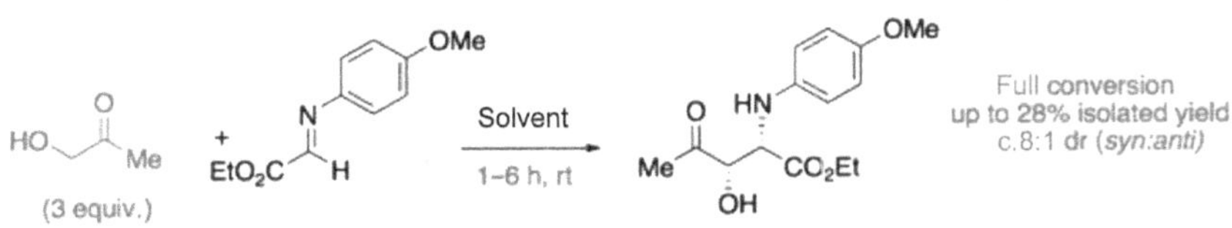

Scheme 13.12 Mannich reaction between hydroxyacetone and an α-iminoethylglyoxylate in the initial absence of any external catalyst.

by time-resolved ^{1}H NMR [25]. They reported that in the absence of any chiral catalyst, and using three equivalents of hydroxyacetone, the conversion of the imine was complete in a few hours, depending on the reaction conditions (solvent, imine concentration, temperature). The Mannich adduct (*syn:anti* mixture) was isolated in variable yields, and always in racemic form (Scheme 13.12).

The conversion time profile for this reaction, obtained by ^{1}H NMR, yielded the S-shaped curve characteristic of autocatalytic processes, and HPLC monitoring seemed to indicate that in some conditions the *syn* adduct, before full conversion, was being generated in nonracemic form (up to 7% *ee*) [25]. We were intrigued by these results and we considered the possibility of studying the same reaction by Blackmond's reaction progress kinetic analysis (RPKA), using microcalorimetric techniques [26].

First, we performed the same Mannich reaction in a specific set of conditions and we were pleased to find that we could obtain reproducible results that were, moreover, in accordance with the work of Tsogoeva (Scheme 13.13). In $CHCl_3$ (or $CDCl_3$) solution, full conversion was observed after 3.5 h at room temperature, using a 3:1 ketone/imine ratio. The NMR yield of the Mannich adduct (6:1 *syn:anti* mixture) was 54% and, was obtained in racemic form.

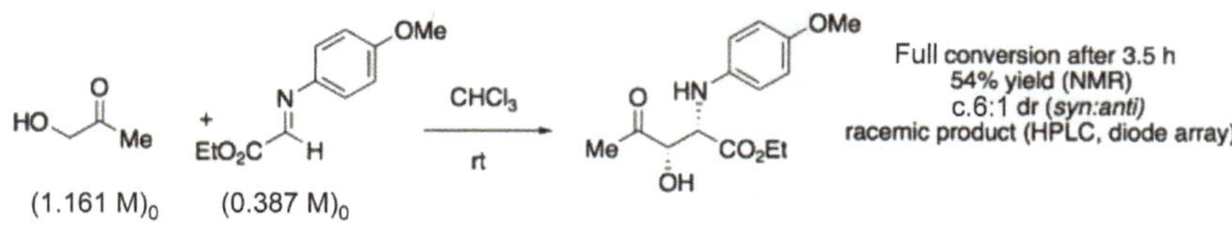

Scheme 13.13 Reaction conditions for Blackmond's reaction progress kinetic analysis study of the autocatalytic Mannich reaction.

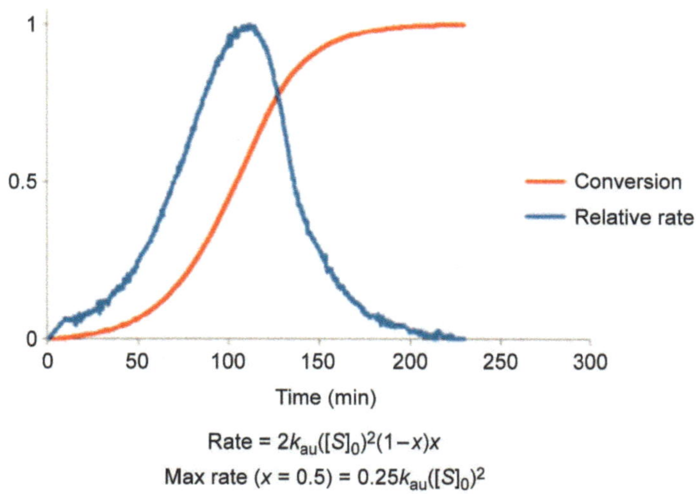

Figure 13.5 Reaction monitoring by microcalorimetry of the Mannich reaction depicted in Scheme 13.13.

With these selected conditions, the microcalorimetric monitoring produced the rate versus time profile expected for a first-order autocatalytic process (Fig. 13.5).

By running the same reaction with different initial concentrations of hydroxyacetone, calorimetric RPKA rapidly established that the reaction order with respect to this reagent was fractionary ($n = 0.4$). This suggests that a reasonable mechanism for this Mannich reaction would involve the reversible formation of an adduct-imine complex, followed by the irreversible reaction with hydroxyacetone to give two molecules of the Mannich product. In addition to that, we have found that when the reaction is run in the presence of a substoichiometric amount (10 mol%) of the preformed Mannich adduct 15:1 (*syn:anti*), the rate is independent of the initial enantiomeric purity of this compound, a result that rules out the possibility of mutual inhibition between both enantiomers of the

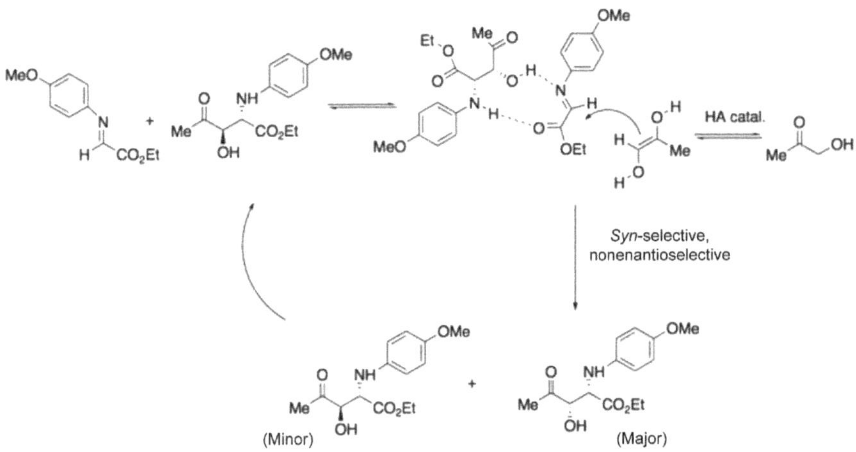

Scheme 13.14 Proposed mechanism for the Mannich reaction depicted in Scheme 13.13, based on the results of RPKA.

product. For a first order autocatalytic system, this would imply that SMSB cannot take place. In fact, when we monitored the Mannich reaction (in the absence of any external catalyst) by chiral HPLC (UV-VIS Diode array detector), we found that the Mannich adduct was racemic throughout the course of the reaction, contrary to the previous report of Tsogoeva [25]. We have no satisfactory explanation for this discrepancy, except that in our hands the most reliable results were obtained for a wavelength of 220 nm, instead of 254 nm.

Finally, our preliminary results seem to indicate that the actual catalyst is the *anti*-diastereomer of the Mannich adduct, which would catalyze the reaction in a racemic fashion, while the diastereomerically pure *syn* adduct does not increase the rate of the reaction [27]. In summary, we propose that the autocatalytic Mannich reaction between hydroxyacetone and the imine derived from *p*-anisidine and ethyl glyoxylate takes place through the mechanism depicted in Scheme 13.14.

13.3 CONCLUSIONS

Reaction networks in which first-order asymmetric autocatalysis is coupled with other reactions in a nonlinear fashion may lead to SMSB. A clear-cut example is apparently provided by the Soai reaction.

The report of SMSB by Mauksch and Tsogoeva in purely organoautocatalytic reactions has drawn considerable attention, and we have reexamined in detail some of these processes.

In the uncatalyzed aldol reaction between acetone and *p*-nitrobenzaldehyde, SMSB may take place at the beginning of the reaction, but most of the aldol product is formed by a racemic mechanism that does not involve autocatalysis. A reversible reaction network featuring enantioselective catalysis by a bis(aldol) adduct is probably involved in the observed SMSB, according both to experiment and to numerical simulations.

The Mannich reaction between hydroxyacetone and a glyoxylate-derived imine is clearly a first-order autocatalytic process, as evinced by microcalorimetric RPKA. In this case, however, no mutual inhibition between product enantiomers takes place, and no SMSB can be observed even in the early stages of the reaction. Also in this case we have proposed a mechanism that is in accordance to experimental results.

ACKNOWLEDGMENTS

Guillem Valero thanks the Spanish Ministry of Economy and Competitivity (MINECO) for a FPI predoctoral fellowship. Financial support from MINECO (Project CTQ2013-47401-C2-1-P) and the participation in the COST Action CM0905 are gratefully acknowledged. We also thank Prof. Josep M. Ribó (Emer.) for the numerical simulations with the *Mathematica* package of programs.

REFERENCES AND NOTES

[1] Plasson, R.; Brandenburg, A.; Jullien, L.; Bersini, H. Autocatalyses. *J. Phys. Chem. A* **2011,** *115*, 8073–8085.

[2] Bissette, A. J.; Fletcher, S. P. Mechanisms of Autocatalysis. *Angew. Chem., Int. Ed.* **2013,** *52*, 12800–12826.

[3] Alberts, A. H.; Wynberg, H. The Role of the Product in Asymmetric C–C Bond Formation: Stoichiometric and Catalytic Enantioselective Autoinduction. *J. Am. Chem. Soc.* **1989,** *111*, 7265–7266.

[4] Frank, F. C. On Spontaneous Asymmetric Synthesis. *Biochim. Biophys. Acta* **1953,** *11*, 459–463.

[5] Crusats, J.; Hochberg, D.; Moyano, A.; Ribó, J. M. Frank Model and Spontaneous Emergence of Chirality in Closed Systems. *ChemPhysChem* **2009,** *10*, 2123–2131.

[6] Ribó, J. M.; Blanco, C.; Crusats, J.; El-Hachemi, Z.; Hochberg, D.; Moyano, A. Absolute Asymmetric Synthesis in Enantioselective Autocatalytic Reaction Networks: Theoretical Games, Speculations on Chemical Evolution and Perhaps a Synthetic Option. *Chem. Eur. J.* **2014,** *20*, 17250–17271.

[7] Soai, K.; Shibata, T.; Morioka, H.; Choji, K. Asymmetric Autocatalysis and Amplification of Enantiomeric Excess of a Chiral Molecule. *Nature* **1995,** *378*, 767–768.

[8] Sato, I.; Urabe, H.; Ishiguro, S.; Shibata, T.; Soai, K. Amplification of Chirality from Extremely Low to Greater than 99.5% ee by Asymmetric Autocatalysis. *Angew. Chem., Int. Ed.* **2003,** *42*, 315–317.

[9] Singleton, D. A.; Vo, L. K. Enantioselective Synthesis without Discrete Optically Active Additives. *J. Am. Chem. Soc.* **2002,** *124*, 10010–10011.

[10] Soai, K.; Sato, I.; Shibata, T.; Komiya, S.; Hayashi, M.; Matsueda, Y., et al. Asymmetric Synthesis of Pyrimidyl Alkanol without Adding Chiral Substances by the Addition of Diisopropylzinc to Pyrimidine-5-carbaldehyde in Conjunction with Asymmetric Autocatalysis. *Tetrahedron: Asymmetry* **2003,** *14*, 185–188.

[11] Blackmond, D. G.; McMillan, C. R.; Ramdeehul, S.; Schorm, A.; Brown, J. M. Origins of Asymmetric Amplification in Autocatalytic Alkylzinc Additions. *J. Am. Chem. Soc.* **2001,** *123*, 10103–10104.

[12] Schiaffino, L.; Ercolani, G. Unraveling the Mechanism of the Soai Asymmetric Autocatalytic Reaction by First-Principles Calculations: Induction and Amplification of Chirality by Self-Assembly of Hexamolecular Complexes. *Angew. Chem., Int. Ed.* **2008,** *47*, 6832–6835.

[13] Calvin, M. *Chemical Evolution;* Oxford University Press: Oxford, 1969.

[14] Avetisov, V. A.; Goldanskii, V. Mirror Symmetry Breaking at the Molecular Level. *Proc. Natl. Acad. Sci. U. S. A.* **1996,** *93*, 11435–11442.

[15] Kondepudi, D. K.; Asakura, K. Chiral Autocatalysis, Spontaneous Symmetry Breaking, and Stochastic Behavior. *Acc. Chem. Res.* **2001,** *34*, 946–954.

[16] Mauksch, M.; Tsogoeva, S. B.; Martynova, I. M.; Wei, S. Evidence of Asymmetric Autocatalysis in Organocatalytic Reactions. *Angew. Chem., Int. Ed.* **2007,** *46*, 393–396.

[17] Mauksch, M.; Tsogoeva, S. B.; Wei, S.; Martynova, I. M. Demonstration of Spontaneous Chiral Symmetry Breaking in Asymmetric Mannich and Aldol Reactions. *Chirality* **2007,** *19*, 816–825.

[18] Valero, G.; Ribó, J. M.; Moyano, A. A Closer Look at Spontaneous Mirror Symmetry Breaking in Aldol Reactions. *Chem. Eur. J.* **2014,** *20*, 17395–17408.

[19] Mauksch, M.; Wei, S.; Freund, M.; Zamfir, A.; Tsogoeva, S. B. Spontaneous Mirror Symmetry Breaking in the Aldol Reaction and its Potential Relevance in Prebiotic Chemistry. *Orig. Life Evol. Biosph.* **2010,** *40*, 79–91.

[20] Gryko, D.; Lipinski, R. Asymmetric Direct Aldol Reaction Catalysed by L-Prolinethioamides. *Eur. J. Org. Chem.* **2006,** 3864–4876.

[21] Kawasaki, T.; Matsumura, Y.; Tsutsumi, T.; Suzuki, K.; Ito, M.; Soai, K. Asymmetric Autocatalysis Triggered by Carbon Isotope ($^{13}C/^{12}C$) Chirality. *Science* **2009,** *324*, 492–495.

[22] Kawasaki, T.; Okano, Y.; Suzuki, E.; Takano, S.; Oji, S.; Soai, K. Asymmetric Autocatalysis: Triggered by Chiral Isotopomer Arising from Oxygen Isotope Substitution. *Angew. Chem., Int. Ed.* **2011,** *50*, 8131–8133.

[23] Amedjkouh, M.; Brandberg, M. Asymmetric Autocatalytic Mannich Reaction in the Presence of Water and its Implication in Prebiotc Chemistry. *Chem. Commun.* **2008,** 3043–3045.

[24] Wang, X.; Zhang, Y.; Tan, H.; Wang, Y.; Han, P.; Wang, D. Z. Enantioselective Organocatalytic Mannich Reactions with Autocatalysts and their Mimics. *J. Org. Chem.* **2010,** *75*, 2403–2406.

[25] Held, F. E.; Fingerhut, A.; Tsogoeva, S. B. Insights into the Spontaneous Emergence of Enantioselectivity in an Asymmetric Mannich Reaction Carried out without External Catalyst. *Tetrahedron: Asymmetry* **2012,** *23*, 1663–1669.

[26] Blackmond, D. G. Reaction Progress Kinetic Analysis: A Powerful Methodology for Mechanistic Studies of Complex Catalytic Reactions. *Angew. Chem., Int. Ed.* **2005,** *44*, 4302–4320.

[27] The non-selective (auto) catalysis by an *anti*-Mannich adduct had been also observed by Amedjkouh and Brandberg; see Ref. [23], note 22.

CHAPTER 14

Isotope Chirality in the Reagents and Intermediates of the Soai Reaction*

Béla Barabás[1], Róbert Kurdi[2] and Gyula Pályi[3]
[1]Institute of Mathematics, Department of Stochastics, Budapest University of Technology and Economics, Budapest, Hungary
[2]Institute of Environmental Engineering, University of Pannonia, Veszprém, Hungary
[3]Department of Life Sciences, University of Modena and Reggio Emilia, Modena, Italy

Abstract

Some fragments of the reagents and of the (hypothetic) intermediates of the asymmetric autocatalysis (Soai reaction) are *prochiral* in the sense of isotopic substitution. The structures as well as the probability of formation of such enantioisotopomeric molecules/groups were identified and calculated by combinatorial and stochastic methods, based on the natural abundances of stable isotopes of H, C, O, Si, and Zn. The expected concentration of isotopically chiral structures can reach ~11 mol%, which should be taken into consideration in mechanistic speculations concerning the molecular events in the Soai reaction.

Keywords: Chirality, by isotopes; diastereoisotopomers; enantioisotopomers; isotope substitution; Soai reaction, intermediates of

14.1 INTRODUCTION

The majority of elements occur as a mixture of stable isotopes in Nature [3,4]. These display only slight differences in their physical features and chemical reactivity, but even so, a considerable body of experience has been accumulated about the role of stable isotopes in chemistry and biochemistry/biology [5,6]. Isotopes may be distributed asymmetrically in molecules, thus causing the formation of enantiomers, also called *enantioisotopomers* (similarly also *diastereoisotopomers* can be defined). These highly interesting structural species in the past were rarely studied [7], even if they occur much more ferquently in Nature or in "pure" laboratory chemicals than is generally recognized [8]. This is, most probably, due to the slight differences in the reactivity of isotopically substituted molecules

* Some parts of this paper were presented at a Symposium [1] and published in an article [2].

Advances in Asymmetric Autocatalysis and Related Topics
DOI: http://dx.doi.org/10.1016/B978-0-12-812824-4.00014-9

and to the technical (analytical) difficulties in the exploration of the stereochemistry of such molecules, as well as the observation of the minor differences in reactivities [9]. These difficulties can be in great part overcome thanks to the discovery of *asymmetric autocatalysis* (Soai reaction) [10,11] in 1995 by the Group of the Tokyo University of Science, headed by Kenso Soai. It became known, shortly after the discovery, that this particular reaction provides a very sensitive tool for the observation of the chirality of other compounds [12], including *cryptochiral* species [13], which otherwise present enormous difficulties for structural studies. The Soai Group discovered as early in 2000 [14], that the asymmetric autocatalysis also "responds" to enantioisotopomers obtained by $^{2}D/^{1}H$ substitution. In recent years these observations could be extended to the observation of chiral induction due to enantioisotopomers obtained by $^{13}C/^{12}C$, $^{15}N/^{14}N$, and $^{16}O/^{17/18}O$ substitution [15]. These exciting results raised the question in our Group, as to whether it is possible to identify isotopically chiral structural elements in the *reagents* and/or *intermediates* of the Soai reaction. Some of the results of these studies were published in a Symposium [1] and in a preliminary publication [2]. Here we summarize these earlier results and report on additional considerations.

14.2 RESULTS AND DISCUSSION

The Soai reaction (Scheme 14.1) was used for a waste choice of substrates [11,16] and speculations about possible reaction intermediates present a rich intellectual challenge [17,18]. We shall discuss first some representative reagents, together with their products (autocatalysts!) and then some selected (hypothetic) intermediates.

Looking through the structures of these species, we find that they are composed of H, C, N, O, Si, and Zn. Prochiral groups contain H, C, O, Si, Zn. The naturally abundant stable isotopes of these elements are shown in Table 14.1. At first sight one can see that the isotopic composition of these elements, with the exception of Zn, is fairly asymmetric, showing a large prevalence of the lightest isotopes. This asymmetry also generates a "technical" hindrance to the formation of larger quantities of enantioisotopomers, but in some cases the richness of variability favors (relatively) the asymmetric structures.

One of the reagents in the Soai reaction is always $i\mathrm{Pr}_2\mathrm{Zn}$ [16]. The aldehydic component may be of more variable structure. Some important and/or representative examples are shown in Table 14.2.

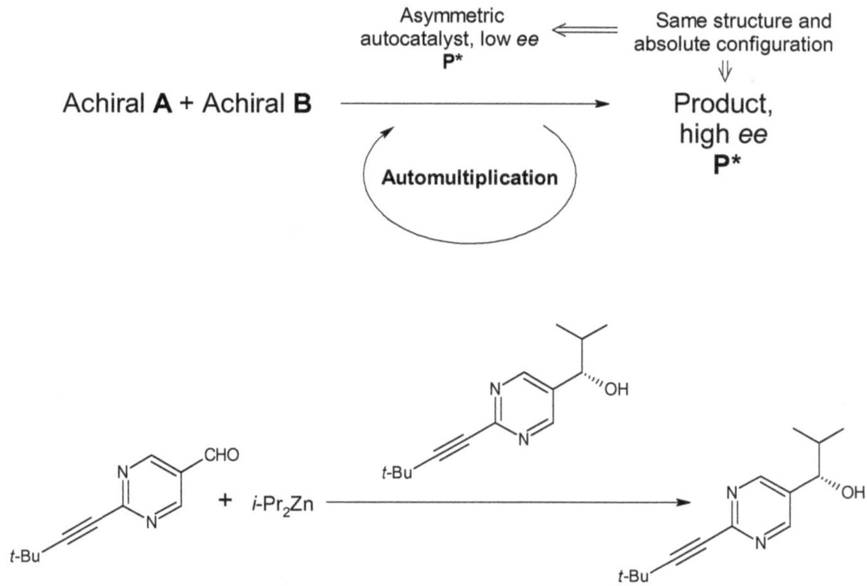

1 2 3

Scheme 14.1 Schematic representation of asymmetric autocatalysis (Soai reaction) [16].

Table 14.1 Natural abundance of stable isotopes in prochiral groups involved in the Soai reaction [3]

Element	Isotopes, natural abundance (%)
Hydrogen	^{1}H 99.985; ^{2}H 0.015
Carbon	^{12}C 98.889; ^{13}C 1.111
Oxygen	^{16}O 99.759; ^{17}O 0.037; ^{18}O 0.204
Silicon	^{28}Si 92.21; ^{29}Si 4.70; ^{30}Si 3.09
Zinc	^{64}Zn 48.89; ^{66}Zn 27.81; ^{67}Zn 4.11; ^{68}Zn 18.57; ^{70}Zn 0.62

14.2.1 Di(iso-propyl)zinc

iso-Propyl group(s) is(are) present in all species (reactant, intermediates, product) of the Soai reaction, with the exception of the starting aldehyde reactant. These groups, however, are in very different chemical environments: while the *i*Pr groups linked to the Zn atom are in a typical organomelallic bond, the *i*Pr groups in the molecules *after* the C—C bonding step are in sp^3—sp^3 covalent bonding environment.

The easiest way to generate (isotopic) chirality in the highly symmetric *i*Pr group (or in its derivatives) is the substitution of either one of the terminal

Table 14.2 Some representative examples of the substituents of aldehydic substrates, **1**, (and the corresponding products/autocatalysts, **3**) in the Soai reaction

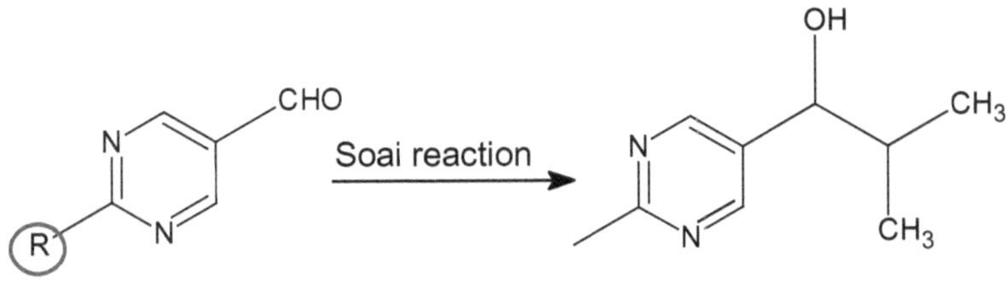

1 **3**

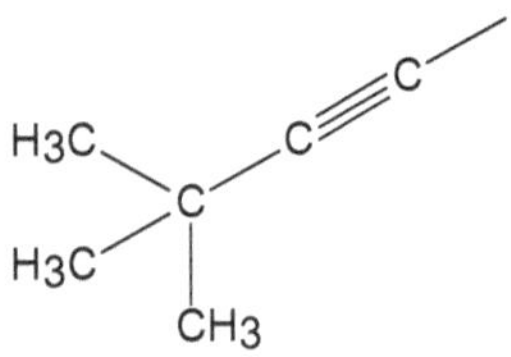

a Ref. [11d]

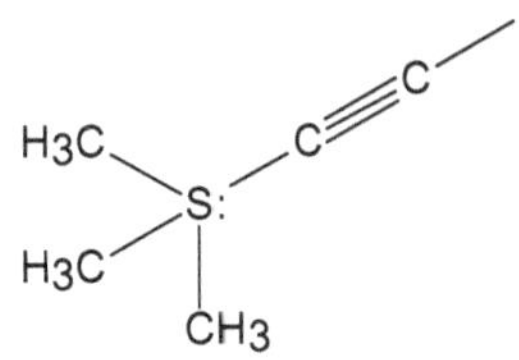

b Ref. [11d]

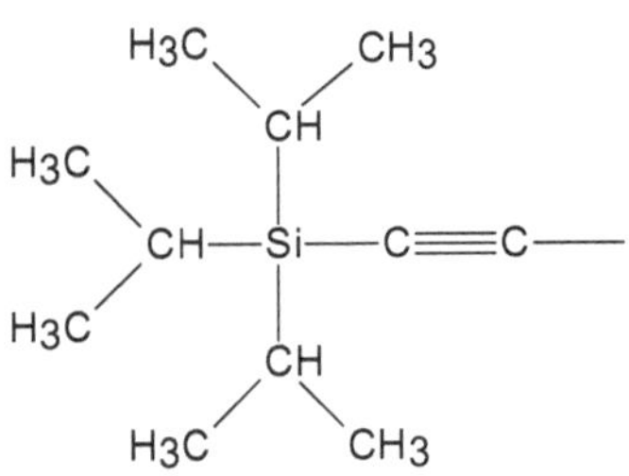

c Ref. [11d]

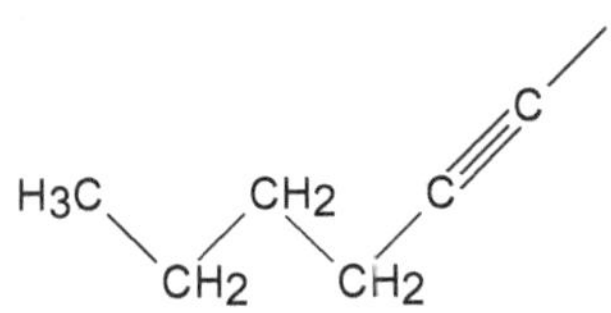

d Ref. [11d]

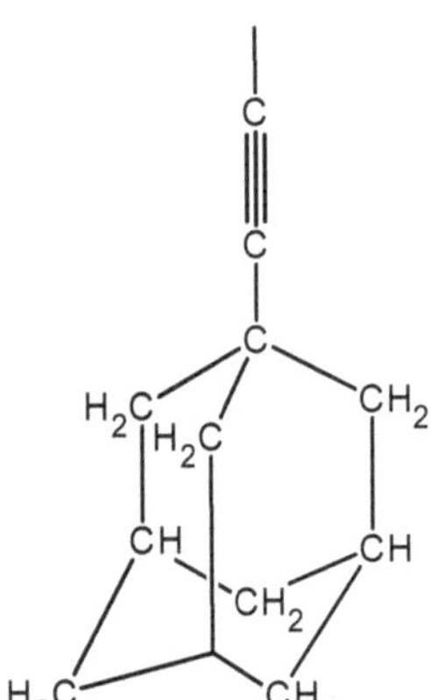

e Ref. [16h]

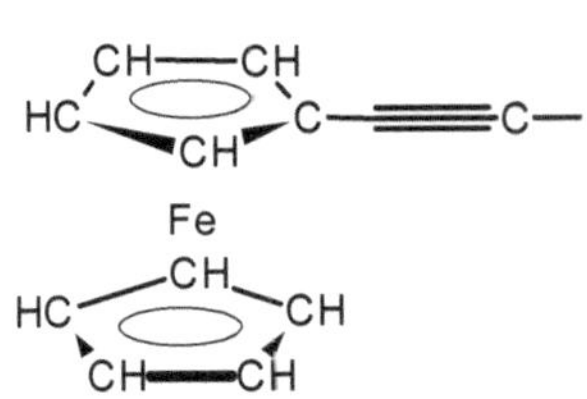

f Ref. [16f]

^{1}H atoms by ^{2}D or exchanging one of the methyl-^{12}C atoms with a ^{13}C. The results of such substitutions can be calculated by elementary combinatorial and probabilistic methods (details are described in Ref. [2]). Taking into account all *combinations* of isotopic substitution one obtains $2 \times 64 = 128$ cases if the isotopic substitution of the central CH group is disregarded and $2 \times 128 = 256$ cases if this latter possibility is also considered. In the following calculations we shall confine ourselves to the former case (128). Not all of these are, however, chiral. The chiral combinations make up 112 cases, while the symmetric ones make up 16. On the other hand, the *probability* of the achiral (symmetric) cases represent an overwhelming majority: 97.7131%, and only 2.2869% remains for the much more numerous chiral combinations.

The alkylation reagent in the Soai reaction, however, contains *two* *i*Pr groups. This considerably increases the number of combinations: $128 \times 128 = 16{,}384$, if only one isotope exchange in the central CH is considered and $4 \times 128 \times 128 = 65{,}536$ if both isotope exchange possibilities in the methylene group is taken in regard. We used the former value in our calculations. The two *i*Pr groups can be regarded in two different ways. If each of the two *i*Pr groups can be "internally" symmetric, the number of achiral combinations amounts to $(8+8)^2 = 16^2 = 256$ cases, while the chiral ones are $16{,}384 - 256 = 16{,}128$. The probability of these chiral combinations amounts to 4.5215% (~4.5%). If the symmetry between the two *i*Pr groups is also considered, the number of the achiral combinations is 128 and that of the chiral structures is $128^2 - 128 = 16{,}384 - 128 = 16{,}256$ and the probability of the chiral structures increases to 6.6200% (~6.6%).

The whole quantity of the chiral components does not represent, however, enantiomeric excess (*ee*) of one of the two enantiomers. The statistically expectable *ee* values can be calculated by the Pars-Mills equation [8,19]. These depend on the sample size. We calculated the $ee_{\text{expectable}}$, 50% values (with 50% confidence) for usual to very small microchemical preparative sample sizes and obtained *ee* values [2] that might be significant in the milimol to femtomol sample size range. It should be mentioned here that these values, especially those with the smaller sample size, are well over the sensitivity threshold of the Soai reaction, as has been demonstrated thoroughly by experimental [20] and theoretical [21] studies. We shall come back to this point later.

Studies concerning the mechanism of the Soai reaction [18] generally conclude that the key intermediates are of fairly complicated structure and contain at least two (or more) $Zn(iPr)_n$ ($n = 1{,}2$) unit pro molecules. This means that the combinatorial analysis of such structures yields very

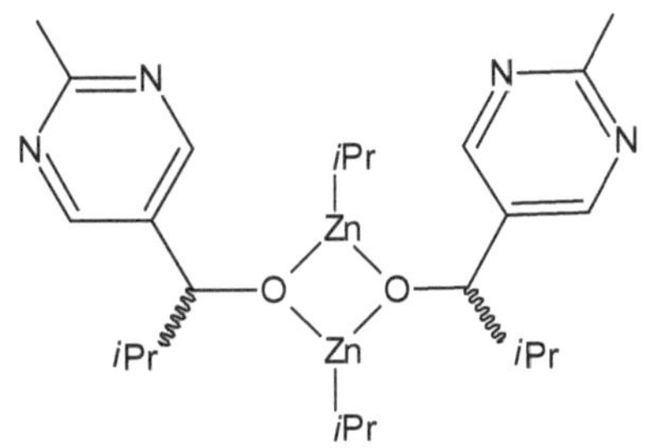

Figure 14.1 Schematic structure of the dimeric intermediate of the Soai reaction, according to Schiaffino and Ercolani [18e].

high numbers for the isotopically substituted derivatives. Here we show only a fairly "simple" case, intermediate **A** [18e] (Fig. 14.1).

This molecule contains two kinds of *i*Pr groups, two of these in organometallic and an additional two in an organic–covalent bonding environment. We considered only one kind of isotopic exchange in the "central" CH groups. According to this approach the total number of combinations amounts to $128^4 = 268{,}435{,}456$. (Several of these are equivalent to each other—"degenerated.") Intermediate **A** can be analyzed as follows:

i. Chirality is due only to Zn-linked *i*Pr groups. This yields $128^2 = 16{,}384$ for all structures. For the chiral combinations one proceeds by substracting the number of the symmetric cases ($16^2 = 256$) from this value and multiplying the rest with the number of the C-bound symmetric cases, which yields: $(16{,}384 - 256) \times 128 = 2{,}064{,}384$. The probability of the chiral structures is 4.5215% (~4.5%) (as obtained for the two *i*Pr groups above).

ii. Chirality is due only to carbon-bound *i*Pr groups. This approach is similar to the second case discussed above, it gives $128 \times 127 \times 16^2 = 4{,}161{,}536$ chiral combinations, with a probability of 6.620% (~6.6%).

iii. If both of these approaches are considered, one obtains for the chiral cases the fairly high number of $(128^2 - 16^2) \times 128 \times 127 = 262{,}176{,}768$. The probability can be obtained by multiplying the probabilities of the former two cases with each other: $0.045215 \times 0.06620 = 0.00300\%$ (~0.3%).

The total of the probabilities of obtaining chiral structures by isotopic substitution in intermediate **A** is the sum of the values obtained under (i)–(iii): that is 4.5% + 6.6% + 0.3% = **11.4%**.

It should be noted that while the number of achiral combinations is much lower, $128 \times 256 = 32{,}768$, these structures represent 100% − 11.4% = **88.6%** of the probabilities.

The very high number of combinations with only four *i*Pr groups (268,435,456) in molar units corresponds to an approximately 0.45 femtomol size sample, if one takes only *one* molecule from each combination. Thus, when approaching such a situation not all combinations can be formed because of statistical reasons, as has been pointed out by Eschenmoser et al. [22] in the context of speculations concerning the origin of DNA or other information carrier molecules originating from the dawn of terrestrial life. This problem associated with intermediates of higher "nuclearity" also becomes significant for the Soai reaction [18], even at sample sizes much greater than a femtomol.

14.2.2 Aldehyde Substrates

The Soai reaction is essentially a C—C coupling reaction of N-heterocyclic carbaldehydes (Scheme 14.1. Refs. [11,16]). The most efficient variants were reported for 2-alkynylpyrimidine-5-carbaldehyde derivatives, bearing various (mostly bulky) substituents on the alkynyl group. We analyzed some of these from the viewpoint of isotope chirality developing by natural abundance $^{1}H/^{2}H$ and/or $^{12}C/^{13}C$ substitution (Table 14.2).

The substituents in these aldehydic substrates are relatively far from the most important reaction center (the aldehyde sp^{2}—carbon atom), but evidently may have a decisive role in the outcome (and enantioselectivity) of the reaction, as has been demonstrated experimentally [16]. Moreover, these organic groups are present not only in the substrate, but also in the product, which also acts as autocatalyst and chirality-inductor in this particular reaction.

In the following analysis we deal only with those groups that form statistically with a probability (P) $>10^{-6} = 0.0001\%$.

Substrates **1a** and **1b** (as well as products **3a** and **3b**) may become chiral by isotopic substitution in the methyl groups. The number of achiral combinations amounts to 144, while that of the chiral structures is 440. Isotopically substituted methyl groups comprise eight kinds, but only three of these correspond to the condition of abundance $>10^{-6}$:P_1 ($^{12}C^{1}H_3$) = 0.98844 = 98.844%; P_2($^{13}C^{1}H_3$) = 0.01109 = 1.109%; P_3($^{12}C^{1}H_2{}^{2}H$) = 0.00046 = 0.046%. In the *t-Bu* and *Me₃Si* groups the three methyl groups generate by isotopic substitution $8^3 = 512$ combinations, but only 27 of these correspond to the condition of $P > 10^{-6}$. The central atom (C or Si) becomes chiral if all three methyl groups are different. The probability of this is $P_{chiral} = 6 \times P_1 \times P_2 \times P_3 = 0.000030254 = 0.0030254\%$ (~0.003%), allowing $P_{achiral} = 0.999969745 = 99.9969745\%$ (~99.997%) for the more-or-less symmetric species. These values are fairly near to the corresponding probabilities calculated by a different approach in our earlier communication [2]. The combinatorial picture of these groups becomes even more rich, if the isotopes of the central C or Si atoms are taken into consideration, but since these do not influence the chirality of the *t-Bu* or Me_3Si groups we did not analyze these options.

Substrate **1c** (and product **3c**) provide very interesting structural features. The total number of achiral combinations is 8424, while the number of chiral isotopomers is 4,262,700. Obviously, due to the C_{3v} symmetry of the (*i*-Pr)$_3$Si group several of these are "degenerated" (equivalent) and others are of very low probability. Once again we analyzed only the most probable three combinations, which are more probable than $P > 10^{-3} = 0.1\%$. These are: $P_4(^{12}C^1H_3{-}^{12}C^1H{-}^{12}C^1H_3) = 0.966159012 = 96.6159012\%$ ($\sim$96.62%); $P_5(^{12}C^1H_3) = 1.0854\%$ ($\sim$1.09%). Already these probabilities show clearly that the majority of the structures is achiral. With this group, however, the possibility emerges again that not the "whole" group, but a fragnent becomes chiral on its "own right", e.g., P_5 or *i*Pr$_2$Zn. If this gets combined with the chirality evolved on the central Si atom, *diastereomeric* structures will be formed, opening new reaction possibilities, since these species are no longer limited by the equal formation energy/formation probability rule valid for *enantiomers* [23].

The on-Si-chirality evolves if the three *i*-Pr groups are different. Thus the corresponding probability is obtained by $P^*_{Si} = 6 \times P_4 \times P_5 \times P_6 = 6 \times 0.966159012 \times 0.021680 \times 0.010854 = 0.001364 = 0.1364\%$ ($\sim$0.14%). Consequently the probability of the structures achiral-on-Si are formed by the probability $P_{Si,achiral} = 1 - P^*_{Si} = 0.998636 = 99.8636\%$ ($\sim$99.86%).

Taking into regard the "own right" chirality of one of the above-mentioned *i*Pr groups, for comodity we write $\mathbf{S_4}$ for structure corresponding to probability P_4 and similarly $\mathbf{S_5}$ and $\mathbf{S_6}$ for the species with P_5 and P_6 respectively, we obtain six combinations: not only $\mathbf{S_4}$; $\mathbf{S_5}$ and $\mathbf{S_6}$ as discussed above, but also five additional cases. These appear with the following probabilities:

$P(\mathbf{S_4\ S_4\ S_5}) = 3 \times 0.966159012^2 = 0.060712449 = 6.0712449\%$

$P(\mathbf{S_4\ S_5\ S_5}) = 0.0011362349 = 0.11362349\%$

$P(\mathbf{S_5\ S_5\ S_5}) = 0.0000102 = 0.00109\%$

$P(\mathbf{S_5\ S_5\ S_6}) = 0.0000153 = 0.00153\%$

$P(\mathbf{S_5\ S_6\ S_6}) = 0.00000766232 = 0.000766232\%$

Thus the sum probabilities of all chiral cases amounts to $P_{\sum chiral} = 0.063471954 = 6.3471954\%$ ($\sim$6.35%). Consequently, the probability of the achiral structures is:

$$P_{\Sigma achiral} = 1 - P_{\Sigma chiral} = 1 - 6.3471954 = 0.936528046\% \ (\sim 93.65\%).$$

The presence of the ~6.35% chiral species in substrate **1c** and in product/autocatalyst **3c** can by no means neglected. It should be noted, however, that this quantity of chiral species is a mixture of several enantio- and diastereoisotopomers, which could have a different effect on the critical stereochemical steps of the corresponding Soai reaction. The size of the influence of these isotopomers could be checked either by the synthesis of isotopically enriched model compounds (which appears to be a very laborious approach) or by the use of isotopically pure reagents (containing only ^{1}H and ^{12}C derivatives, which also could be realized from commercially available starting compounds).

The *n*-Bu group in substrate **1d** (and product/autocatalyst **3d**) can become chiral by mono-deuterium substitution in the CH_2 groups. Combinatorial analysis gives 7776 achiral and 116,640 chiral isotopomeric structures. Thus the probability of formation of these mono-deuterated groups is $P(C^{1}H^{2}H) = 2 \times 0.999844 \times 0.000156 = 0.0003012 = 0.3012\%$. The probability of the achiral homo-isotopic methylene groups is $P(C^{1}H_2$ or $C^{2}H_2) = 0.999688$. Thus achiral is $P_{achiral} = 0.999688^2 = 0.999064 = 99.9064\%$ (~99.91%) and $P_{chiral} = 0.000936 = 0.0936\%$ (~0.094%). Substitution by ^{13}C does not generate chiral center in this group.

Adamantyl derivatives **1e** and **3e** represent a more complicated challenge from the viewpoint of isotopic chirality. The CH_x ($x = 0,1,2$) fragments in the adamantyl group can be classified in four sets (Fig. 14.2). The first of these sets is the "naked" C atom (C−1 in Fig. 14.2), the second set is the regular triangle of the three CH_2 groups directly linked to the quaternary carbon (C−2, C−8, and C−9 in Fig. 14.2), the third set

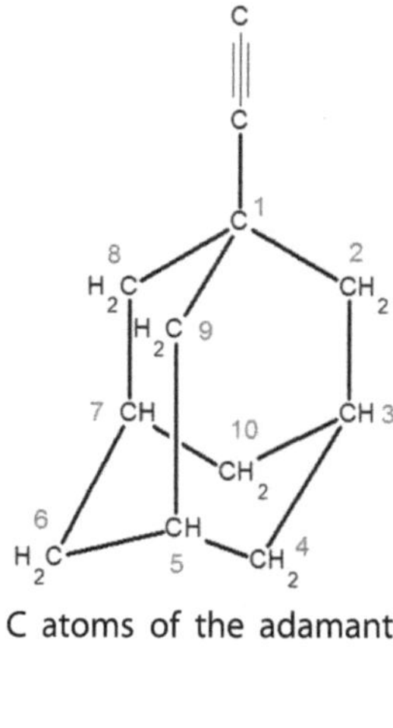

Figure 14.2 Notation of the C atoms of the adamantyl group in substrate **1e** (and product/autocatalyst **3e**).

is the regular triangle consisting of the three methylene groups, parallel to the former one (C−3, C−5, and C−7 in Fig. 14.2) and the fourth set is again a regular triangle, parallel to the previous ones, consisting of the remaining three methylene groups (C−4, C−6, and C−10). The geometry allows various symmetry elements, e.g., a C_{3v} axis through the quaternary carbon and the middle points of these triangles, but we shall use only the symmetry plane containing the C−1/C−2 and C−2/C−3 lines and the middle point of the C−8/C−9 distance. We shall calculate the probability of the symmetric (achiral) array with respect to this symmetry plane. By combinatorial analysis 708,588 achiral and 13,427,640 chiral structures can be identified.

For the first plane, the probalities are as follows: $P(^{12}C^{1}H_2) = 0.98858 = 98.858\%$; $P(^{13}C^{1}H_2) = 0.011106 = 1.1106\%$; the di-deutarated cases can be neglected with probabilities of the order of 10^{-8}. For the symmetry towards the plane defined above, one needs to have the *same* groups in positions C−8 and C−9, that is the probability decreases: $P_{8/9\ \mathrm{achiral}} = 0.98858^2 + 0.011106^2 = 0.97741 = 97.741\%$ (~97.74%), while the probability of the chiral cases is: $P_{8/9\ \mathrm{chiral}} = 1 - P_{8/9\ \mathrm{achiral}} = 0.02259 = 2.259\%$ (~2.26%).

For the third set (second triangle) the following cases were taken into account: $^{12}C^{1}H$, $^{12}C^{2}H$, and $^{13}C^{1}H$ (case $^{13}C^{2}H$ was neglected because of low probability), this results in $P_{5/7\ \mathrm{achiral}} = P^2(^{12}C^{1}H) + P^2(^{13}C^{1}H) + P^2(^{12}C^{2}H) = 0.988735^2 + 0.11108^2 + 0.00015^2 = 0.97772 = 97.772\%$ (~97.78%), consequently the probability of the chiral cases is $P_{5/7\ \mathrm{chiral}} = 1 - P_{5/7\ \mathrm{achiral}} = 0.02228 = 2.228\%$ (~2.23%).

The fourth set, C−4, C−6, C−10 gives numerically the same results as the second set (first plane).

The probability of satisfaction of all symmetry conditions gives:

$$P_{\sum \mathrm{achiral}} = 2 \times (0.97741^2) \times 0.97772 = 0.892322 = 89.2322\%\ (\sim 89.23\%)$$

for the achiral (symmetric) structures and:

$$P_{\sum \mathrm{chiral}} = 1 - P_{\sum \mathrm{achiral}} = 0.107678 = 10.7678\%\ (\sim 10.77\%)$$

for the chiral isotopomers.

The organometallic derivatives **1f** (substrate) and **3f** (product/autocatalyst) represent a very interesting stereochemical situation.

First of all, if any of the C or H atoms from the four CH groups in the substituted ring gets isotopically substituted, the Fe atom becomes chiral (having four different substituents at the points of a deformed tetrahedron). The probability of such a situation can be directly calculated from the natural abundances of the corresponding atoms (Table 14.1):

$$P_{\mathrm{Fe,chiral}} = 4 \times 0.015 + 4 \times 1.111 = 4.504\% \quad (\sim 4.5\%);$$

while the probability of achiral Fe is:

$$P_{\mathrm{Fe,achiral}} = 100 - P_{\mathrm{Fe,chiral}} = 100 - 4.504 = 95.496\% \quad (\sim 95.5\%).$$

Due to the fluxional structure of ferrocene and most of its derivatives [24] we did not analyze the conditions of the C_5H_5 ring, only the substituted C_5H_4 moiety.

The probabilities of the isotopically substituted species in this ring are as follows:

$P(^{12}C^{1}H) = 0.988736$; $P(^{12}C^{2}H) = 0.00015427$; $P(^{13}C^{1}H) = 0.01110827$; $P(^{13}C^{2}H) = 0.0000017332$, where the last one will be omitted again because of very low probability. Circumventing the ring, starting from the "naked" C atom (C−1), this can be one of the two C isotopes, proceeding further one finds at each angle three possibilities, giving a total of cases $2 \times 3^4 = 162$. From these, those will result achiral structures (9), where the neighbors are strictly equal. This will give a probability $P_{\mathrm{achiral}} = 0.95594 = 95.594\%$ and consequently $P_{\mathrm{chiral}} = 1 - 0.95594 = 0.04406 = 4.406\%$. The probabilities of Fe-chirality and ring-chirality are independent of each other, therefore the sum of the probabilities of getting chirality in the ferrocenyl group by isotopic substitution is:

$$P_{\mathrm{ferrocenyl}} = 4.504 + 4.406 = 8.91\%.$$

The above analysis has shown that in most of the substituents of the pyrimidyl ring the most popular Soai-substrates (and consequently in the product/autocatalyst molecules), various enantio- and diastereoisotopomers, can be formed on the basis of natural abundance of H and C stable isotopes, with quantities reaching ~10%. This fact should be taken in consideration in future interpretations of the Soai reaction.

14.2.3 Zinc Isotopic Chirality in Intermediates of the Soai Reaction

Almost all speculations about the mechanism of Soai's asymmetric autocatalysis [18] hypotheses intermediates with more than one Zn atoms. Some of these appear to contain structural fragments where isotopic chirality may develop on the basis of stable isotopes of the Zn atoms. We analyzed [2] one of these (hypothetic) molecules, resulting from the first-principle calculations

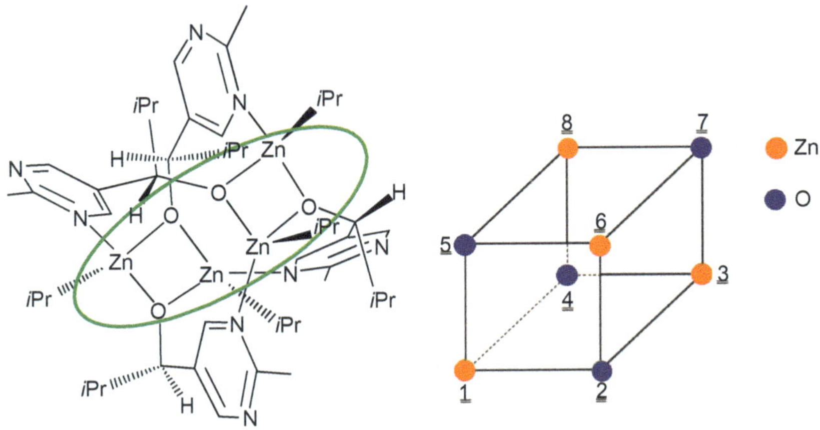

Figure 14.3 Schematic structures of intermediate **B-1** [18e] and its Zn_4O_4 fragment (**B-2**) [2].

of Schiaffino and Ercolani [18e], confirmed most recently by the X-ray structure diffraction study of Soai, Matsumoto et al. [25]. Schematic structures of intermediate (**B-1**) and the Zn_4 isotopically prochiral unit (**B-2**) are shown in Fig. 14.3.

The interesting fragment in this "tetranuclear" intermediate is a Zn_4O_4 polyhedron. In this "cubic" structure Zn and O atoms are distributed in the points of two immaginary tetrahedron, therefore both the stable natural isotopes of Zn and O can lead to chirality of this structure. Combinations of isotopes of these two elements could result in 4690 different chiral distributions, the sum of probabilities of these structures is $P_{\sum} = 0.002672827$ (~0.27%).

The above analysis showed *molecular* (fragment) enantioisotopomeric chirality. In some of the (hypothetic) Soai intermediates, however, there is an other type of metal-centered chirality: that is, if the metal in an approximately tetrahedral bonding environment is linked to four different substituents. This is, for example, the case in three of the seven intermediates suggested by Schiaffino and Ercolani [18e], including the compound discussed above (**B**) (as shown in Fig. 14.3). In this structure the Zn atoms become chiral if the oxygen atoms are different isotopes. We calculated the relative abundance of these Zn-centered enantioisotopomers. This was the result for one of the four Zn atoms

$P_{chiral} = 0.0048099$ (~0.5%) probability, that is for the four Zn atoms in the "cube" shown in Fig. 14.3 **B-2** the sum of the probabilities of metal chiralities is $P_{\text{"cube",chiral}} = 4 \times P_{chiral} = 0.0192396$ (~1.9%).

The probabilities obtained by the above calculations indicate that the metal-linked isotope chirality cannot be neglected in asymmetric autocatalysis. Suitable model experiments, however, are still lacking for the determination of the strength and chiral direction of these effects. Our preliminary calculations [2] concerning the expectable enantiomeric excesses [8,19] demonstrated, that the quantity of excess chirality in these systems is well over the sensitivity threshold of the Soai reaction [11,20].

14.3 CONCLUSIONS

In the present study we summarize our earlier [1,2] and more recent calculations about possible isotopic chirality in the reagents and products, as well as intermediates of the Soai reaction (asymmetric autocatalysis). In most of the cases we found ~0.001%–10% probabilities of the formation of chiral components. This level is high enough to merit further investigation and experimental control. These quantities of chiral enantio- and diastereoisotopomers are the sum of the quantities of several compounds. The general picture is that a relatively low number of isomers are symmetric (achiral), and these provide the overwhelming majority of the chiral components and a very large number of combinations produce the asymmetric (chiral) components, which, however, are at most a fraction (max. 11.4%) or the total. Approximate calculations using the Pars-Mills equation [2,8,19] indicate that the expectable enantiomeric excesses deriving from the chiral components, in the range of milimol to femtomol size samples are of an order of magnitude that is much higher than the sensitivity threshold of the Soai reaction [16,20,21]. It remains, however an open question, whether the enantiomeric excesses composed of several different chiral isotopomers have the same influence as the same quantity of *one* chiral additive (or autocatalyst)?

Let us add a philosophical note. The Soai reaction 20 years ago was an unexpected and truly great discovery. As happens to almost all new discoveries, it answered a few important questions (e.g., provided a reasonable explanation for the problem of the origin of the biological chirality), but also created several new questions, such as, for example, those discussed in the present paper.

REFERENCES AND NOTES

[1] Barabás, B.; Kurdi, R.; Pályi, G. Natural Abundance Isotopic Chirality int he Reagents of the Soai Reaction. In *Third International Symposium on the Soai Reaction and Related Topic*; Felsőmocsolád, Hungary, September 2–5, 2015, Abstracts; p 26.
[2] Barabás, B.; Kurdi, R.; Pályi, G. Natural Abundance Isotopic Chirality in the Reagents of the Soai Reaction. *Symmetry* **2016,** *8*, 2.
[3] Griffith, H. *Stable Isotopes;* BIOS Scientific Publisher Ltd.: Oxford, 1998.
[4] Several natural or artificially generated radioactive isotopes occur in Nature or in modern technology, but these will not be considered in the present paper.
[5] Some recent reviews:
(a) Cleland, W. W. The Use of Isotope Effects to Determine Enzyme Mechanisms. *Arch. Biochem. Biophys.* **2005,** *433*, 2–12.
(b) Kohen, A.; Limbach, H.-H., Eds. *Isotope Effects in Chemistry and Biology;* Taylor & Francis – CRC Press: Boca Raton, FL, 2006.
(c) Galimov, E. N. Isotope Organic Geochemistry. *Org. Geochem.* **2006,** *37*, 1200–1262.
(d) Cook, P. F.; Cleland, W. W. *Enzyme Kinetics and Mechanism;* Garland Science – Taylor & Francis: London, 2007.
(e) Wolfsberg, M.; Van Hook, W. A.; Paneth, P.; Rebelo, L. P. N. *Isotope Effects in the Chemical, Geological and Bio Sciences;* Springer: Dordrecht, 2010.
(f) Van Hook, W. A. Isotope Effects in Chemistry. *Nukleonika* **2011,** *56*, 217–240.
(g) Gu, H.; Zhang, S. Advances in Kinetic Isotope Effect Measurement Techniques for Enzyme Mechanism Study. *Molecules* **2013,** *18*, 9278–9292.
(g) Mosin, O.; Ignatov, I. Studying Isotope Effects of Deuterium in Biological Objects. *Eur. Rev. Chem. Res.* **2015,** *3*, 25–42.
[6] There is a broad range of applications, recent examples:
(a) Lesot, P.; Serhan, Z.; Billault, I. Recent Advances in the Analysis of the Site-Specific Isotopic Fractionation of Metabolites such as Fatty Acids Using Anisotropic Natural-Abundance 2H-NMR Spectroscopy: Application to Conjugated Linolenic Methyl Esters. *Anal. Bioanal. Chem.* **2011,** *399*, 1187–1200.
(b) Umpleby, A. M. (Hormone Measurement Guidelines) Tracing Lipid Metabolism: The Value of Stable Isotopes. *J. Endocrinol.* **2015,** *226*, G1–G10.
(c) Hyodo, F. Stable Carbon and Nitrogen Isotopes in Insect Ecology. *Entomol. Sci.* **2015,** *18*, 295–312.
(d) Gentile, N.; Siegwolf, R. T. W.; Esseiva, P.; Doyle, S.; Zollinger, K.; Delémont, O. Isotope Ratio Mass Spectrometry as a Tool for Source Inference in Forensic Science: A Critical Review. *Forensic. Sci. Int.* **2015,** *251*, 139–158.
[7] (a) Pracejus, H. Steric Isotope Effects as Cause of Asymmetric Synthesis. *Tetrahedron Lett.* **1966,** 7, 3809–3813.
(b) Arigoni, D.; Eliel, E. L. Chirality Due to the Presence of Hydrogen Isotopes at Noncyclic Positions. *Top. Stereochem.* **1969,** *4*, 127–243.
(c) Green, M. M.; Park, J.-W.; Sato, T.; Lifson, S.; Selinger, R. L. B.; Selinger, J. V. The Macromolecular Route to Chiral Amplification. *Angew. Chem., Int. Ed.* **1999,** *38*, 3138–3154.
(d) Cantekin, S.; Balkende, D. W. R.; Smudlers, M. M. J.; Palmans, A. R. A.; Meijer, E. W. The Effect of Isotopic Substitution on the Chirality of a Self-Assembled Helix. *Nat. Chem.* **2011,** *3*, 42–46.
[8] Barabás, B.; Caglioti, L.; Micskei, K.; Zucchi, C.; Pályi, G. Isotope Chirality and Asymmetric Autocatalysis: A Possible Entry to Biological Chirality. *Orig. Life Evol. Biosph.* **2008,** *38*, 317–327.

[9] Observation of chirality in isotopomers.
(a) Lesot, P.; Lafon, O.; Courtieu, J.; Berdagué, P. Analysis of the 13C NMR Spectra of Molecules Chiral by Isotopic Substitution, Dissolved in a Chiral Oriented Environment: Towards the Absolute Assignment of the pro-R/pro-S Character of Enantiotopic Ligands in Prochiral Molecules. *Chem. Eur. J.* **2004,** *10*, 3741–3746.
(b) Baillif, V.; Robins, R. J.; Billault, I.; Lesot, P. Assignment of Absolute Configuration of Natural Abundance Deuterium Signals Associated with (R)- and (S)-Enantioisotopomers in a Fatty Acid Aligned in a Chiral Liquid Crystal: Enantioselective Synthesis and NMR Analysis. *J. Am. Chem. Soc.* **2006,** *128*, 11180–11187.
(c) Lesot, P.; Courtieu, J. Natural Abundance Deuterium NMR Spectroscopy: Developments and Analytical Applications in Liquids, Liquid Crystals and Solid Phases. *Prog. Nucl. Magn. Reson. Spectrosc.* **2009,** *55*, 128–159.
(d) Serhan, Z.; Martel, L.; Billault, J.; Lesot, P. Complete Determination of Natural Site-Specific Enantio-Isotopomeric Excesses in Linolenic Acid Using Natural Abundance Deuterium 2D NMR in Polypeptide Metaphases. *Chem. Commun.* **2010,** *46*, 6599–6601.
(e) Lesot, P.; Lafon, O. Experimental Detection of Achiral and Chiral Naturally Abundant 13C-2H Isotopomers by 2D-NMR in Liquids and Chiral Oriented Solvents. *Anal. Chem.* **2012,** *84*, 4569–4573.
(f) Lesot, P.; Aroulanda, C.; Zimmermann, H.; Luz, Z. Enantiotopic Discrimination in the NMR Spectrum of Prochiral Solutes in Chiral Liquid Crystals. *Chem. Soc. Rev.* **2015,** *44*, 2330–2375.
(g) Berdagué, P.; Lesot, P.; Jacob, J.; Terwillinger, V. J.; Le Milbeau, C. Contribution of NAD 2D NMR in Liquid Crystals to the Determination of Hydrogen Isotope Profile of Methyl Group sin Miliacin. *Geochim. Cosmochim. Acta* **2016,** *173*, 337–351.
[10] Asymmetric autocatalysis is such reaction, where in an achiral-to-chiral reaction the product accelerates its own formation (conventional catalysis) and promotes the formation of product molecules with its enantiomeric configuration (similarly to chiral "auxiliaries").
[11] (a) Soai, K.; Shibata, T.; Morioka, H.; Choji, K. Asymmetric Autocatalysis and the Amplification of Enantiomeric Excess of a Chiral Molecule. *Nature* **1995,** *378*, 767–768.
(b) Shibata, T.; Morioka, H.; Hayase, T.; Choji, K.; Soai, K. Highly Enantioselective Catalytic Asymmetric Automultiplication of Chiral Pyrimidyl Alcohol. *J. Am. Chem. Soc.* **1996,** *118*, 471–472.
(c) Soai, K.; Shibata, T.; Kowata, Y. Asymmetric Synthesis of Enantioenriched Alkanol by Spontaneous Asymmetric Synthesis. *Jpn. Kokai Tokkyo Koho* **1997,** *9* (268), 179.
(d) Shibata, T.; Yonekubo, S.; Soai, K. Practically Perfect Asymmetric Autocatalysis with (2-Alkynyl-5-pyrimidyl)alkanols. *Angew. Chem., Int. Ed.* **1999,** *38*, 659–661.
[12] (a) Sato, I.; Osanai, S.; Kadowaki, K.; Sugiyama, T.; Shibata, T.; Soai, K. Asymmetric Autocatalysis of Pyrimidyl Alkanol Induced by Optically Active 1,1′-Binaphthyl an Atropisomeric Hydrocarbon Generated from Spontaneous Resolution on Crystallization. *Chem. Lett.* **2002,** 168–169.
(b) Kawasaki, T.; Hatase, K.; Fujii, Y.; Yo, K.; Soai, K.; Pizzarello, S. The Distribution of Chiral Asymmetry in Meteorites: An Investigation Using Asymmetric Autocatalytic Chiral Sensors. *Geochim. Cosmochim. Acta* **2006,** *70*, 5395–5402.

(c) Sato, I.; Ohgo, Y.; Igarashi, H.; Nishiyama, D.; Kawasaki, T.; Soai, K. Determination of Absolute Configurations of Amino Acids by Asymmetric Autocatalysis of 2-Alkynylpyrimidyl Alkanol as a Chiral Sensor. *J. Organomet. Chem.* **2007,** *692*, 1783–1787.

(d) Shibata, T.; Iwahashi, K.; Kawasaki, T.; Soai, K. Chiral Secondary Alcohol-Induced Asymmetric Autocatalysis: Correlation between the Absolute Configuration of the Chiral Initiators and the Product. *Tetrahedron: Asymmetry* **2007,** *18*, 1759–1762.

[13] (a) Kawasaki, T.; Tanaka, H.; Tsutsumi, T.; Kasahara, T.; Sato, I.; Soai, K. Chiral Discrimination of Cryptochiral Saturated Quaternary and Tertiary Hydrocarbons by Asymmetric Autocatalysis. *J. Am. Chem. Soc.* **2006,** *128*, 6032–6033.

(b) Kawasaki, T.; Hohberger, C.; Araki, Y.; Hatase, K.; Beckerle, K.; Okuda, J., et al. Discrimination of Cryptochirality in Chiral Isotactic Polystyrene by Asymmetric Autocatalysis. *Chem. Commun.* **2009,** 5621–6523.

[14] Sato, I.; Omiya, D.; Saito, T.; Soai, K. Highly Enantioselectie Synthesis Induced by Chiral Primary Alkanols Due to Deuterium Substitution. *J. Am. Chem. Soc.* **2000,** *122*, 11739–11740.

[15] (a) Kawasaki, T.; Shimizu, M.; Nishiyama, D.; Ito, M.; Ozawa, H.; Soai, K. Asymmetric Autocatalysis Induced by Meteoritic Amino Acids with Hydrogen Isotope Chirality. *Chem. Commun.* **2009,** 4396–4398.

(b) Kawasaki, T.; Matsumura, Y.; Tsutsumi, T.; Suzuki, K.; Ito, M.; Soai, K. Asymmetric Autocatalysis Triggered by Carbon Isotope (13C/12C) Chirality. *Science* **2009,** *324*, 492–495.

(c) Kawasaki, T.; Okano, Y.; Suzuki, E.; Takano, S.; Oji, S.; Soai, K. Asymmetric Autocatalysis Triggered by Chiral Isotopomer Arising from Oxygen Isotope Substitution. *Angew. Chem., Int. Ed.* **2011,** *50*, 8131–8133.

(d) Kawasaki, T.; Soai, K. Asymmetric Induction Arising from Enantiomerically Enriched Carbon-13 Isotopomers and Highly Selective Chiral Discrimination by Asymmetric Autocatalysis. *Bull. Chem. Soc. Jpn.* **2011,** *84*, 879–892.

(e) Kawasaki, T.; Soai, K. Asymmetric Autocatalysis Triggered by Chiral Crystals Formed from Achiral Compounds and Chiral Isotopomers. *Isr. J. Chem.* **2012,** *52*, 582–590.

(f) Matsumoto, A.; Oji, S.; Takano, S.; Tada, K.; Kawasaki, T.; Soai, K. Asymmetric Autocatalysis Triggered by Isotopically Chiral Glycerin. *Org. Biomol. Chem.* **2013,** *11*, 2928–2931.

(g) Ozaki, H.; Tada, K.; Harada, S.; Ayugase, T.; Ozawa, H.; Kawasaki, T.; et al. Asymmetric Autocatalysis Triggered by Nitrogen Isotope Chirality. In *Third International Symposium on the Soai Reaction and Related Topic*; Felsőmocsolád, Hungary, September 2–5, 2015, Abstracts; pp 24–25.

[16] (a) Soai, K.; Shibata, T. Asymmetric Autocatalysis and Biomolecular Chirality. In *Advances in BioChirality;* Pályi, G., Zucchi, C., Caglioti, L., Eds.; Elsevier: Amsterdam, 1999; pp 125–136.

(b) Soai, K.; Shibata, T.; Sato, I. Enantioselective Automultiplication of Chiral Molecules by Asymmetric Autocatalysis. *Acc. Chem. Res.* **2000,** *33*, 382–390.

(c) Soai, K. Asymmetric Autocatalysis and the Origin of Chiral Homogeneity of Biologically Relevant Molecules. In *Fundamentals of Life;* Pályi, G., Zucchi, C., Caglioti, L., Eds.; Éditions Scientifiques et Médicales Elsevier SAS: Paris, 2002; pp 427–435.

(d) Sato, I.; Yanagi, T.; Soai, K. Highly Enantioselective Autocatalysis of 2-Alkenyl- and 2-Vinyl-5—pyrimidyl Alkanols with Significant Amplification of Enantiomeric Excess. *Chirality* **2002,** *14*, 166–168.

(e) Soai, K. Asymmetric Autocatalysis, Absolute Asymmetric Synthesis and the Origin of Homochirality in Biomolecules. In *Progress in Biological Chirality;* Pályi, G., Zucchi, C., Caglioti, L., Eds.; Elsevier: Oxford, 2004; pp 355–364.
(f) Lutz, F.; Kawasaki, T.; Soai, K. Asymmetric Autocatalysis of a Ferrocene-Containing Chiral Compound with Amplification of chirality. *Tetrahedron: Asymmetry* **2006,** *17*, 486–490.
(g) Blackmond, D. G. Mechanistic Study of the Soai Autocatalytic Reaction Informed by Kinetic Analysis. *Tetrahedron: Asymmetry* **2006,** *17*, 584–589.
(h) Busch, M.; Schlageter, M.; Weigand, D.; Gehring, T. Systematic Studies Using 2-(1-Adamantylethynyl)pirimidine-5-carbaldehyde as a Starting Material in Soai's Asymmetric Autocatalysis. *Chem. Eur. J.* **2009,** *15*, 8251–8258.
(i) Stiller, J.; Tse, E. S. Y.; Schlageter, M.; Gehring, T. Systematic and Mechanistic Studies on the Soai Reaction Using 2-(1-Adamantylethynyl-)pyrimydyne-5-carbaldehyde. In *The Soai Reaction and Related Topic;* Pályi, G., Zucchi, C., Caglioti, L., Eds.; Edizioni Artestampa: Modena, 2012.
(j) Soai, K.; Kawasaki, T.; Matsumoto, A. The Origins of Homochirality Examined Using Asymmetric Autocatalysis. *Chem. Rec.* **2014,** *14*, 70–83.
(k) Soai, K.; Kawasaki, T.; Matsumoto, A. Asymmetric Autocatalysis of Pyrimidyl Alcohol and Its Application to the Study of the Origin of Homochirality. *Acc. Chem. Res.* **2014,** *47*, 3643–3654.
[17] It should be pointed out, that in spite of several interesting experimental and theoretical efforts the mechanism of the Soai reaction is actually far to be cleared up, yet(?).
[18] See, for example:
(a) Gridnev, I. D.; Serafimov, J. M.; Brown, J. M. Solution Structure and Reagent Binding of the Zinc Alkoxide Catalyst in the Soai Asymmetric Autocatalytic Reaction. *Angew. Chem., Int. Ed.* **2004,** *43*, 4884–4887.
(b) Gridnev, I. D.; Brown, J. M. Asymmetric Autocatalysis: Novel Structures, Novel Mechanism. *Proc. Natl. Acad. Sci. U. S. A.* **2004,** *101*, 5727–5731.
(c) Gridnev, I. D. Chiral Symmetry Breaking in Chiral Crystallization and Soai Autocatalytic Reaction. *Chem. Lett.* **2006,** 148–153.
(d) Brown, J. M.; Gridnev, I. D.; Klankermayer, J. Asymmetric Autocatalysis with Organozinc Complexes; Elucidation of the Reaction Pathway. *Top. Curr. Chem.* **2008,** *284*, 35–65.
(e) Schiaffino, L.; Ercolani, G. Unraveling the Mechanism of the Soai Asymmetric Autocatalytic Reaction by First-Principles Calculations: Induction and Amplification of Chirality by Self-Assembly of Hexanuclear Complexes. *Angew. Chem., Int. Ed.* **2008,** *47*, 6832–6835.
(f) Schiaffino, L.; Ercolani, G. Mechanism of the Asymmetric Autocatalytic Soai Reaction Studied by Density Functional Theory. *Chem. Eur. J.* **2010,** *16*, 3147–3156.
(g) Quaranta, M.; Gehring, T.; Odell, B.; Brown, J. M.; Blackmond, D. G. Unusual Inverse Temperature Dependence of Reaction Rate in the Asymmetric Autocatalytic Alkylation of Pyrimidyl Aldehydes. *J. Am. Chem. Soc.* **2010,** *132*, 15104–15107.
(h) Gridnev, I. D.; Vorobiev, A. K.; Raskatov, J. A. Role of Oligomerization in Soai Reaction: Structures, Energies, Possible Reactivity. In *The Soai Reaction and Related Topic;* Pályi, G., Zucchi, C., Caglioti, L., Eds.; Edizioni Artestampa: Modena, 2012; pp 79–122.
(i) Schiaffino, L.; Ercolani, G. Mechanistic Insights Into the Soai Reaction from Formal Kinetics and Density Functional Theory Calculations. In *The Soai Reaction and Related Topic;* Pályi, G., Zucchi, C., Caglioti, L., Eds.; Edizioni Artestampa: Modena, 2012; pp 331–346.

[19] Mills, W. H. Some Aspects of Stereochemistry. *Chem. Ind. (London)* **1932,** *51,* 750–759.
[20] (a) Shibata, T.; Yonekubo, S.; Soai, K. Practically Perfect Asymmetric Autocatalysis with (2Alkynyl-5-pyrimidyl)alkanols. *Angew. Chem., Int. Ed.* **1999,** *38,* 659–661.
(b) Soai, K.; Sato, I.; Shibata, T.; Komiya, S.; Hayashi, M.; Matsueda, Y., et al. Asymmetric Synthesis of Pyrimidyl Alkanol without Adding Chiral Substances by the Addition of Diisopropylzinc to Pyrimidine-5-carboxaldehyde in Conjunction with Asymmetric Autocatalysis. *Tetrahedron: Asymmetry* **2003,** *14,* 185–188.
(c) Sato, I.; Urabe, H.; Ishiguro, S.; Shibata, T.; Soai, K. Amplification of Chirality from Extremely Low Level to Greater than 99.5 % ee. *Angew. Chem., Int. Ed.* **2003,** *42,* 315–317.
(d) Kawasaki, T.; Suzuki, K.; Shimizu, M.; Ishikawa, K.; Soai, K. Spontaneous Asymmetric Synthesis int he Presence of Achiral Silica Gel in Conjunction with Asymmetric Autocatalysis. *Chirality* **2006,** *18,* 479–482.
[21] (a) Micskei, K.; Maioli, M.; Zucchi, C.; Caglioti, L.; Pályi, G. Generalization Possibilities of Autocatalytic Absolute Enantioselective Synthesis. *Tetrahedron: Asymmetry* **2006,** *17,* 2960–2962.
(b) Caglioti, L.; Micskei, K.; Pályi, G. Chirality of the Very First Molecule in Absolute Enantioselective Synthesis. *Viva Orig.* **2007,** *35,* 82–84.
(c) Micskei, K.; Rábai, G.; Gál, E.; Caglioti, L.; Pályi, G. Oscillatory Symmetry Breaking in the Soai Reaction. *J. Phys. Chem. B* **2008,** *112,* 9196–9200.
(d) Caglioti, L., et al. First Molecules, Biological Chirality, Origin(s) of Life. *Chirality* **2011,** *23,* 65–68.
(e) Caglioti, L.; Pályi, G. Single Chiral Molecule as Possible Starting Element of Complex Chiral Systems. *Rend. Lincei, Sci. Fis. Nat.* **2013,** *24,* 191–196.
[22] Micura, R.; Bolli, M.; Eschenmoser, A. Pyranosyl-RNA: Chiroselective Self-Assembly of Base Sequences by Ligative Ologomerization of Tetranucleotide-2′,3′-cyclophosphates (with a Commentary Concerning the Origin of Biomolecular Homochirality). *Chem. Biol.* **1997,** *4,* 309–320.
[23] Caglioti, L.; Hajdu, C.; Hoczknecht, O.; Zékány, L.; Zucchi, C.; Micskei, K., et al. The Concept of Racemates and the Soai Reaction. *Viva Orig.* **2006,** *34,* 62–80.
[24] Gielen, M.; Willem, R.; Wrackmeyer, B., Eds. *Fluxional Organometallic and Coordination Compounds;* Wiley: Hoboken, NJ, 2004Chapter 7.8.5. Ferrocene.
[25] Matsumoto, A.; Abe, T.; Hara, A.; Tobita, T.; Sasagawa, T.; Kawasaki, T., et al. Crystal Structure of the Isopropylzinc Alkoxide of Pyrimidyl Alkanol: Mechanistic Insights for Asymmetric Autocatalysis with Amplification of Enantiomeric Excess. *Angew. Chem., Int. Ed.* **2015,** *54,* 15218–15221.

CHAPTER 15

Chiral Polythiophenes

Francesca Parenti[1], Luisa Schenetti[2] and Francesco Tassinari[2]
[1]Dipartimento di Scienze Chimiche e Geologiche, Università di Modena e Reggio Emilia, Modena, Italy
[2]Dipartimento di Scienze della Vita, Università di Modena e Reggio Emilia, Modena, Italy

Abstract

This review reports the synthesis and the properties of chiral polythiophenes (PTs) and in particular it is focused on PTs bearing a sulfur atom directly bonded to the thiophene ring. These polymers show remarkable properties, related to their chirality and electrical conductivity and to the ability to interact with chiral molecules opening the possibility for using them in new potential applications, like chiral sensors, chiral resolution and electrodes.

Keywords: Polythiophene; chirality; synthesis; spin filter

15.1 INTRODUCTION

Chiral biopolymers, such as protein nucleic acids and DNA, are well known and widespread in nature. The majority of these natural biopolymers possess helical conformations with a preferable twist. This is related to the fact that the constituent molecules possess a specific homochirality; e.g., sugars are D and aminoacids are L. For this reason a polypeptide formed by L-aminoacid residues has a right-handed helix. The presence of only a one-handed helical conformation is of great importance when biospecific interactions in enzymatic reactions are concerned.

In recent decades, many efforts have been made in the design and synthesis of chiral polymers with the aim to mimic the behavior of biological polymers and chiral conductive polypyrrole (PPy), polythiophene (PT) and polyaniline (PAn) have attracted particular attention due to their relative ease of synthesis and good environmental stability, besides their chiroptical properties. Starting from the discovery of polyacetylene in the late 1970s [1], the research around organic conducting polymers grew increasingly.

We have to wait till 1985 to have the first examples of chiral conductive polymers [2], showing some form of chirality in their structure. The field has grown over time and today many other polymers have been developed and studied that show not only a high electrical conductivity

Advances in Asymmetric Autocatalysis and Related Topics
DOI: http://dx.doi.org/10.1016/B978-0-12-812824-4.00015-0

but also other interesting properties, thus opening a whole new frontline of possible applications.

15.2 RESULTS AND DISCUSSION

15.2.1 Chiral Conducting Polythiophenes

15.2.1.1 Synthesis

Chiral PT may be synthesized using the same general procedures applied for the synthesis of PTs.

There are several methods for the synthesis of PTs. As the properties of polymers and also the chirality strongly depend on their regiochemistry, it is crucial to choose appropriate synthetic pathways in generating regioregular PTs. A regioregular polymer is characterized by the same repetitive sequence of junctions, and the defects can be analyzed on the basis of four different triads. When the starting material for polymerization is a beta-substituted thiophene, the regiochemistry of the polymer can be described through the four triads arising from the three possible α,α'-coupling of three substituted thiophenes as reported in Fig. 15.1.

When HH coupling is concerned (e.g., regioisomers TT–HH and HT–HH) the strong steric interaction between R groups causes a deviation from planarity and a twisting of the polymer chain is observed. The effect of this twist is a decrease in conjugation length together with a worse electrical conductivity, crystallinity and chirality. In contrast, the regioisomers showing HT–HT junctions, where the substituents are far

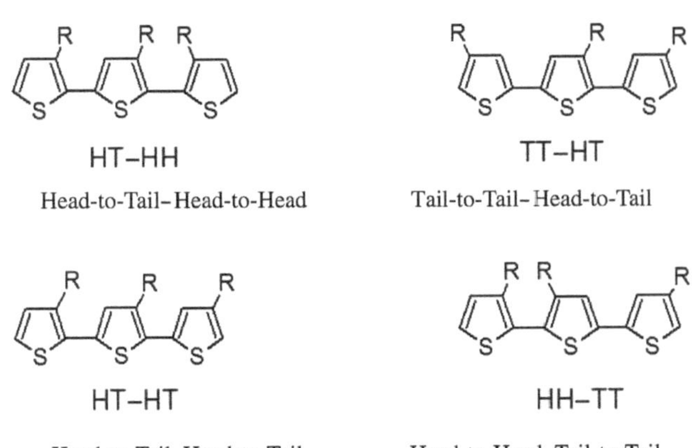

Figure 15.1 The four triads.

apart, are less twisted and can adopt an almost planar arrangement of the polymer backbone.

Regioregular PTs can be generated starting from properly functionalized thiophenes or oligothiophenes, in view of the desired regiochemistry. Different synthetic approaches are reported in literature, which give products with different degrees of regioregularity.

15.2.1.1.1 Coupling of Grignard Reagents with α-Thienyl Bromides in the Presence of a Ni(II) Catalyst

Kumada cross-coupling [3] (Fig. 15.2) affords regiorandom poly(3-alkylthiophenes) starting from 2,5-dibromo-3-alkylthiophene [4].

The monomer is treated with one equivalent of Mg in THF, forming a mixture of Grignard species. A catalytic amount of Ni(dppp)Cl_2 is then added and the polymer is generated through halo-Grignard coupling reaction. This has been the first successful method used in preparing poly(3-alkylthiophenes).

Polyalkylthiophenes obtained through the McCullough method [5] (Fig. 15.2) have nearly 100% HT–HT couplings. In this method, 2-bromo-3-alkylthiophene after treatment with LDA followed by addition of $MgBr_2.OEt_2$ forms 2-bromo-5-(bromomagnesio)-3-alkylthiophene, which is polymerized with catalytic amounts of Ni(dppp)Cl_2 using the Kumada cross-coupling method to give regioregular polyalkylthiophene. The key feature of the synthesis is the selective metallation with LDA of

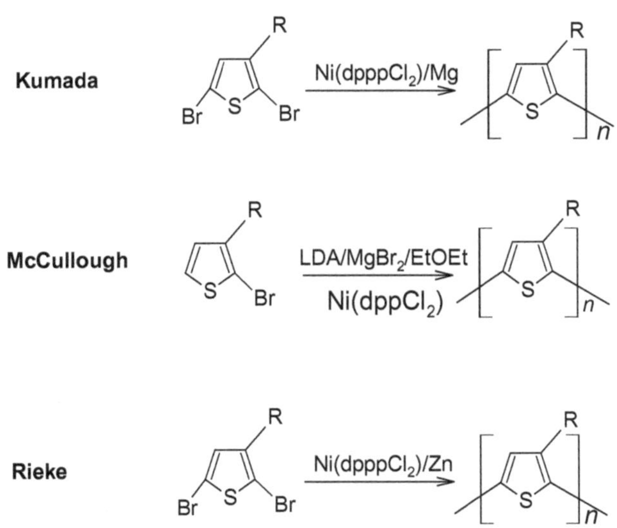

Figure 15.2 Coupling of Grignard reagents with thienyl bromides.

the monomer, which is stable and doesn't undergo halogen-metal exchange at $-78^{\circ}C$. This method has been modified by the use of the Grignard reagent generated by metathesis of 2,5-dibromo-3-alkylthiophene with MeMgBr [6], which gives polymers with 98% regioregularity, although the Grignard reagents have limited functional group tolerance.

In the Rieke method [7], (Fig. 15.2) a 2,5-dibromo-3-alkylthiophene is added to a solution of highly reactive in situ generated "Rieke Zinc" (Zn*). This metal reacts quantitatively to form a mixture of the isomers, 2-bromo-3-alkyl-5-(bromozincio)thiophene and 2-(bromozincio)-3-alkyl-5-bromothiophene. The ratio between these two isomers depends on the reaction temperature and on the steric hindrance of the alkyl group. Although there is no risk of metal-halogen exchange, cryogenic conditions must still be employed because the ratio of isomers produced is greatly affected by temperature. The addition of $Ni(dppe)Cl_2$ as catalyst leads to the formation of a regioregular HT–HT polyalkylthiophene, whereas addition of $Pd(PPh_3)_4$ as a catalyst will result in the formation of a completely regiorandom polyalkylthiophene. Regioregular HT–HT PT is formed starting from 2-iodo-5-bromo-3-alkylthiophene derivative, because the Rieke Zinc reacts only with an iodine atom.

15.2.1.1.2 Coupling of Boronic Acids With α-Thienyl Bromides in the Presence of a Pd(0) Catalyst (Suzuki Coupling)

The mild conditions of this type of reaction [8] are compatible with a lot of functional groups, while the more strong conditions required for the McCullough and Rieke methods are not. Although an extensive application of this method is found in the synthesis of oligothiophenes and in several copolymerization processes, the use of this synthetic approach to generate PTs has proven more difficult both owing to deboronation of thiophene boronates and to the aryl group transfer from phosphine ligands occurring in the termination of the polymerization [9].

Guillerez et al. [10] describe the synthesis of a regioregular (96%–97% HT–HT) PT starting from a boronate derivative of 2-iodo-3-octylthiophene (Fig. 15.3).

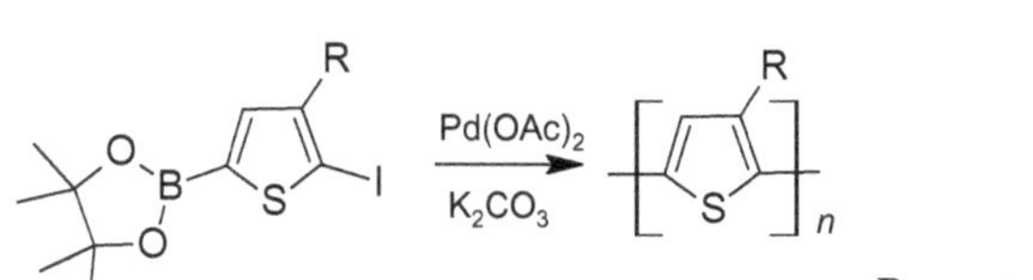

Figure 15.3 Example of polythiophene obtained by Suzuki coupling.

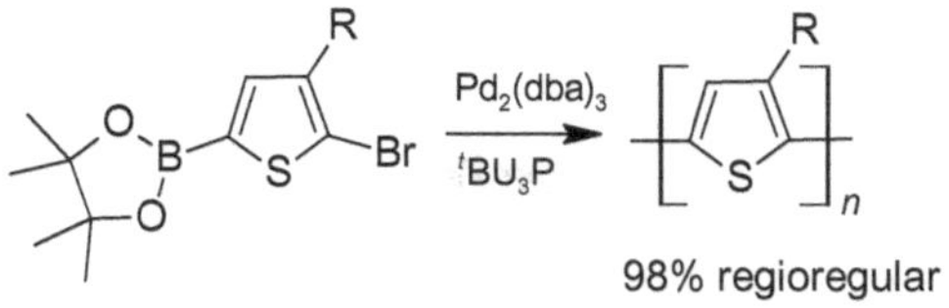

Figure 15.4 Example of polythiophene obtained by Suzuki coupling.

More recently Liversedge et al. [11] obtained a highly HT–HT regioregular polyalkylthiophene using an electron-rich phosphane ligand as a catalyst for the Suzuki coupling (Fig. 15.4).

15.2.1.1.3 Coupling of Stannyl-Derivatives with α-Thienyl Bromides in the Presence of a Pd(0) Catalyst (Stille Coupling)

The Stille reaction [12] involves the coupling of organic halides with organotin compounds catalyzed by a Pd(0) catalyst. The reaction tolerates many functional groups, such as amines, aldehydes, esters, ethers, and nitro groups, but suffers from the use of a toxic stannyl-derivative. Furthermore, it is well known that bromine and the trimethylstannyl group can interchange generating different species in solution that may be responsible of defects in the polymer.

The Stille method has been successfully applied to the catalyzed reaction of β-disubstituted or unsubstituted thiophenes [13]. Moreover, it was found [14] that Pd(0) catalyzed cross coupling reaction of 2-iodo-3-hexyl–5-tri-*n*-butylstannylthiophene lead to poly(3-hexylthiophene) with an HT–HT coupling percentage exceeding 96% (Fig. 15.5). Today the Stille coupling is one of the most widely used polymerization methods for the obtainment of different thiophene-based copolymers, although new tin-free direct heteroarylation polymerization methods are merging [15].

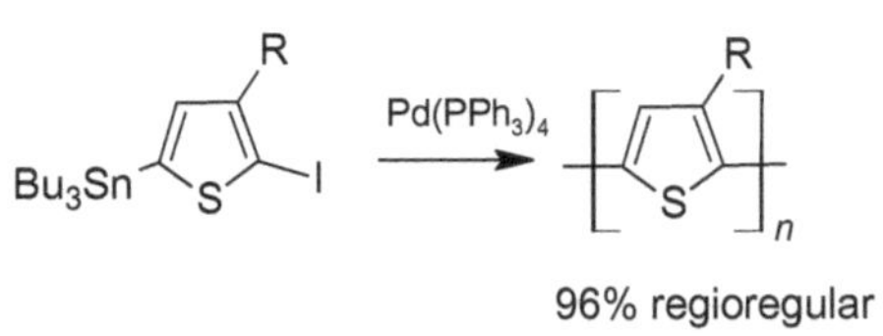

Figure 15.5 Example of polythiophene obtained by Stille coupling.

15.2.1.1.4 Oxydative Coupling With $FeCl_3$

It is assumed that, according to the mechanism of the electrochemical polymerization, the first step in the oxidative coupling is the oxidation of the neutral monomer to radical cation [16] (see Fig. 15.6).

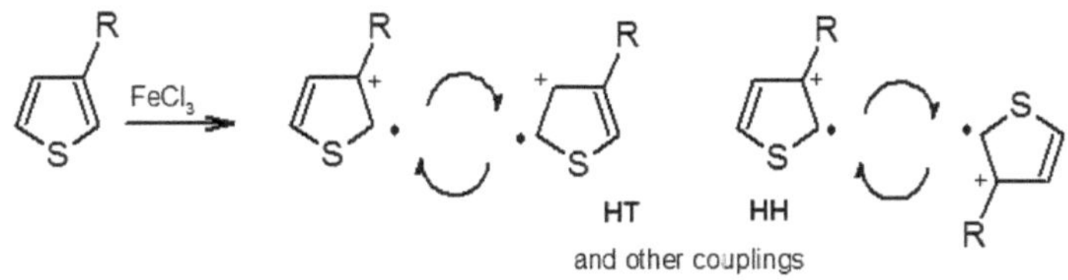

Figure 15.6 Oxidation of neutral monomer to radical cation and possible couplings to generate dimers.

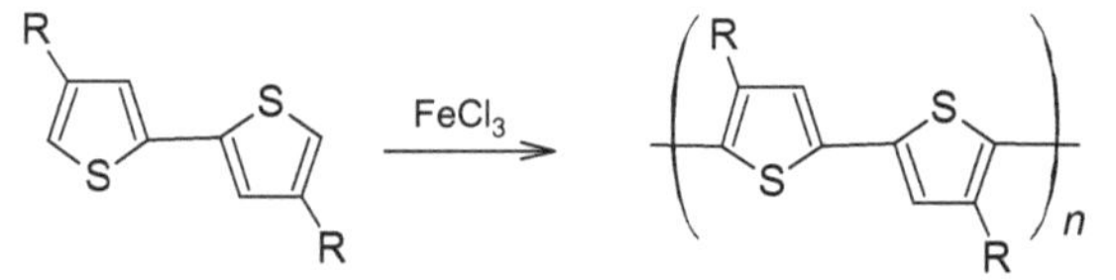

Figure 15.7 Regioregular HH–TT polyalkylthiophenes.

Radicals in positions 2 and 5 are preferably formed and therefore the coupling involves mainly α-positions and only rarely the β ones. A drawback in this type of polymerization is that if the starting monomer is a β-substituted thiophene, regiorandom PTs may be generated. Nevertheless, examples of oxidative polymerization of 3-octylthiophene producing 84% of HT–HT couplings have been reported [17]. The problem of the regio-irregularity can be solved by using bithiophenes with TT or HH junctions as starting materials (Fig. 15.7).

One of the major problems in dealing with $FeCl_3$ is that the oxidative method gives variable results. The reproducibility of the reaction has been examined [18]. The polymerization of 3-octylthiophene with $FeCl_3$ repeated under identical reaction conditions can give polymers with different molecular weights and different contents in terms of Fe impurities. It has been reported that unwanted incorporation of even a fraction per million of $FeCl_3$ into the reaction products could alter their electric properties. The Fe impurities affect PTs' device performance [19] in field effect transistors and in LEDs. Nevertheless, other important physical properties of these materials, such as linear and nonlinear optical properties, should not be altered by slight contamination with $FeCl_3$. Moreover, it should be noted that unwanted contamination by the metals used in the reactants or in the catalysts is always possible depending on the way in which OTs and PTs are synthesized. Today the fine engineerization of the physical chemical properties of the materials, require the introduction of several functional groups that are incompatible with iron chloride.

Despite limitations and drawbacks, the $FeCl_3$ method has been widely used to prepare PTs and their derivatives because it is easy, cheap and gives high weight materials in good yields.

15.2.1.1.5 Electrochemical Polymerization

Chemical synthesis, when dealing with soluble PTs, presents some advantages mainly due to the high yield in polymer, the easy purification process and to the possibility of obtaining regioregular PTs. Nevertheless, PTs also can be prepared by electrochemical polymerization [20]. Since a film is produced on the anode during polymerization, this method is suitable for the preparation of polymers such as PT and poly(3-methylthiophene), which are not greatly soluble in the organic solvents commonly used in chemical polymerization.

On of the drawbacks of electrochemical polymerization is the low yield and another is that the regiochemistry of the polymer is uncontrollable and not well defined. It is generally accepted that electrochemical polymerization proceeds via radical-cationic mechanism (Fig. 15.8) and this seems reasonable since it is an anodic reaction.

The first step is the oxidation of the monomer to the corresponding radical cation, the second step involves the coupling of two radicals to

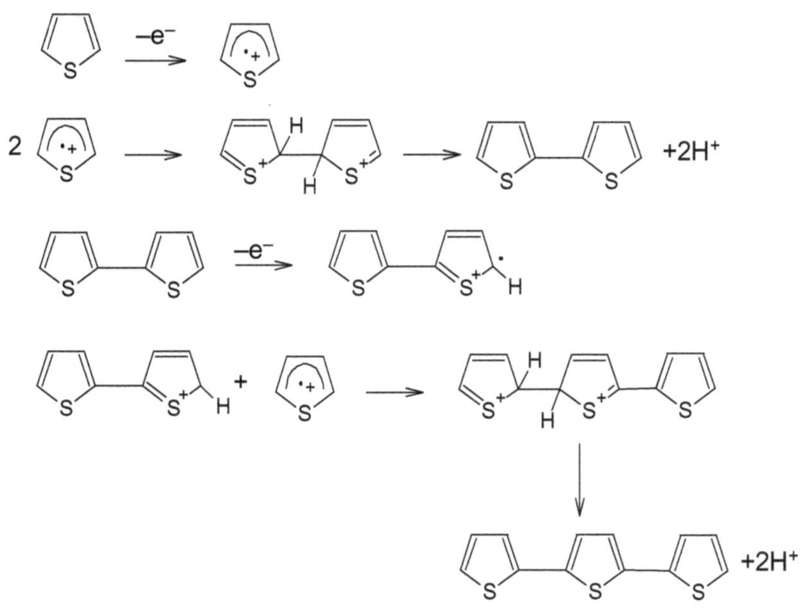

Figure 15.8 Mechanism of electrochemical polymerization.

produce a di-hydro dimer dication, which re-aromatizes through the loss of two protons.

This last process is the driving force of the whole chemical step. The oxidation potential of the dimer is lower than the monomeric one so it is oxidized first and generates a species that can couple with a monomer or with another dimer. Electropolymerization proceeds through electrochemical and chemical steps until the oligomer becomes insoluble and precipitates onto the electrode surface. Thanks to the easy synthesis of the starting monomer, PTs obtained from electrochemical polymerization of 3-alkylthiophenes are well studied [21], but also a series of regiochemically defined dialkoxy-substituted thiophene oligomers and co-polymers were generated electrochemically [22].

15.2.1.2 The Origin of Chirality

The first chiral PTs were prepared by Lemaire and co-workers in 1988 via electrochemical polymerization [23]. Starting from chiral thiophene monomers bearing in the side chain (*S*)- or (*R*)-2-phenylbutyl groups, bonded to the thiophene ring through an alkyloxy spacer (Fig. 15.9), they obtained chiral polymers whose neutral, undoped films exhibited large optical rotations compared to the corresponding monomers.

The authors attribute this high optical activity to the formation of a one-handed helical conformation of the PT backbone related to the configuration (*S*) or (*R*) of the substituent. These polymers were used as electrode modifiers and showed chiral discrimination towards the two optical forms of camphorsulfonic acid.

This hypothesis of helical conformations was supported by theoretical calculations in which the presence of transoid or cisoid helical chains was supposed [24] whereas studies conducted by Meijer and coworkers [25] demonstrate that the chirality arises from a helical packing of predominantly planar chains, rather than from a helical intrachain conformation.

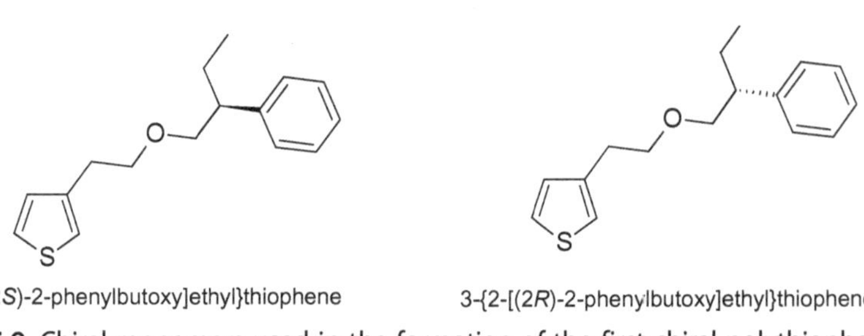

3-{2-[(2*S*)-2-phenylbutoxy]ethyl}thiophene 3-{2-[(2*R*)-2-phenylbutoxy]ethyl}thiophene

Figure 15.9 Chiral monomers used in the formation of the first chiral polythiophenes.

In literature there are different examples of chiral PTs, such as the one reported by Inganas and co-workers [26] where the optical activity is present even in the absence of the aggregates. In particular, an aminoacid functionalised chiral PT [27] with high pH-dependence properties (CD, UV-visible and fluorescence emission spectra) showed that at the isoelectric point of the aminoacid the PT chains are separated and adopt a non-planar helical conformation, while increasing the pH leads to a more planar conformation of the backbone and aggregation of the polymer chains occurs.

There are several hypotheses on the nature of the chirality in PTs:

1. Skewed π-stacking of planar strands, transoid in nature [25]
2. A two-step process, where the strands first organize in 2D sheets (possibly twisted) and subsequently give rise to a one-handed π-stacked packet [28]
3. Organization of one-handed linear chains into an asymmetric crystal, i.e., without inversion center [29].

Recently [30], it has been clearly demonstrated that a block polymer can show chirality even if an excess of chiral centers is not present. The authors claim that the chirality originates from a specific order of events such as pH alteration and solubility that can impart a particular chiral aggregation without having any enantiomeric excess.

The principal difficulty in studying PT systems where chirality is due to the presence of a chiral substituent in the lateral chain is related to the fact that chirality appears only in well-determined conditions, such as in solution of mixed solvents or in specially prepared films, and is not inherent to the molecular core. A recent study [31] shows that both enantiomers of inherently chiral 2,2′-bis(2,2′-bithiophene-5-yl)-3,3′-bithianaphthene have a large optical rotation in solution, which is also reproducible after several oxidative–redox cycles thereby showing the potentiality of introducing chirality inherently in the repeating units.

15.3 ALKYLSULFANYL AND RELATED POLYTHIOPHENES

PTs bearing a sulfur atom directly bonded to the thiophene ring are important in relation to the lowering of the oxidation potential, the stabilization of the oxidized and reduced forms and the narrowing of the band gap between ground and excited state. They can form free-standing films of different thickness that can be easily manipulated and cut. We have synthesized achiral PT bearing different moieties (ionic and not ionic in three positions), but also chiral PTs. The first one [32] was obtained by oxidative polymerization of (+)-4,4′-bis[(*S*)-2-methylbutylsulfanyl]-2,2′-bithiophene (P1, Fig. 15.10). The interesting, small band gap of **P1** allows for easy *p*- and *n*-electrochemical doping.

The CD spectra of P1 in solutions in which poor solvents are present show interesting features and allow the presence of different optically active species to be distinguished (Fig. 15.11).

In Fig. 15.11 the CD/UV-vis spectra of **P1** in mixture of $CHCl_3$/methanol or $CHCl_3$/hexane are shown, whereas in Fig. 15.12 the CD/UV-vis spectra of films obtained by casting from different solvents under different evaporation conditions are shown.

Slow evaporation from a good solvent favors an aggregate phase characterized by a bisignate CD spectrum with a first **positive CE**, whereas fast evaporation from chloroform or slow evaporation from THF favors an aggregate phase with a first **negative CE**. This finding underlines the importance of the solvent polarity and of the evaporation conditions in determining the formation of the aggregates, resulting in different CD spectra.

Two thiophene-based chiral co-polymers co-PT1 and co-PT2 bearing a cysteine moiety in the beta position [33] were synthesized by Stille coupling (Fig. 15.13). A PT functionalized with a cysteine moiety is appealing for numerous reasons: it possesses three potential sites for molecular interaction and metal-ion detection (the thioether, the carboxyl and the amino group), it is able to self-assemble through hydrogen bond formation, and it is suitable for chiral recognition and can induce chain helicity in aggregate phases, thanks to the presence of a stereocenter.

The GPC and UV–vis characterization of the *co*-PTs (Fig. 15.14) evidences the presence in solution of both free and aggregated forms, but their CD is low or absent in solution.

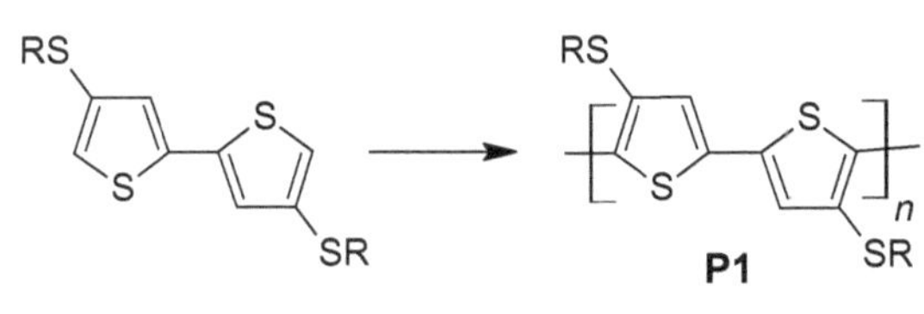

Figure 15.10 Alkylsulfanyl polythiophenes from the functionalized bithiophenes.

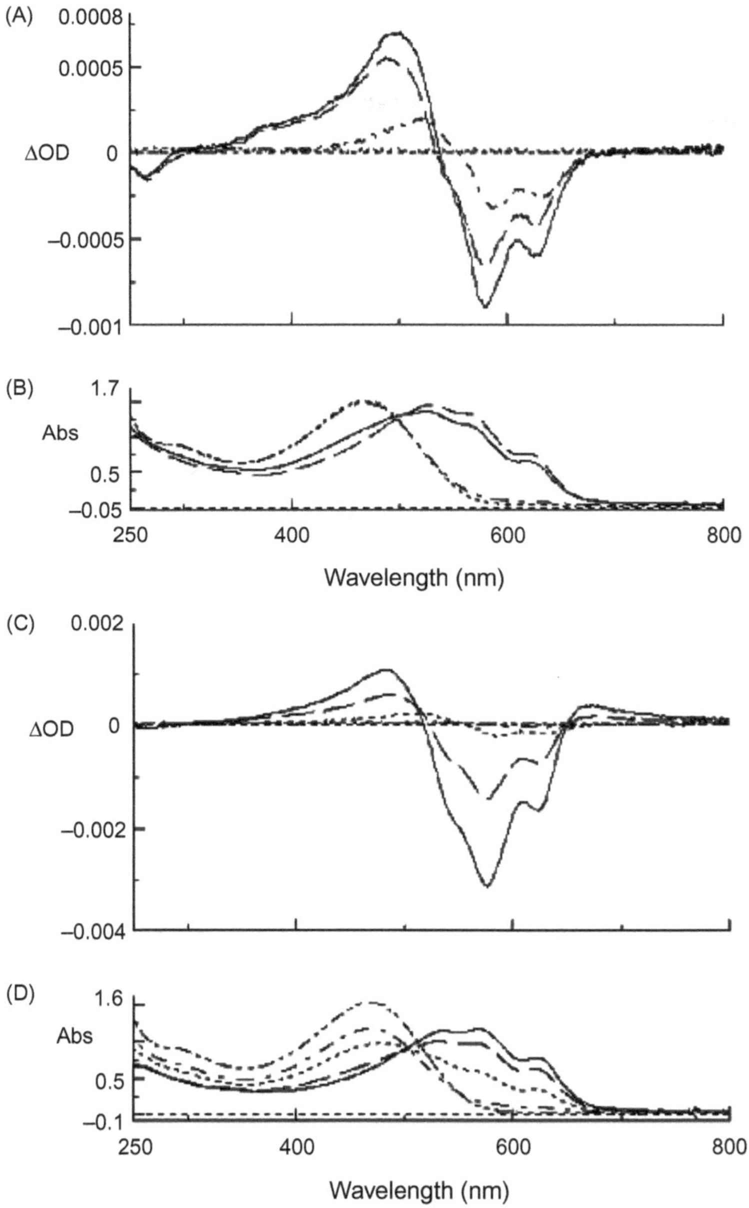

Figure 15.11 (A) CD and (B) UV/Vis spectra of **P1** in chloroform/methanol mixtures: 10:1 (· · · ·), 8:2 (·--·-), 7:3 (——), 6:4 (- - - -); (C) CD and (D) UV/Vis spectra of **P1** in chloroform/*n*-hexane mixtures: 10:1 (· ·-· · ·-), 5:5 (·--·-), 4:6 (· · · ·), 3:7 (- - - -), 1:9 (——).

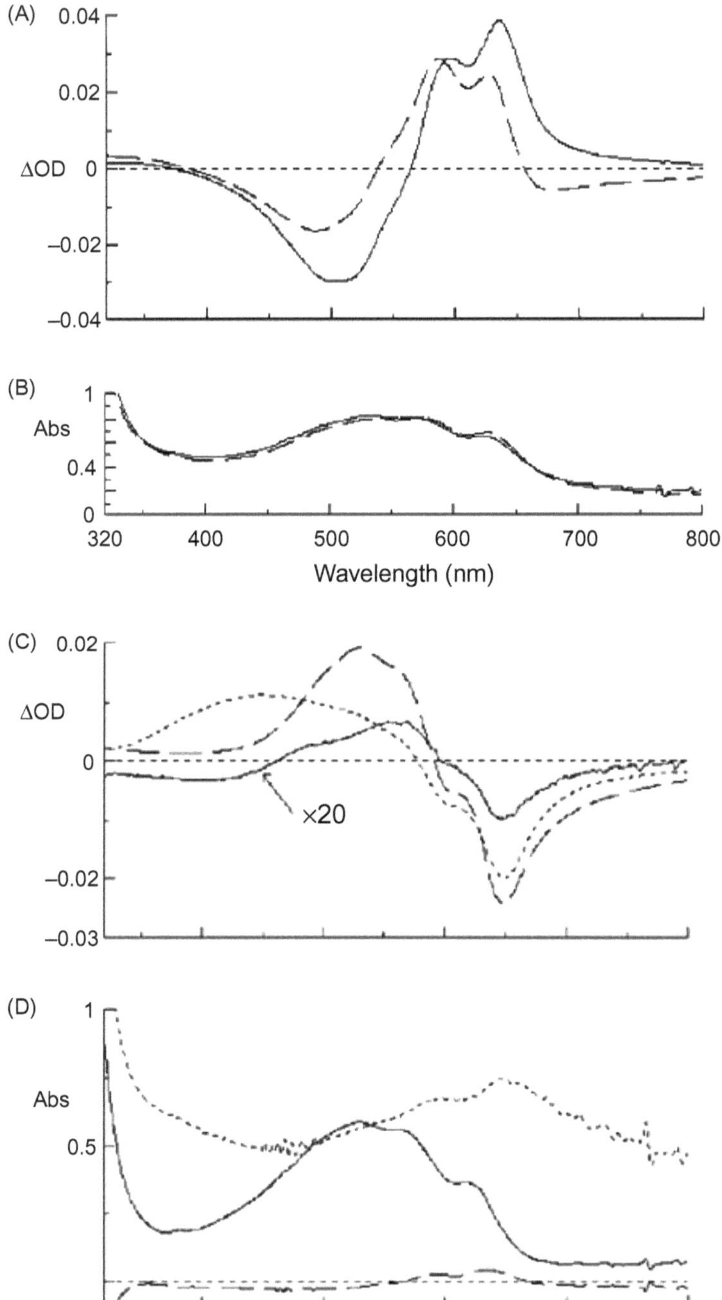

Figure 15.12 (A) CD and (B) UV/Vis spectra of **P1** films cast from chloroform under slow evaporation conditions; (C) CD and (D) UV/Vis spectra of film fast cast from chloroform solution (——); film cast slowly from THF solution (· · · ·); difference between the spectra of Fig. 15.12A and B (- - - -).

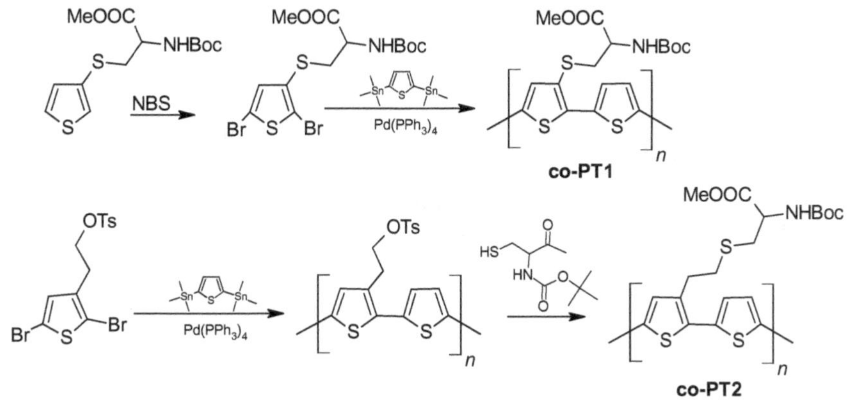

Figure 15.13 Synthetic route to co-PT1 and co-PT2.

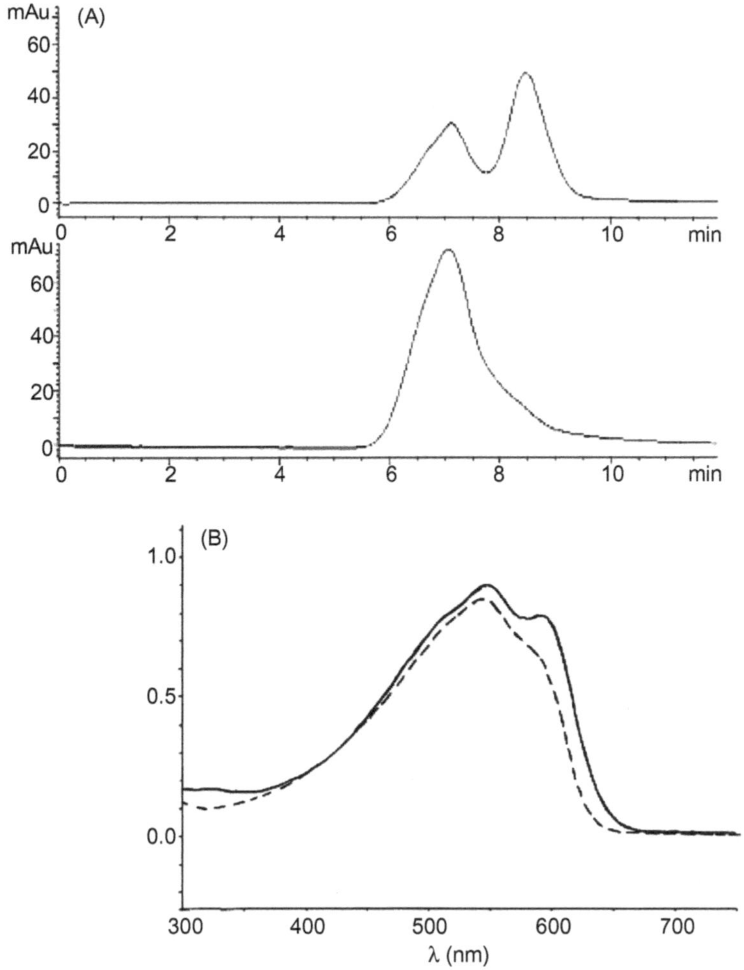

Figure 15.14 (A) GPC of **co-PT1** and **co-PT2**; (B) UV-vis of solutions of co-PT1 (solid line) and co-PT2 (dashed line).

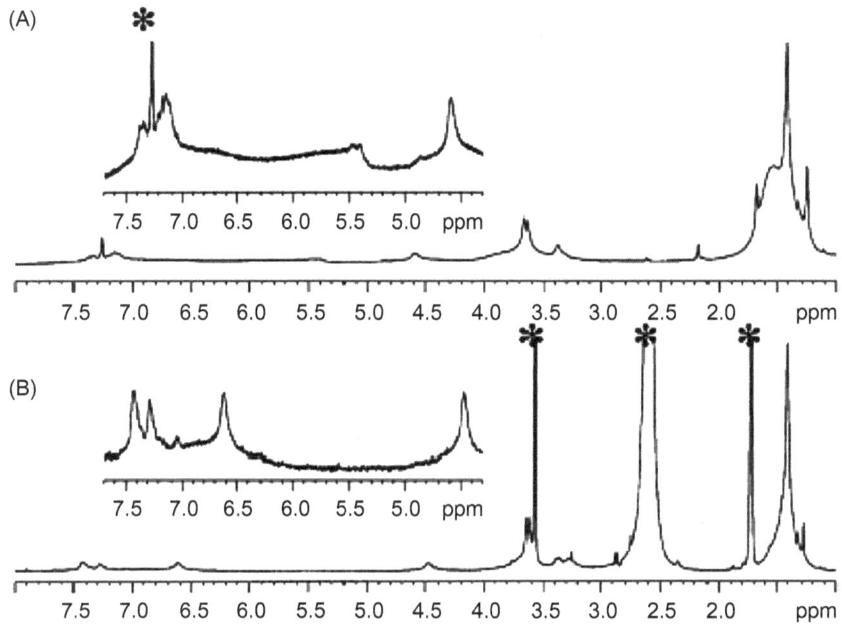

Figure 15.15 ^{1}H nuclear magnetic resonance spectra of **co-PT1** (A) in $CDCl_3$; (B) in THF-d8.

The nuclear magnetic resonance (NMR) spectra of these polymers (Fig. 15.15) display the presence of more or less aggregated forms depending on the solvent utilized.

The NOESY NMR experiments on **co-PT1** and **co-PT2** indicate a spatial proximity of the methoxy and the *t*-butoxy groups in the aggregates, which can be related to the formation of interresidue hydrogen bonds between cysteine units belonging to different polymeric chains, even though intrachain bonds cannot be excluded.

The self-assembling behavior is favored not only by the cysteinic side chains, but also by the alternation of β-substituted and β-unsubstituted thiophene rings, which favors π-stacking. The hydrogen bonding between the aminoacids in the side chains is the main driving force for the chiral aggregation (Fig. 15.16).

Both co-*PTs* exhibit optical activity in the solid state (Fig. 15.17), but their Cotton effect is low or absent in solution (despite the presence of aggregates). Even though aggregation cannot be considered, in principle, a necessary condition for chirality, the main chain chirality induced by the cysteine residues is amplified in the solid state for these *co*-PTs.

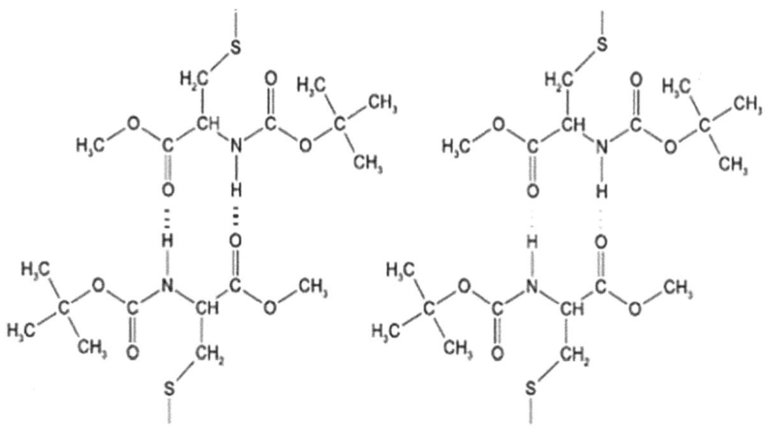

Figure 15.16 Interresidue hydrogen bonds between two different polymeric chains that would explain the formation of aggregates, even though intrachain bonds cannot be excluded.

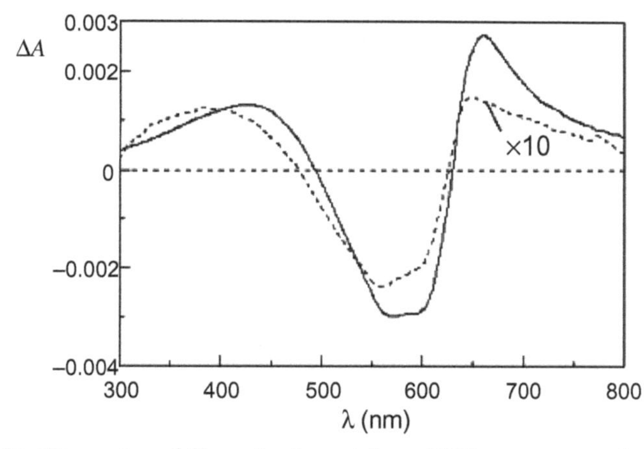

Figure 15.17 CD spectra of films slowly cast from $CHCl_3$ on quartz of co-PT1 (solid line) and co-PT2 (dotted line, 10 times enhanced).

Eventually, *co*-PT1 seems to be more promising than *co*-PT2, on the base of its higher solubility, filmability and optical activity. These results are not surprising, because *co*-PT1 belongs to a class of materials (oligo- and PTs carrying the sulfur atom directly bonded to the thiophene backbone) that possess very interesting physical properties.

15.4 APPLICATIONS OF CHIRAL POLYTHIOPHENES

Conducting organic polymers that possess chirality present some unique features when used as chiral substrates or as chiral electrode materials and they have application potential in electrochemical chiral sensing [34] or electrochemical asymmetric synthesis [35]. Their ability to be processed as particles, membranes or micro- and nano-dimensional fibers opens up possibilities in the design and development of specific molecular recognition/purification systems.

Although a number of chiral PTs have been prepared so far, most studies focused on their unusual chiroptical behavior in dependence on the solvent (solvatochromism) and temperature (thermochromism), and only a few reports are related to their chiral-recognition abilities.

The first chiral recognition by a chiral PT was demonstrated by Roncali et al. [23]. They investigated the chiral-recognition ability of regiorandom chiral PTs with an achiral detector (cyclic voltammetry) on the basis of changes in the shape of the voltammograms of the chiral PTs in the presence of chiral doping agents such as camphorsulfonic acid.

Nilsson et al. [36] showed that a PT having a side-chain of either a D- or L-aminoacid salt can interact in solution with a peptide molecule. Interestingly, based on the conformation of the peptide, the PT adopts different structures, that are probed with multiphoton spectroscopy. The polymer is not only capable of discriminating the structure of the peptide, but the two enantiomers bind differently to the peptide, thus granting chiral recognition.

Another example of chiral recognition from PTs is found in the work of Fukuhara et al. [37], who synthesized a chiral PT that undergoes a structural change upon inclusion of chiral substrates. Using an enantiomeric binaphthyl unit, linked as a bridge between two adjacent thiophenes, they were able to induce a twist in the polymer backbone upon complexation of the binaphthyl with amino-acids derivatives. The complexation affects not only the circular dichroism of the polymer, but also the UV-vis absorbance, and this chiral host–guest interaction can also be analyzed without relying on CD spectrophotometry.

The same authors observed significant conformational changes of the polymer backbone upon inclusion of chiral analyte of a binaphthocrown ether fused PT [38]. Recently Kameta et al. [39] demonstrated that, after nanotubes encapsulation, achiral PT boronic acids became chiral and exhibited chiral recognition abilities for D,L-sugars through preferential complexation. This chiral recognition works only if the polymers are encapsulated inside the nanotubes: even if the polymers maintain their induced chirality for a brief time once they are desorbed from the nanotubes (due to retention of the aggregates structure) they no longer discriminate the chiral sugars, suggesting a cooperative effect between the polymer and the nanotube in the chiral recognition.

Both enantiomers of co-PT1, bearing a cysteine moiety directly bonded to the thiophene ring, found a recent application in the demonstration of the existence of the chiral-induced spin selectivity (CISS) effect in the microscale dimension. These co-polymers are able to organize themselves in chiral superstructures, controlled by the sign of the cysteine chiral center, which may be used as spin filters [40]. Since the magnitude of the spin filtering effect is influenced by the orientation of the helical aggregates in the film, the alignment of the polymer chains in relation to the substrate surface is important. The orientation of the helical superstructures through an applied electric field was studied as a way to modify the properties of the polymer thin film, with the aim of maximizing the spin filtering effect.

The properties of the polymer films are influenced not only by the electric field, but also by the deposition technique used (spincoating or dropcasting) and the rate of solvent evaporation. In particular, the application of an electric field during spincoating promotes a circular dichroism enhancement and a resistivity decrease in the films, due to an increased molecular orientation.

Despite the chain alignment, these films did not show any spin selective effect. This is due to the excessive thickness of the polymer films that these deposition techniques can create, which gives extensive electron scattering during the conduction, resulting in spin randomization.

From these first findings, it was evident that the deposition procedures are essential to obtain thinner films for measuring spin selectivity. Thus, the polymer was deposited on nickel electrodes by electrochemical deposition. By controlling the direction of the magnetization of the nickel electrode, it is possible to inject electrons that have mainly one spin orientation; their selective transport through the polymer was verified by monitoring the cyclic voltammetry (CV) curve of a nonchiral redox couple in an electrochemical cell (Fig. 15.18).

Direct proof that the polymer deposited on the surface retained a chiral structure, despite the extremely thin layer, was found by electrochemical experiments (Fig. 15.19). The CV curves were recorded for nonmagnetic electrodes (Au) coated either with one of the two different enantiomers of co-PT1 (having either a D- or L-cysteine in the lateral chain), using either the R or S enantiomer of *N,N*-dimethyl-1-ferrocenylethylamine as a chiral redox probe. An enantio-selective interaction was found between the adsorbed polymer and the chiral redox probes in

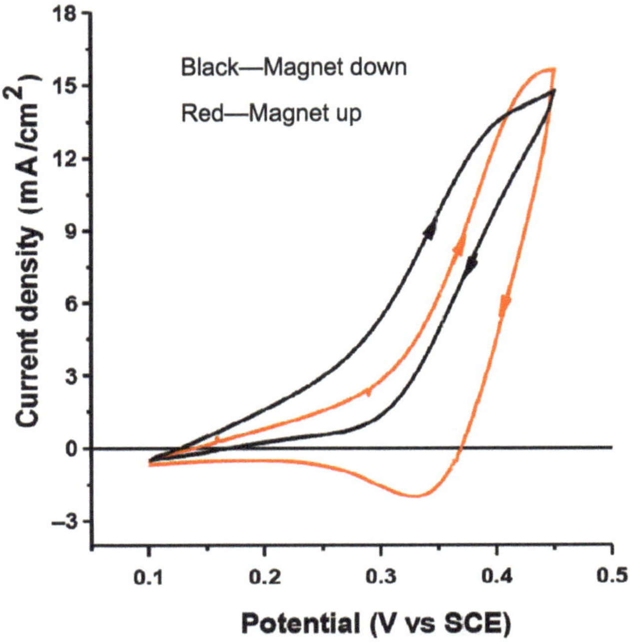

Figure 15.18 Cyclic voltammetry curves obtained with a nickel working electrode coated with conductive PT-LPT1, when achiral ferrocene (Fc) serves as the redox couple in methanol solution. When the nickel is magnetized up (red curve), the current flips its sign upon changing from oxidation to reduction, whereas when the nickel is magnetized down (black), the current does not change direction, indicating a much higher barrier for electron conduction from the nickel electrode towards the solution.

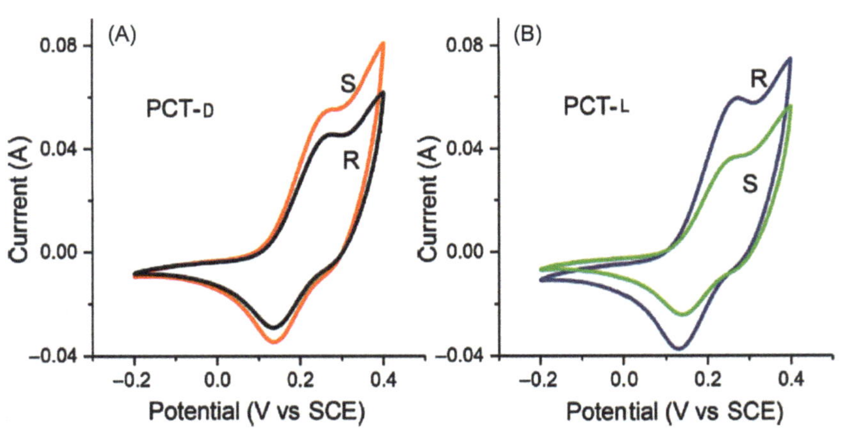

Figure 15.19 (A) Cyclic voltammetry curves obtained with **co-PT1**-D when the redox couple is either S-ferrocene (red) or R-ferrocene (black). (B) CV curves obtained when **coPT1**-L is used with S-ferrocene (green) and R-ferrocene (blue). Working electrode: gold.

the solution. The adsorbed co-PT1-D interacts strongly with S-ferrocene (red curve), whereas co-PT1-L has a stronger interaction with R ferrocene (blue curve).

REFERENCES

[1] Shirakawa, H.; Louis, E. J.; MacDiarmid, A. G.; Chiang, C. K.; Heeger, A. J. Synthesis of Electrically Conducting Organic Polymers: Halogen Derivatives of Polyacetylene, (CH)x. *J. Chem. Soc. Chem. Commun.* **1977,** (16), 578–580.

[2] Elsenbaumer, R. L.; Eckhardt, H.; Iqbal, Z.; Toth, J.; Baughman, R. H. Chiral Metals: Synthesis and Properties of a New Class of Conducting Polymers. *Mol. Cryst. Liq. Cryst.* **1985,** *118* (1), 111–116.

[3] Tamao, K.; Sumitani, K.; Kumada, M. Selective Carbon-Carbon Bond Formation by Cross-Coupling of Grignard Reagents with Organic Halides. Catalysis by Nickel-Phosphine Complexes. *J. Am. Chem. Soc.* **1972,** *94* (12), 4374–4376.

[4] Jen, K. Y.; Oboodi, R.; Elsenbaumer, R. L. Processable and Environmentally Stable Conducting Polymers. *Polym. Mater. Sci. Eng.* **1985,** *53*, 79.

[5] (a) McCullough, R. D.; Lowe, R. D.; Jayaraman, M.; Anderson, D. L. Design, Synthesis, and Control of Conducting Polymer Architectures: Structurally Homogeneous Poly(3-alkylthiophenes). *J. Org. Chem.* **1993,** *58* (4), 904–912.
(b) McCullough, R. D.; Lowe, R. D. Enhanced Electrical Conductivity in Regioselectively Synthesized Poly(3-alkylthiophenes). *J. Chem. Soc. Chem. Commun.* **1992,** (1), 70.

[6] Loewe, R. S.; Ewbank, P. C.; Liu, J.; Zhai, L.; McCullough, R. D. Regioregular, Head-to-Tail Coupled Poly(3-alkylthiophenes) Made Easy by the GRIM Method: Investigation of the Reaction and the Origin of Regioselectivity. *Macromolecules* **2001,** *34* (13), 4324–4333.

[7] Chen, T. A.; Rieke, R. D. The First Regioregular Head-to-Tail Poly(3-hexylthiophene-2,5-diyl) and a Regiorandom Isopolymer: Nickel Versus Palladium Catalysis of 2(5)-Bromo-5(2)-(bromozincio)-3-hexylthiophene Polymerization. *J. Am. Chem. Soc.* **1992,** *114* (25), 10087–10088.

[8] Miyaura, N.; Yanagi, T.; Suzuki, A. The Palladium-Catalyzed Cross-Coupling Reaction of Phenylboronic Acid with Haloarenes in the Presence of Bases. *Synth. Commun.* **1981,** *11* (7), 513–519.

[9] Jayakannan, M.; van Dongen, J. L. J.; Janssen, R. A. J. Mechanistic Aspects of the Suzuki Polycondensation of Thiophenebisboronic Derivatives and Diiodobenzenes Analyzed by MALDI–TOF Mass Spectrometry. *Macromolecules* **2001,** *34* (16), 5386–5393.

[10] Guillerez, S.; Bidan, G. New Convenient Synthesis of Highly Regioregular Poly(3-octylthiophene) Based on the Suzuki Coupling Reaction. *Synth. Met.* **1998,** *93* (2), 123–126.

[11] Liversedge, I. A.; Higgins, S. J.; Giles, M.; Heeney, M.; McCulloch, I. Suzuki Route to Regioregular Polyalkylthiophenes Ising Ir-catalysed Borylation to Make the Monomer, and Pd Complexes of Bulky Phosphanes as Coupling Catalysts for Polymerisation. *Tetrahedron Lett.* **2006,** *47* (29), 5143–5146.

[12] Stille, J. K. Palladium-katalysierte Kupplungsreaktionen Organischer Elektrophile mit Organozinn-Verbindungen. *Angew. Chem.* **1986,** *98* (6), 504–519.

[13] Bao, Z.; Chan, W. K.; Yu, L. Exploration of the Stille Coupling Reaction for the Synthesis of Functional Polymers. *J. Am. Chem. Soc.* **1995,** *117* (50), 12426–12435.

[14] Iraqi, A.; Barker, G. W. Synthesis and Characterisation of Telechelic Regioregular Head-to-Tail Poly(3-alkylthiophenes). *J. Mater. Chem.* **1998,** *8* (1), 25–29.
[15] (a) Bura, T.; Morin, P. O.; Leclerc, M. En Route to Defect-Free Polythiophene Derivatives by Direct Heteroarylation Polymerization. *Macromolecules* **2015,** *48* (16), 5614–5620.
(b) Gobalasingham, N. S.; Noh, S.; Thompson, B. C. Palladium-Catalyzed Oxidative Direct Arylation Polymerization (Oxi-DArP) of an Ester-Functionalized Thiophene. *Polym. Chem.* **2016,** 7 (8), 1623–1631.
[16] Sugimoto, R.; Takeda, S.; Gu, H. B.; Yoahino, K. Preparation of Soluble Polythiophene Derivatives Utilizing Transition Metal Halides as Catalysts and their Property. *Chem. Express* **1986,** *1* (11), 635–638.
[17] Järvinen, H.; Lahtinen, L.; Näsman, J.; Hormi, O.; Tammi, A. L. A New Method to Prepare 3-Octylthiophene and Poly-(3-octylthiophene). *Synth. Met.* **1995,** *69* (1–3), 299–300.
[18] Pomerantz, M.; Tseng, J. J.; Zhu, H.; Sproull, S. J.; Reynolds, J. R.; Uitz, R., et al. Processable Polymers and Copolymers of 3-Alkylthiophenes and their Blends. *Synth. Met.* **1991,** *41* (3), 825–830.
[19] (a) Abdou, M. S. A.; Lu, X.; Xie, Z. W.; Orfino, F.; Deen, M. J.; Holdcroft, S. Nature of Impurities in π-Conjugated Polymers Prepared by Ferric Chloride and their Effect on the Electrical Properties of Metal-Insulator-Semiconductor Structures. *Chem. Mater.* **1995,** 7 (4), 631–641.
(b) Chen, F.; Mehta, P. G.; Takiff, L.; McCullough, R. D. Improved Electroluminescence Performance of Poly(3-alkylthiophenes) Having a High Head-to-Tail (HT) Ratio. *J. Mater. Chem.* **1996,** *6* (11), 1763–1766.
[20] (a) Maior, R. M. S.; Hinkelmann, K.; Eckert, H.; Wudl, F. Synthesis and Characterization of Two Regiochemically Defined Poly(dialkylbithiophenes): A Comparative Study. *Macromolecules* **1990,** *23* (5), 1268–1279.
(b) Zagórska, M.; Kulszewicz-Bajer, I.; Proń, A.; Firlej, L.; Bernier, P.; Galtier, M. Spectroscopic Studies of Poly(4,4′-dialkyl-2,2′-bithiophenes): The "Head-to-Head" Analogues of Poly(3-alkylthiophenes). *Synth. Met.* **1991,** *45* (3), 385–393.
(c) Zagórska, M.; Krische, B. Chemical Synthesis and Characterization of Soluble Poly(4,4′-dialkyl-2,2′-bithiophenes). *Polymer* **1990,** *31* (7), 1379–1383.
(d) Krische, B.; Hellberg, J.; Lilja, C. Conducting Polymers from Dimethyl-2,2′-bithiophenes. *J. Chem. Soc. Chem. Commun.* **1987,** (19), 1476–1478.
[21] (a) Sato, M.; Morii, H. Configurational Feature of Electrochemically Prepared Poly(3-dodecylthiophene). *Polym. Commun.* **1991,** (32), 42.
(b) Sato, M.; Morii, H. Nuclear Magnetic Resonance Studies on Electrochemically Prepared Poly(3-dodecylthiophene). *Macromolecules* **1991,** *24* (5), 1196–1200.
[22] (a) Zotti, G.; Gallazzi, M. C.; Zerbi, G.; Meille, S. V. Conducting Polymers from Anodic Coupling of Some Regiochemically Defined Dialkoxy-Substituted Thiophene Oligomers. *Synth. Met.* **1995,** *73* (3), 217–225.
(b) Triantou, D.; Soulis, S.; Koureli, S.; De Sio, A.; von Hauff, E. Thiophene-Based Copolymers Synthesized by Electropolymerization for Application as Hole Transport Layer in Organic Solar Cells. *J. Appl. Polym. Sci.* **2013,** *127* (1), 585–592.
[23] Lemaire, M.; Delabouglise, D.; Garreau, R.; Guy, A.; Roncali, J. Enantioselective Chiral Poly(thiophenes). *J. Chem. Soc. Chem. Commun.* **1988,** (10), 658.
[24] Cui, C. X.; Kertesz, M. Two Helical Conformations of Polythiophene, Polypyrrole, and their Derivatives. *Phys. Rev. B* **1989,** *40* (14), 9661–9670.
[25] Langeveld-Voss, B. M. W.; Janssen, R. A. J.; Meijer, E. W. On the Origin of Optical Activity in Polythiophenes. *J. Mol. Struct.* **2000,** *521* (1–3), 285–301.
[26] Andersson, M.; Ekeblad, P.; Hjertberg, T.; Wennerstrom, O.; Inganas, O. Polythiophene with a Free Amino-Acid Side-Chain. *Polym. Commun.* **1991,** *32* (18), 546–548.

[27] Nilsson, K. P. R.; Olsson, J. D. M.; Konradsson, P.; Inganäs, O. Enantiomeric Substituents Determine the Chirality of Luminescent Conjugated Polythiophenes. *Macromolecules* **2004,** *37* (17), 6316–6321.

[28] Goto, H.; Okamoto, Y.; Yashima, E. Solvent-Induced Chiroptical Changes in Supramolecular Assemblies of an Optically Active, Regioregular Polythiophene. *Macromolecules* **2002,** *35* (12), 4590–4601.

[29] Catellani, M.; Luzzati, S.; Bertini, F.; Bolognesi, A.; Lebon, F.; Longhi, G., et al. Solid-State Optical and Structural Modifications Induced by Temperature in a Chiral Poly-3-alkylthiophene. *Chem. Mater.* **2002,** *14* (11), 4819–4826.

[30] Monnaie, F.; Van Den Eede, M. P.; Koeckelberghs, G. Expression of Chirality in a Conjugated Polymer without any Excess of Chiral Centers. *Macromolecules* **2015,** *48* (22), 8121–8127.

[31] Longhi, G.; Abbate, S.; Mazzeo, G.; Castiglioni, E.; Mussini, P.; Benincori, T., et al. Structural and Optical Properties of Inherently Chiral Polythiophenes: A Combined CD-Electrochemistry, Circularly Polarized Luminescence, and TD-DFT Investigation. *J. Phys. Chem. C* **2014,** *118* (29), 16019–16027.

[32] Iarossi, D.; Mucci, A.; Parenti, F.; Schenetti, L.; Seeber, R.; Zanardi, C., et al. Synthesis and Spectroscopic and Electrochemical Characterisation of a Conducting Polythiophene Bearing a Chiral β-Substituent: Polymerisation of (+)-4,4′-Bis[(*S*)-2-methylbutylsulfanyl]-2,2′-bithiophene. *Chem. Eur. J.* **2001,** 7 (3), 676–685.

[33] Cagnoli, R.; Lanzi, M.; Mucci, A.; Parenti, F.; Schenetti, L. Polymerization of Cysteine Functionalized Thiophenes. *Polymer* **2005,** *46* (11), 3588–3596.

[34] (a) Tiwari, M. P.; Prasad, A. Molecularly Imprinted Polymer Based Enantioselective Sensing Devices: A Review. *Anal. Chim. Acta* **2015,** *853*, 1–18.
(b) Trojanowicz, M. Enantioselective Electrochemical Sensors and Biosensors: A Mini-Review. *Electrochem. Commun.* **2014,** *38*, 47–52.
(c) Hu, D.; Lu, B.; Zhang, K.; Sun, X.; Xu, J.; Duan, X., et al. Synthesis of Novel Chiral L-phenylalanine Grafted PEDOT Derivatives with Electrochemical Chiral Sensor for 3,4-Dihydroxyphenylalanine Discrimination. *Int. J. Electrochem. Sci.* **2015,** (10), 3065–3081.
(d) Chahma, M.; McTiernan, C. D.; Abbas, S. A. Characterization of Phenomena Occurring at the Interface of Chiral Conducting Surfaces. *New J. Chem.* **2014,** *38* (8), 3379.

[35] Onomura, O. New Development of Asymmetric Reactions in Organic Electrolytic Synthesis. *Electrochemistry* **2010,** (78), 194–202.

[36] Nilsson, K. P. R.; Olsson, J. D. M.; Stabo-Eeg, F.; Lindgren, M.; Konradsson, P.; Inganäs, O. Chiral Recognition of a Synthetic Peptide Using Enantiomeric Conjugated Polyelectrolytes and Optical Spectroscopy. *Macromolecules* **2005,** *38* (16), 6813–6821.

[37] Fukuhara, G.; Inoue, Y. Peptide Chirality Sensing by a Cyclodextrin-Polythiophene Conjugate. *Chem. Eur. J.* **2012,** *18* (36), 11459–11464.

[38] Fukuhara, G.; Inoue, Y. Chirality-Sensing Binaphthocrown Ether-Polythiophene Conjugate. *Chem. Eur. J.* **2010,** *16* (26), 7859–7864.

[39] Kameta, N.; Masuda, M.; Shimizu, T. Soft Nanotubes Acting as Confinement Effecters and Chirality Inducers for Achiral Polythiophenes. *Chem. Commun.* **2016,** *52* (7), 1346–1349.

[40] (a) Tassinari, F.; Mathew, S. P.; Fontanesi, C.; Schenetti, L.; Naaman, R. Electric-Field-Driven Alignment of Chiral Conductive Polymer Thin Films. *Langmuir* **2014,** *30* (16), 4838–4843.
(b) Mondal, P. C.; Kantor-Uriel, N.; Mathew, S. P.; Tassinari, F.; Fontanesi, C.; Naaman, R. Chiral Conductive Polymers as Spin Filters. *Adv. Mater.* **2015,** *27* (11), 1924–1927.

CHAPTER 16

Abiotic Emergence of Biological Homochirality

Josep M. Ribó[1,2], Joaquim Crusats[1,2], Zoubir El-Hachemi[1,2], David Hochberg[3] and Albert Moyano[1]
[1]Department Organic Chemistry, University of Barcelona, Catalonia, Spain
[2]Institute of Cosmos Science (IEEC-UB), University of Barcelona, Catalonia, Spain
[3]Department of Molecular Evolution, Centro de Astrobiología (CSIC-INTA), Madrid, Spain

Abstract

The different types of network leading to spontaneous mirror-symmetry breaking (SMSB) are briefly discussed, as well as the specific characteristics of the thermodynamic systems that allow the achievement of a final chiral stationary state. The recent SMSB model based in enantioselective autocatalysis coupled to enantioselective cross-catalysis aid (hypercyclic replicators) is specifically discussed. Such a model suggests that the RNA-world models could comprise simultaneous emergence of enantioslective replicators and SMSB. The selection of one of two final chiral signs in SMSB in abiotic scenarios is proposed as a primordial Darwinian selection.

Keywords: Enantioselective autocatalysis; enantioselective cross-catalysis; hypercyclic replicators; primordial Darwinian selection; symmetry breaking, spontaneous

16.1 INTRODUCTION

Chemists have simplified the emergence of biological homochirality (BH) [1] to a punctuated accidental kinetic-controlled phenomenon in some enantioselective reaction, under the effect of some external chiral polarization, and followed by chirality transfer reactions. Subsequently, the amplification of chirality through enantioselective reactions showing nonlineal correlations between the initial and final enantiomeric excesses (*ee*) [2] was included as a further hypothesis in this scenario. More recently, the experimental reports on enantioselective autocatalytic reactions leading to spontaneous mirror symmetry breaking (SMSB) [3] have given a new twist to old and new theoretical reports on the topic. However, a reasonable translation of this to reasonable abiotic chemistry scenarios is lacking. In this respect, the study of the phenomenon is of minor interest to synthetic chemists, but an important obstacle to further progress is that the understanding of the

Advances in Asymmetric Autocatalysis and Related Topics
DOI: http://dx.doi.org/10.1016/B978-0-12-812824-4.00016-2

phenomenon necessitates the involvement of different scientific disciplines, and that some reports on SMSB contain basic thermodynamic errors or chemical misunderstandings. Theoretical biologists and biophysicists consider the emergence of BH as an overly complex phenomenon to be taken into account during the latter stages of abiotic evolution. In biophysics and in systems biology, it is implicitly assumed that the emergence of BH occurred in abiotic evolution stages prior to those involving the formation a simple abiotic replicators and that, in consequence, it is not necessary to take the emergence of BH into account in their models, which are already assumed to be homochiral [4]. We summarize how SMSB scenarios can be supported by sound chemical concepts and give a personal point-of-view on how these concepts can be translated to theoretical abiotic models for the selectivity/survival of prebiotic replicators.

Fig. 16.1 displays a widely accepted description of the different stages of abiotic evolution. There is no consensus as to at which stage the emergence of BH may have occurred. The "experimental" evidence for the homochirality of present life is extrapolated back to the final stages of abiotic evolution assuming that pools of homochiral building blocks have yielded the proto-life biopolymers. This means that homochirality was

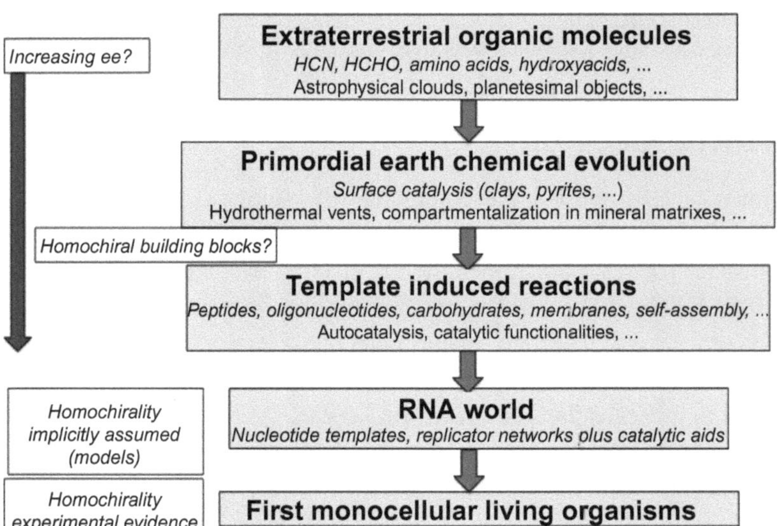

Figure 16.1 Different stages of abiotic chemical evolution correlated to the question of homochiral composition. *Inspired by a figure of Schuster, P. Chemistry on the Early Earth.* Germany-Japan Round Table *[Online]; Heidelberg, November 1–3, 2011. <http://www.tbi.univie.ac.at/~pks/Presentation/heidelberg-11.pdf> [5].*

obtained in a stage prior to the emergence of organocatalytic functionalities and replicator synthesis (Fig. 16.2).

The emergence on Earth of the organic building blocks [6] of chiral biopolymers in $ee \neq 0$ high *ee* should nowadays be considered as an experimental fact [7]. Therefore the most extended point of view is that the formation of enantiopure building blocks occurred in progressive steps of chiral amplification. However, in our opinion, this cannot explain BH because of: (1) thermodynamic controlled bias of the racemic composition yielding $ee \approx 100\%$ needs unreasonably large chiral inductions, i.e., either an already pure chiral media or strong natural chiral forces; (2) kinetically controlled enantioselective synthesis needs to yield high *ee* values of diastereomeric chiral intermediates or to activate complexes that can only be obtained by enantiopure chiral co-reagents; (3) nonlinear amplification effects can only play a secondary role in abiotic evolution because of the necessary repetition of the same processes, such as has already been noted [2] the repetition of the same reaction using the previous outcomes leads one towards the racemic composition (e.g., one transformation yielding an 80% *ee*, its input for a new reaction yields also 80%, this means a decrease of the *ee* of the first input); and (4) chiral building

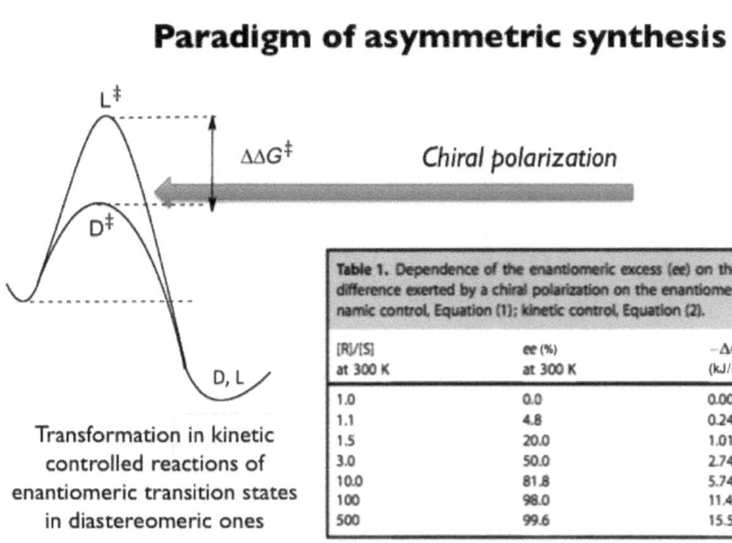

Table 1. Dependence of the enantiomeric excess (*ee*) on the free energy difference exerted by a chiral polarization on the enantiomers: thermodynamic control, Equation (1); kinetic control, Equation (2).

[R]/[S] at 300 K	*ee* (%) at 300 K	$-\Delta G^{\circ}$ or $-\Delta\Delta G^{\circ}$ (kJ/mol)
1.0	0.0	0.00
1.1	4.8	0.24
1.5	20.0	1.01
3.0	50.0	2.74
10.0	81.8	5.74
100	98.0	11.49
500	99.6	15.50

Figure 16.2 Standard free energy differences between initial and final states in thermodynamically controlled $D \leftrightarrows L$ equilibrium or between the transition states necessary to obtain a bias ($ee \neq 0$) from the racemic composition, in the asymmetric induction on enantioselective irreversible transformations.

blocks of biopolymers slowly racemize, therefore an enantiopure composition of chiral building blocks is expected to decrease its initial *ee* value during abiotic evolution. In summary, common effects that are useful in asymmetric synthesis cannot explain BH. However, they have surely played a role in the production of nonracemic mixtures of small chiral compounds in astrophysical scenarios, but cannot explain the question of homochirality in a terrestrial scenario of abiotic evolution. Reaction networks placed in systems unable to equilibrate with their surroundings, such as living beings accomplish, can only explain BH yielding SMSB, i.e., nonracemic outcomes. This is possible when the autocatalysis, today assumed to be the support of life processes [8], is assumed/imagined to be enantioselective [1,9].

16.2 THE SPONTANEOUS MIRROR-SYMMETRY BREAKING SCENARIO

In an isolated system or from an applied chemistry point of view, in any system able to come into equilibrium with its surroundings, and in the absence of any chiral polarization, all possible enantioselective reaction networks must eventually yield the racemic mixture state. However, in some reaction networks showing nonlinear kinetic dependences on the concentrations of the enantiomers, as occurs in enantioselective autocatalysis, it is possible to achieve nonequilibrium thermodynamic final states for which the racemic mixture is less stable than the two degenerate chiral states. This is the so-called SMSB scenario, which cannot be confused with the simple symmetry breaking in the formation of chiral compounds from achiral compounds where the output is always the racemic: there is a decrease of symmetry but there is no SMSB. For example, under thermodynamic control, the (1) crystallization of racemic conglomerate mixtures from achiral compounds, the (2) nucleophilic addition to a prochiral carbonyl group, or (3) the formation of chiral spiral domains in the Belusov-Zhabotinsky oscillatory reaction, all yield a racemic mixture of chemical compounds or chiral domains. In these cases the law of large numbers yields the racemic mixture of the chiral species. Furthermore, SMSB cannot be confused with obtaining chiral mixtures by kinetic control instead of racemic ones. This can be a process useful in bench chemistry, but it cannot play a significant role in chemical evolution, where very long time scales are implied. In SMSB the whole system breaks mirror/chiral symmetry leading to a bias from the racemic mixture triggered by any accidental/random bias from the ideal racemic composition. This can be described as a bifurcation scenario where the racemic mixture is a metastable state and the chiral states correspond to free energy wells of minimum entropy production [10]: chiral reaction noise suffices to take the system out from the racemic composition and towards one of the two stationary enantiomeric states, i.e., parity is broken in the entire system. Obviously, this is only possible in systems prevented from equilibrating with their surroundings, as is the case in the chemical processes of living systems. In this respect, a diversity of thermodynamic system configurations (we call them here *architectures*) can be imagined, for example: (1) systems open to matter transfer of some of the chemical species; (2) compartmentalization of specific reactions, for example by virtue of the anchoring of specific catalysts; (3) closed systems with permanent nonisotropic temperature distributions; or (4) specific energy uptake by only some of the chemical species (e.g., the case of photochemical reactions). However, only enantioselective reaction networks showing sufficiently high nonlinear dependences on the enantiomer concentrations may lead to SMSB, and this only for a certain range of reaction parameters. Such reaction networks that can achieve SMSB in a closed system that is able to equilibrate with its surroundings can show temporary chiral amplifications whose temporal window range increases with exergocity, which is thus of potential interest to applied chemistry.

In the following we discuss, separately: (1) the stoichiometric reactions that may show enantiomeric increases that can lead to the extinction of one of the two enantiomeric catalysts/products, i.e., to SMSB; and (2) the architecture of the thermodynamic system that allows the former reaction networks to achieve a final chiral stationary state as a stable state of the system.

16.3 REACTION NETWORK

Enantioselective autocatalysis is a necessary but not sufficient condition for SMSB [11]. SMSB requires additional conditions beyond those implied by the first-order enantioselective autocatalysis (Eq. 16.1) (quadratic nonlinearity with respect to the enantiomers) represented by the stoichiometric reaction:

$$\mathrm{A_{achiral}} + \mathrm{D} \underset{k_{-a}}{\overset{k_a}{\rightleftharpoons}} 2\,\mathrm{D} \qquad \mathrm{A_{achiral}} + \mathrm{L} \underset{k_{-a}}{\overset{k_a}{\rightleftharpoons}} 2\,\mathrm{L} \tag{16.1}$$

Notice, that this assumes that the direct transformation also exists, albeit as a slower transformation, and it must have the same equilibrium constant:

$$A_{achiral} \underset{k_{-d}}{\overset{k_d}{\rightleftharpoons}} L \qquad A_{achiral} \underset{k_{-d}}{\overset{k_d}{\rightleftharpoons}} L \tag{16.2}$$

$$K_a = K_d = \frac{k_a}{k_{-a}} = \frac{k_d}{k_{-d}} \tag{16.3}$$

Reaction (16.1) in a system able to come into thermodynamic equilibrium with its surroundings yields the racemic mixture. In open systems, and in the best of the scenarios, only the initial *ee* value can be maintained [11]. The set of reactions (16.1) is autocatalytic in the production or total chiral matter, but lacks the selectivity necessary for the survival/extinction of one of the enantiomers. Notice that SMSB is an extreme case for achieving chemical selectivity because the enantiomers are degenerate chemical species [12]. For such a selectivity it is necessary that the kinetic evolution of the enantiomers show a growth dynamic greater than the exponential one of reaction (16.1) in order to lead to the survival/depletion of one of the enantiomers [13]. This is the general problem of selection of chemical species in autocatalytic systems used in chemical evolution models and for enantiomeric species, because their energetic degeneracy poses an extreme case to deal with.

It has been claimed that in the absence of the backward reaction, i.e., the so-called irreversibility of a chemical transformation ($0 = k_{-a} = k_{-d}$), the reaction (16.1), or the combined set (16.1) + (16.2), due to the underlying stochastics, the very first few molecular reactions may lead to high *ee* values. Such an *ee* can be transmitted further when the autocatalytic rate is high, in other words, a chiral kinetically controlled output may be obtained [14]. When Eq. (16.1) is irreversible (a highly exergonic transformation!) in a closed system deterministic kinetics can only lead to the final racemic mixture, but stochastic kinetics may give a better description of how the kinetically controlled *ee* values are obtained. The probability of a large *ee* is higher when the initial concentration of catalysts/products is low. However, the absence of backward reactions is an approximation valid only for relatively high exergonic reactions, but in our opinion it is an invalid approximation on the time scale of chemical evolution where very low return rates will be detected. Notice that stochastic and deterministic kinetics (chemical kinetics model) must both lead to the same result in the thermodynamic limit. Stochastic kinetics applied to scenarios of very small volumes and very low concentrations, is the adequate tool for the study of the first stages of approximately individual transformations; e.g., in protein folding or in the actual mechanisms of enzymatic behavior. However, only the results implied by the thermodynamic limit can arrive at the correct conclusions regarding the emergence of BH [15]. This is because the abiotic evolution towards life is the emergence of complexity through the cooperation and competition between a large number of entities. In summary, only deterministic kinetic, or stochastic but in the thermodynamic limit, are reliable frameworks for the description and simulation of evolution models, which concern the description of stable nonequlibrium thermodynamic states.

Higher order autocatalysis (Eq. (16.4) for $n > 1$), in contrast to the enantioselective autocatalysis of first order (Eq. (16.1), $n = 1$: quadratic dependence with respect to the chiral compounds), may show the growth kinetics able to select one of the enantiomers:

$$A_{\text{achiral}} + n\mathrm{D}(n\mathrm{L}) \leftrightarrows (n+1)\mathrm{D}[(n+1)L] \qquad (16.4)$$

In a closed system the final stage of the reaction is that of a racemic mixture, but temporary windows of amplification of the initial *ee* may occur. In systems unable to come into equilibrium with their surroundings, SMSB may be possible for certain ranges of reaction parameters (reaction rate constants and total chemical mass). The presence of the direct reaction (16.2) contributes obviously to the racemization. In addition, for high n values at low concentrations, the direct reaction (16.2), despite having smaller reaction rate constants, can show overall rates higher than that of the autocatalysis. However, the significant point is that, in the framework of linear thermodynamics (not very high concentrations, mean field assumption, etc.), the stoichiometric reaction (16.4) is expected to show molecularities ≤ 2. Chemical transformations of high stoichiometry show reaction mechanisms of sequences of mono- and bimolecular reactions; therefore they have molecularity of orders ≤ 1. However, when first-order autocatalysis is coupled to reactions implying interactions between the chiral catalysts/products, the total growth of the enantiomers can be high enough to lead to the survival/extinction of one of the enantiomers. In this respect, as reported in Ref. [16], autocatalysis can be defined not as a reaction network that shows an *ee* increase when chiral products/catalysts are added, but as a reaction showing a kinetic signature $\propto [X]^n$.

In summary, the question on the emergence of SMSB has to do with which enantioselective reactions a first-order autocatalysis can be coupled to in order to show the autocatalytic signature in the time evolution of the *ee* value. In the following we summarize in which coupled reaction networks the first-order enantioselective autocatalysis (Eq. 16.1) has been shown to achieve the kinetic growth necessary for SMSB.

Frank-like systems [16]: As first proposed by Frank [17] the coupling of first-order autocatalysis to a heterochiral reaction between enantiomers yielding an achiral compound (Eq. 16.5) may lead to SMSB:

$$\mathrm{D} + \mathrm{L} \underset{k_{-I}}{\overset{k_I}{\rightleftharpoons}} \mathrm{P}_{\mathrm{achiral}} \tag{16.5}$$

When the reaction rate constant of this mutual inhibition between enantiomers is faster than that of the autocatalysis, the disappearance of chiral substance in a racemic ratio determines an increase of the *ee* that is amplified by the autocatalytic reaction. In systems unable to achieve equilibrium with their surroundings, SMSB may occur and in closed systems temporary *ee* amplification windows can be observed before coming into thermodynamic equilibrium. This is the case of the Soai reaction [18], a highly exergonic addition of a zinc organometallic nucleophile to a prochiral carbonyl group: the complex reaction network arising in the reaction is still a matter of study and discussion, but the characteristics that determine its ability for SMSB can be summarized as follows: (1) the actual nucleophiles are oligomeric complexes of the organo-Zn derivative with the final reaction enantiomers, which in fact are the chiral Zn-alkoxydes; (2) the heterochiral interaction between both Zn-alkoxyde enantiomers forms an achiral *meso*-dimer or oligomer, that corresponds to the mutual inhibition reaction (16.5) of the Frank model. Hydrolysis of the reaction crude determines the disappearance of the actual species acting on the reaction mechanisms. The distinction of simple nonlinear chiral effects and the SMSB of the Soai reaction is demonstrated by the chiral *ee* amplification resulting from its characteristic procedure of consecutive reactions at the same experimental conditions, but using the outcome of the previous reaction. This excludes the classical nonlinear effects of classical asymmetric synthesis, which should lead to a decrease of the *ee* value. As has been previously noted, such a procedure is a step-by-step approximation to an open flow system [3a], i.e., the Soai reaction in a continuous flow reactor should lead to a final chiral stationary state.

Limited enantioselective system (LES) [19]: A reaction network that should lead to SMSB is the so-called limited enantioselective inhibition model composed of the enantioselective autocatalysis (Eq. 16.1) coupled to:

$$\mathrm{L} + \mathrm{D} \underset{k_{-li}}{\overset{k_{li}}{\rightleftharpoons}} \mathrm{A}_{\text{achiral}} + \mathrm{D} \qquad \mathrm{L} + \mathrm{D} \underset{k_{-li}}{\overset{k_{li}}{\rightleftharpoons}} \mathrm{A}_{\text{achiral}} + \mathrm{L} \tag{16.6}$$

Notice that reaction (16.6) is the nonenantioselective autocatalysis counterpart of the enantioselective autocatalysis (Eq. 16.1). Therefore both reactions must have the same equilibrium constant, i.e., $k_{-li}/k_{li} = k_a/k_{-a}$. The stability analysis of the LES model shows that only when this constraint is broken can the system show SMSB [20]. This is an example of how overlooking the constraints imposed by the second law of thermodynamics (the principle of microreversibility) can lead to theoretical models of artifactual SMSB [21]. The simple use of common chemical reasoning relating the reaction rate constants to the equilibrium constants would have avoided erroneous reports on new models for SMSB, and in the interpretation of the experimental results in SMSB. However, the consideration of the necessary conditions for LES [18] has allowed us to predict in which type of systems can SMSB be expected [22,23]. For example, when a system composed by autocatalysis (16.1) + (16.6) + (16.2) changes the equilibrium constant ($K_{\text{eq}} = [\mathrm{D}][\mathrm{L}]/[\mathrm{A}]$) from >1 to <1 at two different temperatures, the compartmentalization of Eqs. (16.1) and (16.6) at different temperatures may lead to SMSB [22]. Therefore despite the lack of experimental examples of LES, it is an option for specific abiotic scenarios such as those of deep-ocean hydrothermal vents (Fig. 16.3).

Cross catalytic coupled enantioselective autocatalysis [24]: Recently it has been shown that SMSB may occur through the enantioselective cross-catalytic aid between first-order enantioselective autocatalysis. For example the simple case of two replicators:

$$\left.\begin{array}{ll} \mathrm{A}_{\text{achiral}} + {}^{1}\mathrm{D} + {}^{2}\mathrm{D} \underset{k_{-h1}}{\overset{k_{h1}}{\rightleftharpoons}} 2\,{}^{1}\mathrm{D} + {}^{2}\mathrm{D} & \mathrm{A}_{\text{achiral}} + {}^{1}\mathrm{L} + {}^{2}\mathrm{L} \underset{k_{-h1}}{\overset{k_{h1}}{\rightleftharpoons}} 2\,{}^{1}\mathrm{L} + {}^{2}\mathrm{L} \\ \mathrm{B}_{\text{achiral}} + {}^{2}\mathrm{D} + {}^{1}\mathrm{D} \underset{k_{-h2}}{\overset{k_{h2}}{\rightleftharpoons}} 2\,{}^{2}\mathrm{D} + {}^{1}\mathrm{D} & \mathrm{B}_{\text{achiral}} + {}^{2}\mathrm{L} + {}^{1}\mathrm{L} \underset{k_{-h2}}{\overset{k_{h2}}{\rightleftharpoons}} 2\,{}^{2}\mathrm{L} + {}^{2}\mathrm{L} \end{array}\right] \tag{16.7}$$

${}^{i}\mathrm{D}({}^{i}\mathrm{L})$ are biopolymers, capable of self-reproduction (replicators), formed by achiral building blocks (summarized in A and B).

The role of cross-catalytic coupling to autocatalytic replicators was proposed to solve the question of the error threshold [13a] formulated by

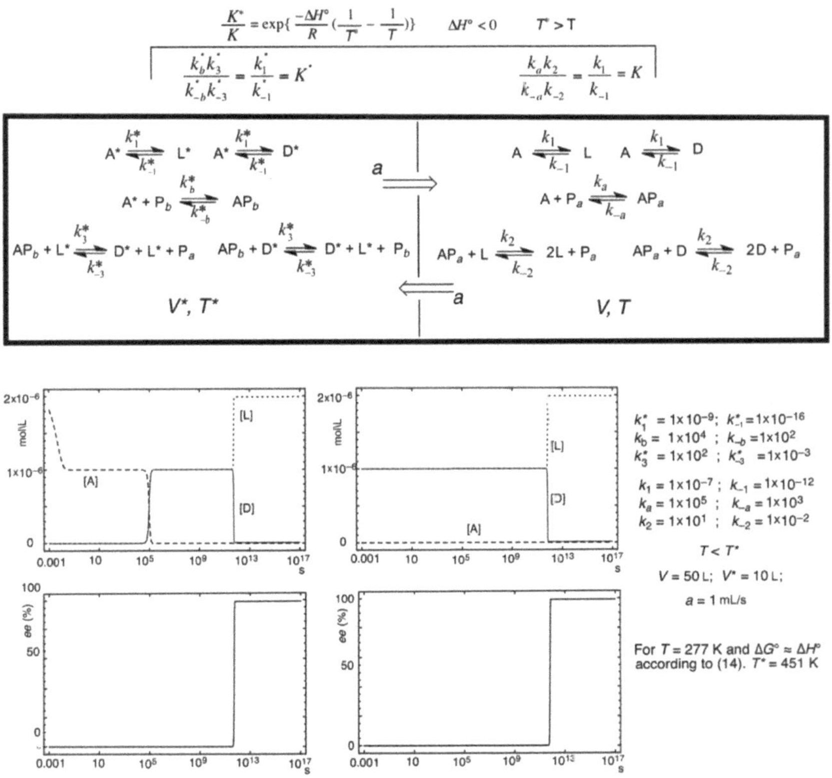

Figure 16.3 Limited enantioselective system reaction network in a two-compartment system at different temperatures and immobilized autocatalysis as a speculative model for the spontaneous mirror-symmetry breaking (SMSB) in abyssal hydrothermal vents. Where A is achiral, D and L are enantiomeric compounds, Pa and Pb enzyme-like immobilized mediators for the enantioselective and nonenantioselective autocatalysis processes, respectively. The simulations of SMSB (evolution of concentrations in the compartment of higher volume at temperature *T*, but after the initial equalization of concentrations the evolution of the species concentration is the same in both compartments). Left column: low reactant concentration and initial conditions for the transformation of A to L and D; $[A]_o = 2 \times 10^{-6}$: $[L]_o = 1 \times 10^{-10} + 1 \times 10^{-20}$: $[D]_o = 1 \times 10^{-10}$: $[Pa]_o = 1 \times 10^{-2}$: $[APa]_o = 0$: $[A^*]_o = 1 \times 10^{-6}$: $[L^*]_o = 1 \times 10^{-10}$: $[D^*]_o = 1 \times 10^{-10}$: $[Pb]_o = 1 \times 10^{-2}$: $[APb]_o = 0$. Right column: low concentrations and starting conditions of equilibrium concentrations between A and D/L; $[A]_o = 1 \times 10^{-10}$: $[L]_o = 1 \times 10^{-6} + 1 \times 10^{-20}$: $[D]_o = 1 \times 10^{-6}$: $[Pa]_o = 1 \times 10^{-2}$: $[APa]_o = 0$: $[A^*]_o = 1 \times 10^{-10}$: $[L^*]_o = 1 \times 10^{-6}$: $[D^*]_o = 1 \times 10^{-6}$: $[Pb]_o = 1 \times 10^{-2}$: $[APb]_o = 0$. For discussion and more details see Ref. [22a]. *Reproduced with permission from Ribó, J.M.; Crusats, J.; El-Hachemi, Z.; Moyano, A.; Blanco, C.; Hochberg, D. Spontaneous Mirror Symmetry Breaking in the Limited Enantioselective Autocatalysis Model: Abyssal Hydrothermal Vents as Scenario for the Emergence of Chirality in Prebiotic Chemistry. Astrobiology 2013, 13, 132–142.*

Eigen in 1971, in order to allow selection between stable replicator sets. This relates to the former discussion on the impossibility of getting selection in first-order autocatalysis. The hypercycle model [13b] consists of first-order autocatalytic replicators forming cycles of sequential cross-catalysis (two replicators in the case of Ref. [7]). It has been a seminal proposal to justify the coexistence, mutual stabilization and growth of self-reproducing chemical species. Although the system has received criticism, with respect to the explanation of evolutive behavior in present-day living systems, it is an adequate description of the role of chemical autocatalysis during the final abiotic stages (Fig. 16.1). However, the ability of the same system for SMSB has been overlooked. Fig. 16.4 shows an example of this. Notice that the ability of the hypercyclic replicators for SMSB strongly suggests that SMSB occurred in the same stages of replicator network formation and of the RNA-world. Further, it justifies that the chiral machinery of the chemical processes of present-day living beings is based on the evolution of hypercyclic networks yielding SMSB: enantioselective autocatalytic coupled replicators in conditions yielding SMSB compose the chiral life machinery.

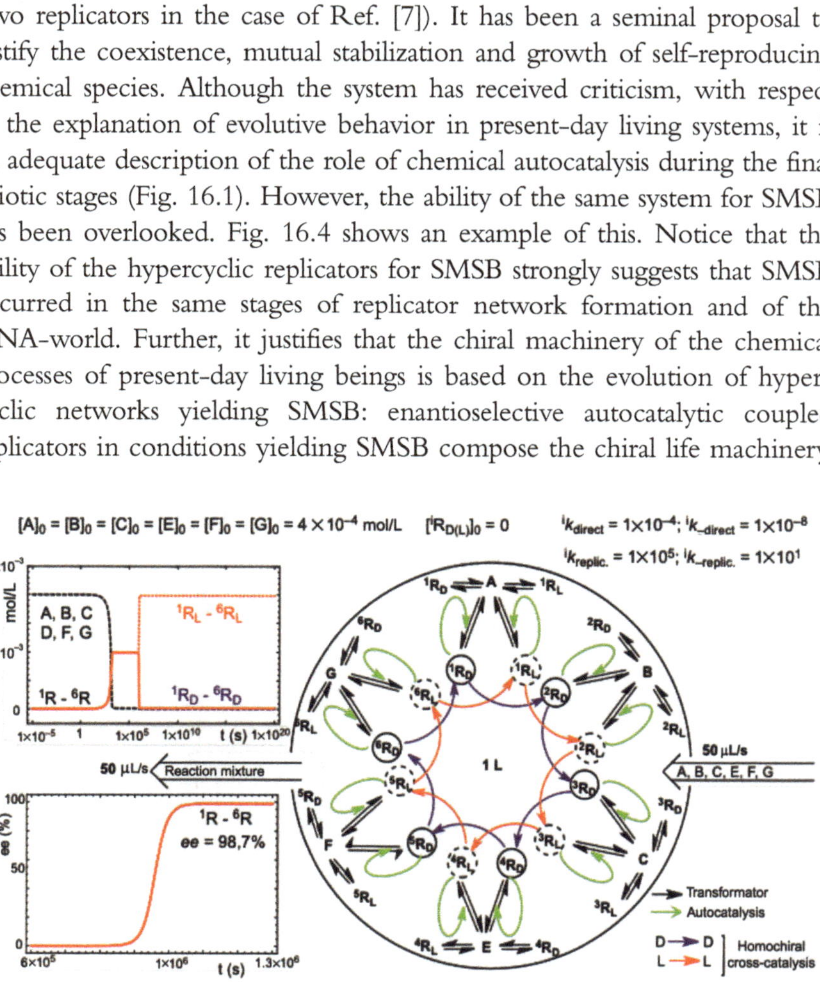

Figure 16.4 Spontaneous mirror-symmetry breaking in a open flow reactor of the reaction network of a six hypercyclic chiral replicator of homochiral cross-catalysis with direct replicator synthesis and no replicator concentration for the initial conditions. A chiral fluctuation in any of the replicators leads to an *ee* near to the 100% value (chiral fluctuation simulated by an initial concentration of 1×10^{-25} mol/L in any of the L-enantiomers). *Reproduced with permission from Blanco, C.; Crusats, J.; El-Hachemi, Z.; Moyano, A.; Hochberg, D.; Ribó, J. M. Spontaneous Emergence of Chirality in the Limited Enantioselectivity Model: Autocatalytic Cycle Driven by an External Reagent. Chem. Phys. Chem.* ***2013****, 14, 2432–2440.*

This nicely justifies how life preserves homochirality and it is resilient to racemization: when death occurs racemization begins.

A significant aspect of this new type of SMSB reaction (16.7) is that it is based on homochiral interactions in contrast to the Frank and LES systems, whose inhibition stages correspond to heterochiral interactions. This is a significant point because heterochiral coupling is undesirable with respect to the achievement of stability and the ability to evolve into biological replicators. In fact, the assumption that SMSB could only occur through Frank-like systems has led to the assumption of homochirality during the final stages of chemical evolution.

The ability of hypercyclic replicators to lead to SMSB cannot overlook the necessity, in order to have a full justification of the emergence of SMSB reaction networks, of how a polymer can be converted into a self-replicating system.

SMSB in polymerizations? Frank-like systems have been reported for SMSB in homochiral polymers (e.g., Ref. [25]). This assumes that the homochiral chain is the output of the polymerization. In biopolymers the diastereoselective selection towards homochiral chains seems to be confirmed by experimental reports on the polymerization of amino acids and nucleotides from racemic mixtures where, after a certain chain length, the diastereoselective formation of homochiral chains occurs preferentially [26]. In consequence, it is difficult to understand how such a system can, where homochiral interactions dominate, at the same time include heterochiral inhibition reactions? Other systems for SMSB in polymers probably exist. In this respect, the SMSB mechanism in the Viedma deracemization [27] (obtained by the crystal grinding of conglomerate racemic mixtures of crystals of achiral or racemic compounds, in their saturated solutions) is not related to Frank-like or LES-type SMSB systems. In our opinion the SMSB mechanism in the Viedma deracemization [11,28] is an example of how SMSB can occur in homochiral polymerizations. The significance of all this with respect to BH is that, assuming that the emergence of catalytic functionalities [4b] is necessary to convert biopolymeric replicators in the hypercyclic replicators as discussed above, such an event has a reasonable probability of occurring when a high number of different homochiral polymers (composed by different residues!) are formed and that very probably homochiral purity is an advantage for its selection. In summary, and in our opinion, this prior SMSB in polymerization is necessary for the emergence of hypercyclic replicators. As the mechanisms of Viedma are relatively well described [27], more work is now necessary for translating these mechanisms into models of SMSB in polymerization reactions and avoiding models based on heterochiral reactions for the inhibition stage of the reaction network.

16.4 SYSTEM THERMODYNAMIC ARCHITECTURES

All SMSB reaction networks can only achieve a chiral final stable state within a range of specific reaction parameters (reaction rate constants and total chemical mass). In addition to this, it is necessary that the reactions occur in a system unable to come into thermodynamic equilibrium with its surroundings. The characteristics of this can also depend on the reaction network. Such systems represent different chemical scenarios; e.g., continuous flow reactors, systems open to matter transfer of specific reactants, closed systems of nonhomogeneous temperature or energy distribution with/without compartmentalized reactions. Sometimes the discussions of SMSB are garbled because the lack of differentiation and the distinction between the reaction network itself and the thermodynamic characteristics of the system where the reactions take place.

The LES example of Fig. 16.3 is an example of how the thermodynamic constraints that hinder the LES model can be overcome by the compartmentalization of different reactions of the SMSB network at different temperatures. This can also be achieved when the inhibition stage of the LES reaction network is driven by external reactants in a system open to them (see Fig. 16.5) [23].

The input of energy to only some species of the system is a well known method in chemistry to obtain nonequilibrium thermodynamic states (stationary), e.g., in photochemistry in the formation of photo stationary states. In SMSB an example of this is the Viedma deracemization, where the mechanical energy of grinding is only supplied to crystals above a certain size range [29].

To design experiments on SMSB in such thermodynamic scenarios should not be an obstacle for experimental synthesis. In fact, most chemical synthesis is performed in conditions leading to nonequilibrium thermodynamic states: the flask reflux process corresponds to a system unable to achieve thermodynamic equilibrium because it has a permanent temperature gradient (between the heated walls and the center of the surface collecting the solvent dropped from, for example, the continuous flow reactor conditions can be worked at the laboratory condensation). Today the new methodologies of micro flow reactors and the anchoring of reagents and catalysts on solid supports aid in this endeavor.

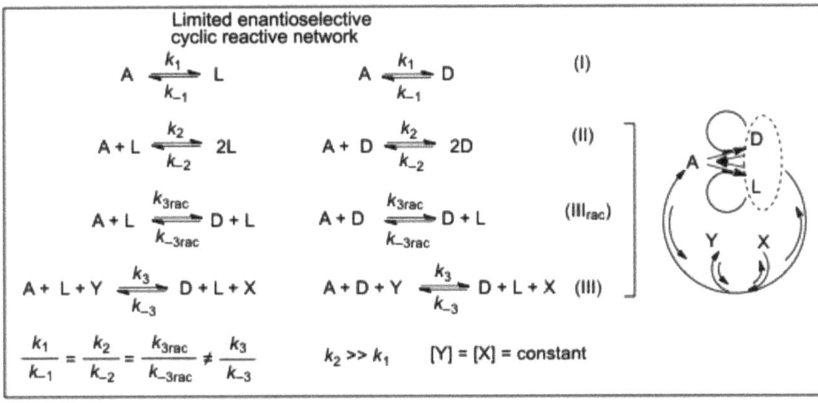

Figure 16.5 Limited enantioselective system (LES) cyclic reaction network with the inhibition heterochiral reaction driven/open by a external reactant X/Y. The cyclic LES network studied herein is that composed by the reactions (16.1), (16.2), and (16.6), but being (Eq. 16.6) promoted by the reactant pair X/Y. Simulations show a wide range of reaction parameters leading to SMSB (see Ref. [23]). *Adapted with permission from Blanco, C.; Crusats, J.; El-Hachemi, Z.; Moyano, A.; Hochberg, D.; Ribó, J. M. Spontaneous Emergence of Chirality in the Limited Enantioselectivity Model: Autocatalytic Cycle Driven by an External Reagent.* Chem. Phys. Chem. ***2013***, 14, *2432–2440.*

16.5 FROM STOCHASTIC TO SETERMINISTIC SPONTANEOUS MIRROR-SYMMETRY BREAKING

An SMSB system in the absence of any chiral polarization breaks parity but the chiral sign shows a stochastic distribution between successive experiments. Therefore for a relatively high number of experiments and in an abiotic scenario of a high number of sites where the reaction occurs, the total net outcome would be the racemic mixture. However, the presence of a very weak chiral polarization determines the final chiral sign [30]. This is a characteristic of the bifurcation scenario [10,30] and in consequence should not be described as an asymmetric induction, that leads to a bias from the racemic composition, but as a chiral sign selection. In this respect, numerical simulations show that the small differences in the enantiomeric reaction rate constants, such as those expected from the charge-parity violation energy differences (PVED) in formation of the atom nuclei, suffice to transform the stochastic final chiral sign into a deterministic one [18b]. This can lead to the temptation to attribute the sign of BH to the *deus ex machina* of PVED; however, the experimental examples of SMSB of the Soai reaction and the Viedma deracemization yield stochastic chiral sign distributions. This is because the stochastic chiral fluctuations of the reaction noise dominate clearly upon the effect of the permanent chiral polarization of PVED. For the simulation of the effect of statistical chiral fluctuations stochastic methods should be used; however, chemists want to have a sound chemical explanation of the origin of chiral chemical noise. In this respect, it has been recently shown how in the Soai reaction the nonhomogeneous isotopic natural composition leading to chirality at the very first steps of the reaction (low concentrations!), the statistical chiral fluctuations have already led to significant *ee* values [31]. This initial *ee* value, of stochastic sign distribution between experiments, would be able to lead the system away from the metastable racemic state.

The easy chiral sign selection in SMSB justifies how a deterministic chiral sign can be achieved in an abiotic scenario of multiple "reactors" in a spatially extended region when we speculate on the action of a natural chiral force upon an extended region or, which is much more likely, when we assume that the chiral compound mixtures forming the resources of SMSB systems already show a small *ee* value. Here the role of asymmetric induction in astrophysical scenarios of abiotic evolution must be taken into account. Such experimental conditions will select the chiral sign, i.e., they will convert the SMSB from a stochastic to a deterministic one.

Notice that the role of SMSB systems is to avoid racemization and the existing chiral machinery of life is probably an improved system of the primordial SMSB reaction networks of the hypercyclic replicator type discussed above. When we consider them as a form abiotic primordial genotype, the chiral sign selection, by the chiral interaction with the surroundings will represent a phenotypic selection; *the less fit* [4d] *of one of the two chiral signs becomes extinct* [32].

REFERENCES AND NOTES

[1] Guijarro, A.; Yus, M. *The Origin of Chirality in the Molecules of Life: A Revision from Awareness to the Current;* RSC Publishing: Cambridge, 2009.

[2] (a) Girard, C.; Kagan, H. B. Effects in Asymmetric Synthesis and Stereoselective Reactions: Ten Years of Investigation. *Angew. Chem., Int. Ed.* **1998,** *37,* 2922–2959.

(b) Feringa, B. L.; van Delden, R. A. Absolute Asymmetric Synthesis: The Origin, Control and Amplification of Chirality. *Angew. Chem. Int. Ed.* **1999,** *38,* 3418–3438.

[3] (a) Plasson, R.; Kondepudi, D. K.; Bersini, H.; Commeyras, A.; Asakura, K. Emergence of Homochirality in Far-From-Equilibrium Systems: Mechanisms and Role in Prebiotic Chemistry. *Chirality* **2007,** *19*, 589–600.
(b) Blackmond, D. G. The Origin of Biological Homochirality. In *The Origins of Life;* Deamer, D., Szostak, J. W., Eds.; Cold Spring Harbor Laboratory Press: New York, 2010.
[4] (a) Gánti, T. *A Theory of Biochemical Supersystems and its Application to Problems of Natural and Artificial Biogenesis*, Vol. 17. University Park Press: Baltimore, MD, 1979.
(b) Kauffman, S. A. Autocatalytic Sets of Proteins. *J. Theor. Biol.* **1986,** *119*, 1–24.
(c) Eigen, M.; McCaskill, J.; Schuster, P. Molecular Quasi-Species. *J. Phys. Chem.* **1988,** *92*, 6881–6891.
(d) Lifson, S.; Lifson, H. A Model for Prebiotic Replication: Survival of the Fittnest vs Extinction of the Unfittnest. *J. Theor. Biol.* **1999,** *199*, 425–433.
(e) Smith, J. M.; Szathmary, E. *The Major Transitions in Evolution;* Oxford University Press: Oxford, 1995.
(f) Orgel, L. E. Prebiotic Chemistry and the Origin of the RNS World. *Crit. Rev. Biochem. Mol. Biol.* **2004,** *39*, 99–123.
[5] Schuster, P. Chemistry on the Early Earth. *Germany-Japan Round Table* [Online]; Heidelberg, November 1–3, 2011. <http://www.tbi.univie.ac.at/~pks/Presentation/heidelberg-11.pdf>
[6] Oro, J. Comets and the Formation of Biochemical Compounds on the Primitive Earth. *Nature* **1961,** *190*, 389–390.
[7] (a) Cronin, J. R.; Pizzarello, S. Enantiomeric Excesses in Meteoric Amino Acids. *Science* **1997,** *275*, 951–955.
(b) Pizzarello, S.; Yarnes, C. T. Enantiomeric Excesses of Chiral Amines in Ammonia-Rich Carbonaceous Meteorites. *Earth Planet. Sci. Lett.* **2016,** *443*, 176–184.
[8] (a) Steel, M. The Emergence of a Self-Catalysing Structure in Abstract Origin-of-Life Models. *Appl. Math. Lett.* **2000,** *13*, 91–95.
(b) Hordik, W.; Hein, J.; Steel, M. Autocatalytic Sets and the Origin of Life. *Entropy* **2010,** *12*, 1733–1742.
[9] Avalos, M.; Babiano, R.; Cintas, P.; Jiménez, J. L.; Palacios, J. C. Chiral Autocatalysis: Where Stereochemistry Meets the Origin of Life. *Chem. Commun.* **2000,** *29*, 887–892.
[10] Kondepudi, D. K.; Nelson, G. Chiral Symmetry Breaking in Nonequilibrium Systems. *Phys. Rev. Lett.* **1983,** *50*, 1023.
[11] Ribó, J. M.; Blanco, C.; Crusats, J.; El-Hachemi, Z.; Hochberg, D.; Moyano, A. Absolute Asymmetric Synthesis in Enantioselective Autocatalytic Reaction Networks. *Chem. Eur. J.* **2014,** *20*, 1–23.
[12] Ribó, J. M.; Hochberg, D. Competitive Exclusion Principle in Ecology and Absolute Asymmetric Synthesis in Chemistry. *Chirality* **2015,** *27*, 722–723.
[13] (a) Eigen, M. Selforganization and the Evolution of Biological Macromolecules. *Naturwissenschaften* **1971,** *58*, 465–523.
(b) Eigen, M.; Schuster, P. *The Hypercycle: A Principle of Natural Self-Organization;* Springer: Berlin; Kiedrowski, Lifston, 1979.
[14] Lente, G. Homogeneous Chiral Autocatalysis: A Simple, Purely Stochastic Kinetic Model. *J. Phys. Chem. A* **2004,** *108*, 9475–9478.
[15] Stich, M.; Ribó, J. M.; Blackmond, D. G.; Hochberg, D. Necessary Conditions for the Emergence of Homochirality via Autocatalytic Self-Replication, submitted for publication.

[16] Plasson, R.; Brandenburg, A.; Jullient, L. H.; Bersini, H. Autocatalyses. *J. Phys. Chem. A* **2011,** *115*, 8073–8085.

[17] Frank, F. C. On Spontaneous Asymmetric Synthesis. *Biochim. Biophys. Acta* **1953,** *11*, 459–463.

[18] (a) Soai, K.; Kawasaki, T.; Matsumoto, A. Asymmetric Autocatalysis of Pyrimidyl Alkanol and its Application to the Study of the Origin of Homochirality. *Acc. Chem. Res.* **2014,** *47*, 3643–3654.

(b) For a Discussion on the SMSB Mechanisms in the Soai Reaction see.; Crusats, J.; Hochberg, D.; Moyano, A.; Ribó, J. M. Frank Model and Spontaneous Emergence of Chirality in Closed Systems. *Chem. Phys. Chem.* **2009,** *10*, 2123–2131.

[19] Avetisov, V. A.; Goldanskii, V. Mirror Symmetry Breaking at the Molecular Level. *Proc. Natl. Acad. Sci. U. S. A.* **1996,** *93*, 11435–11442.

[20] Ribó, J. M.; Hochberg, D. Stability of Racemic and Chiral Steady States in Open and Closed Chemical Systems. *Phys. Lett. A* **2008,** *373*, 111–122.

[21] Blackmond, D. G.; Matar, O. K. Re-examination of Reversibility in Reaction Models for the Spontaneous Emergence of Chirality. *J. Phys. Chem. B* **2008,** *112*, 5098–5104.

[22] (a) Ribó, J. M.; Crusats, J.; El-Hachemi, Z.; Moyano, A.; Blanco, C.; Hochberg, D. Spontaneous Mirror Symmetry Breaking in the Limited Enantioselective Autocatalysis Model: Abyssal Hydrothermal Vents as Scenario for the Emergence of Chirality in Prebiotic Chemistry. *Astrobiology* **2013,** *13*, 132–142.

(b) Blanco, C.; Ribó, J. M.; Crusats, J.; El-Hachemi, Z.; Moyano, A.; Hochberg, D. Mirror Symmetry Breaking with Limited Enantioselective Autocatalysis and Temperature Gradients: a Stability Survey. *Phys. Chem. Chem. Phys.* **2013,** *15*, 1546–1556.

[23] Blanco, C.; Crusats, J.; El-Hachemi, Z.; Moyano, A.; Hochberg, D.; Ribó, J. M. Spontaneous Emergence of Chirality in the Limited Enantioselectivity Model: Autocatalytic Cycle Driven by an External Reagent. *Chem. Phys. Chem.* **2013,** *14*, 2432–2440.

[24] Ribó, J. M.; Crusats, J.; El-Hachemi, Z.; Moyano, A.; Hochberg, D. Spontaneous Mirror Symmetry Breaking in Heterocatalytically Coupled Enantioselective Replicators, submitted for publication.

[25] Sandars, P. G. H. A Toy Model for the Generation of Homochirality During Polymerization. *Orig. Life Evol. Biosph.* **2003,** *33*, 575–587.

[26] (a) Hitz, T.; Luisi, P. L. Chiral Amplification of Oligopeptides in the Polymerization of α-Amino Acid n-Carboxyanhydrides in Water. *Helv. Chim. Acta* **2003,** *86*, 1423–1434.

(b) Lee, D. H.; Granja, J. R.; Martinez, J. A.; Severin, K.; Ghadiri, M. R. A Self-Replicating Peptide. *Nature* **1996,** *383*, 525–528.

(c) Illos, R. A.; Bisogno, F. R.; Clodic, G.; Bolbach, G.; Weissbuch, I.; Lahav, M. Oligopeptides and Copeptides of Homochiral Sequence, via β-sheets, from Mixtures of Racemic α-Amino Acids, in a One-Pot Reaction in Water; Relevance to Biochirogenesis. *J. Am. Chem. Soc.* **2008,** *130*, 8651–8659.

(d) Szepanski, T.; Joyce, G. F. A Cross-Chiral RNS Polymerase Ribozyme. *Nature* **2014,** *515*, 440–442.

[27] Viedma, C. Chiral Symmetry Breaking during Crystallization: Complete Chiral Purity Induced by Nonlinear Autocatalysis and Recycling. *Phys. Rev. Lett.* **2005,** *94*, 065504.

[28] Blanco, C.; Crusats, J.; El-Hachemi, Z.; Moyano, A.; Veintemillas-Verdaguer, S.; Hochberg, D.; Ribó, J. M. The Viedma Deracemization of Racemic Conglomerate

Mixtures as a Paradigm of Spontaneous Mirror Symmetry Breaking in Aggregation and Polymerization. *Chem. Phys. Chem.* **2013,** *14*, 3982–3993.

[29] It is a common place the statement that through the income of energy to a system it is possible to achieve a non-equilibrium thermodynamic state. This is misleading because when the system is able to equilibrate with the surroundings there is a change of the equilibrium constants and of the reaction rate constants but the system evolves simply towards a new thermodynamic equilibrium. Only in the case that the given energy is giving only to some of the chemical species of the system, e.g., in photochemically, a disipative state of minimum entroy production may be achieved.

[30] (a) Kondepudi, D. K.; Prigogine, I.; Nelson, G. Sensitivity of Branch Selection in Nonequilibrium Systems. *Phys. Lett. A* **1985,** *111*, 29–32.
(b) Avetisov, V. A.; Kuz'min, V. V.; Anikin, S. A. Sensitivity of Chemical Chiral Systems to Weak Asymmetric Factors. *Chem. Phys.* **1987,** *112*, 179–187.

[31] Barabás, B.; Kurdi, R.; Pályi, G. Natural Abundance Isotopic Chirality in the Reagents of the Soai Reaction. *Symmetry* **2016,** *8*, 2.

[32] We use here the indication given in Ref. [4d], that in fact the driving force of the replicator selection is the extinction of the less fit.

CHAPTER 17

Catalytic Applications of Supported Ionic Liquid Phases

Csaba Fehér, Máté Papp, Béla Urbán and Rita Skoda-Földes
University of Pannonia, Institute of Chemistry, Department of Organic Chemistry, Veszprém, Hungary

Abstract

Supported ionic liquids (SILPs), prepared via the immobilization of ionic liquids on solid material, can efficiently be used in catalytic reactions as catalysts, or as supports for transition metal catalysts. The SILP concept combines the advantages of homogeneous catalysis with those of heterogeneous systems. In the present paper, this is demonstrated with two examples. In the first application, acidic SILPs are shown to be selective and recyclable catalyst for the trimerization of isobutene. In the second example, neutral SILPs are proved to be excellent supports stabilizing palladium catalysts for carbonylation reactions, as well as ensuring effective catalyst recycling.

Keywords: Ionic liquids; heterogeneous catalysts; recycling; acid catalysis; palladium catalysts

17.1 INTRODUCTION

Over the past 10–15 years there has been a great boom in the number of publications dealing with ionic liquids (ILs). The interest in these compounds seems to persist and several reports appear about new applications of ILs in electrochemistry, materials science, separation technologies, etc. [1].

ILs are salts consisting of bulky organic cations and inorganic or organic anions. They melt at relatively low temperature, usually below 100°C, they have negligible vapor pressure and they are not flammable. They dissolve both polar organic molecules and inorganic salts so they can replace volatile organic solvents.

In organic chemistry ILs have become alternative reaction media for a great variety of transformations, especially for reactions involving acid, base and transition metal catalysis [2]. Low vapor pressure of ILs does not render them green solvents per se, since there are several examples for both toxic and nontoxic derivatives [3,4], but still, their nonvolatility makes them safer to handle and their high polarity facilitates extraction of

Advances in Asymmetric Autocatalysis and Related Topics
DOI: http://dx.doi.org/10.1016/B978-0-12-812824-4.00017-4

products with less polar solvents. They can be reused in most of the organic applications and this evidently reduces the amount of wastes. However, the main advantage of these compounds lies in the great variability in the structure of both cation and anion, which theoretically leads to an infinite number of possible derivatives.

Since ILs are known to stabilize organometallic complexes with metals in low oxidation state, their use as reaction media for transition metal-catalyzed reactions has been thoroughly investigated in recent years [5]. By this methodology, the advantages of homogeneous catalysis can be retained that results in high activity and selectivity. At the same time, the application of IL solvents may help to circumvent problems arising from difficulties in catalyst recovery and recycling.

With the introduction of a properly selected functionality into the cation or anion, task-specific derivatives [6] can be produced that may serve not just as recyclable reaction media, but also as catalysts in acid-, base- or organocatalyzed reactions [7,8].

The use of ILs as solvents and/or catalysts usually makes efficient catalyst recovery and recycling possible. At the same time, because of the high viscosity of some ILs, their handling is often cumbersome and catalytic reactions are limited by diffusion processes. In addition, the use of ILs in their liquid form is inconvenient in an industrial continuous system. Biphasic IL—organic systems require large amounts of the expensive ILs, which also hinders industrial applications. To overcome these difficulties, a new approach, the so-called "supported ionic liquid phase" (SILP) technology that combines the advantages of ILs with those of heterogeneous supports, was developed [9].

In SILP catalysts, a thin film of IL is adsorbed on the surface of a support, usually silica or other inorganic or organic polymers. The bonding of the IL to the support is mainly attributed to hydrogen bonds between surface, anion, and cation. A further possibility for immobilizing ILs on a solid surface is the grafting method. Hence, trialkoxysilylated ILs are condensed with surface hydroxyl groups or co-condensed in a sol–gel process, e.g., with tetraethoxysilane. The supported ILs may function as catalysts themselves, or may stabilize transition metal complexes or catalytically active nanoparticles on the surface of the solid support. In both cases, the catalysts can easily be removed from the reaction mixture by simple filtration and can be recycled.

In the past few years, application of SILPs as an acidic catalyst for oligomerization of isobutene and as supports for palladium catalyzed carbonylations has been explored in our group. Some of the results are summarized in the present paper.

17.2 RESULTS AND DISCUSSION

17.2.1 Acidic Supported Ionic Liquid Phases as Catalysts for Oligomerization of Isobutene

The oligomerization of light olefins, such as isobutene, is an important alternative for the production of higher molecular weight hydrocarbon mixtures useful as fuels (e.g., gasoline or diesel). By the hydrogenation of trimerization products, excellent blending components for jet fuel can be produced [10]. The main challenges in the design of catalysts for this reaction are focused to reach high conversion and high selectivity. Both Brønsted and Lewis acids have been used as oligomerization catalysts in either homogeneous (H_2SO_4, organometallic complexes, etc.) or heterogeneous (mixed oxides, zeolites, clays, ion exchange resins, etc.) phase [11].

Several solid-acid catalysts, less harmful than traditional liquid acids due to their nonvolatility, have been developed [10]. As another approach, Lewis acidic ILs were shown to catalyze isobutene oligomerization with high selectivity toward diisobutene and triisobutene [12]. Imidazolium ionic liquids with SO_3H functionalized side chains were also found to be excellent and reusable catalysts in oligomerization of various alkenes [13]. As supported acidic ILs had been shown to be efficient catalysts in a number of other organic reactions [14], we decided to explore the possibility of the use of similar systems in the oligomerization of isobutene.

17.2.2 Preparation of Acidic Supported Ionic Liquid Phase Catalysts

Two Brønsted acidic ionic liquids (**IL-1** and **IL-2**, Scheme 17.1) were synthesized and used for the impregnation of silica supports with different morphology (Table 17.1) to give catalysts **SILP-1** — **SILP-3** [15,16].

The values of the specific surface area (S_{BET}) and pore volumes of supports (**S1**, **S2**) and SILP catalysts (Table 17.1) were determined from the experimental data of their nitrogen adsorption/desorption isotherms.

S2 had substantially higher BET surface area than **S1**. There was also a great difference in pore size distribution of the supports. **S1** had mesopores (mainly in the 2–10 nm region) and had no measurable micropore surface area. At the same time, **S2** had a considerable amount of

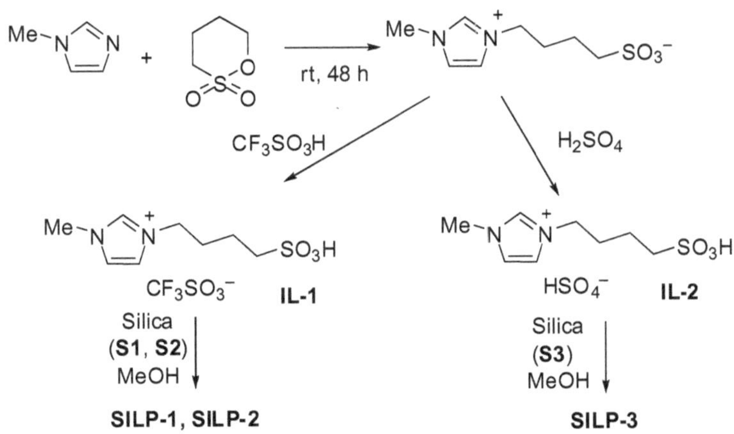

Scheme 17.1 Synthesis of ionic liquids **IL-1** and **IL-2** and preparation of supported ionic liquids **SILP-1—SILP-3.**

Table 17.1 Properties of supports and supported ionic liquid phase catalysts

Support or catalyst	Particle size (mm)	S_{BET} (m^2/g)	Micropore surface area (m^2/g)	Pore volume (cm^3/g)	Amount of IL (mmol IL/g silica)	Acid capacity (mmol NH_3/g)
S1	0.040–0.063	480	0	0.74		
S2	3–6	664	247	0.13		
SILP-1		238	0	0.38	0.86	0.683
SILP-2		196	10	0.05	0.42	0.360

micropores that greatly contributed to its BET surface, but the total volume of its mesopores was low (Fig. 17.1).

As expected, there was a loss of BET surface when the supports were impregnated with IL (Table 17.1). The difference between the S_{BET} of support and SILP catalyst was greater in the case of support with micropores (**S2** vs **SILP-2**). This is in accordance with the observation of Rodriguez, who found that in microporous and mesoporous materials, ionic liquids fill the micropores first [17].

It can be assumed that support **S1** has pear-shaped pores that retain their shape during the catalyst preparation process, although the total pore volume values decreases because of the active ionic liquid film on the wall of the pores. On the other hand, the microporous **S2** contains narrow-shaped pores that become almost totally filled with ionic liquid film (Fig. 17.1). The acid capacity of **SILP-1** is higher than **SILP-2**, in accordance with the different amounts of adsorbed ionic liquid.

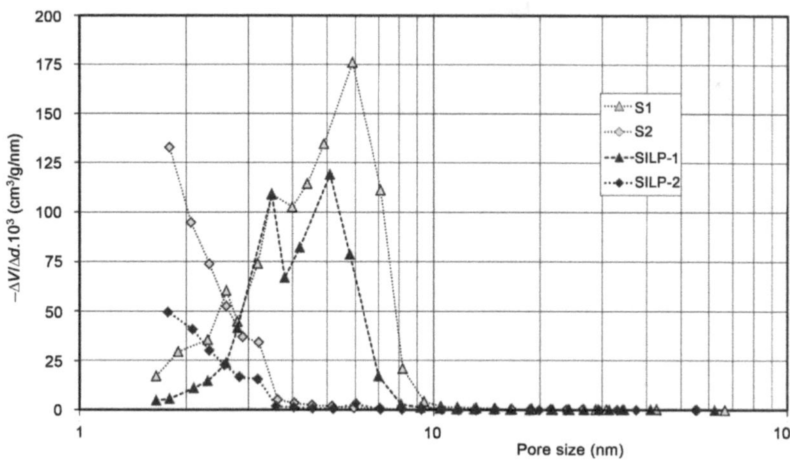

Figure 17.1 Differential pore volume distribution of silica supports and supported ionic liquid phase catalysts.

17.2.3 Catalytic Experiments

All of the supported ionic liquid catalysts (**SILP-1—SILP-3**) were found to be active in oligomerization of isobutene. (It should be mentioned that silica gel itself had no catalytic activity in this reaction.)

The decisive effect of the morphology of the support on the activity and selectivity of the catalysts can be demonstrated upon comparing the results obtained with **SILP-1** and **SILP-2** (Figs. 17.2 and 17.3). Although a twofold amount of the catalyst was used in case of **SILP-2**, in order to keep the isobutene/ionic liquid ratio constant (isobutene/IL = 68), experiments carried out at lower temperatures or with shorter reaction times showed that it was a less active catalyst than **SILP-1** (e.g., TOFs of **SILP-1** and **SILP-2** are 9.3 and 7.4 h^{-1}, at 60°C, respectively).

In case of **SILP-2**, the diffusion of isobutene to the micropores is blocked by the ionic liquid that fills the pore, so the contact surface between the ionic liquid phase and the organic phase is lower than that of the SILP catalyst with mesopores (**SILP-1**). The lower contact surface results in lower catalytic activity.

With the exception of the reaction carried out at 60°C, the use of **SILP-1** led to the C_{12} compounds as the main products. As expected, a shift toward the heavier oligomers was observed in longer reaction times. The same tendency was observed with an increase in the temperature.

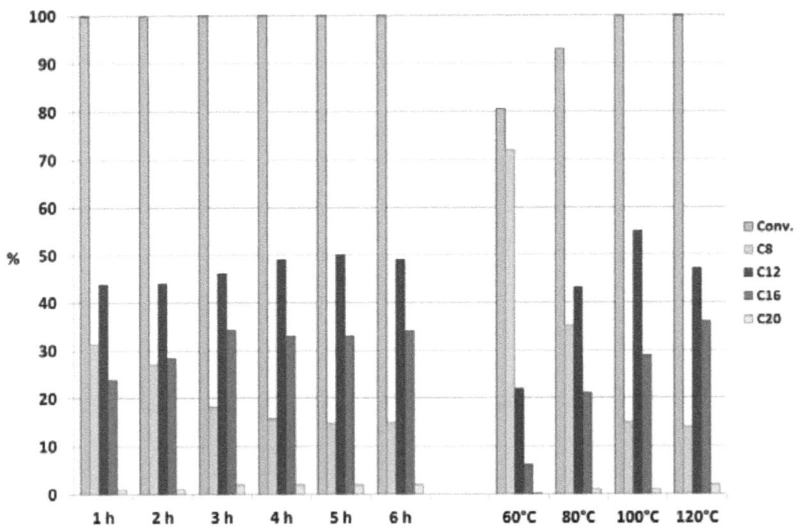

Figure 17.2 Oligomerization in the presence of **SILP-1** with different reaction time (temperature: 100°C) and at different temperatures (reaction time: 5 h).

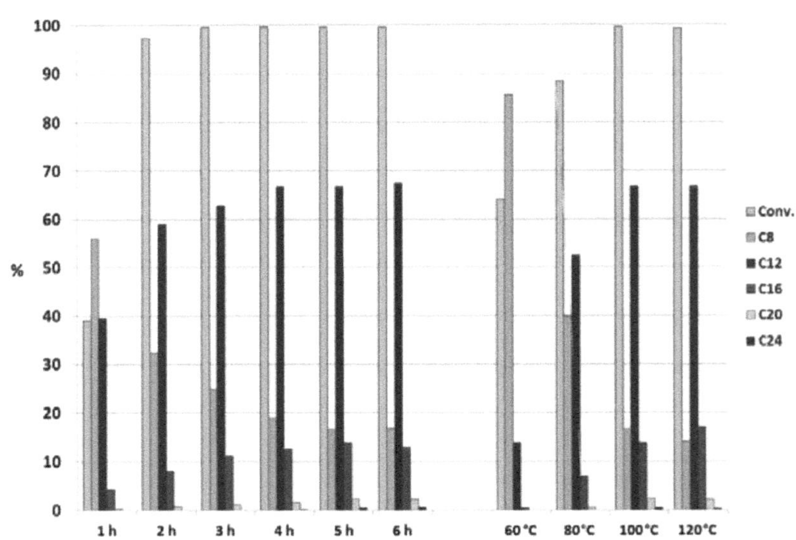

Figure 17.3 Oligomerization in the presence of **SILP-2** with different reaction time (temperature: 100°C) and at different temperatures (reaction time: 5 h).

Best C_{12} selectivity was obtained at 100°C (reaction time: 5 h). The lower activity of **SILP-2** results in an increased ratio of lower oligomers under identical conditions. At the same time, an improved selectivity for trimers could be observed at 100–120°C compared to the results obtained with **SILP-1**.

The possibility of the reuse of the SILP catalysts was also investigated. **SILP-1** and **SILP-2** could be recycled several times without loss of activity or remarkable change in selectivity (Figs. 17.4 and 17.5). The main products were the C_{12} isomers obtained in a 49%–59% yield with **SILP-1**, while **SILP-2** showed an even better C_{12} selectivity (67%–70%).

Excellent C_8 selectivity (83%) was achieved with the imidazolium hydrogensulfate catalyst (**SILP-3**), but unfortunately a great drop of activity was observed upon reuse (from 85% conversion to 21% in the second run). This shows a decisive effect of the anion on the stability of the catalyst. In case of **SILP-1**, an overall 2% loss of the catalyst was observed during the eight runs. Leaching was observed to be approximately double with catalysts incorporating the hydrogensulfate anion. The higher rate of catalyst loss might be one of the factors that lead to inactivation of **SILP-3**, but the difference in the leaching of the two catalysts is not great enough to account for the great loss of activity of **SILP-3** alone.

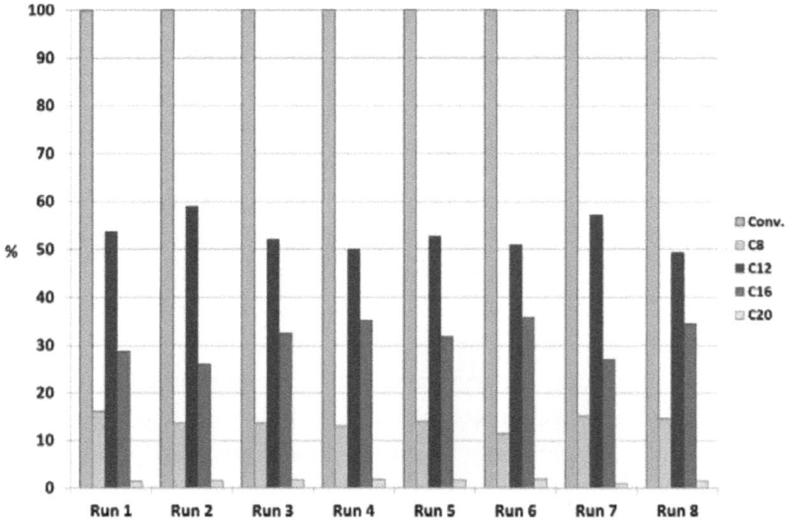

Figure 17.4 Recyclability of **SILP-1** in oligomerization of isobutene (100°C, 5 h).

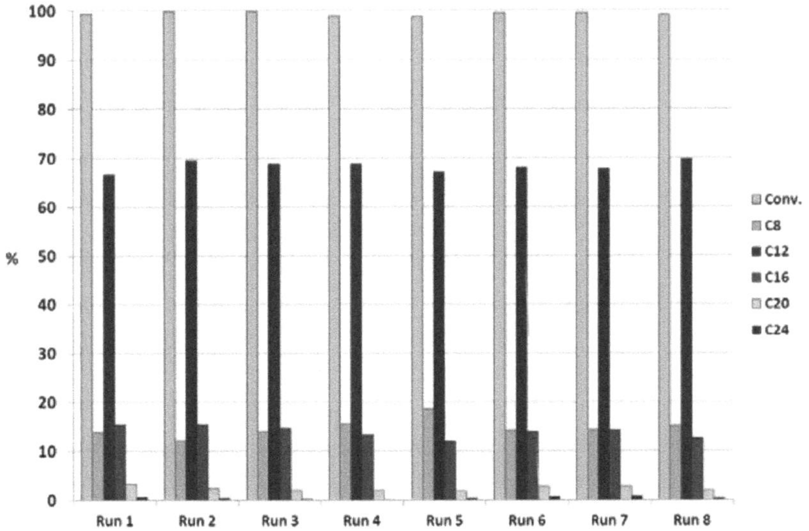

Figure 17.5 Recyclability of **SILP-2** in oligomerization of isobutene (100°C, 5 h).

As a conclusion it can be stated that both **SILP-1** and **SILP-2** are stable and recyclable catalysts in the oligomerization of isobutene. They lead to formation of C_{12} products, starting materials for blending components for jet fuel, with good to excellent selectivity.

17.2.4 The Use of Palladium Catalysts Immobilized on Supported Ionic Liquid Phases in Aminocarbonylation of Aryl Halides

Heterogeneous palladium catalysts obtained by the immobilization of metal nanoparticles or complexes on SILPs have been used efficiently for C—C coupling reactions, but examples for carbonylations are rare [18]. Palladium catalyzed carbonylation in the presence of organic nucleophiles serves as a powerful tool for the conversion of aryl/alkenyl halides or halide equivalents to carboxylic acid derivatives [19], so the development of stable, recyclable catalysts may facilitate industrial applications.

The reaction of amines, aryl halides and carbon monoxide in the presence of a palladium catalyst and a base leads to amides and α-ketoamides, via mono- and double carbonylation processes, respectively. Both types of the products have practical importance: the amide functionality is an important motif in a lot of biologically active molecules [20] and some α-ketoamides also show interesting pharmacological properties, such as anti-HIV activity [21] or HCV protease inhibition [22].

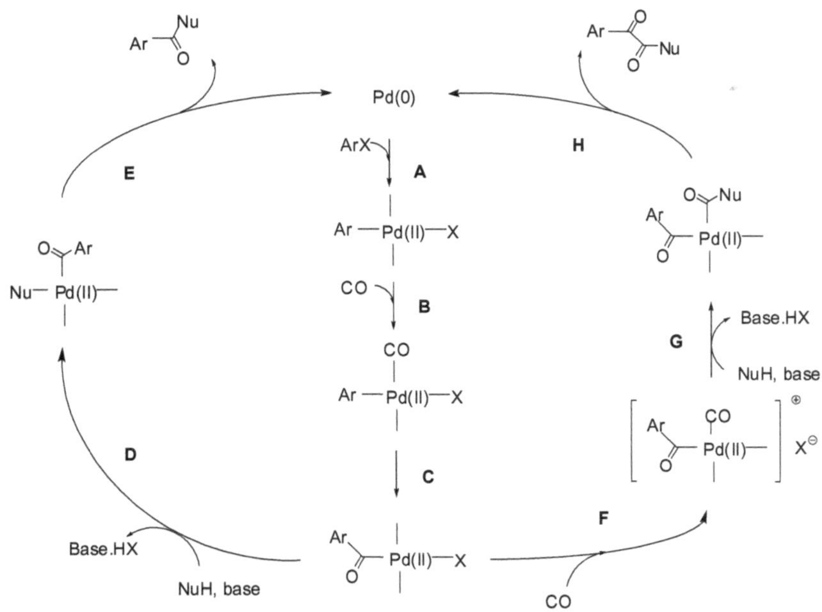

Figure 17.6 Mechanism of palladium-catalyzed carbonylation.

Under homogeneous conditions the mechanism, depicted in Fig. 17.6, is more or less explored [19b,23]. The catalytically active palladium(0) species undergoes oxidative addition in the presence of the aryl halide (**A**), followed by coordination (**B**) and migratory insertion (**C**) of CO, nucleophilic attack (**D**) and reductive elimination (**E**).

In the case of heterogeneous catalysts, the catalytic systems can be very complicated, as palladium either adsorbed on the surface of the support or dissolved in the reaction mixture (or both) may play a role in the carbonylation. Moreover, both palladium nanoparticles and palladium complexes formed under catalytic conditions may be involved. According to our own results, modification of the support, that of the palladium precursor or conditions of immobilization may affect considerably the selectivity and recyclability of the catalysts.

In the past few years, a detailed investigation of the influence of immobilization methodology on the activity, selectivity and recyclability of the catalysts has been carried out in our group.

17.2.5 Preparation of the Catalysts

Supported ionic liquid phases were obtained either by impregnation of silica gel with the IL (**SILP-4**, Fig. 17.7) or by anchoring the IL

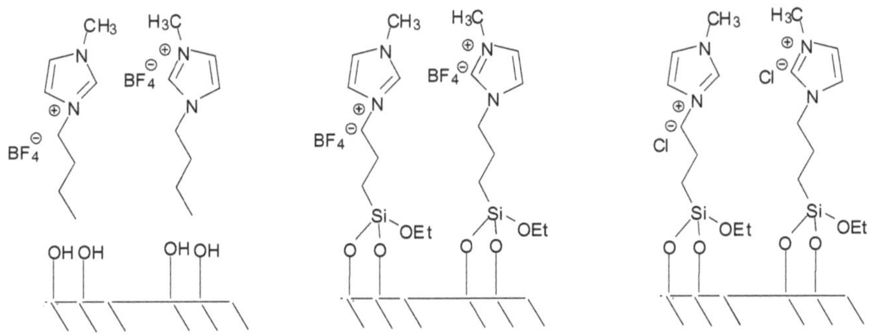

SILP-4 SILP-5 SILP-6

Figure 17.7 Supported ionic liquid phases used as supports in Pd-catalyzed carbonylations.

Table 17.2 Preparation of supported palladium catalysts used in the aminocarbonylation reactions

Catalyst	Support	Pd-precursor	Solvent	Additive/ ligand	Pd-content (m/m%)
CAT-1	Silica	$Pd_2(dba)_3.CHCl_3$	THF, CH_3CN	–	0.55
CAT-2	**SILP-4**	$Pd_2(dba)_3.CHCl_3$	THF, CH_3CN		0.40
CAT-3	**SILP-5**	$Pd_2(dba)_3.CHCl_3$	THF, CH_3CN		0.29
CAT-4	**SILP-5**	$Pd_2(dba)_3.CHCl_3$	EtOH		0.44
CAT-5	**SILP-5**	$Pd(OAc)_2$	THF, CH_3CN	DPPBA	0.58
CAT-6	**SILP-6**	$Pd(OAc)_2$	EtOH	KO*t*Bu	0.60

covalently to the support (grafting, **SILP-5** and **SILP-6**) [24–26]. Immobilization of ILs on silica was proved by ^{13}C-, and ^{29}Si cross-polarization magic angle spinning nuclear magnetic resonance (CP-MAS NMR) and Fourier transform infrared spectroscopy. In order to explore the effect of the IL on the effectiveness of the catalyst, unmodified silica was also used as a support (**CAT-1**, Table 17.2).

Palladium was immobilized on these phases using different Pd(0) or Pd(II) precursors, additives, ligands and solvents (Table 17.2). $Pd(OAc)_2$ may be reduced to catalytically active Pd(0) during the aminocarbonylation reaction with the help of CO, Et_3N, by the solvent (e.g., dimethylformamide (DMF)) or phosphine ligand. In case of **CAT-4** and **CAT-6** heterogenization was carried out in EtOH to facilitate reduction of Pd(II) to Pd(0) [27] prior to carbonylation.

A catalysts system composed of $Pd(OAc)_2$ and DPPBA (4-diphenylphosphino-benzoic acid) (**CAT-5**) was also prepared as it had been proved to be an efficient and recyclable catalyst in aminocarbonylations carried out in ionic liquids as solvents under homogeneous conditions [28]. The ^{13}C CP-MAS NMR spectrum of **CAT-5** proved the presence of both IL and DPPBA on the solid support. The ^{31}P CP-MAS NMR spectrum supported the formation of a palladium(II)-phosphine complex.

Imidazolium ILs with halide anions are known to produce carbene complexes that can be active catalysts themselves or may serve as a reservoir for palladium nanoparticles. As a consequence, **CAT-6** palladium was immobilized on **SILP-6** in the presence of KO*t*Bu to facilitate the formation of a carbene complex.

Palladium-content of the catalysts was determined by inductively coupled plasma (ICP). In some cases, the catalysts were characterized by transmission electron microscope (TEM).

17.2.6 Double Carbonylation Reactions

Aminocarbonylation of iodobenzene (**1**) with morpholine (**2**) was chosen as the model reaction (Scheme 17.2) to test the selectivity and recyclability of the catalysts. According to literature examples [23] and our previous results [29], under homogeneous conditions the ratio of mono- to double-carbonylation products decreases with CO pressure and increases with temperature. The use of polar solvents favors the formation of α-ketoamides.

Since there had been no precedent for a double carbonylation reaction carried out with a SILP catalyst, the reaction was investigated first at 30 bar CO pressure and in DMF in order to push the reaction towards the formation of the α-ketoamide **4**.

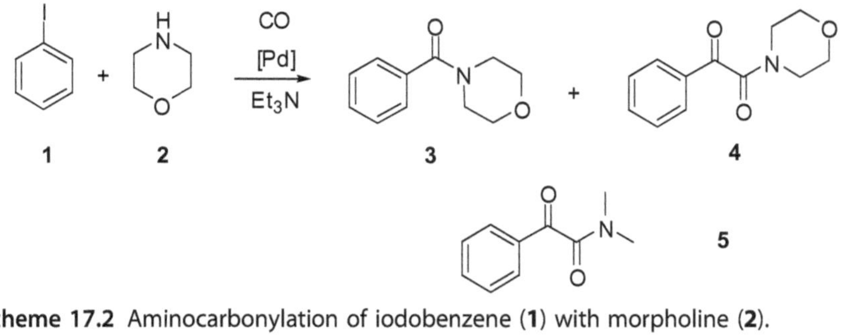

Scheme 17.2 Aminocarbonylation of iodobenzene (**1**) with morpholine (**2**).

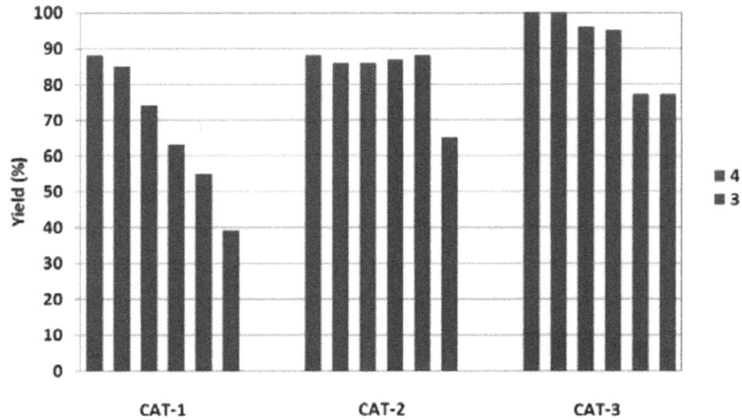

Figure 17.8 Recycling experiments in aminocarbonylation of iodobenzene (**1**) with morpholine (**2**). (1.8 mol% Pd, DMF, 30 bar, 100°C, 3 h).

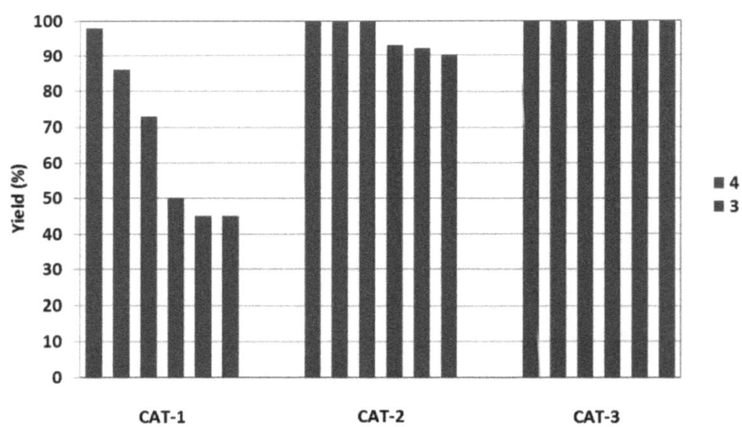

Figure 17.9 Recycling experiments in aminocarbonylation of iodobenzene (**1**) with morpholine (**2**). (1.8 mol% Pd, DMF, 30 bar, 100°C, 8 h.)

The modification of the support with ionic liquids, even by physisorption, resulted in more active catalysts (**CAT-2**, **CAT-3**) than that obtained from silica itself (**CAT-1**). Although the selectivity of the latter catalyst is comparable to that of **CAT-2** and **CAT-3**, it showed the greatest drop of activity in recycling experiments (Figs. 17.8 and 17.9).

The catalyst (**CAT-2**) obtained from the support prepared by simple impregnation of silica with the IL (**SILP-4**) showed better selectivity for the double carbonylation product **4**, unfortunately together with a somewhat greater decrease of activity upon reuse. The latter phenomenon can

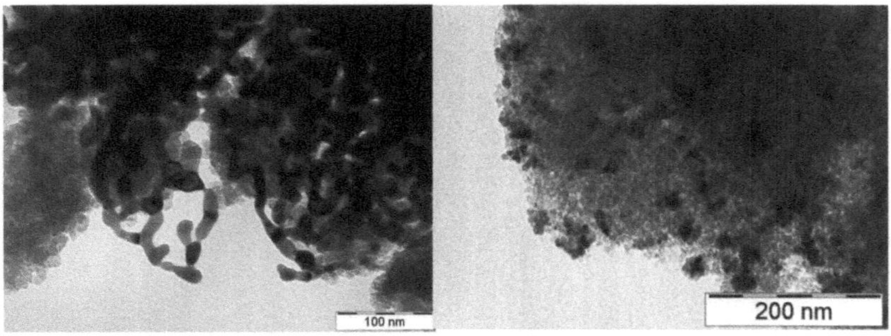

Figure 17.10 TEM images of spent **CAT-2** (left) and **CAT-3** (right) after two runs (with 8 h reaction time).

be explained by a higher loss of palladium: an average of 5.3% in experiments carried out for 3 h. As a comparison, Pd leaching was 3.9% under similar conditions with **CAT-3**. The greatest difference in palladium loss was observed in the first runs, being 23.7% and 7.6% with **CAT-2** and **CAT-3**, respectively. This can be explained by the fact that due to the presence of the polar solvent, DMF in the carbonylation experiments, most of the adsorbed IL becomes dissolved in the reaction mixture, that promote dissolution of palladium. TEM image (Fig. 17.10) of spent **CAT-2** proves the presence of palladium nanoparticles and shows some residual ionic liquid despite the great loss of IL in the first cycle detected by ^{1}H NMR. It should be mentioned that a lower loss of palladium was observed in case of **CAT-3** in the recycling experiments too (an average of 1.6 compared to that of 3.2% observed with **CAT-2**).

CAT-3, prepared with **SILP-5** as support, turned out to be the most efficient catalyst with excellent recyclability (Figs. 17.8 and 17.9). Total conversion of iodobenzene (**1**) to ketoamide **4** could be achieved even in the 10th run in experiments carried out for 8 h.

The improved performance of catalysts in carbonylations carried out for 8 h instead of 3 h can be explained not only by the better conversion that can evidently be achieved in a longer reaction time, but also by a dissolution/reprecipitation mechanism, proposed by Köhler [30]. Even if soluble palladium complexes or palladium nanoparticles are involved in the catalytic cycle, they can get back to the surface of the support in the form of nanoparticles after the reaction is completed. The TEM image of the spent catalyst proved again the presence of palladium nanoparticles (Fig. 17.10).

The reprecipitation of palladium is supported by the smaller leaching of the catalysts in case of longer reactions (3.9% and 3.1% of the original

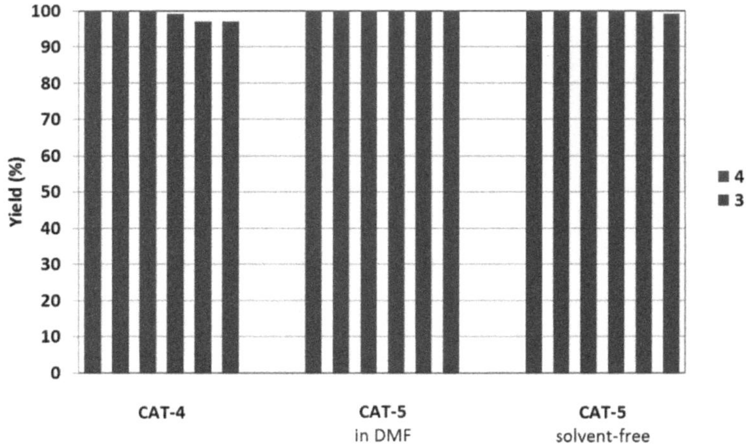

Figure 17.11 Catalyst recycling experiments in aminocarbonylation of iodobenzene (**1**) with morpholine (**2**) (left: **CAT-4**, middle: **CAT-5**, 1.8 mol% Pd, DMF, 30 bar, 100°C, 8 h, right: **CAT-5**, 0.9 mol% Pd, solvent-free, 5 bar, 120°C, 8 h).

load of **CAT-3** in experiments carried out for 3 h and 8 h, respectively). The higher activity of the recycled catalyst **CAT-3** in the longer experiments could also be proved. Mixtures, obtained after a 1 h reaction time in the second run with spent catalysts of the 3-h-long and 8-h-long experiments, revealed 38% and 47% conversion, respectively. Hot filtration and mercury tests showed that only a low conversion of iodobenzene (**1**) could be observed after the removal of the heterogeneous catalyst and the reaction could be completely stopped by the addition of mercury. This means that a small amount of catalytically active palladium nanoparticles have leached to the reaction mixture.

A catalyst with even higher activity could be obtained when palladium was deposited on **SILP-5** in EtOH (**CAT-4**). In this case, the presence of a side product (**5**, Scheme 17.2) was also observed due to the formation of dimethylamine by decomposition of DMF [31]. A change in the solvent of the reaction to acetonitrile results in a very selective reaction producing **4** in a 92%–97% yield (GC) (Fig. 17.11).

In order to show the efficiency of the latter catalyst, aminocarbonylation of iodobenzene in the presence of other aliphatic amines as well as the reaction of other aryl iodides with morpholine were investigated. A total conversion of the aryl iodides and good selectivity towards ketoamide products were observed in each case even after catalyst recycling (Fig. 17.12).

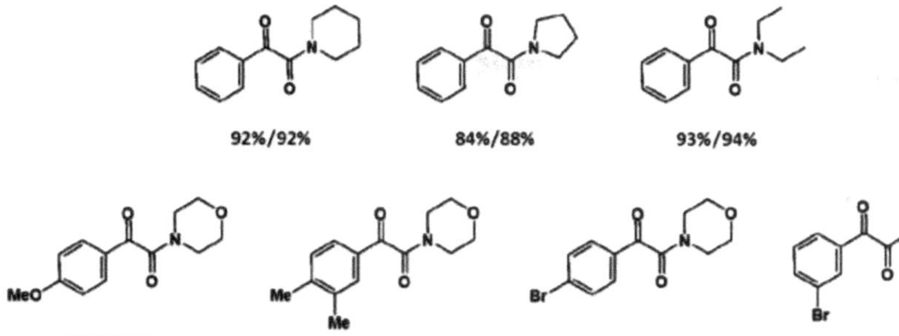

Figure 17.12 GC yields of α-ketoamides obtained via aminocarbonylation in the presence of **CAT-4** in DMF (1.8 mol% Pd, 30 bar, 100°C, 8 h).

Efficient double carbonylation also could be carried out in the presence of a catalyst obtained from $Pd(OAc)_2$ and DPPBA immobilized on **SILP-5** (**CAT-5**). Comparison of the FT-IR spectra of fresh and spent catalysts shows a loss of DPPBA ligand during the first run that accounts for the change in selectivity after the first experiment (Fig. 17.11). At the same time, the palladium–phosphine complex serves as a good precursor for the active species that leads to ketoamide **4** with excellent selectivity after the second run.

17.2.7 Monocarbonylation Reactions Leading to Carboxamides

In order to explore the possibility of the synthesis of carboxamides with the same catalysts, carbonylation was carried out at lower pressure (1–10 bar) and higher temperature (120°C). Although some change of selectivity towards **3** could be observed during the model reaction (Scheme 17.2), the results were not satisfactory. Besides, the use of lower pressure led to a decrease in the recyclability of the catalysts.

At the same time, when carbonylation was performed in the absence of a solvent and the products were extracted by toluene before reuse of the catalyst, monocarbonylation was the prevailing reaction in the presence of **CAT-1**, **CAT-2**, **CAT-3**, and **CAT-5**. Catalytic activity was even enhanced, so a reduced amount of catalysts (0.9 mol% instead of 1.8 mol%) could be used under solvent-free conditions. It can be assumed that due to the higher concentration of the reacting molecules, nucleophilic attack of the amine (Fig. 17.6, **D**) is very fast and there is not enough time for the coordination of a second CO (Fig. 17.6, **F**) that would lead to the

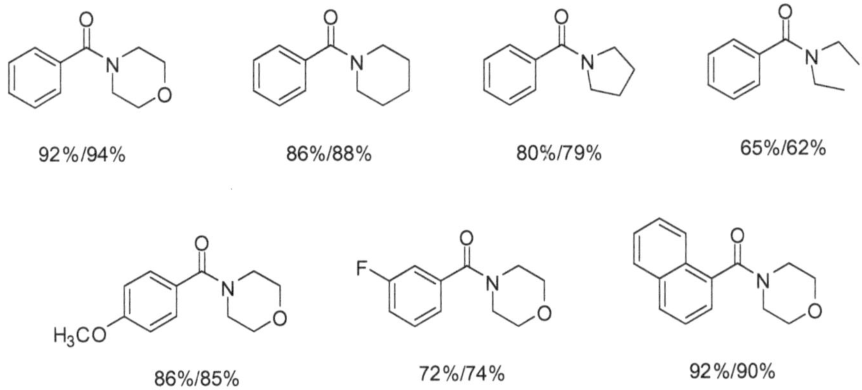

Figure 17.13 Isolated yields of carboxamides obtained via aminocarbonylation in the presence of **CAT-5** under solvent-free conditions in two successive runs (0.9 mol% Pd, 5 bar, 120°C, 8 h).

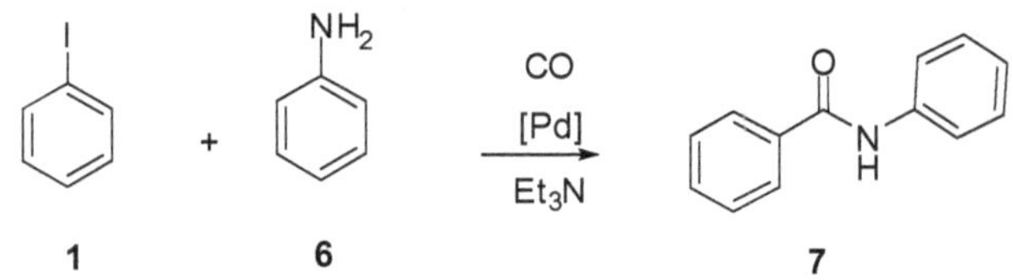

Scheme 17.3 Aminocarbonylation of iodobenzene (**1**) with aniline (**6**).

formation of a ketoamide. Among the catalysts investigated, **CAT-5** showed the best performance leading to the amide **3** with 98%–99% selectivity (Fig. 17.11), and retaining its activity in at least eight successive runs.

A loss of only 2.2% of the original palladium load was observed after the first run. Hot filtration and mercury tests showed that palladium nanoparticles are once again the main catalytically active species to have leached into the reaction mixture.

Under the same conditions, carboxamides could be isolated in good-to-excellent yields in aminocarbonylation of iodobenzene in the presence of other aliphatic amines, as well as in the reaction of other aryl iodides with morpholine (Fig. 17.13).

Carbonylation of aryl iodides with aromatic amines as nucleophiles uniformly led to the selective formation of monocarbonylated products (Scheme 17.3) even under high CO pressure. At 30 bar CO, catalysts **CAT-2** and **CAT-4** showed the best performance, though some loss of activity was observed with the latter catalyst (Fig. 17.14).

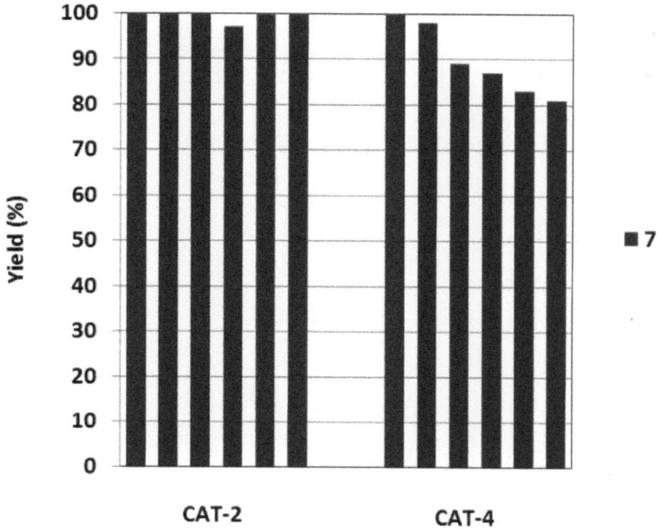

Figure 17.14 Catalyst recycling experiments in aminocarbonylation of iodobenzene (**1**) with aniline (**6**) (left: **CAT-2**, right: **CAT-4**, 3 mol% Pd, DMF, 30 bar, 100°C, 4 h).

Although under homogeneous conditions carbonylation of iodobenzene can be achieved even in atmospheric conditions [32], the use of all of the supported catalysts mentioned above (**CAT-1** to **CAT-5**) resulted in low conversion under 1 bar CO. Moreover, a great loss of activity was observed upon reuse. To our satisfaction we have found that **CAT-6**, obtained from $Pd(OAc)_2$ in ethanol with KO*t*Bu and immobilized on the chloride-containing support **SILP-6**, performed high activity and good recyclability in the reaction of iodobenzene with a number of aromatic amines, as well as in the carbonylation of substituted iodobenzenes with aniline (Fig. 17.15).

It should be mentioned, however, that the same catalyst showed a poor performance in aminocarbonylation with aliphatic amines under 1 bar CO. Lower conversion of iodobenzene, compared to that in the reaction of aniline, and poor selectivities were observed in each case.

As a summary, several recyclable heterogeneous catalysts were prepared, via immobilization of palladium on SILPs. **CAT-5** is the first SILP catalyst that can be used either for selective mono- or double carbonylation, depending on the reaction conditions. **CAT-6** is an effective catalyst for the conversion of aromatic amines, even under atmospheric conditions.

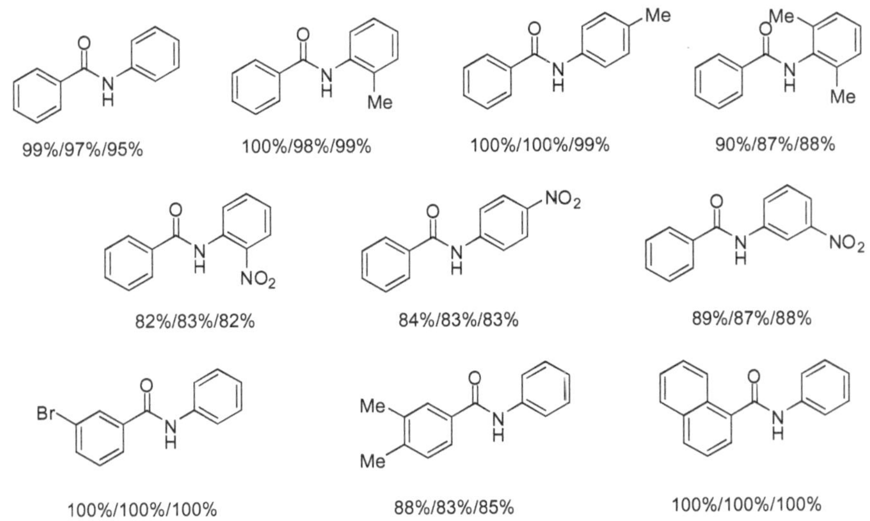

Figure 17.15 GC yields of carboxamides obtained via aminocarbonylation in the presence of **CAT-6** in DMF (3 mol% Pd, 1 bar, 100°C, 8 h).

17.3 CONCLUSIONS

The efficiency of SILPs in the preparation of recyclable heterogeneous catalysts was demonstrated in two very different reactions. A high variability of the ionic liquids and a diverse structure of possible supports make it possible to prepare task-specific SILPs. There is still a great potential in as yet unexplored applications, such as in asymmetric synthesis. Ionic liquids tagged with chiral groups were shown to catalyze enantioselective reactions [33]. Development of supported versions may lead to recyclable catalysts that are easier to handle and that can also be used in continuous flow reactors.

ACKNOWLEDGMENTS

The authors thank the Hungarian National Science Foundation for financial support (OTKA K105632), E. Kriván and J. Hancsók for their contribution in oligomerization experiments and E. Drotár for the TEM measurements.

REFERENCES

[1] (a) Kokorin, A., Ed. *Ionic Liquids: Theory, Properties, New Approaches;* InTech Open: Rijeka, 2011.

(b) Torimoto, T.; Tsuda, T.; Okazaki, K.; Kuwabata, S. New Frontiers in Materials Science Opened by Ionic Liquids. *Adv. Mater.* **2010,** *22*, 1196–1221.

[2] Qureshi, Z. S.; Deshmukh, K. M.; Bhanage, B. M. Applications of Ionic Liquids in Organic Synthesis and Catalysis. *Clean Technol. Environ. Policy* **2014,** *16*, 1487–1513.

[3] Ventura, S. P. M.; de Barros, R. L. F.; Sintra, T.; Soares, C. M. F.; Lima, A. S.; Coutinho, J. A. P. Simple Screening Method to Identify Toxic/Non-toxic Ionic Liquids: Agar Diffusion Test Adaptation. *Ecotoxicol. Environ. Saf.* **2012,** *83*, 55–62.

[4] Petkovic, M.; Seddon, K. R.; Rebelo, L. P. N.; Silva Pereira, C. Ionic Liquids: A Pathway to Environmental Acceptability. *Chem. Soc. Rev.* **2011,** *40*, 1383–1403.

[5] (a) Pârvulescu, V. I.; Hardacre, C. Catalysis in Ionic Liquids. *Chem. Rev.* **2007,** *107*, 2615–2665.

(b) Liu, Y.; Wang, S. S.; Liu, W.; Wan, Q. X.; Wu, H. H.; Gao, G. H. Transition-Metal Catalyzed Carbon-Carbon Couplings Mediated with Functionalized Ionic Liquids, Supported-Ionic Liquid Phase, or Ionic Liquid Media. *Curr. Org. Chem.* **2009,** *13*, 1322–1346.

(c) Olivier-Bourbigou, H.; Magna, L.; Morvan, D. Ionic Liquids and Catalysis: Recent Progress from Knowledge to Applications. *Appl. Catal. A* **2010,** *373*, 1–56.

(d) Navalon, S.; Alvaro, M.; Garcia, H. Polymer- and Ionic Liquid-Containing Palladium: Recoverable Soluble Cross-Coupling Catalysts. *ChemCatChem* **2013,** *5*, 3460–3480.

[6] Yue, C.; Fang, D.; Liu, L.; Yi, T. F. Synthesis and Application of Task-Specific Ionic Liquids Used as Catalysts and/or Solvents in Organic Unit Reactions. *J. Mol. Liq.* **2011,** *163*, 99–121.

[7] Ratti, R. Ionic Liquids: Synthesis and Applications in Catalysis. *Adv. Chem.* **2014,** *2014*, 1–16.

[8] Kuchenbuch, A.; Giernoth, R. Ionic Liquids Beyond Simple Solvents: Glimpses at the State of the Art in Organic Chemistry. *Chem. Open* **2015,** *4*, 677–681.

[9] (a) Riisager, A.; Fehrmann, R.; Haumann, M.; Wasserscheid, P. Supported Ionic Liquids: Versatile Reaction and Separation Media. *Top. Catal.* **2006,** *40*, 91–102.

(b) Gu, Y.; Li, G. Ionic Liquids-Based Catalysis with Solids: State of the Art. *Adv. Synth. Catal.* **2009,** *351*, 817–847.

(c) Giacalone, F.; Gruttadauria, M. Covalently Supported Ionic Liquid Phases: An Advanced Class of Recyclable Catalytic Systems. *ChemCatChem* **2016,** *8*, 664–684.

[10] Jhung, S. H.; Chang, J. S. Trimerization of Isobutene Over Solid Acid Catalysts. *Catal. Surv. Asia* **2009,** *13*, 229–236.

[11] Olah, G. A.; Molnár, Á. Oligomerization and Polymerization. In *Hydrocarbon Chemistry;* 2nd ed. John Wiley & Sons, Inc.: Hoboken, NJ, 2003; pp 723–806.

[12] Yang, S.; Liu, Z.; Meng, X.; Xu, C. Oligomerization of Isobutene Catalyzed by Iron(iii) Chloride Ionic Liquids. *Energy Fuels* **2009,** *23*, 70–73.

[13] Gu, Y.; Shi, F.; Deng, Y. SO_3H-functionalized Ionic Liquid as Efficient, Green and Reusable Acidic Catalyst System for Oligomerization of Olefins. *Catal. Commun.* **2003,** *4*, 597–601.

[14] Skoda-Földes, R. The Use of Supported Acidic Ionic Liquids in Organic Synthesis. *Molecules* **2014,** *19*, 8840–8884.

[15] Fehér, C.; Kriván, E.; Hancsók, J.; Skoda-Földes, R. Oligomerisation of Isobutene with Silica Supported Ionic Liquid Catalysts. *Green Chem.* **2012,** *14*, 403–409.

[16] Fehér, C.; Kriván, E.; Kovács, J.; Hancsók, J.; Skoda-Földes, R. Support Effect on the Catalytic Activity and Selectivity of SILP Catalysts in Isobutene Trimerization. *J. Mol. Catal. A* **2013,** *372*, 51–57.

[17] Lemus, J.; Palomar, J.; Gilarranz, M. A.; Rodriguez, J. J. Characterization of Supported Ionic Liquid Phase (SILP) Materials Prepared from Different Supports. *Adsorption* **2011,** *17*, 561–571.
[18] Gadge, S. T.; Bhanage, B. M. Recent Developments in Palladium Catalysed Carbonylation Reactions. *RSC Adv.* **2014,** *4*, 10367–10389.
[19] (a) Kollár, L., Ed. *Modern Carbonylation Methods;* Wiley-VCH: Weinheim, 2008.
(b) Grigg, R.; Mutton, S. P. Pd-catalysed Carbonylations: Versatile Technology for Discovery and Process Chemists. *Tetrahedron* **2010,** *66*, 5515–5548.
[20] Roy, S.; Roy, S.; Gribble, G. W. Metal-Catalyzed Amidation. *Tetrahedron* **2012,** *68*, 9867–9923.
[21] Wang, T.; Zhang, Z.; Wallace, O. B.; Deshpande, M.; Fang, H.; Yang, Z., et al. Discovery of 4-Benzoyl-1-[(4-methoxy-1H-pyrrolo[2,3-b]pyridin-3-yl)oxoacetyl]-2-(R)-methylpiperazine (BMS-378806): A Novel HIV-1 Attachment Inhibitor that Interferes with CD4-gp120 Interactions. *J. Med. Chem.* **2003,** *46*, 4236–4239.
[22] Chatel-Chaix, L.; Baril, M.; Lamarre, D. Hepatitis C Virus NS3/4A Protease Inhibitors: A Light at the End of the Tunnel. *Viruses* **2010,** *2*, 1752–1765.
[23] Barnard, C. F. J. Palladium-Catalyzed Carbonylations: A Reaction Come of Age. *Organometallics* **2008,** *27*, 5402–5422.
[24] Papp, M.; Skoda-Földes, R. Phosphine-Free Double Carbonylation of Iodobenzene in the Presence of Reusable Supported Palladium Catalysts. *J. Mol. Catal. A* **2013,** *378*, 193–199.
[25] Urbán, B.; Papp, M.; Srankó, D.; Skoda-Földes, R. Phosphine-Free Atmospheric Carbonylation of Aryl Iodides with Aniline Derivatives in the Presence of a Reusable Silica-Supported Palladium Catalyst. *J. Mol. Catal. A* **2015,** *397*, 150–157.
[26] Papp, M.; Urbán, B.; Drotár, E.; Skoda-Földes, R. Mono-and Double Carbonylation of Iodobenzene in the Presence of Reusable Supported Palladium Catalysts. *Green Proc. Synth.* **2015,** *4*, 103–115.
[27] Kume, Y.; Qiao, K.; Tomida, D.; Yokoyama, C. Selective Hydrogenation of Cinnamaldehyde Catalyzed by Palladium Nanoparticles Immobilized on Ionic Liquids Modified-Silica Gel. *Catal. Commun.* **2008,** *9*, 369–375.
[28] Skoda-Földes, R.; Takács, E.; Horváth, J.; Tuba, Z.; Kollár, L. Palladium-Catalysed Aminocarbonylation of Steroidal 17-Iodo-androst-16-ene Derivatives in N,N′-dialkyl-imidazolium-type Ionic Liquids. *Green Chem.* **2003,** *5*, 643–645.
[29] Szarka, Z.; Skoda-Földes, R.; Kuik, Á.; Berente, Z.; Kollár, L. Synthesis of Ferrocene Amides and Alpha-Ketoamides via Palladium-Catalyzed Homogeneous Carbonylation Reaction. *Synthesis* **2003,** 545–550.
[30] Heidenreich, R. G.; Krauter, J. G. E.; Pietsch, J.; Köhler, K. Control of Pd Leaching in Heck Reactions of Bromoarenes Catalyzed by Pd Supported on Activated Carbon. *J. Mol. Catal. A* **2002,** *182–183*, 499–509.
[31] Muzart, J. N,N-Dimethylformamide: Much More than a Solvent. *Tetrahedron* **2009,** *65*, 8313–8323.
[32] Takács, E.; Varga, C.; Skoda-Földes, R.; Kollár, L. Facile Synthesis of Primary Amides and Ketoamides via a Palladium-Catalysed Carbonylation–Deprotection Reaction Sequence. *Tetrahedron Lett.* **2007,** *48*, 2453–2456.
[33] Xu, D. C.; Liu, Y.; Shi, S.; Wang, Y. Chiral Quaternary Alkylammonium Ionic Liquid [Pro-dabco][BF_4]: As a Recyclable and Highly Efficient Organocatalyst for Asymmetric Michael Addition Reactions. *Tetrahedron: Asymmetry* **2010,** *21*, 2530–2534.

CHAPTER 18

Asymmetric Autocatalysis Initiated by Crystal Chirality of Achiral Compounds

Arimasa Matsumoto[1,2], Yoshiyasu Kaimori[1], Tsuneomi Kawasaki[2,3] and Kenso Soai[1,2]
[1]Department of Applied Chemistry, Tokyo University of Science, Tokyo, Japan
[2]Research Center for Chirality, Research Institute for Science and Technology (RIST), Tokyo University of Science, Tokyo, Japan
[3]Department of Materials Science and Engineering, University of Fukui, Bunkyo, Fukui, Japan

Abstract

Chiral crystallization of achiral compounds is attracting much attention, especially in areas of chirality research such as chiral separation and generation. Spontaneous chiral crystallization generates chirality and is a possible origin of homochirality. Although various organic and inorganic achiral compounds form chiral crystals, there are few examples of the use of this chirality for the control of enantioselective reactions of organic compounds. In this chapter, we describe how asymmetric autocatalysis can successfully utilize crystal chirality of achiral organic compounds to afford highly enantioenriched products with the absolute structure corresponding to the chirality of the crystals used.

Keywords: Asymmetric autocatalysis; amplification of ee; chirality; chiral crystal

18.1 INTRODUCTION

Crystal structures can be classified by periodic pattern and point group symmetry. Although there are 230 space group symmetries for 3D crystal structures, the most important space group classification from a chirality perspective is whether the crystal structure has inversion and mirror type symmetry. There are 65 Sohncke space groups that have only rotation or screw axis symmetry and no inversion or mirror type symmetry. This means that crystal structures that can be assigned to this Sohncke space group have "handedness" and their mirror image cannot be superimposed on the original structure [1]. Enantiopure chiral compounds can crystallize only in one of these 65 Sohncke space groups. However, crystallization of achiral compounds in these Sohncke space groups is also possible (Fig. 18.1) [2]. This chiral crystallization of achiral compounds is a

DOI: http://dx.doi.org/10.1016/B978-0-12-812824-4.00018-6

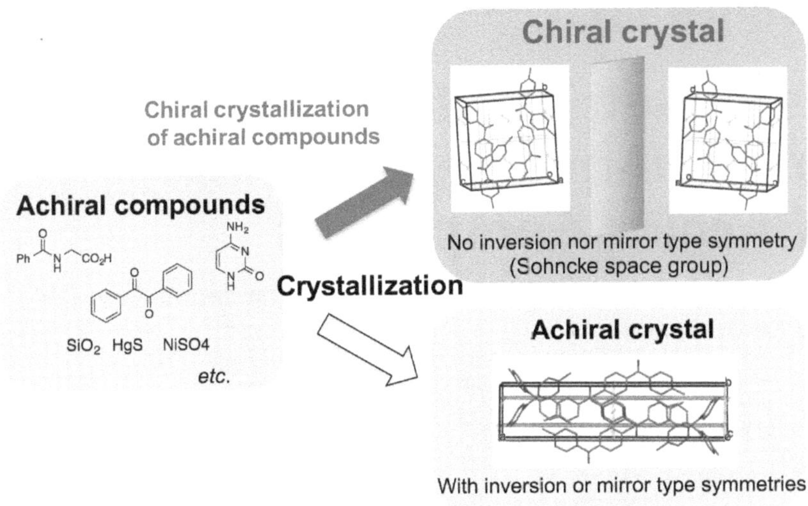

Figure 18.1 Chiral crystallization of achiral compounds.

spontaneous chirality generation process that has attracted much attention with respect to both chirality generation and separation [3,4].

If we can correlate the crystal chirality of achiral compounds to the chirality of organic molecules, this spontaneous chiral crystallization becomes a possible candidate for the origin of homochirality. There are some examples of using the crystal chirality of a reaction substrate to mediate selective conversion of the reactant into the enantioenriched product under solid-state photoreaction [5] or low-temperature reaction conditions [6]. However, the use of this crystal chirality to control the reaction selectivity as a catalyst or additive remains a highly challenging topic.

We have been studying asymmetric autocatalysis of pyrimidyl alkanol **1**. In this system, the product alkanol can act as an asymmetric catalyst for the addition reaction of diisopropylzinc to pyrimidine-5-carbaldehyde **2** to give the same pyrimidyl alkanol with high enantiomeric excess (*ee*) with the same absolute configuration (Scheme 18.1) [7–10]. The most remarkable point of this reaction is the asymmetric amplification that occurs during the reaction process, which generates the product with high *ee* even when the *ee* of the initial catalyst is undetectably low.

This high amplification of *ee* makes it possible to recognize the minute chiral environment, which traditionally has not been considered as a chiral initiator for the organic reaction. If the crystal chirality can control the reaction selectivity, asymmetric autocatalysis of pyrimidyl alkanol can

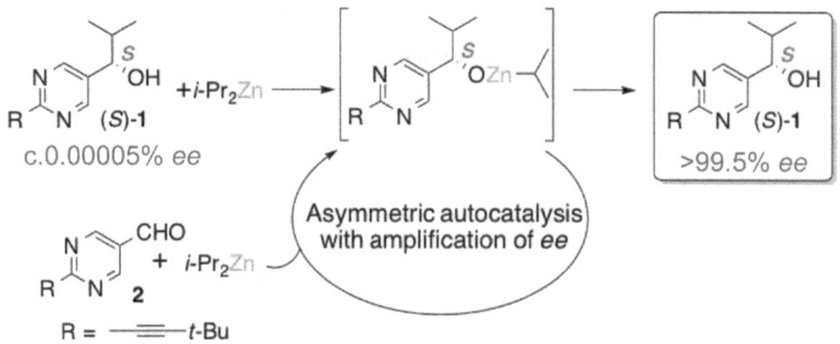

Scheme 18.1 Asymmetric autocatalysis of pyrimidyl alkanol.

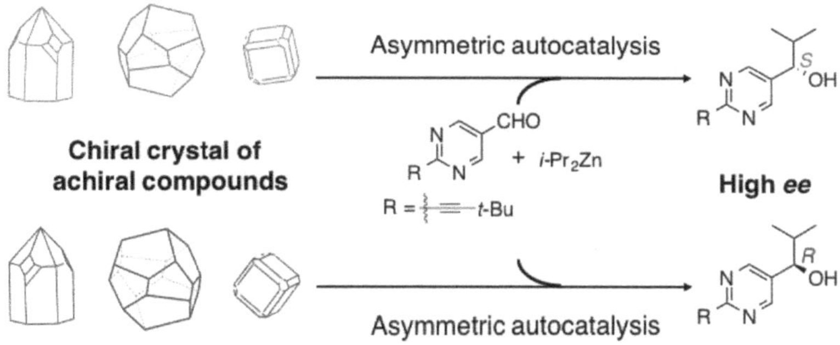

Scheme 18.2 Asymmetric autocatalysis initiated by the crystal chirality of achiral compounds.

amplify the *ee* through the enantioselectivity induced by crystal chirality. In this chapter, we show that asymmetric autocatalysis can be initiated by a range of chiral crystals that are constructed from achiral inorganic [11] and organic [12] compounds (Scheme 18.2).

18.2 RESULTS AND DISCUSSION

18.2.1 Asymmetric Autocatalysis Initiated by Inorganic Crystals

Given that inorganic minerals existed on Earth before the first live organisms, asymmetric induction of organic molecules from inorganic crystals, especially inorganic minerals, is an extremely interesting topic for the study of the origin of biological homochirality. We have demonstrated that the chirality of various inorganic crystals can cause enantioselectivity in the asymmetric autocatalysis reaction of pyrimidyl alkanol.

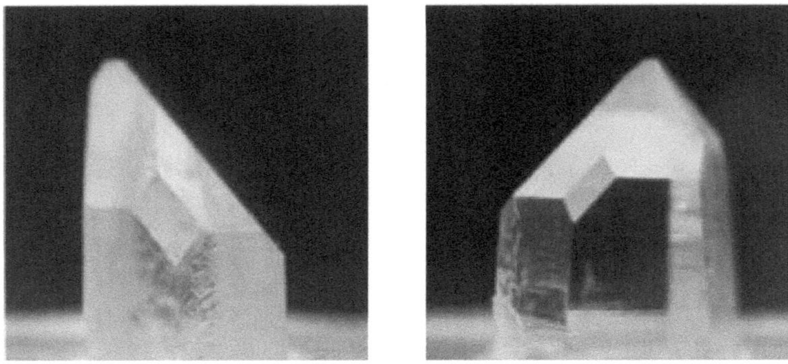

Figure 18.2 Hemihedral structures of quartz crystals.

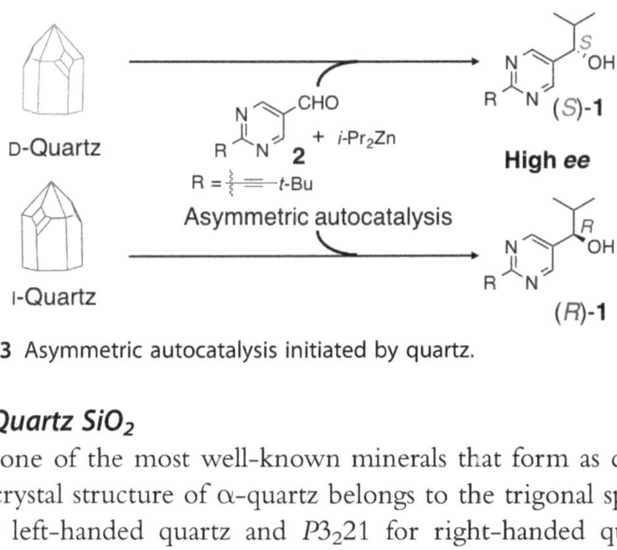

Scheme 18.3 Asymmetric autocatalysis initiated by quartz.

18.2.1.1 Quartz SiO_2

Quartz is one of the most well-known minerals that form as chiral crystals. The crystal structure of α-quartz belongs to the trigonal space group $P3_121$ for left-handed quartz and $P3_221$ for right-handed quartz. The crystal handedness of well-grown crystals can sometimes be recognized by morphological differences, as shown in Fig. 18.2. More accurately, crystal handedness is determined by observation of optical rotation (OR). An important note to avoid confusion is that left or right assignment of OR is not the same as left or right packing structure symmetry. In the case of quartz, *levorotatory* L-quartz has a right-handed 3_1 screw axis in the packing structure and D-quartz has a left-handed 3_2 screw axis.

Asymmetric autocatalysis reaction of pyrimidine-5-carbaldehyde **2** and diisopropylzinc was performed in the presence of a fine powder of chiral crystals of quartz [11a]. In the presence of D-quartz, (*S*)-alkanol was obtained with high *ee*, whereas the use of L-quartz afforded the opposite enantiomer, (*R*)-alkanol (Scheme 18.3). This was the first example to

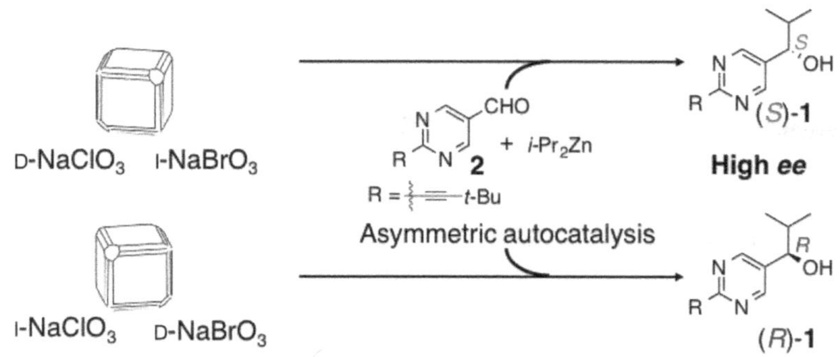

Scheme 18.4 Asymmetric autocatalysis initiated by $NaClO_3$ and $NaBrO_3$.

demonstrate that the chirality of a mineral affected the enantioselectivity of a reaction.

18.2.1.2 Sodium Chlorate and Bromate

Sodium chlorate is also a famous achiral compound that crystallizes in chiral structures. Although both D- and L-crystals belong to the cubic $P2_13$ space group, the packing structures of D- and L-crystals are mirror images of each other. Crystals of sodium chlorate can be obtained from solution and crystal chirality can be recognized by observation with a microscope under polarized light illumination. Studies on such crystals led to several interesting findings on chiral crystallization such as spontaneous chiral crystallization under stirring [3a] and enantioenrichment of chiral crystals under stirring [4a].

We examined asymmetric autocatalysis with both sodium chlorate and bromate crystals and found that D-$NaClO_3$ and L-$NaBrO_3$ both afforded (*S*)-alkanol with high *ee*. On the other hand, the reaction catalyzed by L-$NaClO_3$ and D-$NaBrO_3$ both give (*R*)-alkanol (Scheme 18.4) [11b,c]. Note that the replacement of Cl ions with Br ions in the crystal packing structure of D-$NaClO_3$ changed the sign of the OR to give the L-$NaBrO_3$ crystal [13].Thus, this result shows that the same crystal-packing structure gives the same selectivity for both $NaClO_3$ and $NaBrO_3$.

18.2.1.3 Cinnabar HgS

Cinnabar HgS is another chiral mineral, the crystal structure of which is composed of Hg and S atoms in threefold spiral chains. The crystal belongs to trigonal $P3_121$ or $P3_221$ space groups for (*P*)-HgS or (*M*)-HgS, respectively. The crystal structure, including crystal handedness, was characterized by anomalous dispersion of single-crystal X-ray

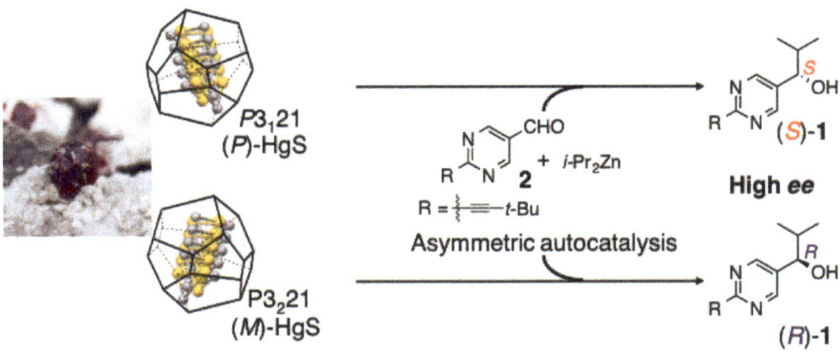

Scheme 18.5 Asymmetric autocatalysis initiated by cinnabar.

diffraction analysis and used for asymmetric autocatalysis (Scheme 18.5). In the presence of (*P*)-HgS, (*S*)-alkanol **1** was obtained with high *ee* and the opposite enantiomer was obtained in the presence of (*M*)-HgS.

Regarding this crystal, the absorbance of pyrimidyl alkanol on the HgS crystal surface could be observed by atomic force microscopy, and the same technique was used to show that the characteristics of the crystal growth difference of pyrimidyl alkanol on the surface determines the absolute chirality of alkanol [11d]. This result clearly supports the conclusion that the chiral crystal surface of HgS interacts enantioselectively with the alkanol.

18.2.1.4 Retgersite

A recent example of asymmetric autocatalysis initiated by a chiral inorganic crystal involves retgersite (nickel sulfate hexahydrate; $NiSO_4 \cdot 6H_2O$). Retgersite is a sulfate mineral that is ubiquitous on Earth. The sulfate ion itself is achiral, but the ion placed around a fourfold spiral axis in the retgersite crystal structure belongs to either the $P4_12_12$ or $P4_32_12$ space group. The characteristic absorption in the visible light region means that the crystal chirality can be determined by solid-state circular dichroism (CD) measurements in Nujol mull [14]. The $P4_12_12$ crystal has a negative cotton effect at a wavelength of approximately 390 nm (CD(−)390_{Nujol}) and $P4_32_12$ shows a positive effect in the same region CD(+)390_{Nujol}. The crystal chirality was distinguished by both X-ray diffraction analysis and CD spectrum and the crystal acts as a chiral initiator for the asymmetric autocatalysis. The crystal with CD(+)390_{Nujol} afforded (*S*)-alkanol in the asymmetric autocatalysis of pyrimidyl alkanol, whereas the CD(−) 390_{Nujol} crystal afforded (*R*)-alkanol. This result shows that even highly

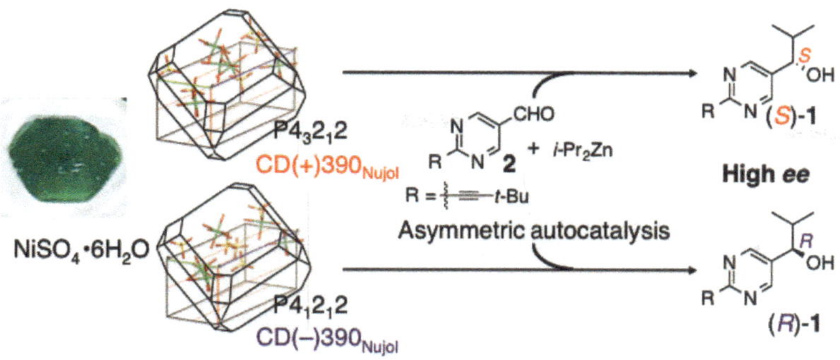

Scheme 18.6 Asymmetric autocatalysis initiated by $NiSO_4 \cdot 6H_2O$.

hydrated minerals can act as chiral initiator for the asymmetric autocatalysis (Scheme 18.6) [11e].

18.2.2 Asymmetric Autocatalysis Initiated by Organic Crystals

Achiral organic compounds can also form enantiomorphic crystals. Various kinds of achiral organic molecules have chirality because of restricted rotation of bonds and/or spiral packing in the structure. This spontaneous chiral crystallization of achiral organic compounds is not so rare and is considered to be a possible origin of biological homochirality. We have examined these kinds of chiral crystals constructed with achiral organic compounds as a chiral initiator in asymmetric autocatalysis and demonstrated that the crystal chirality can act as an origin of chirality of organic compounds.

An initial example of asymmetric autocatalysis induced by chiral crystals of achiral organic compounds is found for the cobalt complex. Some metal complexes of achiral organic ligands have chirality with respect to their conformation. The octahedral cobalt complexes with edta **3** and trdta **4** have chirality originating from the mode of coordination [15]. Although these complexes can racemize in solution, the chirality is fixed in the crystal state. The chirality of these complexes can be characterized by OR measured at 546 nm. In the presence of these chiral crystals of $K[Co(edta)] \cdot 2H_2O$ or $K[Co(trdta)] \cdot 2H_2O$, the asymmetric autocatalysis afforded (*S*)-alkanol in the presence of crystals of complexes with (+)OR, and (*R*)-alkanol was formed in the presence of crystals with (−)OR (Scheme 18.7) [12a].

Asymmetric autocatalysis initiated by crystals composed of only organic compounds has also been achieved. Indole amine and indole

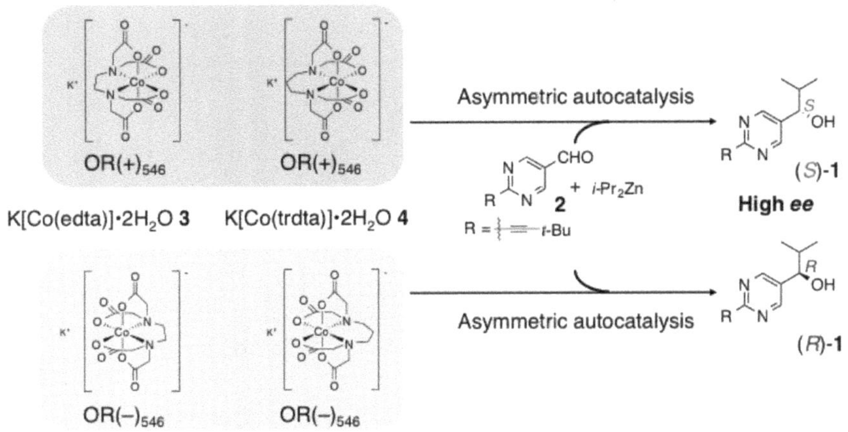

Scheme 18.7 Asymmetric autocatalysis initiated by a chiral cobalt complex.

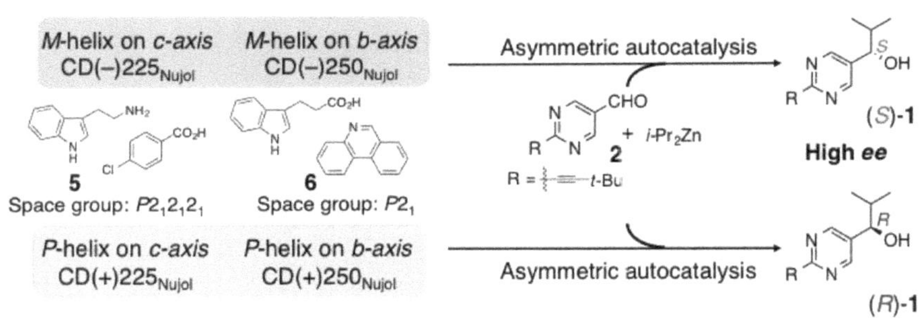

Scheme 18.8 Asymmetric autocatalysis initiated by chiral co-crystals of achiral compounds.

carboxylate are achiral organic compounds; however, co-crystallization of these indole derivatives with chlorobenzoic acid and phenanthridine afforded chiral crystals **5** and **6** with space groups $P2_12_12_1$ and $P2_1$, respectively. Their crystal chirality can be distinguished by their CD spectrum in Nujol mull recorded at 225 nm for crystal **5** and at 250 nm for crystal **6** [16]. The addition reaction of diisopropylzinc to pyrimidine-5-carbaldehyde was performed in the presence of these chiral crystals of achiral organic compounds. As a result, the pyrimidyl alkanol was obtained with high *ee* by asymmetric autocatalysis and the absolute structure of the obtained product had a reproducible correlation to the crystal chirality of the organic co-crystals employed (Scheme 18.8) [12b].

Benzil **7** also is an achiral compound that crystallized in a chiral structure. In the crystal state, the molecular torsion of the benzil moiety fixed

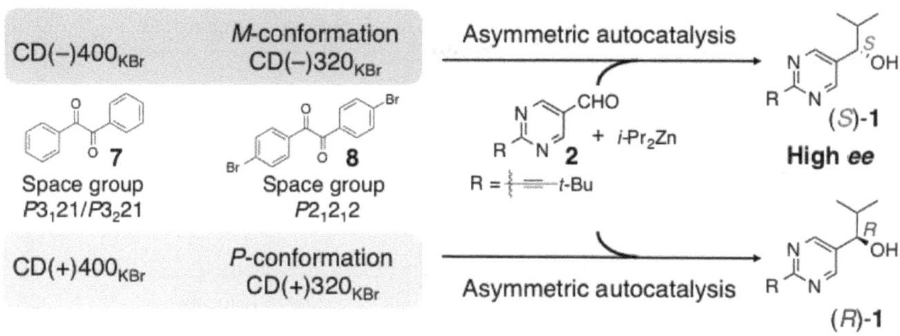

Scheme 18.9 Asymmetric autocatalysis initiated by chiral crystals of achiral benzil derivatives.

the *P* or *M* form of the crystal and led to the observed CD spectrum. This benzil crystal also acts as a chiral initiator for asymmetric autocatalysis and the selectivity of the reaction was correlated with its CD signals: (*S*)-alkanol for CD(−)400_{KBr}-**7** and (*R*)-alkanol for CD(+)400_{KBr}-**7**. The correlation between molecular torsion and CD spectrum was investigated for bromo-substituted benzil **8**, the high anomalous dispersion effects that allowed the absolute structure to be easily distinguished by X-ray diffraction analysis. Bromo-substituted benzil **8**, with *P* torsion structure, was found to exhibit CD(−)320_{KBr} signals and afforded (*S*)-alkanol in the asymmetric autocatalysis (Scheme 18.9) [12e].

Simple achiral hydrocarbons such as tetraphenylethylene **9** also form chiral crystals of the space group $P2_1$, the crystal chirality of which was distinguished based on the CD spectrum. We also found that this chiral crystal has a well-developed hemihedral shape that depends on the crystal chirality. Chloro and bromo substituted tetraphenylethylene **10** and **11**, respectively, also form chiral crystals containing *p*-xylene with space group $P2_12_12_1$. The crystal chirality of **10** and **11** can be determined by X-ray diffraction analysis, and crystals of **10** also exhibit a CD spectrum after removal of *p*-xylene. Crystals of **10** and **11** with *M*-conformation afforded (*S*)-alkanol **1**, whereas crystals with *P*-conformation afforded (*R*)-alkanol **1** in asymmetric autocatalysis (Scheme 18.10) [12g].

N-benzoylglycine **12** and *N*-thiophenecarbonyl glycine **13** are achiral derivatives of achiral amino acid glycine. Their crystal structure has chirality with space group $P2_12_12_1$. These crystals show CD signals and act as a chiral initiator for the asymmetric autocatalysis. Crystals of CD(+)260_{Nujol}-**12** and CD(+)300_{KBr}-**13** both give the (*S*)-alkanol and CD(−)260_{Nujol}-**12** and CD(−)300_{KBr}-**13** had the opposite selectivity

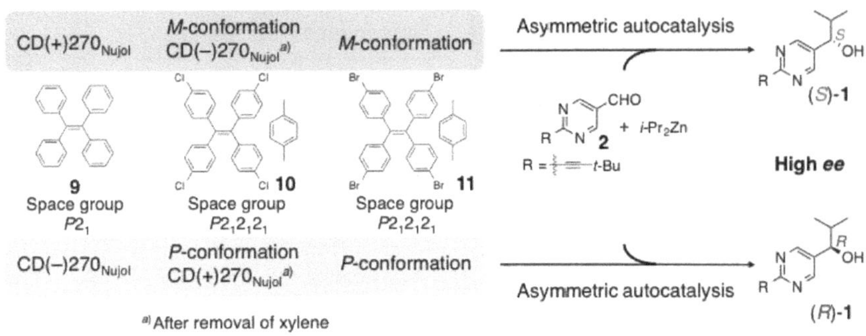

Scheme 18.10 Asymmetric autocatalysis initiated by chiral crystals of achiral tetraphenylethylene and its derivatives.

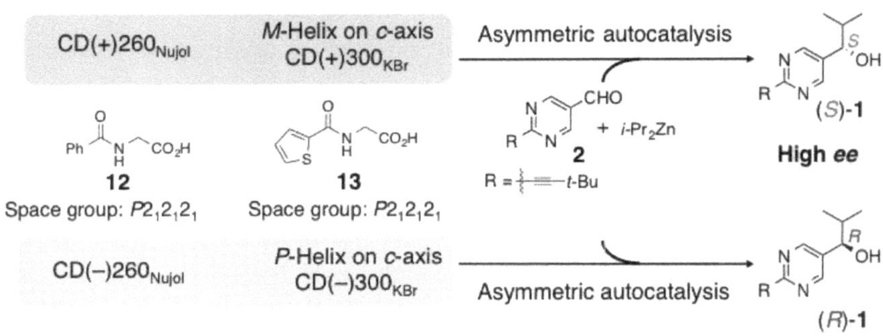

Scheme 18.11 Asymmetric autocatalysis initiated by chiral crystals of achiral *N*-benzoylglycine and *N*-thiophenecarbonyl.

(Scheme 18.11) [12c,i]. For the *N*-thiophenecarbonyl glycine, a computational study was also performed to predict the selectivity of the asymmetric autocatalysis with the crystal [12i,17].

Cytosine and adenine are achiral nucleobases and crucial components of DNA. These achiral nucleobases also form chiral crystals. Cytosine dehydrated crystal **14** and adenine dinitrate crystal **15** both have chirality and both show CD signals in Nujol at 250 and 310 nm, respectively. The asymmetric autocatalysis initiated by these chiral crystals of achiral nucleobases CD(−)250_{Nujol}-**14** or CD(−)310_{Nujol}-**15** afforded (*S*)-alkanol in high *ee* (Scheme 18.12) [12d,j]. It should be noted that the CD spectra of cytosine crystals depend on the measurement conditions, and CD(−)250_{Nujol}-**14** shows a positive cotton effect in KBr disk around 310 nm.

Furthermore, we have found that control of chirality of cytosine crystal **14** can be achieved by dehydration of cytosine monohydrate achiral

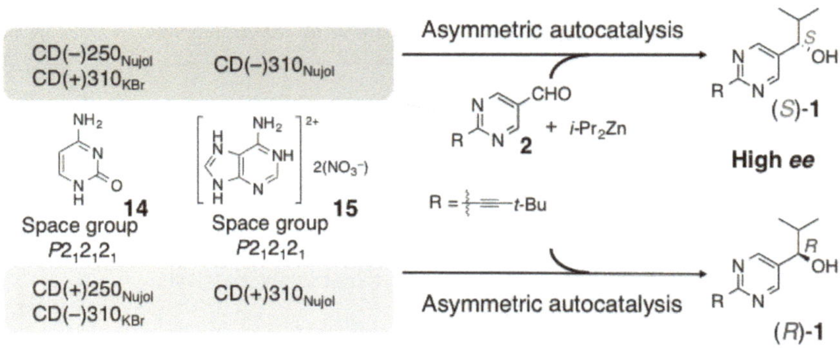

Scheme 18.12 Asymmetric autocatalysis initiated by chiral crystals of cytosine and adenine.

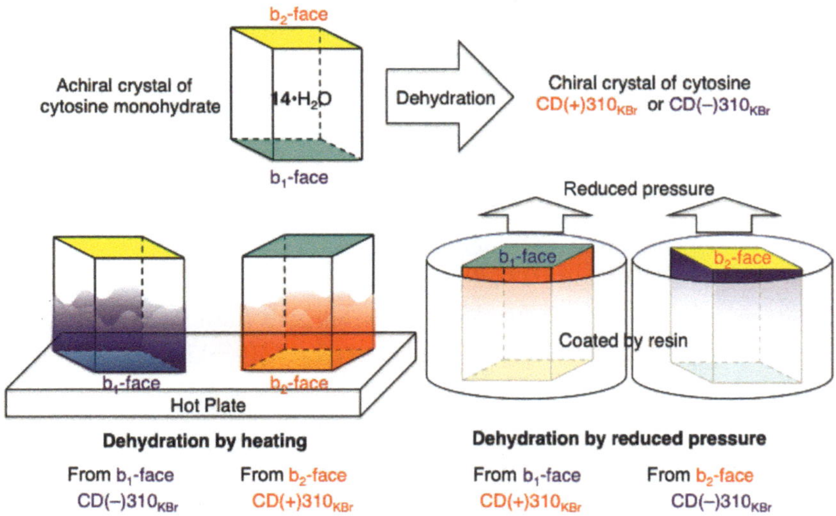

Figure 18.3 Chirality control of cytosine crystals by dehydration.

crystal **14** · H_2O (Fig. 18.3). Cytosine forms the achiral monohydrate crystal with space group $P2_1/c$ when crystallized from water. The crystallization water can be removed either by heating or under reduced pressure to give a chiral cytosine dehydrated crystal. In this dehydration process, the resulting chirality of the crystal of cytosine can be controlled by choosing the dehydration method and the direction of dehydration. When dehydration was performed by heating on a hot plate from the b_1-face, the CD(−)310_{KBr} cytosine crystal was obtained; CD(+)310_{KBr} was obtained by heating from the b_2-face. In contrast, in the case of dehydration under reduced pressure, the crystal dehydrated from the

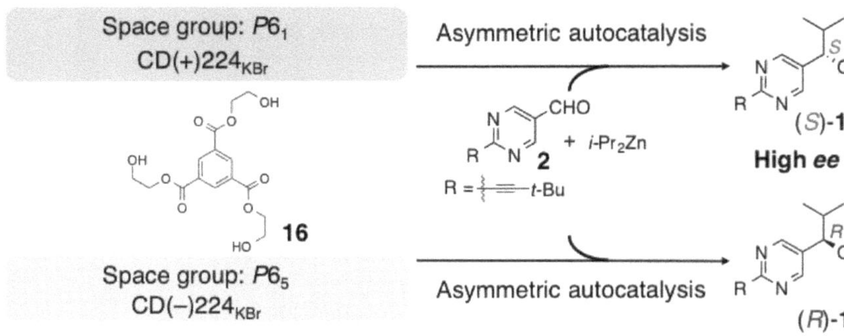

Scheme 18.13 Asymmetric autocatalysis initiated by chiral crystals of benzene triester.

b_1-face by exposure to reduced pressure to afford CD(+)310_{KBr}, and CD(−)310_{KBr} crystal was obtained from dehydration by exposing the b_2-face to reduced pressure (Fig. 18.3) [12f,l].

One recent example of asymmetric autocatalysis initiated by chiral crystals of achiral organic compounds is benzene triester **16**. The benzene triester **16** form right- or left-handed helical stacking of benzene rings with space group $P6_1$ or $P6_5$ in the crystal structure [18]. The structures can be recognized from the solid-state CD spectra as CD(+)224_{KBr} and CD(−)224_{KBr} for $P6_1$ and $P6_5$, respectively. In the presence of this chiral crystal, the asymmetric autocatalysis reaction give the (*S*)-alkanol **1** with CD(+)224_{KBr}-**16** crystal and (*R*)-alkanol for CD(−)224_{KBr}-**16** crystal (Scheme 18.13) [12k].

Another recent example is ethylenediamine(ethylenediammonium) sulfate **17**, which is an achiral diamine that is widely used as a ligand for metals. However, the crystal structure of its sulfate salt has chirality. The sulfate crystals have tetragonal structures with $P4_12_12$ or $P4_32_12$ symmetry. The crystal chirality can be easily determined with a microscope under polarized light [19]. Asymmetric autocatalysis with ethylenediamine sulfate afforded (*R*)-alkanol in the presence of (+)-crystal **17** and (*S*)-alkanol **1** in the presence of (−)-crystal **17** (Scheme 18.14) [12m].

18.2.3 Chiral Crystal of Racemic Compounds

Crystals of the amino acid serine also provide some interesting examples of autocatalysis. A racemic mixture of serine crystallized from sulfuric acid to afford chiral crystals of racemic DL-serine sulfate monohydrate.

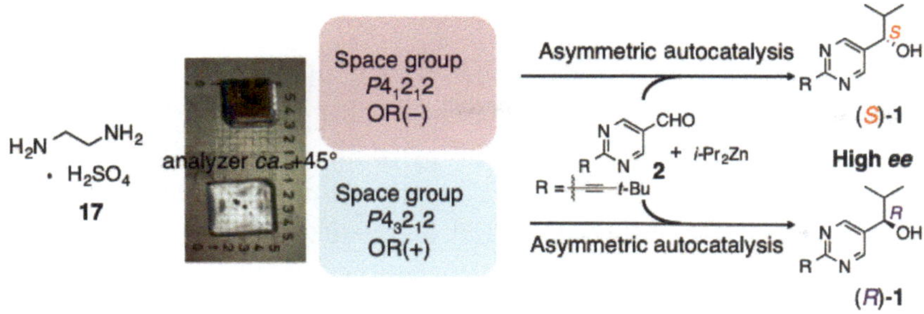

Scheme 18.14 Asymmetric autocatalysis initiated by chiral crystals of ethylenediamine sulfate.

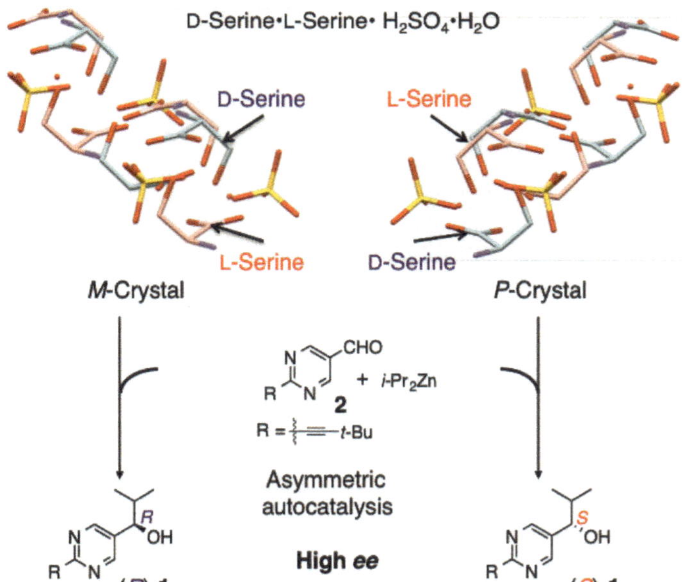

Scheme 18.15 Asymmetric autocatalysis initiated by chiral crystals of racemic serine sulfate.

The crystal structure has chirality even though the crystal itself contains both D- and L-serine in 1:1 ratio (Scheme 18.15) [12h]. The crystal chirality can be determined by single-crystal X-ray diffraction analysis and the helicity can be established from the hydrogen-bond network for the *c*-axis direction. The asymmetric autocatalysis initiated with this DL-serine sulfate crystal afforded (*R*)-alkanol in the presence of the *M*-crystal and (*S*)-alkanol in the presence of the *P*-crystal.

18.3 CONCLUSION

We have demonstrated various chiral crystals that are constructed from achiral organic and inorganic components that act as a chiral initiator for the asymmetric autocatalysis of pyrimidyl alkanol. These studies reveal that the chirality of the crystal structure can undoubtedly affect the enantioselectivity of the reaction and represent a possible candidate for the origin of homochirality. Typical organic compounds contained only light atoms, which are weakly absorbed at X-ray wavelengths and thus required careful measurements to determine their absolute crystal structure. Asymmetric autocatalysis of pyrimidyl alkanol **1** can be used to detect various types of crystal chirality of achiral organic compounds, which provides useful methods to determine the crystal chirality of an achiral compound.

ACKNOWLEDGMENTS

This work has been financially supported by Grant-in-Aid for Scientific Research from Japan Society for the Promotion of Science (JSPS KAKENHI Grant Numbers 23685012, 26810026, and 15H03781) and MEXT-Supported Program for the Strategic Research Foundation at Private Universities, 2012–16.

We thank the enthusiastic graduate students whose names appear in references.

REFERENCES

[1] Flack, H. D. Chiral and Achiral Crystal Structures. *Helv. Chim. Acta* **2003,** *86*, 905–921.

[2] (a) Matsuura, T.; Koshima, H. Introduction to Chiral Crystallization of Achiral Organic Compounds Spontaneous Generation of Chirality. *J. Photochem. Photobiol. C Photochem. Rev.* **2005,** *6*, 7–24.
(b) Pidcock, E. Achiral Molecules in Non-centrosymmetric Space Groups. *Chem. Commun.* **2005,** 3457–3459.

[3] (a) Kondepudi, D. K.; Kaufman, R. J.; Singh, N. Chiral Symmetry Breaking in Sodium Chlorate Crystallizaton. *Science* **1990,** *250*, 975–976.
(b) McBride, J. M.; Carter, R. L. Spontaneous Resolution by Stirred Crystallization. *Angew. Chem., Int. Ed.* **1991,** *30*, 293–295.
(c) Kondepudi, D. K.; Asakura, K. Chiral Autocatalysis, Spontaneous Symmetry Breaking, and Stochastic Behavior. *Acc. Chem. Res.* **2001,** *34*, 946–954.

[4] (a) Viedma, C. Chiral Symmetry Breaking During Crystallization: Complete Chiral Purity Induced by Nonlinear Autocatalysis and Recycling. *Phys. Rev. Lett.* **2005,** *94*, 065504.
(b) McLaughlin, D. T.; Nguyen, T. P. T.; Mengnjo, L.; Bian, C.; Leung, Y. H.; Goodfellow, E., et al. Viedma Ripening of Conglomerate Crystals of Achiral Molecules Monitored Using Solid-State Circular Dichroism. *Cryst. Growth Des.* **2014,** *14*, 1067–1076.

(c) Xiouras, C.; Van Aeken, J.; Panis, J.; Ter Horst, J. H.; Van Gerven, T.; Stefanidis, G. D. Attrition-Enhanced Deracemization of $NaClO_3$: Comparison Between Ultrasonic and Abrasive Grinding. *Cryst. Growth Des.* **2015,** *15*, 5476–5484.

[5] (a) Elgavi, A.; Green, B. S.; Schmidt, G. M. J. Reactions in Chiral Crystals. Optically Active Heterophotodimer Formation from Chiral Single Crystals. *J. Am. Chem. Soc.* **1973,** *95*, 2058–2059.

(b) Green, B. S.; Lahav, M.; Rabinovich, D. Asymmetric Synthesis via Reactions in Chiral Crystals. *Acc. Chem. Res.* **1979,** *12*, 191–197.

(c) Sakamoto, M. Absolute Asymmetric Synthesis from Achiral Molecules in the Chiral Crystalline Environment. *Chem. Eur. J.* **1997,** *3*, 684–689.

(d) Tanaka, K.; Toda, F.; Mochizuki, E.; Yasui, N.; Kai, Y.; Miyahara, I., et al. Enantioselective Single-Crystal-to-Single- Srystal Photodimerization of Coumarin and Thiocoumarin in Inclusion Complexes with Chiral Host Compounds. *Angew. Chem., Int. Ed.* **1999,** *38*, 3523–3525.

(e) Yagishita, F.; Sakamoto, M.; Mino, T.; Fujita, T. Asymmetric Intramolecular Cyclobutane Formation via Photochemical Reaction of *N,N*-diallyl-2-quinolone-3-Carboxamide Using a Chiral Crystalline Environment. *Org. Lett.* **2011,** *13*, 6168–6171.

[6] (a) Sakamoto, M.; Iwamoto, T.; Nono, N.; Ando, M.; Arai, W.; Mino, T., et al. Memory of Chirality Generated by Spontaneous Crystallization and Asymmetric Synthesis Using the Frozen Chirality. *J. Org. Chem.* **2003,** *68*, 942–946.

(b) Sakamoto, M.; Kato, M.; Aida, Y.; Fujita, K.; Mino, T.; Fujita, T. Photosensitized 2 + 2 Cycloaddition Reaction Using Homochirality Generated by Spontaneous Crystallization. *J. Am. Chem. Soc.* **2008,** *130*, 1132–1133.

(c) Yagishita, F.; Takagishi, N.; Ishikawa, H.; Kasashima, Y.; Mino, T.; Sakamoto, M. Deracemization of Quinolonecarboxamides by Dynamic Crystalline Salt Formation and Asymmetric Photoreaction by Using the Frozen Chirality. *Eur. J. Org. Chem.* **2014,** *2014*, 6366–6370.

[7] (a) Soai, K.; Shibata, T.; Morioka, H.; Choji, K. Asymmetric Autocatalysis and Amplification of Enantiomeric Excess of a Chiral Molecule. *Nature* **1995,** *378*, 767–768.

(b) Shibata, T.; Yonekubo, S.; Soai, K. Practically Perfect Asymmetric Autocatalysis with (2-Alkynyl-5-pyrimidyl)alkanols. *Angew. Chem., Int. Ed.* **1999,** *38*, 659–661.

(c) Sato, I.; Urabe, H.; Ishiguro, S.; Shibata, T.; Soai, K. Amplification of Chirality from Extremely Low to Greater than 99.5 % ee by Asymmetric Autocatalysis. *Angew. Chem., Int. Ed.* **2003,** *42*, 315–317.

[8] (a) Soai, K.; Kawasaki, T. Asymmetric Autocatalysis with Amplification of Chirality. In *Amplification of Chirality. Topics in Current Chemistry;* Soai, K., Ed.; Vol. 284; Springer Berlin Heidelberg: Berlin, Heidelberg, 2008; p 1–31.

(b) Kawasaki, T.; Soai, K. Asymmetric Induction Arising from Enantiomerically Enriched Carbon-13 Isotopomers and Highly Sensitive Chiral Discrimination by Asymmetric Autocatalysis. *Bull. Chem. Soc. Jpn.* **2011,** *84*, 879–892.

(c) Soai, K.; Kawasaki, T. Asymmetric Autocatalysis: Discovery and State of the Art. In *The Soai Reaction and Related Topic;* Palyi, G., Zucchi, C., Caglioti, C., Eds.; Edizioni Artestampa: Modena, 2012; pp 9–34.

(d) Kawasaki, T.; Soai, K. Asymmetric Autocatalysis Triggered by Chiral Crystals Formed from Achiral Compounds and Chiral Isotopomers. *Isr. J. Chem.* **2012,** *52*, 582–590.

(e) Kawasaki, T.; Sato, I.; Mineki, H.; Matsumoto, A.; Soai, K. Approaches Toward the Origin of Homochirality Using the Synthetic Organic Chemistry—Asymmetric Autocatalysis with Amplification of Enantiomeric Excess—. *J. Synth. Org. Chem. Jpn.* **2013,** *71*, 109–123.

(f) Soai, K.; Kawasaki, T. *Asymmetric Autocatalysis of Pyrimidyl Alkanol. Inventing Reactions;* Springer: Berlin and New York, 2013; pp 261–279.
(g) Soai, K.; Kawasaki, T.; Matsumoto, A. The Origins of Homochirality Examined by Using Asymmetric Autocatalysis. *Chem. Rec.* **2014,** 70–83.
(h) Soai, K.; Kawasaki, T.; Matsumoto, A. Asymmetric Autocatalysis of Pyrimidyl Alkanol and Its Application to the Study on the Origin of Homochirality. *Acc. Chem. Res.* **2014,** *47*, 3643–3654.

[9] (a) Avalos, M.; Babiano, R.; Cintas, P.; Jiménez, J. L.; Palacios, J. C. Chiral Autocatalysis: Where Stereochemistry Meets the Origin of Life. *Chem. Commun.* **2000,** 887–892.
(b) Blackmond, D. G. Asymmetric Autocatalysis and Its Implications for the Origin of Homochirality. *Proc. Natl. Acad. Sci. U. S. A.* **2004,** *101*, 5732–5736.
(c) Podlech, J.; Gehring, T. New Aspects of Soai's Asymmetric Autocatalysis. *Angew. Chem., Int. Ed.* **2005,** *44*, 5776–5777.
(d) Brown, J. M.; Gridnev, I. D.; Klankermayer, J. Asymmetric Autocatalysis with Organozinc Complexes; Elucidation of the Reaction Pathway. In *Amplification of Chirality. Topics in Current Chemistry;* Soai, K., Ed.; Springer Berlin Heidelberg: BerlinHeidelberg, 2008; Vol. 284, pp 35–65.
(e) Gehring, T.; Busch, M.; Schlageter, M.; Weingand, D. A Concise Summary of Experimental Facts about the Soai Reaction. *Chirality* **2010,** *22* (Suppl 1), E173–E182.
(f) Barabás, B.; Tóth, J.; Pályi, G. Stochastic Aspects of Asymmetric Autocatalysis and Absolute Asymmetric Synthesis. *J. Math. Chem.* **2010,** *48*, 457–489.
(g) Lente, G. The Role of Stochastic Models in Interpreting the Origins of Biological Chirality. *Symmetry (Basel)* **2010,** *2*, 767–798.
(h) Micheau, J.-C.; Coudret, C.; Buhse, T. Systems Chemistry of the Soai Reaction. In *The Soai Reaction and Related Topic;* Palyi, G., Zucchi, C., Caglioti, C., Eds.; Edizioni Artestampa: Modena, 2012; pp 169–196.
(i) Bissette, A. J.; Fletcher, S. P. Mechanisms of Autocatalysis. *Angew. Chem., Int. Ed.* **2013,** *52*, 12800–12826.
(j) Fülöp, O.; Barabás, B. Impact of the Soai-Autocatalysis on Natural Sciences. *J. Math. Chem.* **2016,** *54*, 10–17.

[10] (a) Sato, I.; Omiya, D.; Igarashi, H.; Kato, K.; Ogi, Y.; Tsukiyama, K., et al. Relationship Between the Time, Yield, and Enantiomeric Excess of Asymmetric Autocatalysis of Chiral 2-Alkynyl-5-Pyrimidyl Alkanol with Amplification of Enantiomeric Excess. *Tetrahedron: Asymmetry* **2003,** *14*, 975–979.
(b) Blackmond, D. G. Asymmetric Autocatalysis and Its Implications for the Origin of Homochirality. *Proc. Natl. Acad. Sci. U. S. A.* **2004,** *101*, 5732–5736.
(c) Gridnev, I. D.; Serafimov, J. M.; Brown, J. M. Solution Structure and Reagent Binding of the Zinc Alkoxide Catalyst in the Soai Asymmetric Autocatalytic Reaction. *Angew. Chem., Int. Ed.* **2004,** *43*, 4884–4887.
(d) Klankermayer, J. J.; Gridnev, I. D.; Brown, J. M. Role of the Isopropyl Group in Asymmetric Autocatalytic Zinc Alkylations. *Chem. Commun.* **2007,** 3151–3153.
(e) Schiaffino, L.; Ercolani, G. Unraveling the Mechanism of the Soai Asymmetric Autocatalytic Reaction by First-Principles Calculations: Induction and Amplification of Chirality by Self-Assembly of Hexamolecular Complexes. *Angew. Chem., Int. Ed.* **2008,** *47*, 6832–6835.
(f) Crusats, J.; Hochberg, D.; Moyano, A.; Ribó, J. M. Frank Model and Spontaneous Emergence of Chirality in Closed Systems. *Chem. Phys. Chem.* **2009,** *10*, 2123–2131.

(g) Quaranta, M.; Gehring, T.; Odell, B.; Brown, J. M.; Blackmond, D. G. Unusual Inverse Temperature Dependence on Reaction Rate in the Asymmetric Autocatalytic Alkylation of Pyrimidyl Aldehydes. *J. Am. Chem. Soc.* **2010,** *132,* 15104–15107.
(h) Ercolani, G.; Schiaffino, L. Putting the Mechanism of the Soai Reaction to the Test: DFT Study of the Role of Aldehyde and Dialkylzinc Structure. *J. Org. Chem.* **2011,** *76,* 2619–2626.
(i) Gehring, T.; Quaranta, M.; Odell, B.; Blackmond, D. G.; Brown, J. M. Observation of a Transient Intermediate in Soai's Asymmetric Autocatalysis: Insights from ^{1}H NMR Turnover in Real Time. *Angew. Chem., Int. Ed.* **2012,** *51,* 9539–9542.
(j) Gridnev, I. D.; Vorobiev, A. K. Quantification of Sophisticated Equilibria in the Reaction Pool and Amplifying Catalytic Cycle of the Soai Reaction. *ACS Catal.* **2012,** *2,* 2137–2149.
(k) Micheau, J.-C.; Coudret, C.; Cruz, J.-M.; Buhse, T. Amplification of Enantiomeric Excess, Mirror-image Symmetry Breaking and Kinetic Proofreading in Soai Reaction Models with Different Oligomeric Orders. *Phys. Chem. Chem. Phys.* **2012,** *14,* 13239–13248.
(l) Gehring, T.; Quaranta, M.; Odell, B.; Blackmond, D. G.; Brown, J. M. Observation of a Transient Intermediate in Soai's Asymmetric Autocatalysis: Insights from ^{1}H NMR Turnover in Real Time. *Angew. Chem., Int. Ed.* **2012,** *51,* 9539–9542.
(m) Ercolani, G. Principles for Designing an Achiral Receptor Promoting Asymmetric Autocatalysis with Amplification of Chirality. *Tetrahedron: Asymmetry* **2014,** *25,* 405–410.
(n) Barabás, B.; Zucchi, C.; Maioli, M.; Micskei, K.; Pályi, G. Stochastic and Empirical Models of the Absolute Asymmetric Synthesis by the Soai-Autocatalysis. *J. Mol. Model.* **2015,** *21,* 33.
(o) Matsumoto, A.; Abe, T.; Hara, A.; Tobita, T.; Sasagawa, T.; Kawasaki, T., et al. Crystal Structure of the Isopropylzinc Alkoxide of Pyrimidyl Alkanol: Mechanistic Insights for Asymmetric Autocatalysis with Amplification of Enantiomeric Excess. *Angew. Chem., Int. Ed.* **2015,** *54,* 15218–15221.

[11] (a) Soai, K.; Osanai, S.; Kadowaki, K.; Yonekubo, S.; Shibata, T.; Sato, I. D- and L-Quartz-Promoted Highly Enantioselective Synthesis of a Chiral Organic Compound. *J. Am. Chem. Soc.* **1999,** *121,* 11235–11236.
(b) Sato, I.; Kadowaki, K.; Soai, K. Asymmetric Synthesis of an Organic Compound with High Enantiomeric Excess Induced by Inorganic Ionic Sodium Chlorate. *Angew. Chem., Int. Ed.* **2000,** *39,* 1510–1512.
(c) Sato, I.; Kadowaki, K.; Ohgo, Y.; Soai, K. Highly Enantioselective Asymmetric Autocatalysis Induced by Chiral Ionic Crystals of Sodium Chlorate and Sodium Bromate. *J. Mol. Catal. A Chem* **2004,** *216,* 209–214.
(d) Shindo, H.; Shirota, Y.; Niki, K.; Kawasaki, T.; Suzuki, K.; Araki, Y., et al. Asymmetric Autocatalysis Induced by Cinnabar: Observation of the Enantioselective Adsorption of a 5-Pyrimidyl Alkanol on the Crystal Surface. *Angew. Chem., Int. Ed.* **2013,** *52,* 9135–9138.
(e) Matsumoto, A.; Ozawa, H.; Inumaru, A.; Soai, K. Asymmetric Induction by Retgersite, Nickel Sulfate Hexahydrate, in Conjunction with Asymmetric Autocatalysis. *New J. Chem.* **2015,** *39,* 6742–6745.

[12] (a) Sato, I.; Kadowaki, K.; Ohgo, Y.; Soai, K.; Ogino, H. Highly Enantioselective Asymmetric Autocatalysis Induced by Chiral Cobalt Complexes Due to the Topology of the Coordination of Achiral Ligands. *Chem. Commun.* **2001,** 1022–1023.

(b) Kawasaki, T.; Jo, K.; Igarashi, H.; Sato, I.; Nagano, M.; Koshima, H., et al. Asymmetric Amplification Using Chiral Cocrystals Formed from Achiral Organic Molecules by Asymmetric Autocatalysis. *Angew. Chem., Int. Ed.* **2005,** *44*, 2774–2777.
(c) Kawasaki, T.; Suzuki, K.; Hatase, K.; Otsuka, M.; Koshima, H.; Soai, K. Enantioselective Synthesis Mediated by Chiral Crystal of Achiral Hippuric Acid in Conjunction with Asymmetric Autocatalysis. *Chem. Commun.* **2006,** *2*, 1869–1871.
(d) Kawasaki, T.; Suzuki, K.; Hakoda, Y.; Soai, K. Achiral Nucleobase Cytosine Acts as an Origin of Homochirality of Biomolecules in Conjunction with Asymmetric Autocatalysis. *Angew. Chem., Int. Ed.* **2008,** *47*, 496–499.
(e) Kawasaki, T.; Harada, Y.; Suzuki, K.; Tobita, T.; Florini, N.; Pályi, G., et al. Enantioselective Synthesis Utilizing Enantiomorphous Organic Crystal of Achiral Benzils as a Source of Chirality in Asymmetric Autocatalysis. *Org. Lett.* **2008,** *10*, 4085–4088.
(f) Kawasaki, T.; Hakoda, Y.; Mineki, H.; Suzuki, K.; Soai, K. Generation of Absolute Controlled Crystal Chirality by the Removal of Crystal Water from Achiral Crystal of Nucleobase Cytosine. *J. Am. Chem. Soc.* **2010,** *132*, 2874–2875.
(g) Kawasaki, T.; Nakaoda, M.; Kaito, N.; Sasagawa, T.; Soai, K. Asymmetric Autocatalysis Induced by Chiral Crystals of Achiral Tetraphenylethylenes. *Orig. Life Evol. Biosph.* **2010,** *40*, 65–78.
(h) Kawasaki, T.; Sasagawa, T.; Shiozawa, K.; Uchida, M.; Suzuki, K.; Soai, K. Enantioselective Synthesis Induced by Chiral Crystal Composed of DL-Serine in Conjunction with Asymmetric Autocatalysis. *Org. Lett.* **2011,** *13*, 2361–2363.
(i) Carter, D. J.; Rohl, A. L.; Shtukenberg, A.; Bian, S.; Hu, C.; Baylon, L., et al. Prediction of Soai Reaction Enantioselectivity Induced by Crystals of *N*-(2-Thienylcarbonyl)glycine. *Cryst. Growth Des.* **2012,** *12*, 2138–2145.
(j) Mineki, H.; Hanasaki, T.; Matsumoto, A.; Kawasaki, T.; Soai, K. Asymmetric Autocatalysis Initiated by Achiral Nucleic Acid Base Adenine: Implications on the Origin of Homochirality of Biomolecules. *Chem. Commun.* **2012,** *48*, 10538–10540.
(k) Kawasaki, T.; Uchida, M.; Kaimori, Y.; Sasagawa, T.; Matsumoto, A.; Soai, K. Enantioselective Synthesis Induced by the Helical Molecular Arrangement in the Chiral Crystal of Achiral Tris(2-Hydroxyethyl) 1,3,5-Benzenetricarboxylate in Conjunction with Asymmetric Autocatalysis. *Chem. Lett.* **2013,** *42*, 711–713.
(l) Mineki, H.; Kaimori, Y.; Kawasaki, T.; Matsumoto, A.; Soai, K. Enantiodivergent Formation of a Chiral Cytosine Crystal by Removal of Crystal Water from an Achiral Monohydrate Crystal under Reduced Pressure. *Tetrahedron: Asymmetry* **2013,** *24*, 1365–1367.
(m) Matsumoto, A.; Ide, T.; Kaimori, Y.; Fujiwara, S.; Soai, K. Asymmetric Autocatalysis Triggered by Chiral Crystal of Achiral Ethylenediamine Sulfate. *Chem. Lett.* **2015,** *44*, 688–690.

[13] (a) Simon, H. J.; Bloembergen, N. Second-Harmonic Light Generation in Crystals with Natural Optical Activity. *Phys. Rev.* **1968,** *171*, 1104–1114.
(b) Xue, D.; Zhang, S. Calculations of Nonlinear Optical Responses of Isomorphous Crystals $NaClO_3$ and $NaBrO_3$ with Natural Optical Activity. *Chem. Phys. Lett.* **1998,** *287*, 503–508.

[14] Stadnicka, K.; Glazer, A. M.; Koralewski, M. Structure, Absolute Configuration and Optical Activity of α-Nickel Sulfate Hexahydrate. *Acta Crystallogr. Sect. B Struct. Sci.* **1987,** *43*, 319–325.

[15] (a) Gyarfas, E. C.; Mellor, D. P. The Resolution and Racemization of Potassium Ethylenediaminetetraacetatocobaltate(III). *J. Phy. Chem.* **1955,** *59*, 296–297.
(b) Ogino, H.; Takahashi, M.; Tanaka, N. Stereochemistry of Trimethylenediaminetetraacetatocobaltate(III). *Bull. Chem. Soc. Jpn.* **1970,** *43*, 424–428.
[16] (a) Koshima, H.; Hayashi, E.; Matsuura, T.; Tanaka, K.; Toda, F.; Kato, M., et al. Preparation, Structure and Discrimination of a Chiral Bimolecular Crystal by the Self-Assembly of 3-Indolepropionic Acid and Phenanthridine. *Tetrahedron Lett.* **1997,** *38*, 5009–5012.
(b) Koshima, H.; Miyauchi, M.; Shiro, M. Helicity Control in the Preparation of Chiral Co-crystals from Tryptamine and Achiral Carboxylic Acids by Pseudo-seeding. *Supramol. Chem.* **2001,** *13*, 137–142.
(c) Koshima, H.; Nagano, M.; Asahi, T. Optical Activity Induced by Helical Arrangements of Tryptamine and 4-Chlorobenzoic Acid in their Cocrystal. *J. Am. Chem. Soc.* **2005,** *127*, 2455–2463.
[17] Carter, D. J.; Kahr, B.; Rohl, A. L. Computational Methodology for Chirality Determination in the Soai Reaction by Crystals: γ-glycine. *Theor. Chem. Acc.* **2012,** *131*, 1125.
[18] Azumaya, I.; Uchida, D.; Kato, T.; Yokoyama, A.; Tanatani, A.; Takayanagi, H., et al. Absolute Helical Arrangement of Stacked Benzene Rings: Heterogeneous Double-Helical Interaction Comprising a Hydrogen-Bonding Belt and an Offset Parallel Aromatic–Aromatic-Interaction Array. *Angew. Chem., Int. Ed.* **2004,** *43*, 1360–1363.
[19] Cuccia, L. A.; Koby, L.; Ningappa, J. B.; Dakessian, M. Chiral Crystallization of Ethylenediamine Sulfate. *J. Chem. Educ.* **2005,** *82*, 1043.

CHAPTER 19

Amino Acids as Food Quality Factors in Parmigiano Reggiano Hard Cheese

Franco Bellesia[1], Adriano Pinetti[1], Livia Simon-Sarkadi[2], Claudia Zucchi[1], János Csapó[3], Bart Weimer[4], Luciano Caglioti[5] and Gyula Pályi[1]

[1]Department of Chemistry, University of Modena and Reggio Emilia, Modena, Italy
[2]Institute of Biochemistry and Food Chemistry, Technical University of Budapest, Budapest, Hungary
[3]Faculty of Animal Science, University of Kaposvár, Kaposvár, Hungary
[4]Nutrition and Food Sciences, Utah State University, Logan, UT, United States
[5]Department of Chemistry and Technology of Biologically Active Compounds, University La Sapienza, Rome, Italy

Abstract

Free and peptide/protein bonded acids were qualitatively and quantitatively determined in parmesan (Parmigiano Reggiano) cheese samples by HPLC. D/L ratio was measured using a chiral derivatizing agent. The diffusion processes of free amino acids occurring in Parmigiano Reggiano hard cheese can be regarded and utilized as a characteristic molecular marker of its ripening.

Keywords: Free amino acids; HPLC analysis; d/l ratio of amino acids; cheese ripening; Parmigiano Reggiano cheese

19.1 INTRODUCTION

One of the most typical products of Italian agricultural/food industry is the grana parmesan (Parmigiano Reggiano) type hard cheese [1]. Two principal kinds are known as Grana Padano and Grana Parmigiano Reggiano (these names are also registered trademarks). The latter is produced in the territories ranging from the Po river to the Apennine mountains in the provinces of Parma, Reggio Emilia, Modena, and, partly, of Bologna and Mantova [2]. A Consortium of small- and medium-sized cheese factories, mainly cooperatives, produces Parmigiano Reggiano exclusively from cows' milk. The producers must obey the rigorously controlled rules of the Consortium (D.M. 17/06/1956), starting from the alimentation of the cows to the production and ripening processes. These

DOI: http://dx.doi.org/10.1016/B978-0-12-812824-4.00019-8

rules are based on centuries-old experience as well as some easily measurable parameters, e.g., density, pH, temperature, bulk size (about 40 kg) [3], and on laws concerning the dairy industry (Italian Law DPR 30/10/1965 no. 1269; EU Regulations no. 2081/82 and 1107/96). We are working on a project with the aim of adding to these parameters some new molecular level features derived from analytical organic chemistry [1c,4].

The free amino acid and short peptide content account for the high nutritional value and easy digestibility of Parmigiano Reggiano cheese [1a,b]. In slow seasoning water evaporation occurs mainly from the external parts and a hydrophilicity gradient builds up. This factor may force many water-soluble compounds and lipids or lipid-derivatives to migrate in the bulk of the "forma", resulting in composition differences with respect to the homogeneous paste of the freshly formed cheese.

In several biogenic protein-containing foods, L-amino acids are often accompanied by D-isomers [5] because of partial racemization owing to spontaneous processes, enzymatic or microbial activity of either the residual bacterial content, or by new microorganisms that could be present in aging cheeses. Because of newly recognized [6] interesting features of D-amino acids, their formation adds nutritional value to cheeses. The D/L ratio could represent a very important process marker in the cheese industry [7].

Besides the overall migration processes involving the bulk of the cheese, some "local" differentiation may occur, accounting for the ease with which it breaks into different sized smaller flat pieces according to internal discontinuity surfaces.

With the aim of evaluating from a molecular point of view the differences between the inner and the outer parts, as well as the local ones, we planned a systematic analysis of free compounds (amino acids, fatty acids and volatile flavoring compounds) that could be subjected to migration inside the Parmigiano Reggiano cheese during ripening. Here we report on the utilization of amino-acid content and amino-acid D/L ratio as molecular markers of ripening.

19.2 AMINO-ACID ANALYSIS

A set of Parmigiano Reggiano cheese samples (aged 28–24 months, 200–300 mg each) of seven cheeses (PR-1–PR-7) from the same factory were collected at regular distances between the center and the rim of the cheese bulk according to Fig. 19.1.

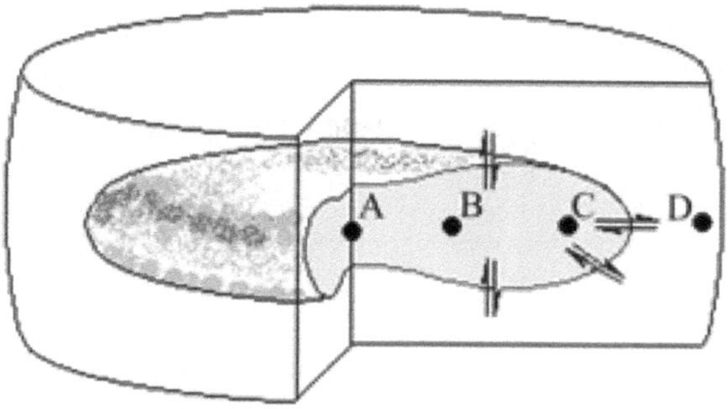

Figure 19.1 Location of sampling in a Parmigiano Reggiano "forma" and an idealized surface of same migration of free amino acids.

Samples were finely ground and treated under argon with deoxygenated distilled water [8], obtaining thin milk-like suspensions. After careful filtration [9], clear solutions suitable for high-performance liquid chromatographic (HPLC) analysis was achieved. The HPLC analysis was performed using a Kromasil octyl C8 column (EKA Nobel AB, Bohus, Sweden). For the determination of the (enantiomeric) D/L ratio, a precolumn manual derivatization was made, using *o*-phthalaldehyde and 2,3,4,6-tetra-*O*-acetyl-1-thio-β-gluco-pyranoside [10].

The solid from the filtration was washed again and suspended in 2 mL of deoxygenated distilled water. Hydrochloric acid (0.01 M, 0.2 mL) was added and the mixture allowed to stand at 50°C overnight. After a further filtration, HPLC analysis was then performed as before.

The result of these analyses is shown in Tables 19.1 and 19.2 and in Figs. 19.2 and 19.3.

19.3 RESULTS AND DISCUSSION

Analytical data for the free amino acids show an analogous distribution pattern of each free amino acid in all Parmigiano Reggiano cheese samples. Calculated mean values, to better represent this behavior, indicate a higher amount of free amino acids from B to C zones, lower quantities in the samples collected near the rim, and, surprisingly, also in the samples from the center of the bulk.

As far as bounded amino acids (proteins) are concerned a reverse behavior can be recognized: higher amounts are measured near the rim (D) and in the center (A) and lower quantities in points B and C. This

Table 19.1 Free amino acid content (mg/g)[a] in samples from seven aged Parmigiano Reggiano cheeses

		Cys	Orn	Met	Gly	Ala	His	Thr	Ile	Arg
PR-1	**A**	0.09	2.62	1.55	1.61	2.04	2.12	2.51	3.39	1.87
	B	0.11	3.27	1.42	1.46	1.78	1.81	1.93	2.95	0.76
	C	0.15	5.45	1.86	1.82	2.23	2.02	2.10	3.16	0.24
	D	0.08	2.50	1.18	1.14	1.25	1.35	1.14	2.58	0.35
PR-2	**A**	0.00	0.43	2.02	2.54	2.37	2.40	2.00	4.53	4.45
	B	0.00	0.42	2.06	2.45	2.31	2.31	2.01	4.53	4.38
	C	0.03	1.30	2.02	2.28	2.11	2.20	1.62	4.50	3.23
	D	0.00	0.46	1.41	1.70	1.48	1.59	1.09	3.49	2.70
PR-3	**A**	0.00	0.44	1.42	1.31	1.49	2.04	1.95	3.03	3.10
	B	0.00	0.22	1.50	1.45	1.55	2.00	2.00	3.36	3.40
	C	0.00	0.13	1.37	1.44	1.40	1.63	1.62	3.20	2.92
	D	0.00	0.33	1.11	1.05	1.04	1.33	1.21	2.52	2.28
PR-4	**A**	0.00	0.32	2.72	1.83	2.48	2.45	4.26	5.28	4.64
	B	0.00	0.33	2.46	1.99	2.58	2.54	4.35	5.08	4.93
	C	0.00	0.17	2.50	1.91	2.40	2.35	3.86	5.22	4.49
	D	0.00	0.22	1.31	1.02	1.25	1.37	1.97	2.98	2.68
PR-5	**A**	0.00	0.61	2.16	2.43	2.44	2.38	2.41	4.84	3.56
	B	0.00	1.05	2.56	2.94	2.94	2.75	2.71	5.61	2.87
	C	0.00	1.05	2.76	2.78	2.76	2.50	2.35	5.51	2.16
	D	0.00	0.78	1.72	1.62	1.51	1.50	1.65	3.99	1.20
PR-6	**A**	0.02	0.18	0.75	0.67	0.65	0.59	2.00	1.48	0.78
	B	0.08	0.43	1.65	1.25	1.47	1.65	2.22	3.20	1.68
	C	0.09	0.52	1.52	1.37	1.49	1.65	1.84	2.96	1.73
	D	0.01	0.13	1.01	1.13	1.20	1.44	1.18	1.87	1.81
PR-7	**A**	0.09	0.24	0.65	0.63	0.63	0.68	1.81	1.44	0.50
	B	0.10	1.87	1.35	1.17	1.43	1.78	2.35	2.85	0.63
	C	0.13	0.41	1.29	1.27	1.42	1.72	1.77	2.84	2.40
	D	0.19	0.36	1.44	1.20	1.33	1.54	1.53	2.92	1.94

(Continued)

Table 19.1 (Continued)

		Val	Phe	Tyr	Ser	Asp	Lys	Leu	Pro	Glu
PR-1	**A**	4.59	3.29	1.72	4.47	3.80	5.64	4.13	8.53	13.24
	B	3.93	2.76	1.50	3.87	3.17	4.78	3.63	7.20	11.24
	C	4.49	3.31	1.82	5.43	4.15	5.38	3.80	9.84	14.67
	D	3.15	2.11	0.86	2.66	2.35	3.87	3.27	5.02	7.74
PR-2	**A**	5.82	3.84	3.03	5.93	4.57	7.33	5.72	10.33	15.20
	B	5.68	3.72	2.65	5.65	4.35	7.12	5.81	9.97	14.59
	C	5.41	3.53	2.00	5.22	4.26	6.90	5.70	8.90	13.90
	D	4.07	2.76	1.22	3.70	3.03	5.30	4.61	6.54	9.20
PR-3	**A**	4.09	2.97	1.42	2.81	1.88	5.78	4.67	6.84	9.07
	B	4.27	2.90	1.29	3.19	2.18	5.78	4.74	7.00	9.51
	C	3.87	2.48	1.32	3.20	2.36	5.31	4.28	6.27	8.74
	D	2.95	2.00	1.01	2.25	1.69	4.27	3.72	4.70	5.94
PR-4	**A**	6.67	4.62	4.66	5.60	4.59	7.32	6.79	10.13	14.09
	B	6.80	4.51	3.15	6.01	5.05	7.75	6.46	10.56	15.00
	C	6.61	4.20	3.02	5.77	4.85	7.46	6.51	9.78	14.39
	D	3.75	2.52	1.39	3.16	2.47	4.72	3.92	5.29	7.52
PR-5	**A**	6.36	4.32	3.01	5.41	4.51	7.46	6.39	9.43	14.91
	B	7.54	5.04	3.77	6.63	5.28	8.72	7.24	11.69	17.51
	C	7.10	4.63	3.53	6.43	5.17	7.95	7.06	10.95	16.59
	D	4.44	2.88	1.36	3.71	2.76	5.34	5.48	6.32	9.24
PR-6	**A**	1.42	1.29	0.82	1.61	1.13	2.46	1.88	1.33	3.20
	B	2.67	2.23	1.47	3.31	2.07	5.34	4.32	2.97	6.29
	C	2.60	2.17	2.40	3.36	2.23	5.48	3.75	2.93	6.17
	D	1.56	1.70	1.09	2.46	1.94	3.79	2.10	2.60	4.14
PR-7	**A**	1.42	1.21	0.57	1.37	0.81	2.22	1.84	1.56	3.26
	B	2.47	2.06	1.32	2.46	1.75	5.03	3.65	3.68	5.98
	C	2.49	2.09	1.79	2.80	1.79	5.18	3.48	3.25	6.15
	D	2.51	2.07	1.48	2.68	1.69	4.91	3.89	3.05	5.98

[a] A 0.00 value means concentrations below the instrument integration limits.

Table 19.2 Bonded amino acid content (mg/g)[a] of five Parmigiano Reggiano cheeses

		Cys	Orn	Met	Gly	Ala	His	Ile	Arg	Val
PR-1	**A**	0.48	4.99	6.19	5.99	9.43	9.76	14.90	13.50	19.70
	B	0.40	6.78	5.28	5.87	9.04	9.42	13.50	10.90	18.30
	C	0.34	11.80	5.29	6.22	9.98	10.00	13.90	10.40	18.90
	D	0.61	5.00	7.46	7.21	10.80	11.80	17.20	16.40	22.60
PR-2	**A**	0.28	0.88	4.08	6.74	9.07	9.17	13.90	14.50	19.30
	B	0.26	0.84	5.43	6.68	9.04	9.15	14.30	14.70	19.20
	C	0.29	0.89	6.07	7.21	9.74	10.00	15.70	16.10	20.80
	D	0.47	1.07	7.87	7.39	10.60	11.40	17.60	18.10	23.20
PR-3	**A**	0.37	1.38	6.60	5.99	9.18	9.77	14.90	15.80	19.80
	B	0.34	1.03	4.46	5.78	8.72	9.10	13.80	14.90	18.10
	C	0.34	0.79	2.70	6.03	8.89	9.19	13.20	14.70	17.80
	D	0.49	1.27	4.71	7.26	10.70	11.70	16.60	18.70	21.90
PR-4	**A**	0.34	1.30	4.96	5.75	9.04	9.37	13.00	15.20	18.80
	B	0.18	1.31	2.39	5.92	9.16	9.66	13.80	15.80	18.90
	C	0.38	1.00	4.79	6.26	9.64	10.10	15.30	17.10	20.10
	D	0.67	1.31	10.90	7.37	11.10	11.90	18.20	19.40	24.90
PR-5	**A**	0.41	2.91	6.24	6.96	10.10	10.10	14.60	14.50	20.40
	B	0.36	3.59	4.73	6.72	9.70	9.81	12.90	13.10	18.70
	C	0.42	3.79	8.88	7.30	10.40	10.40	16.80	14.60	22.20
	D	0.55	3.45	7.88	8.11	12.10	13.40	16.20	17.40	22.00

(Continued)

Table 19.2 (Continued)

		Phe	Tyr	Ser	Asp	Lys	Leu	Pro	Glu
PR-1	**A**	16.90	17.30	17.60	21.10	23.00	25.20	36.20	64.30
	B	15.90	16.00	16.50	20.30	21.50	22.50	34.10	60.70
	C	16.60	16.60	17.50	21.10	22.50	24.10	33.60	63.10
	D	21.10	20.60	20.60	25.80	25.80	27.80	42.80	74.70
PR-2	**A**	16.30	18.00	17.70	22.10	23.00	24.70	40.50	64.20
	B	16.30	17.60	17.70	20.60	22.40	24.50	40.30	63.80
	C	18.00	19.10	18.80	21.80	23.90	26.90	43.30	68.40
	D	21.10	20.40	20.50	24.70	25.90	28.40	49.20	75.60
PR-3	**A**	17.20	18.00	17.90	21.30	24.80	26.30	36.70	66.40
	B	16.00	17.90	17.30	19.30	23.30	25.60	34.70	62.80
	C	15.90	17.00	17.50	23.70	22.90	25.10	34.50	63.10
	D	20.70	21.60	21.30	28.10	27.30	28.40	43.60	76.80
PR-4	**A**	16.70	19.30	18.50	21.30	21.90	25.00	39.30	61.00
	B	17.10	20.00	18.90	21.80	21.90	25.40	39.50	61.50
	C	18.40	19.10	19.70	23.30	22.20	26.00	41.60	62.60
	D	22.60	21.30	22.10	27.10	26.80	29.50	49.50	75.10
PR-5	**A**	18.20	19.40	20.10	23.30	23.30	26.40	44.10	63.90
	B	17.30	20.70	19.30	22.40	22.50	24.80	42.40	61.80
	C	19.30	22.00	20.90	24.50	24.60	29.40	44.30	67.40
	D	23.30	23.70	23.70	29.50	26.40	29.80	54.40	75.60

[a] A 0.00 value means concentrations below the instrument integration limits.

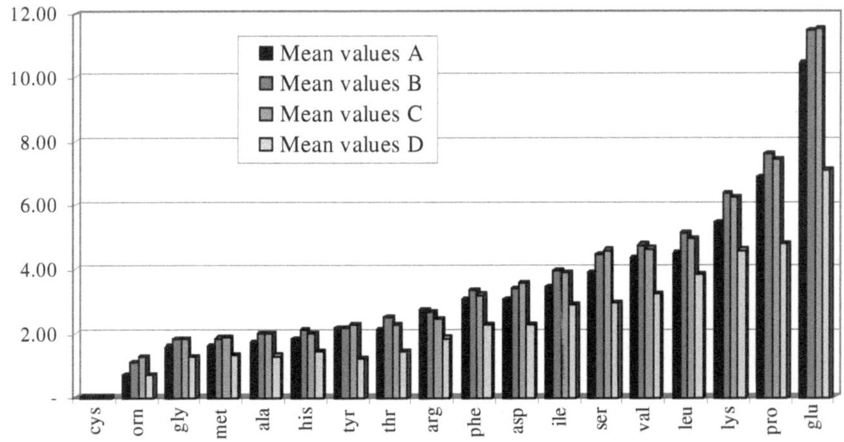

Figure 19.2 Distribution of free amino acids.

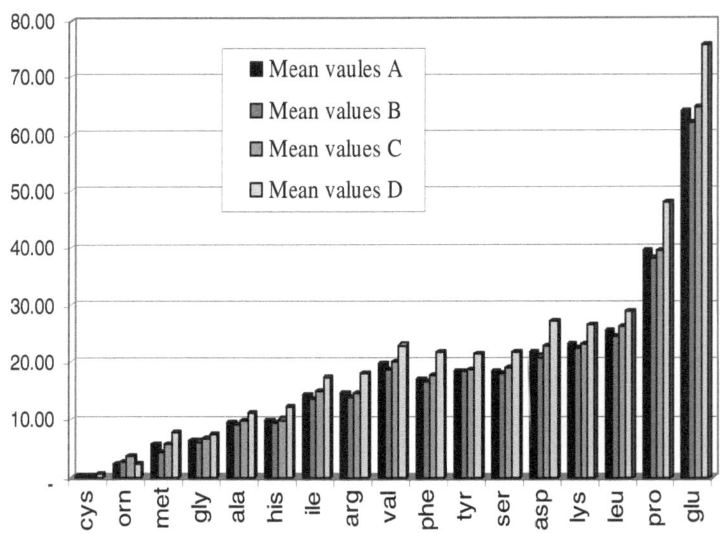

Figure 19.3 Distribution of (protein) bonded amino acids.

tendency can be more clearly seen from the mean values. The observed trends could be accounted for by: (i) the different hydrolytic activity caused by a lower water content in the most lipophilic parts and (ii) migration of free amino acids from A and D zones to B and C owing to their hydrophilic properties. Both processes slowly occur during the ripening period (18–24 months) (Table 19.3).

Table 19.3 D/L ratio for aspartic and glutamic acid in four Parmigiano Reggiano cheeses

	PR-4				PR-5				PR-6				PR-7			
D/L	A	B	C	D	A	B	C	D	A	B	C	D	A	B	C	D
Asp	0.68	0.76	0.76	0.77	0.71	0.8	0.81	0.74	0.77	0.94	0.87	0.82	0.77	0.93	0.94	0.84
Glu	0.19	0.21	0.18	0.18	0.21	0.27	0.27	0.2	0.24	0.25	0.3	0.27	0.26	0.24	0.23	0.24

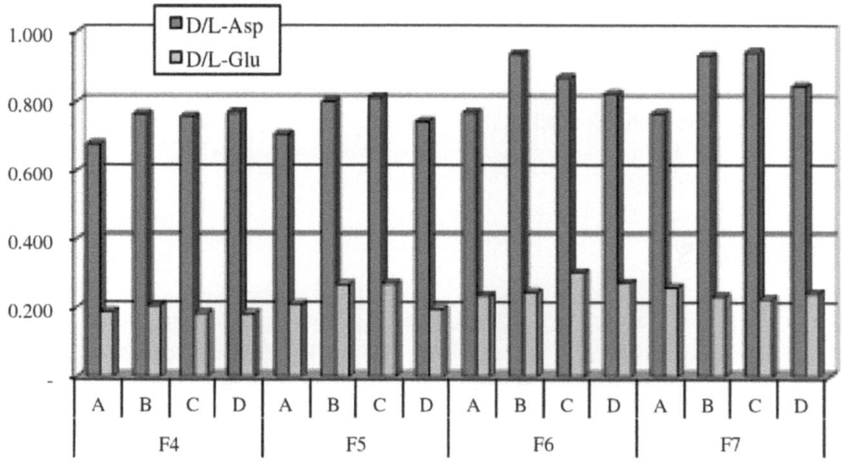

Figure 19.4 L/D ratio for aspartic and glutamic acid in four Parmigiano Reggiano "formas".

Trends of transformation of L- to D-amino acids during the ripening processes were studied for aspartic and glutamic acids in two sets of samples. These measurements showed that the changes in the D/L ratio of these two amino acids are analogous (Fig. 19.4).

The experimental results suggest that an ideal surface, defined by the points of the same concentration values for each free amino acid and of similar L to D conversion intensity, could have a "red blood cell" shape (see Fig. 19.1). Furthermore, the distance of this ideal surface from the rim, corresponding to a particular amino acid concentration, could also represent a valuable marker of ripening for Parmigiano Reggiano or even for any other hard type cheese.

19.4 CONCLUSIONS

The results reported in the present paper on free and protein-bound amino acid content, as well as the D/L enantiomeric ratio of some characteristic amino acids, may provide a valuable tool for the study of the internal inhomogeneities of the Parmigiano Reggiano cheese forms. These parameters could also serve as markers of the ripening stage as well as—to a certain extent—the origin of the cheese sample. Studies on statistical evaluation of these parameters with a major number of samples are in course in our laboratories.

ACKNOWLEDGMENTS

Financial supports received from Ministry of University and Research (MURST), from the National Research Council (CNR, Roma), from the Comunità Montana del Frignano and the Town Council of Serramazzoni (MO) are acknowledged.

REFERENCES

[1] (a) Scott, R. *Cheese Making Practice;* Applied Science Publishers: London, 1981.
(b) Belitz, H.-D.; Grosch, W. *Food Chemistry,* 2nd ed.; Springer: Berlin, 1999; pp 498.
(c) Bellesia, F.; Pinetti, A.; Pagnoni, U. M.; Rinaldi, R.; Zucchi, C.; Caglioti, L., et al. Volatile Components of Grana Parmigiano Reggiano Type Hard Cheese. *Food Chem.* **2003,** *83,* 55–61.
(d) Bellesia, F.; Pinetti, A.; Pagnoni, U. M.; Rinaldi, R.; Caglioti, L. Volatile Compounds in Food Aroma: Biosynthesis and Biotransformations. In *Progress in Biological Chirality;* Pályi, G., Zucchi, C., Caglioti, L., Eds.; Elsevier: Amsterdam, 2004; pp 253–257.

[2] Stein, J. L.; Imhof, K. In *Ullmann's Enc. Tech. Chem.,* Vol. 16. Wiley-VCH: Berlin, 1978; pp 689.

[3] Secchi, G. *I nostri alimenti;* H. Hoepli: Milano, 1979; pp 798–800.

[4] (a) Foissy, H.; Brandl, E. Dairy Research in Austria. *Milchwissenschaft* **1995,** *50* (8), 424–428.
(b) Mayer, H. K. Ripening Control and Quality Assessment of Parmesan Cheese. Electrophoretic Determination of Proteolysis Degree. *Lebensmittelind. Milch.* **1996,** *117* (7), 318–325.
(c) Mayer, H. K.; Rockenbauer, C.; Hannes, M. Evaluation of Proteolysis in Parmesan Cheese Using Electrophoresis and HPLC. *Lait* **1998,** *78,* 425–438.
(d) LeQuere, J. L.; Pierre, A.; Riaublanc, A.; Demaizieres, D. Characterization of Aroma Compounds in the Volatile Fraction of Soft Goat Cheese During Ripening. *Lait* **1998,** *78,* 279–290.
(e) Jaillais, B.; Bertrand, V.; Auger, J. Cryo-trapping/SPME/GC Analysis of Cheese Aroma. *Talanta* **1999,** *48,* 747–753.
(f) Careri, M.; Mangia, A.; Mori, G.; Musci, M. A New Multivariate Approach for the Optimisation of the Simultaneous Distillation-Extraction Technique for Free Fatty Acids Using a Face Centred Cube Experimental Design: Application to Parmigiano Reggiano Cheese. *Anal. Chim. Acta* **1999,** *386,* 169–180.
(g) Sforza, S.; Galaverna, G.; Neviani, E.; Pinelli, C.; Dossena, A.; Marchelli, R. Study of the Oligopeptide Fraction in Grana Padano and Parmigiano Reggiano Cheeses by LC-Electron Spray Ionization MS. *Eur. J. Mass Spectrom.* **2004,** *10,* 421–427.

[5] (a) Nagata, Y. d-Amino Acids in Nature. In *Advances in Bio Chirality;* Pályi, G., Zucchi, C., Caglioti, L., Eds.; Elsevier: Amsterdam, 1999; pp 271–283.
(b) Simon-Sarkadi, L. Occurrence of d-Amino Acids in Food. In *Progress in Biological Chirality;* Pályi, G., Zucchi, C., Caglioti, L., Eds.; Elsevier: Amsterdam, 2004; pp 329–338.
[6] (a) Fuchs, S. A.; Berger, R.; Klomp, L. W. J.; de Koening, T. J. d-Amino Acids in the Central Nervous System in Breathe and Disease. *Mol. Genet. Metab.* **2005,** *85*, 168–180.
(b) Maekawa, M.; Watanabe, M.; Yamaguchi, S.; Konno, R.; Hori, Y. Spatial Learning and Long-Term Potentiation of Mutant Mice Lacking D-Amino Acid Oxidase. *Neurosci. Res.* **2005,** *53*, 34–38.
[7] (a) Man, H.; Bada, J. L. Dietary d-Amino Acids. *Annu. Rev. Nutr.* **1987,** *7*, 209–225.
(b) Brückner, H.; Hausch, M. D-Amino Acids in Dairy Products: Detection, Origin and Nutritional Aspects. II. Ripened Cheeses. *Milchwissenschaft* **1990,** *45* (357–360), 421–425.
(c) Gandolfi, I.; Palla, G.; Delprato, L.; DeNisco, F.; Marchelli, R.; Salvadori, C. d-Amino Acids in Milk as Related to Heat Treatments and Bacterial Activity. *J. Food Sci.* **1992,** *57*, 377–381.
(d) Marchelli, R.; Palla, G.; Dossena, A.; Galaverna, G.; Corradini, R.; Clementi, S. d-Ammino Acidi: Markers Molecolari di Stagionatura e di Tipicità per il Parmigiano Reggiano e il Grana Padano. *Sci. Tec. Latt.-Casearia* **1997,** *48*, 21–32.
[8] (a) Csapó, J.; Csapó-Kiss, Z.; Stefler, J.; Martin, T. G.; Némethy, S. Influence of Mastitis on d-Amino Acid Content of Milk. *J. Dairy Sci.* **1995,** *78*, 2375–2381.
(b) Csapó, J.; Csapó-Kiss, Z.; Schmidt, J.; Martin, T. G. Quantitative Determination of Protein of Bacterial Origin. *Trends Anal. Chem.* **2001,** *20*, 42–48.
[9] (a) Ziarelli, F.; Shintu, L.; Caldarelli, S. Analysis of Cheese: A Practical Speciation Tool on Intact Samples. In Proceedings of the 3rd Italian-French Meeting on Organic Chemistry: Organic Chemistry towards Interfaces, Pisa, **2002**, Abstract OC7.
(b) Zucchi, C.; Bellesia, F.; Pinetti, A.; Simon-Sarkadi, L.; Weimer, B.; Caglioti, L.; et al. Featuring Compounds of the Greasy Base of the Parmesan Cheese Aroma. In Proceedings of the 3rd Italian-French Meeting on Organic Chemistry: Organic Chemistry towards Interfaces, Pisa, **2002,** Abstract P60.
(c) Shintu, L.; Ziarelli, F.; Caldarelli, S. Is High-Resolution Magic Angle Spinning NMR a Practical Speciation Tool for Cheese Samples? Parmigiano Reggiano as a Case Study. *Magn. Reson. Chem.* **2004,** *42*, 396–401.
(d) Vezzani, M.; Bellesia, F.; Caglioti, L.; Pinetti, A.; Zucchi, C.; Pályi, G. ^{1}H-NMR Spectra: A Highly Sensitive Tool in the Analysis of Grana Parmigiano-Reggiano Type Hard Cheese. *Agro Food Ind. Hi-Tech* **2015,** *26* (2), 34–37.
[10] Einarsson, S.; Folestad, S.; Josefsson, B. Separation of Amino Acid Enantiomers Using Precolumn Derivatization with *o*-Phthalaldehyde and 2,3,4,6-Tetra-O-acetyl-1-thio-β-glucopyranoside. *J. Liq. Chromatogr.* **1987,** *10*, 1589–1601.

INDEX

Note: Page numbers followed by "*f*" and "*t*" refer to figures and tables, respectively.

A

Abiotic emergence, of biological homochirality, 299–313
Absolute asymmetric synthesis, 14–16, 72, 74–76, 91, 94
Absolute configuration, inversion of
  retention of, 48
Achiral inorganic compounds, crystals of
  organic compounds, crystals of, 7–12
  substrate, 150, 162
Achiral-to-chiral reactions, 53–54
Adenine dinitrate, chiral crystals of, 9
Adipic acid, 224–225
Agrochemical industry, chiral molecules in, 32
Alcohol, secondary, chiral, 53–54, 167
Aldehyde substrates, isotope chirality in, 265–269
Aldol reaction, asymmetric catalysis in, 244–252
2-Alkenylpyrimidine-5-carbaldehyde, 243
Alkylation, asymmetric, 34–35
  asymmetric induction in, 34–35
  of azaaryl aldehydes, 46–48
Alkylsulfanyl polythiophenes, 285–291, 286*f*
Alkynylation, asymmetric induction in
  of azaaryl aldehydes, 43–45, 44*t*
2-Alkynylpyrimidyl alkanol, 22
Alliot, M.-J., 204
α-Thienyl bromide, coupling with boronic acids
  Grignard reagent, 279–280
  stannyl derivatives, 281
Amines, achiral, 74–75
Amino acid enantiomers, HPLC analysis of
  isotope ratio of, in meteorites, 18
D-Amino acid oxidase, 120
D-Amino acids, 120, 124–128
L-Amino acids, 120
D-Amino acids, analysis of
  in peptides, 120
Amino acids, free, in Parmigiano Reggiano cheese, 359
  protein-bound, in Parmigiano Reggiano cheese, 365–366
Aminocarbonylation, of aryl halides, 324–325
D-Aminolevulinate, 224
Amplification of chirality, remote, 46–48, 46*f*
Aryl halides, aminocarbonylation of, 324–325
D-Aspartate, formation mechanism of
  in human tissues, 120
  influence on protein structure, 123–124
  oxidase, 120
Aspartic acid, D/L ratio in Parmigiano Reggiano cheese, 365*t*
Asymmetric alkylation, 34–35
  alkynylation, 39–46
  autocatalysis, 34–35, 39–46
  autocatalysis, chiral crystals of achiral compounds in, 337
  autocatalysis, mathematical mechanism of, 2
  autocatalysis, on enantiotopic crystal faces, 13
  autocatalyst, X-ray structure of, 186*f*
  catalysis, non-autocatalytic, 2
  physical field, 243–244
  reduction of ketones, 223
Asymmetric synthesis, absolute, 14, 74–76
  electrochemical, 291–292
  spontaneous, 14–16
Asymmetry, general, 203
Autocatalysis, asymmetric, 1–21, 2*f*, 3*f*, 31–32, 33*f*, 75–77, 91, 149–162, 183–194, 242, 246, 253, 261*f*, 337–349, 349*f*
Autocatalysis, cross-coupled
  higher order, 305

Autocatalysis/catalysis, combination of, 45–46
Automultiplication, of chiral compound, 2, 72, 183–184, 242
Avetisov, V. A., 243–244
Azaaryl aldehydes, alkynylation of, 43–45

B

Bakers' yeast, reduction by, 229–230, 230*f*
Balaban, A. T., 134–135
Barabás, B., 19, 94, 100
Belusov [Belousov]-Zhabotinsky oscillating reaction, 302–303
Benzil, chiral crystals of, 344–345, 345*f*
BINAP-derivatives, as ligands for Ru catalysts, 231
Biodiesel additives, 224
Biological chirality, 53
  homochirality, 1–16, 22–23
Biomass, selective conversion of, 223–224, 237*f*
Biomolecules, chiral homogeneity of, 31–32
Bisphenol-A, 224
Blackmond, D. G., 32–33, 60, 86–87
Blackmond's Community, 60, 61*f*, 62*f*
Bolstad, R., 204
Brown, J. M., 32–33, 60, 81
Buhse, T., 33, 88, 171–172
Buono, F. G., 86–87

C

$C(sp^3)$-$C(sp^3)$ bond making, 53–54
Caglioti, L., 113
Calvin, M., 243–244
5-Carbamoyl-3-pyridine, 74*t*
Carbon, isotopes of, 18
Carboxamides, by monocarbonylation, 331–333
4-Carboxy-[2.2]-paracyclophane, 79*t*
Catalysis, in Supported Ionic Liquid Phases (SILP), 317–333
Catalysis/autocatalysis, combination of, 45–46
Catalyst, heterogeneous, recyclable, from SLIP, 333
Cayley, A., 134–135
CD Spectra, 5, 286, 346
Cellulose, 224
Cheese ripening, L-amino acid inversion in, 365
Chemical evolution of life, 243–244
Chiral additives, to Soai reaction mixtures, 77
  amplification, non-linear, 301–302
  crsytals of achiral organic compounds, 7–12, 338–339, 343
  crystals, of achiral compounds, 53–54
  crystals, of achiral inorganic compounds, 338–339
  discrimination, by asymmetric autocatalysis, 16–21
  Induced Spin Selectivity (CISS), 293
  molecules, synthesis of, 32
  product, of asymmetric autocatalysis, 22, 53–54, 183–184
  transmission process, 41–42
Chirality, amplification of, 3–16
  biological, 53, 116
  definition of, 53, 112–113
  degrees of, 203, 205
  descriptors, 218–219
  due to isotopic substitution, 265–266, 269
  in thiophene polymers, 277–295
  measure of, 203–219
  metal centered, 270–271
  molecular, 2
  propagation of, 31–32, 32*f*, 183–184
  remote amplification of, 46–48
  three dimensional, graph theory of, 146–147
  two dimensional, graph theory of, 133
Chitin, 224
Cinnabar (HgS), chiral crystals of, 7*t*, 341–342
Circularly polarized light (CPL), 4–5, 22–23, 76–77
Cobalt, chiral EDTA complex of, 343
Cobalt, chiral TRDTA complex of, 343
Complexity of molecules, 111–112
Configuration space, chemical, 215–218
Cotton effect, in CD spectra, 5, 8
Cryptochiral compounds, 16, 20, 20*f*, 53–54

Cryptochirality, discrimination by, 20
Crystal chirality of inorganic compounds, 184
Crystal chirality of organic compounds, 184, 350
Crystallization conditions, effect of, on crystal structure, 186
Cytosine, crystals of
  dehydrated, 8–9
  monohydrate, chiral crystals of, 8–9, 9*f*

D

Degrees of chirality, measure of, 205
Dermorphin, 120
Di(*iso*propyl)zinc, 13*f*, 72–73
  isotope chirality in, 261–265
Dialkylzinc, 34–35, 41, 84–85
Diastereoisotopomers, 259–260, 271
Dimensions, three, chirality in, 133, 146–147
Dimensions, two, chirality in, 133, 138–142
Dimeric catalyst, 81, 86–87, 185
Diols, chiral, 79*t*
Dissymmetry, molecular, 1–2
DNA, higher order structures of, 120
Double carbonylation, 324, 327–331
d-Sugars, 1–2, 72, 120, 149–150, 243–244
Dubois, J.-E., 204

E

Earth, early, molecules in, 112*f*
Earth, primordial, chemical evolution on, 300*f*
Eigen, M., 307–310
Electric conductivity, of chiral polythiophenes, 278–279
Electrochemical asymmetric synthesis, 291–292
  polymerization of thiophene derivatives, 283–284
Electrodes, chiral, from chiral polythiophenes, 293
Enantioenrichment, 2–4, 22–23, 31–32, 76–77, 151, 158
Enantioisotopomers, 259–260, 270–271
Enantiomeric excess, 2, 72, 183–184, 227–228, 263
  amplification of, 88, 91, 243
Enantiomorph organisms, 53
Enantiopurity, 3, 38–39, 43–45, 85–86
Enantioselective inhibition, limited, 307
Enantiotopic faces, of crystals, asymmetric autocatalysis on, 8–9, 13, 22–23
Energy, prebiotic, solar, 111–112
Ercolani, G., 32–33, 89–90, 269–271
Ethylenediammonium sulfate, chiral crystals of, 9
Euler, L., 134–135
Extraterrestrial origin of organic molecules, 300*f*

F

Fastgreedy Algorithm, 58–59, 62*f*, 63*f*
Ferrocene, fluxionality of, 269
Ferrocenylethynylpyrimidine carbaldehyde, 74*t*
Flavor industry, chiral molecules in, 32
Florini, N., 62–63
Formic acid, 224
Fossil-based chemicals, 223–224
Fragrance industry, chiral molecules in, 32
Frank, F. C., 2, 32–33, 72, 249–250, 306, 310–311
Fructose, 224

G

γ-valerolactone (GVL), 224–228, 228*f*, 230–231
Gehring, T., 74–75
General asymmetry, 203, 205–219
Geodiamolide, synthesis of, 227*f*
Geothermal energy, prebiotic, 111–112
Glucose, 224
Glutamic acid, D/L raito of, in Parmigiano Reggiano cheese, 365, 365*t*
Google Scholar's Database, 56
Graph, articulation points of, 58*f*
  average degree of, 56
  theory, 134–135

Gridnev, I. D., 74–75, 81, 93
Group theory, for symmetry, 204
Gryko, D., 247–248
Guillerez, S., 280
GVL Economy, 224

H

Helical mesoporous silica, chiral discrimination by, 21
Heterogeneous recyclable catalyst from SLIP, 324, 333
Hexakis(5-pyrimidyl alkanol) hexasilane, 152–154, 160–162
Homochirality, biological, 1–2
of living organisms, 1–2, 120
of organic compounds, 1–21
origins of, 1–21
propagation of, 31–32, 32*f*, 34
Horváth, I. T., 224
HPLC, chiral, 153–154
Hydroformylation, in GVL solvent, 224–225
Hydrogen, isotopes of, 16–18
Hydrogenation, enantioselective, 225–227, 227*f*
Hydrogen isotope ratio in meteorites, 18
5-(Hydroxymethyl)furfural (5-HMF), 224
4-Hydroxyvaleric acid (4-HVA), 224–226, 230
Hypercyclic replicators, 307–311, 313

I

Intermediates, dimeric, of the Soai reaction
of the Soai reaction, isotope chirality in
oligomeric, 91
tetrameric, 86–87
Ionic Liquid Phases, Supported (SLIP), 317–333
Iron(III), oxidative coupling by, of thiophene derivatives, 281–283
Isobutene, oligomerization of, by SLIP catalyst, 319
Isodensity contour, of molecules, 205
Isotopes, natural abundance of, 75–76, 100, 261*t*
Isotopomers, chiral, 16–19, 271

K

Kagan, H. B., 32, 86–87
Kawasaki, T., 76–77
Ketones, asymmetric reduction of, 223–236
Kinetic analysis, of the Soai reaction, 86–88
Kinetics, deterministic, 304–305
stochastic, 304–305
Kitaigorodski, A. I., 138–139
Klankermayer, J., 82–83
Kondepudi, D. K., 243–244
Kumada, M., 279–280

L

Lemaire, M., 284
Levulinic acid (LA), 224–232, 234–237
Life, autocatalytic system in
history of, on Earth, 73
origin of, 31–32, 116
Light, circularly polarized, 4–5, 76–77, 150–151, 184
Lignocellulose, 224
Liversedge, I. A., 281
Living organisms, self-replication of, 1–2

M

Macrocyclic intermediates, of the Soai reaction, 81
Macromolecular World, 111–116
Mannich reaction, 101, 243–244, 252–256, 253*f*, 254*f*, 255*f*, 256*f*
Mannich reaction, first order autocatalysis in, 252, 257
Matsumoto, Y., 63–64
Mauksch, M., 243–245, 252–253, 257
McCullough, R. D., 279–280
Metal centered chirality, 270–271
Meteorites, amino acid H/D ratio in, 18
2-Methylpyrimidine-5-carbaldehyde, 14–16, 184
Micheau, J.-C., 33
Microcalorimetry, for study of the Mannich reaction
for study of the Soai reaction, 255*f*
Micromolecular World, 111–116

Micskei, K., 94
Mills, W. H., 75–76
Mirror image, non-superimposability of, 203–204
Mirror-symmetry breaking, in the Soai reaction
organocatalytic, possibility of, 243–244
spontaneous, 312–313
Mislow, K., 14, 20, 204
Models of the Soai reaction, comparison of, 94
Molecule, single, chiral, 116, 168, 215–216
Molecules, biologically important
chemical identity of, 203–204
multifunctionalized, 162
stochastic distribution of, 14–16
Multiasymmetric, stereogenic centers, 162

N

Nature 1995, Soai's publication in, 66*t*
N-Benzoylglycine, chiral crystals of, 346*f*
Nickel(II), catalyst, for thienyl bromide coupling, 279–280
Non-autocatalytic alkylzinc addition, 88, 185
Nonlinear, chiral amplification, 301–302
Norephedrine, dialkyl, 79*t*
Noyori, R., 63–64, 64*f*, 231–232
Noyori's catalyst, 234
Noyori's Community, 62–63, 64*f*

O

Oligomeric catalyst, X-ray structure of, 195
One-pot process, for catalysis/autocatalysis combination, 35, 48
Organic molecules, extraterrestrial origin of, 300*f*
Organocatalytic reactions, spontaneous symmetry breaking in, 101, 241–257
Oxygen, isotopes of, 19

P

Palladium(0) catalyst, for α-thienyl bromide coupling, 280–281
Palladium catalyst, supported by SLIP, 324–325
Panaye, A., 204
Parachirality, definition of
in life science, 120–129
Parity violation energy difference (PVED), 312–313
Parmesan cheese, amino acid enantiomers in, 357–358
Parmigiano Reggiano cheese, amino acids in, 358–359, 365–366, 365*t*
Pars-Mills equation, 263, 271
Pasteur, L., 1–2
1,4-Pentanediol, 224–225
Peptides, D-amino acids in, 299–300
Pharmaceutical industry, chiral molecules in, 32
Platform molecules, chiral, 223–237
Polymerization, first molecule in, 116
Polymers, with mixed stereochemistry
with random sequence, 111
Polystyrene, isotactic, 20
Polythiophene, chiral, CD spectra of
fluoressence emission spectra of, 285
UV-VIS spectra of, 285
Polythiophenes, alkylsulfanyl
chiral, 285–291, 286*f*
Prebiotic chemistry
energy, solar, 111–112
Primitive biology, 111
Propargylic alcohols, enantioenriched, 39
Propylene oxide, chiral, 79*t*
Proteins, higher order structure of, 120
Pyridine-3-carbaldehyde, 2, 73, 184
3-Pyridyl alkanol, 2, 3*f*
Pyrimidinyl aldehydes, alkylation of, 72
2-Pyrimidyl alkanol, 14–16
Pályi, G., 62*f*, 63*f*
Pályi's community, 62*f*, 63*f*

Q

Quartz, 5–7, 22–23
Quinoline carbaldehyde, 73–74

R

Racemases, 120
Racemate, definition of, 115–116
  fluctuation in, 75–76
Racemates, chiral crystals of, 113, 115–116
Racemic mechanism, in aldol reaction, 257
Remote amplification, of chirality, 46–48
Replicators, hypercyclic, 307–311
Retgersite ($NiSO_4.6H_2O$), chiral crystals of, 5–7, 7*t*, 342–343
Ribó, J. M., 33
Rieke, R. D., 280
RNA, higher order structures of
  World, 120
RNA-peptide world, 112*f*
(*R*, *R*)-1,2-diphenyl-ethylenediamine, as chiral auxiliary, 231
Ruthenium/BINAP, reduction catalyst, 228–229

S

Saito, T., 231–232, 234–235
Sato, I., 59, 86–87
Self-improvement
  -replication, 149–162
DL-Serine sulfate, chiral crystal of, 13–14
Silica gel, amorphous
  helical, 22–23
  mesoporous helical, 22–23
Silicon, isotopes of, 261*t*
Single chiral molecule, 113
Singleton, D. A., 74–75, 243
Soai, K
  publication of, in Nature, 1995, 53–54, 56
Soai autocatalyst, 31–49
Soai graph, 54–64
Soai reaction, DFT study of
  enantiomer distribution in, 93–94, 167–177
  graph theoretical analysis of, 53–65
  impact of, on natural sciences, 53–65
  kinetics of, 72–102
  mirror symmetry breaking in, 32–33, 72–102
  models of, 88–95
  NMR study of, 185
  quantum chemistry of, 91–93
  single molecule in, 168
  statistical analysis of, 53–65
  stochastic models of, 93–94
  structural aspects of, 72–102
  theoretical models of, 88–95
Soai's community
  group, 72, 74–75, 88, 100, 167–169, 259–260
Social interactions between Authors, 57
Social Network Analysis (SNA), 54, 56
Sodium bromate, chiral crystals of
  chlorate, chiral crystals of, 6–7
Solar energy, prebiotic, 111–112
Spontaneous asymmetric synthesis, 75–76, 75*t*, 86–87, 90
Spontaneous mirror symmetry breaking, search for
  stochastic, 312–313
  thermodynamics of, 305–307
S-Sulcatol, from S-GVL, 226*f*
Starch, 224
Statistical analysis
  fluctuation of enantiomers, 14
Stereogenic centers, multiasymmetric, 39
Stille, J. K., 281
Stille coupling, 281, 286
Stochastic enantiomer distribution, 167–177
Styrene oxide, chiral, 79*t*
L-Sugars, 120
Suzuki coupling of α-thienyl bromides, 280–281
Symmetries, approximate, 204–205
Symmetry breaking
  deficiency, 208
  group theory for, 242
  imperfect, 205

T

t-butyl ethynyl pyrimidine, 74*t*
t-butyl ethynyl substitution, chiral amplification by, 73–74
2-(*tert*-butyldimethylsilylethynyl) pyrimidine-5-carbaldehyde, 13, 13*f*
Tetrakis(5-pyrimidylalkanol)tetrasilane, 158–162

Tetrameric catalyst, of the Soai reaction, 93
Tetraphenylethylene, chiral crystals of, 9, 345, 346*f*
Tetrathia-[7]-helicene, 79*t*
Thalidomide, enantiomers of, 113
2-Thenoylglycine, chiral crystals of, 9
Thiophene derivatives, electrochemical polymerization of
 oxidative coupling of, 281, 283–284
Three dimensions, chirality in, 133
Transition state, of the Soai reaction, 86–87
Tsogoeva, S. B., 243–245, 252–256
Two dimensions, chirality in, 133

V

Viedma, C., 310–313
2-Vinyl-5-pyrimidine carbaldehyde, 74*t*

W

Walktrap algorithm, 58–59, 59*f*, 61, 61*f*, 62*f*, 63
Waste, biomass-based, 224, 225*f*, 236, 237*f*

X

X-ray crystallography, of Soai reaction ntermediates, 85–86

Z

Zinc alkoxide, X-ray structure of
 alkoxides, crystallographic parameters of, 21, 186–191
 dialkyl reagents, 41, 84–85
 diisopropyl, 187–189, 195
 isotopes, 270
 oxide (Wurtzite), catalytic effect of, 102

FSC
www.fsc.org
MIX
Papier | Fördert
gute Waldnutzung
FSC® C083411

Kolibri 360 GmbH
Ferdinand-Jühlke-Straße 7
99095 Erfurt, Deutschland
produktsicherheit@kolibri360.de